A
ma   3

Advanced mathematics

3

C W Celia

formerly Principal Lecturer in Mathematics,
City of London Polytechnic

A T F Nice

Mathematics Department, Lady Eleanor Holles
School, Hampton; formerly Principal Lecturer
in Mathematics, Middlesex Polytechnic

K F Elliott

formerly Head of the Division of Mathematics
Education, Derby Lonsdale College of Higher
Education

Consultant Editor: Dr C. Plumpton, Moderator
in Mathematics, University of London School
Examinations Board: formerly Reader in
Engineering Mathematics; Queen Mary
College, London

MACMILLAN

First published 1985
Reprinted 1987

Published by
MACMILLAN EDUCATION LTD
Houndmills, Basingstoke, Hampshire RG21 2XS
and London
Companies and representatives
throughout the world

Printed in Hong Kong

British Library Cataloguing in Publication Data
Celia, C.W.
Advance mathematics.
3
1. Mathematics—1961—
I. Title   II. Nice, A.T.F.
III. Elliott, K.F.
510   QA39.2
ISBN 0-333-34827-3

Contents

Preface

This is the third of a series of books written for students preparing for A level Mathematics. Book 1 covers the essential core of sixth-form mathematics now accepted by the GCE Boards, while Book 2 covers the applied mathematics, i.e. the numerical methods, mechanics and probability, contained in most single-subject mathematics syllabuses.

Book 3 covers the additional pure mathematics needed by students taking the double subject Mathematics and Further Mathematics, and by those taking Pure Mathematics as a single subject.

It must be emphasised that the reader is expected to be familiar with the use of a calculator, so that numerical work presents no difficulties.

The material is arranged under well-known headings and is organised so that the teacher is free to follow his or her own preferred order of treatment. The chapter contents are listed and an index is also provided to make it easy for both the teacher and the student to refer back rapidly to any particular topic. For ease of reference, a list of the notation used is given at the back of the book together with a list of formulae.

The approach in Book 3 is the same as in Books 1 and 2, each topic being developed mainly through worked examples. There is a brief introduction to each new piece of work followed by worked examples and numerous simple exercises to build up the student's technical skills and to reinforce his or her understanding. It is hoped that this approach will enable the individual student working on his or her own to make effective use of the books and the teacher to use them with mixed ability groups. At the end of each chapter there are many miscellaneous examples, taken largely from past A level examination papers. In addition to their value as examination preparation, these miscellaneous examples are intended to give the student the opportunity to apply the techniques acquired from the exercises throughout the chapter to a considerable range of problems of the appropriate standard.

We are most grateful to the University of London University School Examinations Board (L), the Associated Examining Board (AEB), the University of Cambridge Local Examination Syndicate (C) and the Joint Matriculation Board (JMB) for giving us permission to use questions from their past examination papers.

We are also grateful to the staff of Macmillan Education for the patience they have shown and the help they have given us in the preparation of these books.

C. W. Celia
A. T. F. Nice
K. F. Elliott

1 The idea of proof

1.1 Propositions

Any mathematical proof depends on valid arguments, and so a discussion of proof must consider logical relations between statements.

A statement which is either true or false is known as a *proposition*. A true proposition has a truth value T; a false proposition has a truth value F.

Two propositions p and q can be combined to form *compound* propositions.

The conjunction $p \wedge q$, read as 'p and q', is true when and only when p and q are both true.

The disjunction $p \vee q$, read as 'p or q', is true when either p or q is true, or when both are true.

The symbol $\sim p$, read as 'not p', stands for the negation of p. When p is true, $\sim p$ is false, and when p is false, $\sim p$ is true.

The truth values of $p \wedge q$, $p \vee q$ and $\sim p$ for different truth values of p and q are shown in the truth table below:

p	q	$p \wedge q$	$p \vee q$	$\sim p$
T	T	T	T	F
T	F	F	T	F
F	T	F	T	T
F	F	F	F	T

Two propositions are said to be equal if they have the same truth values in all possible cases. For example, $p \wedge q$ and $q \wedge p$ are clearly equal.

Consider the propositions $(\sim p) \vee (\sim q)$ and $\sim(p \wedge q)$. When p and q are both true, each of these propositions is false. In all other cases, they are both true. As the following truth table shows, these two propositions are equal:

p	q	$(\sim p) \vee (\sim q)$	$\sim(p \wedge q)$
T	T	F	F
T	F	T	T
F	T	T	T
F	F	T	T

Example

Let p be the proposition that, in the triangle ABC, AB $=$ CA, and let q be the proposition that AB $=$ BC.

When p and q are both true, AB $=$ BC $=$ CA. Hence when $p \wedge q$ is true, the triangle is equilateral.

When either p or q is true, or both are true, at least two of the sides of the triangle are equal. Hence when $p \vee q$ is true, the triangle is certainly isosceles, and possibly equilateral.

The proposition $(\sim p) \vee (\sim q)$ is true when either AB $\neq$ CA or AB $\neq$ BC, or both, i.e. when the triangle is not equilateral.

The proposition $\sim (p \wedge q)$ is the negation of the proposition that AB $=$ BC $=$ CA, i.e. the negation of the proposition that the triangle is equilateral.

Open statements

A statement involving a variable x is called an *open* statement.

An open statement becomes a proposition when x is given a particular value. Thus the equation $x^2 - 5x + 4 = 0$, where $x \in \mathbb{R}$, is an open statement.

This statement is true when $x = 1$ or when $x = 4$, but otherwise it is false. The set $\{1, 4\}$ is known as the truth set of the equation. Similarly, the truth set of the inequality $x^2 < 1$, where $x \in \mathbb{R}$, is the set $\{x : -1 < x < 1\}$.

Let P and Q be the truth sets of the open statements p and q.

Then the truth set of $p \wedge q$ is the intersection $P \cap Q$, and the truth set of $p \vee q$ is the union $P \cup Q$.

An equation in x which is true for all values of x is called an *identity*. Its truth set will be the set $\mathbb{R}$. For example,

$$x^2 - 1 \equiv (x + 1)(x - 1),$$
$$\sin 2x \equiv 2 \sin x \cos x,$$
$$x^4 + 1 \equiv (x^2 + \sqrt{2}x + 1)(x^2 - \sqrt{2}x + 1).$$

Note that the equals sign is replaced by the identity sign.

Exercises 1.1

1 Given that p is the proposition 'Jack is tall' and that q is the proposition 'Jill is slim', express the following propositions in terms of p and q:
   (a) Either Jack is short or Jill is slim (or both),
   (b) Jack is tall and Jill is not slim,
   (c) Jill is slim and Jack is not tall,
   (d) Either Jill is not slim or Jack is tall (or both).

2 Draw up truth tables for $\sim p \wedge q$ and $p \vee \sim q$.

3 Show that the propositions $\sim p \vee q$ and $\sim (p \wedge \sim q)$ are equal.

4 Given that x is real, find the truth sets of the propositions $p \wedge q$ and $p \vee q$ when p and q are the following open statements:

(a) $(x - 1)(x - 2) = 0$, $(x - 2)(x - 3) = 0$
(b) $\sin x = 0$, $\cos x = 1$
(c) $x^2 < 2x$, $x^2 + 3 < 4x$
(d) $|x| < 1$, $|x - 1| < 1$.

1.2 Implication

The proposition 'if p is true, then q is true' is called an *implication*, and is represented by $p \Rightarrow q$.

This can be read as 'p implies q' or as 'if p, then q'. For example,

$$x = 4 \quad \Rightarrow \quad x^2 = 16,$$
$$\cos x = 1 \quad \Rightarrow \quad \sin x = 0,$$
$$\ln x = 2 \quad \Rightarrow \quad x = e^2.$$

Consider the truth value of the proposition $p \Rightarrow q$.
When p is true and also q is true, $p \Rightarrow q$ is true.
When p is true but q is false, $p \Rightarrow q$ is false.
When p is false and q is true, and also in the case when p is false and q is false, the proposition $p \Rightarrow q$ is defined to be true.

Thus the proposition $p \Rightarrow q$ is false only when p is true and q is false.
The converse of the proposition $p \Rightarrow q$ is the proposition $q \Rightarrow p$.
It is important to realise that $p \Rightarrow q$ and its converse are not equal propositions, for it is possible for one to be true while the other is false. This can be seen from the following truth table:

p	q	$p \Rightarrow q$	$q \Rightarrow p$
T	T	T	T
T	F	F	T
F	T	T	F
F	F	T	T

Consider the three statements

p: x is an even integer,
q: $x + 2$ is an even integer,
r: x^2 is an even integer.

The proposition $p \Rightarrow q$ is true, and so is its converse $q \Rightarrow p$. The proposition $p \Rightarrow r$ is true, but its converse $r \Rightarrow p$ is false.

Example
Show that the propositions $\sim p \Rightarrow \sim q$ and $q \Rightarrow p$ are equal.

The proposition $\sim p \Rightarrow \sim q$ is true except in the case when $\sim p$ is true and $\sim q$ is false, i.e. except when p is false and q is true.

The truth table for $q \Rightarrow p$ shows that $q \Rightarrow p$ is true except when p is false and q is true.

It follows that these two propositions are equal.

A common mistake is to assume that $p \Rightarrow q$ and $\sim p \Rightarrow \sim q$ are equal.

To show that they are not equal, consider the proposition p, 'I live in Kent', and the proposition q, 'I live in England'.

Since Kent is part of England, p implies q.

Now $\sim p$ is the proposition 'I do not live in Kent', and $\sim q$ is the proposition 'I do not live in England'. In this case, $\sim p$ does not imply $\sim q$, and so while $p \Rightarrow q$ is true, $\sim p \Rightarrow \sim q$ is false.

Exercises 1.2

1   Use a truth table to show that $p \Rightarrow q$, $q \vee \sim p$ and $\sim q \Rightarrow \sim p$ are all equal.

2   Find whether the proposition $p \Rightarrow q$ and its converse are true or false, where p and q respectively are the following open statements
   (a) $\cos x = 1$, $\sin x = 0$, $x \in \mathbb{R}$
   (b) $\sin x > 0$, $0 < x < \pi/2$, $x \in \mathbb{R}$
   (c) $|x| < 1$, $|x| < 4$, $x \in \mathbb{R}$
   (d) $x^2 \in \mathbb{Q}$, $x \in \mathbb{Q}$.

3   Given that x is real, find whether the following propositions are true. In each case, state the converse proposition and whether it is true.
   (a) $(x > 0) \Rightarrow (x > 1)$
   (b) $(\sin x > 0) \Rightarrow (\cos x > 0)$
   (c) $(\tan x = 1) \Rightarrow (\cot x = 1)$
   (d) $(\ln x > 2) \Rightarrow (e^x > 2)$.

4   Find the truth values of the propositions $p \Rightarrow q$, $q \Rightarrow p$, $\sim p \Rightarrow \sim q$ and $\sim q \Rightarrow \sim p$, when p and q respectively are the following propositions:
   (a) $\tan^2 x = 2$, $\sec^2 x = 3$, $x \in \mathbb{R}$
   (b) $z = 2i$, $z^2 = -4$, $z \in \mathbb{C}$
   (c) $x = \pi/4$, $\tan x = 1$, $x \in \mathbb{R}$
   (d) iz is real, $\arg z = \pi/2$, $z \in \mathbb{C}$.

1.3   Necessary and sufficient conditions

Necessary conditions

When p implies q and q is false, p cannot be true. This means that p can be true only when q is true.

In this case, q is said to be a *necessary condition* for p.

For example,

$$x \in \mathbb{Q} \Rightarrow x \in \mathbb{R},$$

and so $x \in \mathbb{R}$ is a necessary condition for $x \in \mathbb{Q}$.

In words, 'x is rational' implies that 'x is real', and so x cannot be rational if it is not real.

Hence x must be real if it is to be rational.

Note that $x \in \mathbb{Q}$ is not a necessary condition for $x \in \mathbb{R}$, since a number can be real but not rational.

Sufficient conditions

When p implies q, q will certainly be true when p is true.

Then p is said to be a *sufficient condition* for q.

For example, 'x is rational' is a sufficient condition for 'x to be real'.

Consider the inequality $(x - 1)(x - 4) < 0$.

The condition $x < 4$ is necessary for this inequality to be true, but it is not sufficient.

The condition $x = 3$ is sufficient for the inequality to be true, but it is not necessary.

The condition $x > 3$ is neither necessary nor sufficient.

Necessary and sufficient conditions

Two propositions p and q are said to be equivalent when p implies q and q implies p, i.e. when

$$p \Rightarrow q \quad \text{and} \quad q \Rightarrow p.$$

Then p and q are either both true or both false.

This is expressed by $p \Leftrightarrow q$.

When p and q are equivalent, q is true if p is true, and only if p is true. Thus p is a sufficient condition for q, and at the same time p is a necessary condition for q.

Similarly, p is true if and only if q is true, so that q is a necessary and sufficient condition for p.

The phrase 'if and only if' can be abbreviated to 'iff'.

Example 1
Let p be the proposition

the three sides of the triangle ABC are equal,

and let q be the proposition

the three angles of the triangle ABC are equal.

Then $p \Rightarrow q$ and $q \Rightarrow p$, and so p and q are equivalent, i.e. $p \Leftrightarrow q$.

Example 2
Show that $k^2 < 1$ is a necessary and sufficient condition for $x^2 + 2kx + 1$ to be positive for all real values of x.

(a) Since

$$x^2 + 2kx + 1 = (x + k)^2 + (1 - k^2),$$

this expression will be negative or zero when $x = -k$ if $k^2 \geqslant 1$.
   Therefore $k^2 < 1$ is a necessary condition.

(b) When k^2 is less than 1, $(x + k)^2 + (1 - k^2)$ is positive for all real values of x. Therefore $k^2 < 1$ is a sufficient condition.
   This shows that the given condition is both necessary and sufficient.

Summary

When $p \Rightarrow q \begin{cases} p \text{ is a sufficient condition for } q, \\ q \text{ is a necessary condition for } p. \end{cases}$

When $p \Leftrightarrow q \begin{cases} p \text{ is a necessary and sufficient condition for } q, \\ q \text{ is a necessary and sufficient condition for } p. \end{cases}$

Exercises 1.3

1 State a proposition equivalent to the proposition that in the triangle ABC, $a^2 = b^2 + c^2$.

2 Show that $e^x > 1$, where $x \in \mathbb{R}$, if and only if $x > 0$.

3 Show that $b^2 = 4ac$ if and only if the roots of the quadratic equation $ax^2 + bx + c = 0$ are equal.

4 Show that $|z_1 + z_2| = |z_1| + |z_2|$ if and only if z_1/z_2 is real and positive.

5 ABCD is a quadrilateral. Show that the propositions 'AB = CD and BC = DA' and 'opposite sides are parallel' are equivalent.

6 Prove that $P(a) = 0$ is a necessary and sufficient condition for $x - a$ to be a factor of the polynomial $P(x)$.

7 The first term of an infinite geometric series is a and the common ratio is r. Determine which of the following conditions are (i) necessary, (ii) sufficient for the series to be convergent:
   (a) $r < 2$,   (b) $r > -1$,   (c) $r = \frac{1}{2}$,   (d) $a = 1 - r$.

8 For each of the conditions (i) $k > 0$, (ii) $4k > 1$, (iii) $k = 1$, determine whether it is
   (a) only a necessary condition,
   (b) only a sufficient condition,
   (c) a necessary and sufficient condition
   for $x^2 + x + k$ to be positive for all real values of x.

1.4 Relationship analysis

Suppose that two open statements **1** and **2** are given, and that it is required to find the relationship between them. The following possibilities are exhaustive:

A   **1** always implies **2** but **2** does not always imply **1**   [**1** ⇒ **2**, **2** ⇏ **1**]

B   **2** always implies **1** but **1** does not always imply **2**   [**2** ⇒ **1**, **1** ⇏ **2**]

C   **1** always implies **2** and **2** always implies **1**   [**1** ⇔ **2**]

D   **1** always denies **2** and **2** always denies **1**

E   none of the above relationships holds.

Problems of this kind are called *relationship analysis* problems.

Example 1
(a) **1** $x^2 - x - 6 < 0$
    2 $-2 < x < 3$

1 implies that $(x + 2)(x - 3) < 0$, and so one of the factors $(x + 2)$ and $(x - 3)$ must be positive and the other negative.
    This is the case when $-2 < x < 3$.
Hence   **1** ⇒ **2**.
Also, if **2** is true, then **1** is true, and so **2** ⇒ **1**.
Thus **1** ⇔ **2**, and the key is **C**.

(b) **1** $x^2 - x - 6 < 0$
    2 $-2 < x < 2$

Since the set $\{x : -2 < x < 2\}$ is a subset of the set $\{x : -2 < x < 3\}$,

$$2 \Rightarrow 1.$$

However, **1** does not imply **2** when x lies between 2 and 3.
    The key in this case is **B**.

(c) **1** $x^2 - x - 2 < 0$
    2 $-2 < x < 3$

1 implies that $(x + 1)(x - 2) < 0$, i.e. that $-1 < x < 2$. This shows that when **1** is true, **2** is true, i.e. **1** ⇒ **2**. But for $-2 < x \leqslant -1$ and for $2 \leqslant x < 3$, **2** is true and **1** is false.
    Hence **2** does not imply **1**, and the key is **A**.

(d) **1** $x^2 - x - 6 < 0$
    2 $x < -2$ or $x > 3$

Since **1** is equivalent to $-2 < x < 3$, **1** denies **2** and **2** denies **1**, giving the key **D**.

(e) **1** $x^2 - x - 6 < 0$
    2 $x > 0$

1 and **2** are both true when $0 < x < 3$. However, when $-2 < x \leqslant 0$, **1** is true and **2** is false, while when $x \geqslant 3$, **2** is true and **1** is false.

None of the keys **A, B, C, D** is correct, and so the key is **E**.

Example 2
$x, y \in \mathbb{R}, x, y \neq 0$.

1 $x^2 = y^2$
2 $x^3 = y^3$

1 $\Rightarrow x^2 - y^2 = 0 \Rightarrow$ either $x = y$ or $x = -y$. **2** is true when $x = y$, but not when $x = -y$.

Hence **1** does not always imply **2**.

2 $\Rightarrow x^3 - y^3 = 0 \Rightarrow (x - y)(x^2 + xy + y^2) = 0$.
Since $x^2 + xy + y^2 > 0$, this gives **2** $\Rightarrow x = y$.

Hence **2** always implies **1**, and the key is **B**.

Example 3
1 $y = \sec x$

2 $\dfrac{dy}{dx} = y \tan x$

When $y = \sec x$, $\dfrac{dy}{dx} = \sec x \tan x = y \tan x$. Hence **1** $\Rightarrow$ **2**.

Integration of the separable differential equation gives $y = P \sec x$, where P is an arbitrary constant.

Hence **2** does not imply **1**, and the key is **A**.

Exercises 1.4
Directions. Each of the following questions consists of two statements (in some cases following very brief preliminary information). You are required to determine the relationship between these statements and to choose

A   if **1** always implies **2** but **2** does not imply **1**

B   if **2** always implies **1** but **1** does not imply **2**

C   if **1** always implies **2** and **2** always implies **1**

D   if **1** always denies **2** and **2** always denies **1**

E   if none of the above relationships holds.

Directions summarised	
A	**1** $\Rightarrow$ **2**, **2** $\not\Rightarrow$ **1**
B	**2** $\Rightarrow$ **1**, **1** $\not\Rightarrow$ **2**
C	**1** $\Leftrightarrow$ **2**
D	**1** denies **2**, **2** denies **1**
E	none of these

1 1 $\tan x = 1$
   2 $\cos x = 1/\sqrt{2}$

2 1 $x^2 > 4$
   2 $x^2 > 2x$

3 1 $\sin^2 x = 1$
   2 $\cos x = 0$

4 1 $f(x)$ is an odd function
   2 $\displaystyle\int_{-1}^{1} f(x)\,dx = 0$

5 1 The mean value of $f(x)$ for $0 \leqslant x \leqslant 4$ is 2
   2 $\displaystyle\int_{0}^{4} f(x)\,dx = 2$

6 1 $|x - 2| = 2 - x$
   2 $\qquad x > 2$

7 1 $V = \frac{4}{3}\pi r^3$
   2 $\dfrac{\delta V}{V} = 3\dfrac{\delta r}{r}$

8 **a** and **b** are non-zero vectors
   1 $|\mathbf{a}| = |\mathbf{b}|$
   2 $\mathbf{a} = \mathbf{b}$

9 1 $p \Rightarrow q$
   2 $\sim p \Rightarrow \sim q$

10 1 The matrix **X** has 2 rows and 3 columns
   2 The matrix $\mathbf{XX}^\mathrm{T}$ has 2 rows and 2 columns

11 1 $zz^* = 2$
   2 $|z|^2 = 2$

12 1 $\sin x = \cos y$
   2 $\cos x = \sin y$

13 1 $f''(0) = 0$
   2 The curve $y = f(x)$ has a point of inflexion at the origin

14 1 The line $ax + by = 1$ touches the circle $x^2 + y^2 = 1$
   2 $a^2 + b^2 = 1$

15 $a, b \in \mathbb{R}$
   1 $2 - i$ is a root of the equation $z^3 + az + b = 0$
   2 $-2 + i$ is a root of the equation $z^3 + az + b = 0$

16 1 $\cos(x - y) = \cos(x + y)$
   2 $y = 0$

17 $x, y \in \mathbb{R},$ $\quad x, y \neq 0$
  1 $x > y$
  2 $1/x < 1/y$

18 **1** $\overrightarrow{PQ} = \overrightarrow{SR}$
  2 $\overrightarrow{PS} = \overrightarrow{QR}$

19 X and Y are 3×3 matrices
  1 $XY = 0$
  2 $X = 0$ or $Y = 0$

20 M is a 2×2 matrix
  1 $\det M = 1$
  2 $\det(2M) = 2$

21 **1** $x(x + 1)(2 - x) > 0$
  2 $-1 < x < 2$

22 $p \Rightarrow q$
  1 p is false
  2 q is false

23 **1** $\mathbf{a} = \mathbf{b} + \mathbf{c}$
  2 $|\mathbf{a}|^2 = |\mathbf{b}|^2 + |\mathbf{c}|^2$

24 **1** $x > 0$
  2 $x + 1/x \geqslant 2$

25 **1** The lines $y = m_1 x$ and $y = m_2 x$ are at right angles
  2 $m_1 m_2 = 1$

1.5 Data sufficiency

Suppose that a problem is set, accompanied by two statements giving information. The question is asked whether or not the data in the statements are sufficient to solve the problem. The following possibilities are exhaustive:

A EACH statement (i.e. statement **1** ALONE and statement **2** ALONE) is sufficient by itself to solve the problem

B statement **1** ALONE is sufficient but statement **2** alone is not sufficient to solve the problem

C statement **2** ALONE is sufficient but statement **1** alone is not sufficient to solve the problem

D BOTH statements **1** and **2** TOGETHER are sufficient to solve the problem, but NEITHER statement ALONE is sufficient

E statements **1** and **2** TOGETHER are NOT sufficient to solve the problem, and additional data specific to the problem are needed

Problems of this type are called *data sufficiency* problems.

Example 1
In an arithmetic progression the first term is 2. Find the sum of the first 100 terms.

1 The hundredth term is 200
2 The common difference is 2

With the usual notation, the sum S of the series is given by $S = \frac{1}{2}n(a + l)$. It is known that $a = 2$ and $n = 100$.
    Statement **1** gives $l = 200$, and so statement **1** alone is sufficient. The sum is also given by $S = \frac{1}{2}n[2a + (n - 1)d]$.
    Statement **2** gives $d = 2$, so that statement **2** alone is sufficient.
    Hence the key is **A**.

Example 2
Prove that the quadratic equation $ax^2 + bx + c = 0$ has one positive root and one negative root.

1 $a > 0$
2 $c < 0$

The statement $a > 0$ alone is not enough, e.g. the equation $4x^2 - 8x + 3 = 0$ has two positive roots.
    Statement **2** alone is not enough, e.g. the equation $-4x^2 + 8x - 3 = 0$ has two positive roots.
    When $a > 0$ and $c < 0$, $b^2 - 4ac$ will be positive and the roots will be real. The product of the roots equals c/a, and so the product will be negative. Hence one root will be positive and one negative.
    Since both statements together are sufficient, but neither alone is sufficient, the key is **D**.

Example 3
Evaluate $\log_{10} 81$.

1 $\log_{10} 3$ is given
2 $\log_{10} 2$ is given

Since $3^4 = 81$, $\log_{10} 81 = 4 \log_{10} 3$. This shows that statement **1** alone is sufficient.
    Since 81 cannot be expressed as a rational power of 2, statement **2** alone cannot be sufficient. Hence the key is **B**.

Example 4
Evaluate $\log_2 105$.

   1 $\log_2 3$ is given
   2 $\log_2 5$ is given

Since $105 = 3 \times 5 \times 7$, the value of $\log_2 7$ is needed.
   Statements **1** and **2** together are not sufficient, and so the key is **E**.

Exercises 1.5
Each of the following questions consists of a problem and two statements, **1** and **2**, in which certain data are given. You are not asked to solve the problem: you have to decide whether the data given in the statements are *sufficient* for solving the problem. Using the data given in the statements, choose

A if EACH statement (i.e. statement **1** ALONE and statement **2** ALONE) is sufficient by itself to solve the problem

B if statement **1** ALONE is sufficient but statement **2** alone is not sufficient to solve the problem

C if statement **2** ALONE is sufficient but statement **1** alone is not sufficient to solve the problem

D if BOTH statements **1** and **2** TOGETHER are sufficient to solve the problem, but NEITHER statement ALONE is sufficient

E if statements **1** and **2** TOGETHER are NOT sufficient to solve the problem, and additional data specific to the problem are needed

Directions summarised	
A	either
B	1
C	2
D	both
E	neither

1 Show that the roots of the quadratic equation $ax^2 + bx + c = 0$ are real

   1 $b^2 > 4ac$
   2 $ac < 0$

2 PQRS is a parallelogram. Find $\overrightarrow{PS}$

   1 $\overrightarrow{PQ} = 2\mathbf{i} - 3\mathbf{j}$
   2 $\overrightarrow{QR} = 3\mathbf{i} + 2\mathbf{j}$

3 Find the value of $\sin x$

1 $\cos x = 1/\sqrt{3}$
2 $\tan x = -\sqrt{2}$

4 Find the sum of the first ten terms of an arithmetic progression in which the first term is 25

    1 The fifth term is 13
    2 the tenth term is -2

5 Determine the values of the constants P, Q and R in the identity

$$f(x) \equiv P(x^2 + 3x + 7) + (Qx + R)(x - 1)$$

    1 $f(1) = 11$
    2 $f(0) = 0$

6 Prove that $|z| > 3$

    1 $|z - 5| < 1$
    2 $|z - 2| > 3$

7 A circle passes through the point P(4, 0). Find the equation of the tangent to the circle at P

    1 The line $x + 4 = 0$ is a tangent to the circle
    2 The radius of the circle is 4

8 Prove that $|\mathbf{a}| = |\mathbf{b}|$, where $\mathbf{a}$ and $\mathbf{b}$ are non-zero vectors

    1 $(\mathbf{a} - \mathbf{b}) \cdot (\mathbf{a} + \mathbf{b}) = 0$
    2 $|\mathbf{a} + \mathbf{b}| = |\mathbf{a} - \mathbf{b}|$

9 Show that the equation $f(x) = 0$ has a root in the interval $1 < x < 2$

    1 $f(1) > 0$
    2 $f(2) < 0$

10 Show that $\sim p \Rightarrow \sim q$

    1 $p \Rightarrow q$
    2 $q \Rightarrow p$

11 Show that $\sin 2\theta = 0\cdot8$

    1 $\tan \theta = 0\cdot5$
    2 $\cos^2 \theta = 0\cdot6$

12 Show that $\tan 2\theta = \frac{4}{3}$

    1 $\tan \theta = \frac{1}{2}$
    2 $\tan (\pi - 2\theta) = -\frac{4}{3}$

13 Prove that $x^2 - 1$ is a factor of the polynomial P(x)

1 $P(1) = P(-1)$
2 $P(1) + P(-1) = 0$

14 Prove that the point $P(x, y)$ lies in the region $x^2 + y^2 < 1$

1 $x^3 + y^3 < 1$
2 $x^4 + y^4 < 1$

15 Show that $|z| \times |z - 1| < 1$.

1 $|z| < \frac{1}{2}$
2 $|z| \times |z + 1| < 1$

16 Show that $p \Rightarrow q$

1 $\sim p \Rightarrow \sim q$
2 $\sim q \Rightarrow \sim p$

17 Find the value of k

1 $x - 1$ is a factor of $x^3 - 2x + k$
2 the remainder when $x^3 - 2x + k$ is divided by $x + 1$ is 2

18 Show that $|z - 2| > |z|$

1 $\text{Im } z > 0$
2 $\text{Re } z < 1$

19 Find maximum value of $ax^3 + b(x^2 + 1)$

1 $a = 2$
2 $b = -2$

20 Find the non-zero vectors $\mathbf{p}$ and $\mathbf{q}$

1 $\mathbf{p} \times \mathbf{q} = \mathbf{i}$
2 $\mathbf{p} \cdot \mathbf{q} = 1$

21 Prove that p is true

1 q is true and $p \Rightarrow q$
2 r is false and $r \Rightarrow \sim p$

22 Prove that $\sin x = \frac{1}{2}$

1 $\tan x = 1/\sqrt{3}$
2 $0 < x < \pi/2$

23 Prove that $\mathbf{PQ} = \mathbf{R}$, where $\mathbf{P}$, $\mathbf{Q}$ and $\mathbf{R}$ are 2×2 matrices

1 $\mathbf{Q} = \mathbf{P}^{-1}\mathbf{R}$
2 $\mathbf{P} = \mathbf{R}\mathbf{Q}^{-1}$

24 Prove that the roots of the equation $ax^2 + bx + c = 0$ are real, given that a, b and c are real

1 $ac < 0$

2 $a + c < 0$

25 Evaluate the integral $\displaystyle\int_{-1}^{1} (ax^3 + bx^2)\,dx$

1 $a = 4$

2 $b = 5$

1.6 Solution of equations

When solving an equation, it is essential to be wary of any step which is not reversible. If at any stage an equation is replaced by a new equation which is not equivalent to the old equation then the step is not reversible and the truth set is changed.

Squaring each side of an equation is not a reversible step, except in a case such as $|z| = |z - 1|$. This equation is equivalent to the equation $|z|^2 = |z - 1|^2$, because a modulus is essentially positive.

Example 1

Solve the equation $\sqrt{(3x - 2)} + \sqrt{(x + 2)} = 4$.   ... (1)

Squaring each side gives the equation

$$3x - 2 + 2\sqrt{(3x - 2)}\sqrt{(x + 2)} + x + 2 = 16 \qquad \ldots (2)$$
$$\Leftrightarrow \qquad \sqrt{(3x^2 + 4x - 4)} = 8 - 2x \qquad \ldots (3)$$
$$\Rightarrow \qquad 3x^2 + 4x - 4 = 4x^2 - 32x + 64 \qquad \ldots (4)$$
$$\Leftrightarrow \qquad x^2 - 36x + 68 = 0 \qquad \ldots (5)$$
$$\Leftrightarrow \qquad x = 2 \quad \text{or} \quad x = 34. \qquad \ldots (6)$$

It can be checked by substitution that 2 is a root of the given equation, but that 34 is not.

Note that equations (1) and (2) are not equivalent, and that equations (3) and (4) are not equivalent.

The equation

$$\sqrt{(3x - 2)} - \sqrt{(x + 2)} = 4,$$

which is satified when x equals 34, also implies equation (4), but is not equivalent to it.

Example 2

Find the values of θ between 0 and π which satisfy the equation

$$\sin \theta - \cos \theta = \sin 2\theta - 1.$$

The given equation implies, but is not equivalent to, the equation

$$(\sin \theta - \cos \theta)^2 = (\sin 2\theta - 1)^2$$

$\Leftrightarrow$ $$1 - \sin 2\theta = \sin^2 2\theta - 2 \sin 2\theta + 1$$
$\Leftrightarrow$ $$(\sin 2\theta)(1 - \sin 2\theta) = 0$$
$\Leftrightarrow$ $$\sin 2\theta = 0 \text{ or } \sin 2\theta = 1.$$

(a) For $0 < \theta < \pi$, $\sin 2\theta = 1 \Rightarrow \theta = \pi/4$.
When $\theta = \pi/4$,

$$\sin \theta - \cos \theta = 0$$

and $$\sin 2\theta - 1 = 0.$$

Hence $\pi/4$ is a root of the given equation.

(b) For $0 < \theta < \pi$, $\sin 2\theta = 0 \Rightarrow \theta = \pi/2$.
When $\theta = \pi/2$,

$$\sin \theta - \cos \theta = 1$$

and $$\sin 2\theta - 1 = -1.$$

Hence $\pi/2$ is not a root of the given equation, but is a root of the equation

$$\sin \theta - \cos \theta = 1 - \sin 2\theta.$$

Example 3
Solve the equation $\cos x + \sin x + 1 = 0$ for values of x between 0 and 2π.

When $t = \tan (x/2)$, $\cos x = \dfrac{1 - t^2}{1 + t^2}$ and $\sin x = \dfrac{2t}{1 + t^2}$.

The given equation becomes

$$\frac{1 - t^2}{1 + t^2} + \frac{2t}{1 + t^2} + 1 = 0$$
$\Leftrightarrow$ $$1 - t^2 + 2t + 1 + t^2 = 0$$
$\Leftrightarrow$ $$2t + 2 = 0$$
$\Leftrightarrow$ $$t = -1$$
$\Leftrightarrow$ $$\tan (x/2) = -1.$$

The only root of this equation between 0 and 2π is $3\pi/2$.

Since $\cos (3\pi/2) = 0$ and $\sin (3\pi/2) = -1$, the given equation is satisfied. However, it can be seen that the equation is also satisfied when $x = \pi$. This root was lost when the substitution $t = \tan (x/2)$ was made, because t is not defined at $x = \pi$.

Exercises 1.6
1 Solve each of the equations
    (a) $\qquad x + \sqrt{(x + 1)} = 11,$
    (b) $\sqrt{(3x + 4)} - \sqrt{(3 - x)} = \sqrt{(2x + 3)},$
    (c) $\sqrt{(x + 9)} - \sqrt{(x + 4)} = \sqrt{(1 - x)},$
    (d) $\sqrt{(x^2 - 3x)} - \sqrt{(x^2 - 9)} = \sqrt{(4x^2 - 7x - 15)}.$

2 Find the value of $\tan \theta$ given that

$$\cos \theta + 2 \sin \theta + 1 = 0.$$

3 Solve the equation

$$2 \cos^2 \theta + \sin \theta = 2$$

for values of θ in the interval $0 \leqslant \theta \leqslant \pi$,
(a) by the substitution $t = \tan (\theta/2)$,
(b) by solving for $\sin \theta$.

1.7 Methods of proof

The aim of a proof is to show that a given proposition implies another proposition. Definitions do not require proof, since they rely on agreement for their meaning. For example, the sum of the vectors $\overrightarrow{AB}$ and $\overrightarrow{BC}$ equals $\overrightarrow{AC}$ by definition, and no proof is needed.

Direct proof

In a direct proof, the initial proposition or propositions, known as the *premise* or *hypothesis*, is followed by a chain of statements leading to the conclusion.

As an example, consider a proof that the perpendicular bisectors of the sides of a triangle ABC are concurrent.
(a) Each point on the perpendicular bisector of AB is equidistant from A and B.
(b) Each point on the perpendicular bisector of CA is equidistant from C and A.
Let the two perpendiculars meet at O.
(c) Then OA = OB and OA = OC.
(d) Therefore OB = OC.
(e) The point O must lie on the perpendicular bisector of BC.
(f) The three perpendicular bisectors of the sides meet at O, i.e. they are concurrent.

In this proof, the propositions (a) and (b), which are taken to be true, form the premise.

The conjunction of (a) and (b) gives (c).

Then (c) $\Rightarrow$ (d), (d) $\Rightarrow$ (e) and (e) $\Rightarrow$ (f).

The truth of the conclusion (f) depends on the truth of (a) and (b).

Proof by contradiction

Let it be assumed that the conclusion (f) above is false.

Let the perpendicular bisector of AB meet the perpendicular bisector of CA at O_1. Then O_1 is the centre of the circumcircle of the triangle ABC.

Let the perpendicular bisector of AB meet the perpendicular bisector of BC at O_2. Then O_2 also is the centre of the circumcircle of the triangle ABC.

It is absurd for a circle to have two distinct centres. Therefore (f) cannot be false, i.e. (f) is true.

This is an example of a proof by contradiction, or by *reductio ad absurdum*. It is the method used in *Advanced Mathematics 1*, Section 8.5, to prove that $\sqrt{2}$ is not rational.

Proof by induction

Let $P(n)$ be an open statement involving the positive integer n, and let T be the set of positive integers for which $P(n)$ is true. Then the Principle of Mathematical Induction asserts that 'If $(k \in T) \Rightarrow (k + 1 \in T)$ for every $k \geqslant N$, and if $N \in T$, then for every $n \geqslant N$, $n \in T$.'

This principle can also be stated as follows: 'If $P(k + 1)$ is true whenever $P(k)$ is true, and if also $P(N)$ is true, then $P(n)$ is true for $n \geqslant N$.'

In many cases, N can be taken to be 1, as in *Advanced Mathematics 1*, Section 3.4.

The necessity of establishing that $P(n)$ is true for some value of n is shown by the following example.

Let $P(n)$ be the statement that, for any positive integer n, 3^n is an even number.

Assume that $P(k)$ is true.

When 3^k is even, $3^k + 3^k + 3^k$ is even, i.e. 3^{k+1} is even.

This shows that $P(k + 1)$ will be true when $P(k)$ is true,

i.e. $$P(k) \Rightarrow P(k + 1).$$

To clinch the argument, an integer N must be found such that 3^N is an even number. Clearly there is no such integer.

Proof by exhaustion

This method can be used only when the number of possible cases is small. Consider the proposition that the number of permutations of the three letters a, b, c is 6. The truth of this proposition can be demonstrated by listing all the cases:

$$abc, acb, bca, bac, cab, cba.$$

Care must be taken to ensure that every possibility is considered.

Disproof by counter-example

A single counter-example is enough to prove that a general statement is false. This was the method used earlier to show that $p \Rightarrow q$ and $\sim p \Rightarrow \sim q$ were not equal propositions.

Consider the proposition that, for 2×2 matrices, multiplication is commutative.

Let $\mathbf{A} = \begin{pmatrix} 1 & 1 \\ 0 & 0 \end{pmatrix}$ and $\mathbf{B} = \begin{pmatrix} 2 & 1 \\ 3 & 0 \end{pmatrix}$.

Then $\mathbf{AB} = \begin{pmatrix} 5 & 1 \\ 0 & 0 \end{pmatrix}$ and $\mathbf{BA} = \begin{pmatrix} 2 & 2 \\ 3 & 3 \end{pmatrix}$.

Since $\mathbf{AB}$ and $\mathbf{BA}$ are not equal matrices, the proposition is false.

Note that a general statement can never be established by quoting particular examples. Consider the claim that for any integer n the expression $n^2 + n + 41$ is a prime number. This expression is a prime number for each value of n from 1 to 40, but the counter-example provided by $n = 41$ is enough to disprove the claim.

2 Complex numbers

2.1 De Moivre's theorem

Let n be an integer, i.e. let $n \in \mathbb{Z}$. Then de Moivre's theorem states that

$$(\cos \theta + i \sin \theta)^n = \cos n\theta + i \sin n\theta.$$

Case 1 Let n be positive. The result is clearly true when $n = 1$.
Assume that the result is true when $n = k$, where $k \in \mathbb{Z}^+$,

i.e. $\qquad\qquad (\cos \theta + i \sin \theta)^k = \cos k\theta + i \sin k\theta.$

Then $\qquad (\cos \theta + i \sin \theta)^{k+1} = (\cos \theta + i \sin \theta)^k(\cos \theta + i \sin \theta)$
$$= (\cos k\theta + i \sin k\theta)(\cos \theta + i \sin \theta).$$

The real part of this expression is

$$\cos k\theta \cos \theta - \sin k\theta \sin \theta,$$

which equals $\cos (k + 1)\theta$.
The imaginary part is $\sin k\theta \cos \theta + \cos k\theta \sin \theta$, i.e. $\sin (k + 1)\theta$.

Hence $\qquad (\cos \theta + i \sin \theta)^{k+1} = \cos (k + 1)\theta + i \sin (k + 1)\theta.$

This shows that if the result is true for $n = k$ it is true for $n = k + 1$. The result is true for $n = 1$. Hence, by induction, it is true for $n = 2, 3$ or any positive integer.

Case 2 Let n be negative and equal to $- m$, where $m \in \mathbb{Z}^+$.

Since $\quad (\cos \theta + i \sin \theta)(\cos \theta - i \sin \theta) = \cos^2 \theta + \sin^2 \theta = 1,$
$$(\cos \theta + i \sin \theta)^{-1} = \cos \theta - i \sin \theta$$
$$= \cos (- \theta) + i \sin (- \theta).$$

Therefore $\quad (\cos \theta + i \sin \theta)^{-m} = \cos (-m\theta) + i \sin (-m\theta).$

This shows that the result

$$(\cos \theta + i \sin \theta)^n = \cos n\theta + i \sin n\theta$$

is true when n is a negative integer.

Let p and q be two integers with no common factor. It will be seen in Section 2.4 that the expression $(\cos \theta + i \sin \theta)^{p/q}$ has q distinct values. In this case, de Moivre's theorem states that one of these values is $\cos (p\theta/q) + i \sin (p\theta/q)$.
To prove this, let

$$(\cos \theta + i \sin \theta)^{p/q} = \cos \alpha + i \sin \alpha.$$

Note that each side of this equation has modulus 1.

Let each side be raised to the qth power. This gives

$$\cos p\theta + i \sin p\theta = \cos q\alpha + i \sin q\alpha,$$

since p and q are integers. This equation is satisfied by $p\theta = q\alpha$, $\alpha = p\theta/q$. Hence one value of $(\cos \theta + i \sin \theta)^{p/q}$ is $\cos (p\theta/q) + i \sin (p\theta/q)$.

De Moivre's theorem can be used to establish trigonometrical identities.

Cos $n\theta$ and sin $n\theta$

For any positive integer n, the even function $\cos n\theta$ can be expressed as a polynomial of degree n in $\cos \theta$ by equating real parts in the identity

$$\cos n\theta + i \sin n\theta \equiv (\cos \theta + i \sin \theta)^n$$

and then replacing $\sin^2 \theta$ by $(1 - \cos^2 \theta)$.

When $\sin \theta$ is not zero, the even function $\dfrac{\sin n\theta}{\sin \theta}$ can be expressed as a polynomial of degree $(n - 1)$ in $\cos \theta$ by equating imaginary parts in the same identity.

By the same method, the odd function $\sin n\theta$ can be expressed as a polynomial of degree n in $\sin \theta$ when n is an odd positive integer.

Example 1

Express $\cos 6\theta$ and $\dfrac{\sin 6\theta}{\sin \theta}$ in terms of $\cos \theta$.

By de Moivre's theorem,

$$\cos 6\theta + i \sin 6\theta \equiv (\cos \theta + i \sin \theta)^6.$$

The binomial expansion of the right-hand side gives

$$c^6 + 6ic^5s - 15c^4s^2 - 20ic^3s^3 + 15c^2s^4 + 6ics^5 - s^6,$$

where c denotes $\cos \theta$ and s denotes $\sin \theta$.

The real part of this expansion must equal $\cos 6\theta$, giving

$$\begin{aligned}
\cos 6\theta &= c^6 - 15c^4s^2 + 15c^2s^4 - s^6 \\
&= c^6 - 15c^4(1 - c^2) + 15c^2(1 - c^2)^2 - (1 - c^3 \\
&= 32c^6 - 48c^4 + 18c^2 - 1
\end{aligned}$$

$\Rightarrow \qquad \cos 6\theta = 32 \cos^6 \theta - 48 \cos^4 \theta + 18 \cos^2 \theta - 1.$

By equating imaginary parts in the identity above,

$$\begin{aligned}
\sin 6\theta &= 6c^5s - 20c^3s^3 + 6cs^5 \\
&= 6c^5s - 20(c^3 - c^5)s + 6(c - 2c^3 + c^5)s.
\end{aligned}$$

Provided that sin θ is not zero, this gives

$$\frac{\sin 6\theta}{\sin \theta} = 32 \cos^5 \theta - 32 \cos^3 \theta + 6 \cos \theta.$$

Example 2

Express tan 3θ in terms of tan θ.

By de Moivre's theorem,

$$\cos 3\theta + i \sin 3\theta \equiv (\cos \theta + i \sin \theta)^3.$$

The binomial expansion of the right-hand side gives

$$\cos^3 \theta + 3i \cos^2 \theta \sin \theta - 3 \cos \theta \sin^2 \theta - i \sin^3 \theta.$$

By equating the real parts in the identity,

$$\cos 3\theta = \cos^3 \theta - 3 \cos \theta \sin^2 \theta.$$

By equating the imaginary parts,

$$\sin 3\theta = 3 \cos^2 \theta \sin \theta - \sin^3 \theta.$$

Hence

$$\tan 3\theta = \frac{\sin 3\theta}{\cos 3\theta}$$

$$= \frac{3 \cos^2 \theta \sin \theta - \sin^3 \theta}{\cos^3 \theta - 3 \cos \theta \sin^2 \theta}.$$

When numerator and denominator are both divided by $\cos^3 \theta$, this becomes

$$\tan 3\theta = \frac{3 \tan \theta - \tan^3 \theta}{1 - 3 \tan^2 \theta}.$$

Powers of cos θ and sin θ

Let

$$\cos \theta + i \sin \theta = z.$$

Then

$$\cos \theta - i \sin \theta = 1/z.$$

Hence

$$2 \cos \theta = z + 1/z, \qquad \ldots (1)$$

and

$$2i \sin \theta = z - 1/z. \qquad \ldots (2)$$

Also, for any integer p,

$$\cos p\theta + i \sin p\theta = (\cos \theta + i \sin \theta)^p = z^p,$$

and

$$\cos p\theta - i \sin p\theta = (\cos \theta - i \sin \theta)^p = 1/z^p.$$

Hence

$$2 \cos p\theta = z^p + 1/z^p, \qquad \ldots (3)$$

and

$$2i \sin p\theta = z^p - 1/z^p. \qquad \ldots (4)$$

Consider first $\cos^n \theta$, which is an even function of θ. From equation (1),

$$2^n \cos^n \theta = (z + 1/z)^n.$$

The binomial expansion of $(z + 1/z)^n$ contains $(n + 1)$ terms. By pairing each power of z with its reciprocal, and using equation (3) above, $\cos^n \theta$ can be expressed in terms of cosines of multiples of θ.

The corresponding result for $\sin^n \theta$ depends on whether n is even or odd.

When n is even, $\sin^n \theta$ is an even function of θ. Let $n = 2m$. From equation (2),

$$(2i \sin \theta)^{2m} = (z - 1/z)^{2m}.$$

In the binomial expansion of $(z - 1/z)^{2m}$, each power of z can be paired with its reciprocal. Then, by means of equation (3) above, $\sin^{2m} \theta$ can be expressed in terms of cosines of multiples of θ.

When n is odd, $\sin^n \theta$ is an odd function of θ. Let $n = 2m + 1$.

In the binomial expansion of $(z - 1/z)^{2m+1}$, each power of z can be paired with its reciprocal. Then, by means of equation (4) above, $\sin^{2m+1} \theta$ can be expressed in terms of sines of multiples of θ.

The expression $\cos^p \theta \sin^q \theta$, where p and q are positive integers, is an even function of θ when q is even, and can then be expressed in terms of cosines of multiples of θ.

When q is odd, $\cos^p \theta \sin^q \theta$ is an odd function of θ, and can be expressed in terms of sines of multiples of θ.

Example 3
Express $\cos^6 \theta$ in terms of cosines of multiples of θ.

When $z = \cos \theta + i \sin \theta$, $2 \cos \theta = z + 1/z$.

$$
\begin{aligned}
2^6 \cos^6 \theta &= (z + 1/z)^6 \\
&= (z^6 + z^{-6}) + 6(z^4 + z^{-4}) + 15(z^2 + z^{-2}) + 20.
\end{aligned}
$$

Now $z^2 + z^{-2} = (\cos 2\theta + i \sin 2\theta) + (\cos 2\theta - i \sin 2\theta)$
$= 2 \cos 2\theta$.

Similarly $z^4 + z^{-4} = 2 \cos 4\theta$, $z^6 + z^{-6} = 2 \cos 6\theta$.

Hence $64 \cos^6 \theta = 2 \cos 6\theta + 12 \cos 4\theta + 30 \cos 2\theta + 20$
$\Rightarrow \cos^6 \theta = (\cos 6\theta + 6 \cos 4\theta + 15 \cos 2\theta + 10)/32.$

Example 4
Express $\sin^5 \theta$ in terms of sines of multiples of θ.

When $z = \cos \theta + i \sin \theta$, $2i \sin \theta = z - 1/z$.

$$
\begin{aligned}
(2i)^5 \sin^5 \theta &= (z - 1/z)^5 \\
&= (z^5 - z^{-5}) - 5(z^3 - z^{-3}) + 10(z - z^{-1}).
\end{aligned}
$$

Now
$$z - z^{-1} = 2i \sin \theta.$$
$$z^3 - z^{-3} = (\cos 3\theta + i \sin 3\theta) - (\cos 3\theta - i \sin 3\theta)$$
$$= 2i \sin 3\theta.$$
$$z^5 - z^{-5} = 2i \sin 5\theta.$$

Hence
$$(2i)^5 \sin^5 \theta = 2i \sin 5\theta - 10i \sin 3\theta + 20i \sin \theta$$
$$\Rightarrow \qquad \sin^5 \theta = (\sin 5\theta - 5 \sin 3\theta + 10 \sin \theta)/16.$$

Example 5

Express $\cos^3 \theta \sin^2 \theta$ in terms of cosines of multiples of θ.

$$2 \cos \theta = z + 1/z, \quad 2i \sin \theta = z - 1/z,$$

where $z = \cos \theta + i \sin \theta$.

$$(2 \cos \theta)^3 (2i \sin \theta)^2 = (z + 1/z)^3 (z - 1/z)^2$$
$$= (z^5 + z^{-5}) + (z^3 + z^{-3}) - 2(z + z^{-1})$$
$$= 2 \cos 5\theta + 2 \cos 3\theta - 4 \cos \theta$$
$$\Rightarrow \qquad \cos^3 \theta \sin^2 \theta = (2 \cos \theta - \cos 3\theta - \cos 5\theta)/16.$$

Exercises 2.1

1 Simplify (a) $(\sqrt{3} + i)^6$, (b) $(1 + i)^{10}$.

2 Simplify (a) $(3 \cos 30° + 3i \sin 30°)^3$,
   (b) $(2 \cos 40° - 2i \sin 40°)^{-3}$.

3 Express $\sin 3\theta$ and $\sin 5\theta$ in terms of $\sin \theta$.

4 Express $\dfrac{\sin 4\theta}{\sin \theta}$ and $\dfrac{\sin 6\theta}{\sin \theta}$ in terms of $\cos \theta$.

5 Express $\cos 5\theta$ and $\dfrac{\sin 5\theta}{\sin \theta}$ in terms of $\cos \theta$.

6 Express $\cos 7\theta$ and $\dfrac{\sin 7\theta}{\sin \theta}$ in terms of $\cos \theta$.

7 Show that $\tan 4\theta = \dfrac{4 \tan \theta - 4 \tan^3 \theta}{1 - 6 \tan^2 \theta + \tan^4 \theta}$. Use the substitution $x = \tan^2 \theta$ to show that the roots of the equation $x^2 - 6x + 1 = 0$ are $\tan^2 (\pi/8)$ and $\tan^2 (3\pi/8)$.

8 Show that $\tan 5\theta = \dfrac{5 \tan \theta - 10 \tan^3 \theta + \tan^5 \theta}{1 - 10 \tan^2 \theta + 5 \tan^4 \theta}$.

9 By expressing $\tan 7\theta$ in terms of $\tan \theta$, show that the roots of the equation

$$x^3 - 21x^2 + 35x - 7 = 0$$

are $\tan^2 (\pi/7)$, $\tan^2 (2\pi/7)$ and $\tan^2 (3\pi/7)$.

10 Use de Moivre's theorem to prove that

$$\sin 7\theta = 7 \sin \theta - 56 \sin^3 \theta + 112 \sin^5 \theta - 64 \sin^7 \theta.$$

Deduce that the only real solutions of the equation

$$\sin 7\theta = 7 \sin \theta$$

are given by $\theta = n\pi$, where $n \in \mathbb{Z}$.

11 Express (a) $\cos^4 \theta$, (b) $\cos^5 \theta$ in terms of cosines of multiples of θ.

12 Show that $64 \sin^7 \theta = 35 \sin \theta - 21 \sin 3\theta + 7 \sin 5\theta - \sin 7\theta$. Evaluate $\int_0^{\pi/2} \sin^7 \theta \, d\theta.$

13 Show that $8 \sin^4 \theta = \cos 4\theta - 4 \cos 2\theta + 3$. Evaluate $\int_0^{\pi} \sin^4 \theta \, d\theta.$

14 Show that $\cos^2 \theta \sin^3 \theta = (2 \sin \theta + \sin 3\theta - \sin 5\theta)/16$. Evaluate $\int_0^{\pi/2} \cos^2 \theta \sin^3 \theta \, d\theta.$

15 Show that $64 \sin^4 \theta \cos^3 \theta = \cos 7\theta - \cos 5\theta - 3 \cos 3\theta + 3 \cos \theta.$

2.2 Square roots of complex numbers

Consider the equation $z^2 = a$, where $a \in \mathbb{C}$.
  Let $|z| = r$ and $\arg z = \theta$, so that $z = r(\cos \theta + i \sin \theta)$.
  Let $|a| = R$ and $\arg a = \alpha$, so that $a = R(\cos \alpha + i \sin \alpha)$.
  We will adopt the convention that $\arg z$ denotes the principal value of the argument of z.
  Then the angles θ and α are such that

$$-\pi < \theta \leqslant \pi, \qquad -\pi < \alpha \leqslant \pi.$$

Since $z^2 = a$, $\qquad\qquad\qquad |z|^2 = |a|$
$\Rightarrow \qquad\qquad\qquad\qquad r^2 = R$
$\Rightarrow \qquad\qquad\qquad\qquad r = \sqrt{R},$

i.e. r is equal to the positive square root of R.
  From $z^2 = a$,

$$r^2(\cos 2\theta + i \sin 2\theta) = R(\cos \alpha + i \sin \alpha)$$
$\Rightarrow \qquad\quad \cos 2\theta + i \sin 2\theta = \cos \alpha + i \sin \alpha$
$\Rightarrow \qquad\qquad\quad\; \cos 2\theta = \cos \alpha, \qquad \sin 2\theta = \sin \alpha.$

These two equations give

$$2\theta = \alpha + 2k\pi, \qquad \text{where } k \in \mathbb{Z},$$
$\Rightarrow \qquad\qquad\quad \theta = \alpha/2 + k\pi.$

The only suitable values of k are those which give $-\pi < \theta \leqslant \pi$.

When $0 < \alpha \leqslant \pi$, the suitable values of k are 0 and -1.

For $k = 0$, $\theta = \alpha/2$ and $z = \sqrt{R}\left(\cos\dfrac{\alpha}{2} + \text{i}\sin\dfrac{\alpha}{2}\right)$.

For $k = -1$, $\theta = \alpha/2 - \pi$, giving

$$\begin{aligned} z &= \sqrt{R}[\cos(\alpha/2 - \pi) + \text{i}\sin(\alpha/2 - \pi)] \\ &= -\sqrt{R}[\cos(\alpha/2) + \text{i}\sin(\alpha/2)]. \end{aligned}$$

When $-\pi < \alpha \leqslant 0$, the suitable values of k are 0 and 1.

For $k = 0$, $\theta = \alpha/2$, and $z = \sqrt{R}\left(\cos\dfrac{\alpha}{2} + \text{i}\sin\dfrac{\alpha}{2}\right)$.

For $k = 1$, $\theta = \alpha/2 + \pi$, giving

$$\begin{aligned} z &= \sqrt{R}[\cos(\alpha/2 + \pi) + \text{i}\sin(\alpha/2 + \pi)] \\ &= -\sqrt{R}[\cos(\alpha/2) + \text{i}\sin(\alpha/2)]. \end{aligned}$$

Hence in all cases the square roots of the complex number $R(\cos\alpha + \text{i}\sin\alpha)$ are

$$\pm\sqrt{R}[\cos(\alpha/2) + \text{i}\sin(\alpha/2)].$$

In Fig. 2.1, the point A represents the complex number a, with $|a| < 1$ and $0 < \arg a \leqslant \pi$. The points P and Q represent the two square roots of a.

Example

Find the square roots of $21 + 20\text{i}$.

Method 1   Let $21 + 20\text{i} = R(\cos\alpha + \text{i}\sin\alpha)$

$\Rightarrow \qquad\qquad R^2 = 21^2 + 20^2 = 841$

$\Rightarrow \qquad\qquad R = 29.$

The two square roots will be $\pm\sqrt{29}[\cos(\frac{1}{2}\alpha) + \text{i}\sin(\frac{1}{2}\alpha)]$.

By equating real parts,

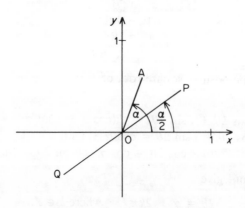

Fig. 2.1   Square roots

$$29 \cos \alpha = 21$$
$$\Rightarrow \qquad \cos \alpha = 21/29.$$

By equating imaginary parts,

$$29 \sin \alpha = 20$$
$$\Rightarrow \qquad \sin \alpha = 20/29.$$

Since $\cos \alpha$ and $\sin \alpha$ are both positive, $0 < \alpha < \pi/2$ and $\cos \frac{1}{2}\alpha$ and $\sin \frac{1}{2}\alpha$ will be both positive.

$$\cos \alpha = 2 \cos^2 \tfrac{1}{2}\alpha - 1 = 21/29$$
$$\Rightarrow \qquad \cos^2 \tfrac{1}{2}\alpha = 25/29$$
$$\Rightarrow \qquad \cos \tfrac{1}{2}\alpha = 5/\sqrt{29}$$
$$\Rightarrow \qquad \sin \tfrac{1}{2}\alpha = 2/\sqrt{29}.$$

Hence the two square roots are $\pm(5 + 2i)$.

Method 2   Let p and q be real numbers such that

$$(p + iq)^2 = 21 + 20i$$
$$\Rightarrow \qquad p^2 - q^2 + 2pqi = 21 + 20i.$$

By equating real parts,

$$p^2 - q^2 = 21.$$

By equating imaginary parts,

$$2pq = 20.$$

These two simultaneous equations for p and q can be solved by substituting $10/p$ for q in the first equation. This gives

$$p^2 - 100/p^2 = 21$$
$$\Rightarrow \qquad p^4 - 21p^2 - 100 = 0$$
$$\Rightarrow \qquad (p^2 - 25)(p^2 + 4) = 0$$

Since p is real, this gives $p = \pm 5$.

From $2pq = 20$, $q = \pm 2$, and so $p + iq = \pm(5 + 2i)$.

The solution can be shortened by the use of the relation $p^2 + q^2 = 29$.

Exercises 2.2

1   Find in the form $p + iq$ (p, q real) the square roots of
   (a) $2i$, (c) $16 + 30i$,
   (b) $3 - 4i$, (d) $5 - 12i$.

2   Find the modulus r and the argument θ of each square root of
   (a) $1 + i$, (c) $-2\sqrt{2} + 2\sqrt{2}i$,
   (b) $3 + i\sqrt{3}$, (d) $-\sqrt{3} - i$.

2.3 Cube roots of complex numbers

Consider the equation $z^3 = a$, where $a \in \mathbb{C}$.

Let $a = R(\cos \alpha + \mathrm{i} \sin \alpha)$, where $R > 0$ and $-\pi < \alpha \leqslant \pi$.

Let $z = r(\cos \theta + \mathrm{i} \sin \theta)$, where $r > 0$ and $-\pi < \theta \leqslant \pi$.

By de Moivre's theorem,

$$z^3 = r^3(\cos 3\theta + \mathrm{i} \sin 3\theta).$$

From $z^3 = a$, $|z^3| = |a|$,

$$\Rightarrow \qquad\qquad r^3 = R$$
$$\Rightarrow \qquad\qquad r = \sqrt[3]{R}, \text{ the positive cube root of } R.$$

Also, $r^3(\cos 3\theta + \mathrm{i} \sin 3\theta) = R(\cos \alpha + \mathrm{i} \sin \alpha)$

$$\Rightarrow \qquad r^3 \cos 3\theta = R \cos \alpha, \ r^3 \sin 3\theta = R \sin \alpha$$
$$\Rightarrow \qquad \cos 3\theta = \cos \alpha, \ \sin 3\theta = \sin \alpha$$
$$\Rightarrow \qquad 3\theta = \alpha + 2k\pi, \text{ where } k \in \mathbb{Z},$$
$$\Rightarrow \qquad \theta = \tfrac{1}{3}\alpha + \tfrac{2}{3}k\pi.$$

$k = 0$ gives $\theta = \tfrac{1}{3}\alpha$.
$k = 1$ gives $\theta = \tfrac{1}{3}\alpha + \tfrac{2}{3}\pi$.
$k = -1$ gives $\theta = \tfrac{1}{3}\alpha - \tfrac{2}{3}\pi$.

These are the only values of k for which $-\pi < \theta \leqslant \pi$.

The cube roots of $R(\cos \alpha + \mathrm{i} \sin \alpha)$ are therefore

$$\sqrt[3]{R}\left(\cos \frac{\alpha}{3} + \mathrm{i} \sin \frac{\alpha}{3}\right),$$

$$\sqrt[3]{R}\left(\cos \frac{\alpha + 2\pi}{3} + \mathrm{i} \sin \frac{\alpha + 2\pi}{3}\right)$$

and

$$\sqrt[3]{R}\left(\cos \frac{\alpha - 2\pi}{3} + \mathrm{i} \sin \frac{\alpha - 2\pi}{3}\right).$$

Since 1 can be expressed as $\cos 0 + \mathrm{i} \sin 0$, the cube roots of 1 are

$$1, \ \cos 120° + \mathrm{i} \sin 120°, \ \cos(-120°) + \mathrm{i} \sin(-120°),$$

i.e.,
$$1, \ -\frac{1}{2} + \mathrm{i}\frac{\sqrt{3}}{2}, \ -\frac{1}{2} - \mathrm{i}\frac{\sqrt{3}}{2}.$$

Let ω denote the complex number $(-1 + \mathrm{i}\sqrt{3})/2$.

Then
$$\omega^2 = (-1 + \mathrm{i}\sqrt{3})^2/4$$
$$= (-1 - \mathrm{i}\sqrt{3})/2,$$

so that the cube roots of 1 are given by 1, ω and ω^2.

Note that

$$1 + \omega + \omega^2 = 1 + (-1 + i\sqrt{3})/2 + (-1 - i\sqrt{3})/2$$
$$\Rightarrow \quad 1 + \omega + \omega^2 = 0.$$

Example

Find the cube roots of $2 + 2i$.

Let
$$2 + 2i = R(\cos \alpha + i \sin \alpha).$$

Then
$$R = |2 + 2i| = 2\sqrt{2},$$
$$\alpha = \arg (2 + 2i) = 45°$$
$$\Rightarrow \quad 2 + 2i = 2\sqrt{2}(\cos 45° + i \sin 45°).$$

The positive cube root of R is $\sqrt{2}$ and $\alpha/3 = 15°$. Hence one cube root is $\sqrt{2}(\cos 15° + i \sin 15°)$.

In Fig. 2.2 the point P represents $2 + 2i$ and the point A represents the cube root with argument 15°.

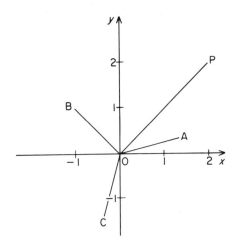

Fig. 2.2   Cube roots of $2 + 2i$

The other two cube roots, represented by B and C, will have the same modulus $\sqrt{2}$, and their arguments will be $(15° + 120°)$ and $(15° - 120°)$ respectively. Note that each of the angles AOB, BOC and COA is 120°.

Thus the cube roots of $2 + 2i$ are

$$\sqrt{2}(\cos 15° + i \sin 15°),$$
$$\sqrt{2}(\cos 135° + i \sin 135°),$$

and
$$\sqrt{2}[\cos (-105°) + i \sin (-105°)].$$

Exercises 2.3

1   Express in the form $p + iq$ (p, q real) the cube roots of
   (a) 8,   (b) 8i,   (c) -8.

2 Find the modulus r and the argument θ of each cube root of the following, and represent them in an Argand diagram.

(a) 27i,   (b) $2 - 2i$,   (c) $4\sqrt{3} + 4i$.

3 Given that $\omega = (-1 + i\sqrt{3})/2$, simplify

(a) $(1 + \omega)^3$,   (b) $(a + b)(a + \omega b)(a + \omega^2 b)$,   (c) $(\omega - 1)^3$,

(d) $(1 + \omega^4 + \omega^8)$,   (e) $(1 + \omega - \omega^2)(1 - \omega - \omega^2)(1 - \omega + \omega^2)$.

2.4 nth roots of unity

Consider the equation $z^n = 1$, where n is a positive integer. Since $|z^n| = 1$, $|z| = 1$, i.e. each root has unit modulus.

Let
$$z = \cos \theta + i \sin \theta.$$

By de Moivre's theorem,

$$z^n = \cos n\theta + i \sin n\theta.$$

From $z^n = 1$, $\cos n\theta + i \sin n\theta = 1$

$\Rightarrow$ $\cos n\theta = 1, \sin n\theta = 0$

$\Rightarrow$ $n\theta = 2k\pi,$ where $k \in \mathbb{Z}$.

The values of k must be such that $-\pi < \theta \leqslant \pi$.

Case 1 n even

The required values of k are

$$0, \pm 1, \pm 2, \ldots, \pm \frac{n - 2}{2} \text{ and } \frac{n}{2}.$$

The corresponding values of θ are

$$0, \pm \frac{2\pi}{n}, \pm \frac{4\pi}{n}, \ldots, \pm \frac{(n - 2)\pi}{n}, \pi,$$

and the n roots of $z^n = 1$ are

$$1, \quad \cos \frac{2\pi}{n} \pm i \sin \frac{2\pi}{n}, \quad \cos \frac{4\pi}{n} \pm i \sin \frac{4\pi}{n}, \quad \ldots, \quad -1.$$

In Fig. 2.3 the points A, B, C, D, E and F on the unit circle represent the roots of the equation $z^6 = 1$.

These roots are

$$1, \quad \cos 60° \pm i \sin 60°, \quad \cos 120° \pm i \sin 120°, \quad -1.$$

Case 2 n odd

The required values of k are

$$0, \quad \pm 1, \quad \pm 2, \quad \ldots, \quad \pm (n - 1)/2.$$

The corresponding values of θ are

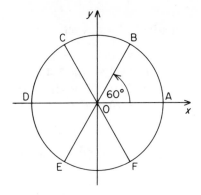

Fig. 2.3 Roots of $z^6 = 1$

$$0, \quad \pm \frac{2\pi}{n}, \quad \pm \frac{4\pi}{n}, \quad \ldots, \quad \pm \frac{(n-1)\pi}{n},$$

and the n roots are

$$1, \quad \cos \frac{2\pi}{n} \pm i \sin \frac{2\pi}{n}, \quad \cos \frac{4\pi}{n} \pm i \sin \frac{4\pi}{n}, \quad \ldots,$$

$$\cos \frac{(n-1)\pi}{n} \pm i \sin \frac{(n-1)\pi}{n}.$$

In Fig. 2.4, the roots of the equation $z^9 = 1$ are represented by the vertices of the nine-sided regular polygon inscribed in the unit circle. These roots are

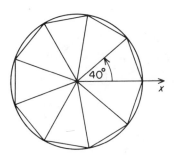

Fig. 2.4 Roots of $z^9 = 1$

$1, \quad \cos 40° \pm i \sin 40°, \quad \cos 80° \pm i \sin 80°, \quad \cos 120° \pm i \sin 120°$
and $\cos 160° \pm i \sin 160°$.

For any value of n, odd or even, the n roots of unity are represented in the Argand diagram by the n vertices of a regular polygon. Each side of this polygon will subtend an angle $2\pi/n$ at the origin.

Example 1
Find the roots of the equation $z^5 = 4 + 4i$.

Let $\qquad\qquad\qquad 4 + 4i = R(\cos \alpha + i \sin \alpha)$.

Then $\qquad\qquad\quad R = |4 + 4i| = 4\sqrt{2}$,

$\qquad\qquad\qquad\qquad \alpha = \arg (4 + 4i) = 45°$.

Let $\qquad\qquad\qquad z = r(\cos \theta + i \sin \theta)$.

By de Moivre's theorem,

$$z^5 = r^5(\cos 5\theta + i \sin 5\theta).$$

From $\qquad\qquad\qquad z^5 = 4 + 4i$,

$\qquad r^5(\cos 5\theta + i \sin 5\theta) = 4\sqrt{2}(\cos 45° + i \sin 45°)$

$\Rightarrow \qquad\qquad\quad r^5 = 4\sqrt{2} \quad \Rightarrow \quad r = \sqrt{2}$.

Also $\qquad\qquad \cos 5\theta = \cos 45°, \sin 5\theta = \sin 45°$

$\Rightarrow \qquad\qquad\quad 5\theta = (360k + 45)°, \qquad$ where $\qquad k \in \mathbb{Z}$,

$\Rightarrow \qquad\qquad\quad \theta = (72k + 9)°$.

The values of k for which $-180° < \theta \leqslant 180°$ are $0, \pm 1$ and ± 2, giving

$$\theta = 9°, \quad 81°, \quad 153°, \quad -63° \quad \text{and} \quad -135°.$$

The five roots of the equation $z^5 = 4 + 4i$ are

$\sqrt{2}(\cos 9° + i \sin 9°), \quad \sqrt{2}(\cos 81° + i \sin 81°), \quad \sqrt{2}(\cos 153° + i \sin 153°),$
$\sqrt{2}(\cos 63° - i \sin 63°) \quad \text{and} \quad \sqrt{2}(\cos 135° - i \sin 135°)$.

These roots are represented in Fig. 2.5 by the vertices of the regular pentagon. Each side of the pentagon subtends an angle of 72° at the origin.

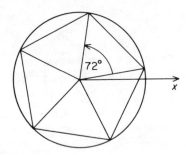

Fig. 2.5   Roots of $z^5 = 4 + 4i$

Example 2
Show that the expression $(\cos \theta + i \sin \theta)^{p/q}$, where p and q are integers with no common factor, has q distinct values (q being assumed positive).

Let $(\cos \theta + i \sin \theta)^{p/q} = r(\cos \alpha + i \sin \alpha)$, where $r > 0$, $-\pi < \alpha \leqslant \pi$. Then by de Moivre's theorem,

$$\cos p\theta + i \sin p\theta = r^q(\cos q\alpha + i \sin q\alpha).$$

Since the modulus of the left-hand side is 1, r must equal 1.

Also,
$$\cos p\theta = \cos q\alpha, \qquad \sin p\theta = \sin q\alpha$$
$$\Rightarrow \qquad q\alpha = p\theta + 2k\pi, \qquad \text{where} \qquad k \in \mathbb{Z},$$
$$\Rightarrow \qquad \alpha = (p\theta + 2k\pi)/q.$$

Consecutive values of k give values of α which differ by $2\pi/q$, so that there will be q values of k which satisfy the condition

$$-\pi < \alpha \leqslant \pi.$$

Hence there will be q values of α in this interval, giving the q distinct values of $(\cos \theta + i \sin \theta)^{p/q}$.

Exercises 2.4

1 Find the arguments of the roots of the equations

(a) $z^4 = 1$,   (b) $z^7 = 1$,   (c) $z^8 = -1$,   (d) $z^5 = -1$.

In each case represent the roots in an Argand diagram.

2 Find the fourth roots of (a) -4, (b) $-2 + i2\sqrt{3}$, giving your answers in the form $p + iq$, where p and q are real.

3 Find the modulus r and the argument θ of each root of the equation

(a) $z^6 = -1$,   (b) $z^8 = 16$,   (c) $z^5 = -243$.

4 Find the roots of the equation $z^4 + a^4 = 0$, where a is real and positive. Express $z^4 + a^4$ as the product of factors with real coefficients.

5 Find the modulus and argument of each root of the equation

$$z^4 + 8 + i8\sqrt{3} = 0.$$

6 Express in the form $r(\cos \theta + i \sin \theta)$ the roots of the equation

$$z^5 - 16 + 16i\sqrt{3} = 0.$$

7 Represent the roots of the equation $z^9 = -1$ in an Argand diagram.

8 Find the roots which the equations $z^8 = 1$ and $z^{12} = -1$ have in common.

9 Show that the roots of the equation $z^{10} = 1$ can be represented in an Argand diagram by the vertices of a regular polygon. By considering similar triangles, or otherwise, show that the length of a side of the polygon is $(\sqrt{5} - 1)/2$.

10 Prove that the sum of the roots of the equation $z^n = 1$, where $n \in \mathbb{Z}^+$, is zero. By taking $n = 5$, show that $1 + 2\cos 72° + 2\cos 144° = 0$.

2.5 Conjugate complex numbers

In Fig 2.6 let P represent z, where $z = x + iy$, and let Q represent the conjugate complex number z^*, where $z^* = x - iy$.

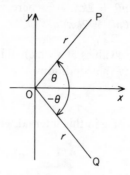

Fig. 2.6 z and z^*

Then $$|z^*| = |z| \;\Rightarrow\; OQ = OP$$
and $$\arg z^* = -\arg z.$$

The sum of two conjugate complex numbers is real, for

$$z + z^* = (x + iy) + (x - iy)$$
$$= 2x.$$

The product of two conjugate complex numbers is real, for

$$(z)(z^*) = (x + iy)(x - iy)$$
$$= x^2 + y^2$$
$$\Rightarrow zz^* = |z|^2 = |z^*|^2.$$

Hence the product of two conjugate complex numbers equals the square of their modulus.

Let $|z| = r$ and let $z = r(\cos\theta + i\sin\theta)$.

Then $$z^* = r(\cos\theta - i\sin\theta)$$
$$\Rightarrow \quad zz^* = r^2(\cos^2\theta + \sin^2\theta) = r^2,$$

giving the same result.

By de Moivre's theorem,

$$z^n = r^n(\cos n\theta + i\sin n\theta), \qquad \text{where } n \in \mathbb{Z},$$
$$\Rightarrow \quad (z^n)^* = r^n(\cos n\theta - i\sin n\theta).$$
Also, $$z^* = r(\cos\theta - i\sin\theta)$$
$$\Rightarrow \quad (z^*)^n = r^n(\cos n\theta - i\sin n\theta).$$

Hence $(z^n)^*$ and $(z^*)^n$ are equal.

Consider the polynomial $P(z)$ given by

$$P(z) = a_n z^n + a_{n-1} z^{n-1} + \ldots + a_1 z + a_0.$$

Provided that the coefficients $a_0, a_1, \ldots, a_n$ are real, then

$$[P(z)]^* = a_n(z^n)^* + a_{n-1}(z^{n-1})^* + \ldots + a_1 z^* + a_0$$
$$= a_n(z^*)^n + a_{n-1}(z^*)^{n-1} + \ldots + a_1 z^* + a_0$$
$$= P(z^*).$$

Let $\quad P(z) = P(x + iy) = a + ib, \quad$ where $\quad a, b \in \mathbb{R}.$

Then $\qquad\qquad [P(z)]^* = a - ib$

$\Rightarrow \qquad\qquad P(z^*) = P(x - iy) = a - ib.$

When $P(z) = 0$, both a and b must be zero, so that $P(z^*) = 0$.

Hence $\qquad\qquad P(z) = 0 \quad \Leftrightarrow \quad P(z^*) = 0.$

This proves that, when α is a complex root of the equation $P(z) = 0$ in which the coefficients are real, α^* will also be a root of this equation.

Thus the complex roots of the equation $P(z) = 0$ will occur in conjugate pairs.

Example
Show that $2 + 3i$ is a root of the equation $z^3 - z^2 + z + 39 = 0$, and find the other roots.

Substitute $2 + 3i$ for z:

$$(2 + 3i)^3 - (2 + 3i)^2 + (2 + 3i) + 39$$
$$= (-46 + 9i) - (-5 + 12i) + (2 + 3i) + 39$$
$$= (-46 + 5 + 2 + 39) + i(9 - 12 + 3)$$
$$= 0.$$

This confirms that $2 + 3i$ is a root of the equation.

Since the coefficients are all real, $2 - 3i$ is also a root.

$$\text{Sum of the roots} = -\frac{\text{coefficient of } z^2}{\text{coefficient of } z^3} = 1 \quad \text{(Section 8.1).}$$

Since $(2 + 3i) + (2 - 3i) = 4$, the third root must be -3.

$$\text{Product of the roots} = -\frac{\text{constant term}}{\text{coefficient of } z^3} = -39.$$

$$(2 + 3i)(2 - 3i)(-3) = (13)(-3) = -39,$$

giving a check on the result.

Exercises 2.5

1 Find zz^* when z equals (a) $3 - 2i$, (b) $1 - i$, (c) $3i$, (d) $4 - 3i$.

2 Solve the equations (a) $z^2 - 2z + 2 = 0$, (b) $z^2 + 4z + 8 = 0$.

3 Given that $z + z^* = 6$ and $zz^* = 25$, find z.

4 Show that $3 + 3i$ is a root of the equation

$$z^3 - 8z^2 + 30z - 36 = 0,$$

and find the other roots.

5 Show that i is a root of the equation

$$z^4 - 4z^3 + 6z^2 - 4z + 5 = 0,$$

and find the other roots.

6 Solve the equation $z^3 + z + 10 = 0$, given that one root is a negative integer.

7 Solve the equation $z^3 + 4z^2 + 4z + 16 = 0$.

8 Show that one root of the equation $z^4 + 12z - 5 = 0$ is $1 + 2i$. Find the other roots, and express $z^4 + 12z - 5$ as the product of two quadratic factors with real coefficients.

9 The product of two roots of the equation

$$z^4 + 56z^2 - 80z + 500 = 0$$

is 10. Solve the equation.

10 The sum of two roots of the equation

$$z^4 + 2z^3 + 2z^2 + 10z + 25 = 0$$

is 2. Solve the equation.

2.6 The exponential function

The function e^z, or $\exp z$, is defined with domain $\mathbb{C}$ by the series

$$e^z = 1 + z + \frac{z^2}{2!} + \frac{z^3}{3!} + \ldots + \frac{z^n}{n!} + \ldots .$$

It can be shown that this series converges for all values of z.

Let $z = x + iy$. When $y = 0$, $z = x$ and the series above becomes the Maclaurin series for e^x.

When $x = 0$, $z = iy$ and series for e^z becomes

$$e^{iy} = 1 + iy + \frac{(iy)^2}{2!} + \frac{(iy)^3}{3!} + \ldots + \frac{(iy)^n}{n!} + \ldots .$$

The real part of the series for e^{iy} is

$$1 - \frac{y^2}{2!} + \frac{y^4}{4!} - \frac{y^6}{6!} + \ldots + (-1)^n \frac{y^{2n}}{(2n)!} + \ldots ,$$

which is the Maclaurin series for $\cos y$.

The imaginary part of the series for e^{iy} is

$$y - \frac{y^3}{3!} + \frac{y^5}{5!} - \frac{y^7}{7!} + \ldots + (-1)^n \frac{y^{2n+1}}{(2n+1)!} + \ldots,$$

which is the Maclaurin series for $\sin y$.

It follows that

$$e^{iy} = \cos y + i \sin y,$$

and, since y is any real number, this gives

$$e^{i\theta} = \cos \theta + i \sin \theta.$$

By means of this relation, known as *Euler's relation*, $\cos \theta$ and $\sin \theta$ can be expressed in terms of the exponential function. The conjugate of $e^{i\theta}$ is given by

$$e^{-i\theta} = \cos \theta - i \sin \theta$$

$\Rightarrow$

$$e^{i\theta} + e^{-i\theta} = 2 \cos \theta$$

and

$$e^{i\theta} - e^{-i\theta} = 2i \sin \theta.$$

Hence

$$\cos \theta = \frac{e^{i\theta} + e^{-i\theta}}{2}, \qquad \sin \theta = \frac{e^{i\theta} - e^{-i\theta}}{2i}.$$

Any complex number can be expressed in the form $re^{i\theta}$.
For when $|z| = r$ and $\arg z = \theta$,

$$z = r(\cos \theta + i \sin \theta)$$

$\Rightarrow$

$$z = re^{i\theta}.$$

For example, the cube roots of 8 can be expressed as

$$2, \quad 2e^{2\pi i/3} \quad \text{and} \quad 2e^{-2\pi i/3}.$$

Example 1
Show that $8 \cos^4 \theta = \cos 4\theta + 4 \cos 2\theta + 3$.

$$2 \cos \theta = e^{i\theta} + e^{-i\theta}$$

$\Rightarrow$

$$16 \cos^4 \theta = (e^{i\theta} + e^{-i\theta})^4$$
$$= (e^{4i\theta} + e^{-4i\theta}) + 4(e^{2i\theta} + e^{-2i\theta}) + 6$$
$$= 2 \cos 4\theta + 8 \cos 2\theta + 6$$

$\Rightarrow$

$$8 \cos^4 \theta = \cos 4\theta + 4 \cos 2\theta + 3.$$

Example 2
Find the sum of the series

$$1 + \cos x + \cos 2x + \ldots + \cos nx,$$

where $x \in \mathbb{R}$ and $\cos x \neq 1$.

The given series is the real part of the complex series

$$1 + e^{ix} + e^{2ix} + e^{3ix} + \ldots + e^{inx}.$$

Let the sum of this geometric series be $C + iS$, where C and S are real:

$$C + iS = (e^{i(n+1)x} - 1)/(e^{ix} - 1)$$

$$= \frac{(e^{i(n+1)x} - 1)(e^{-ix} - 1)}{(e^{ix} - 1)(e^{-ix} - 1)}$$

$$= \frac{e^{inx} - e^{i(n+1)x} - e^{-ix} + 1}{1 - e^{ix} - e^{-ix} + 1}.$$

The denominator is real, since it equals $2 - 2 \cos x$.
The real part of the numerator is $[\cos nx - \cos (n + 1)x - \cos x + 1]$.
Hence, by equating real parts,

$$C = \frac{\cos nx - \cos (n + 1)x - \cos x + 1}{2 - 2 \cos x}.$$

Exercises 2.6

1 Prove that e^{ix} is periodic with period 2π.

2 Show that $e^{2\pi i} = 1$, $e^{\pi i} = -1$ and $e^{\pi i/2} = i$.

3 Using the exponential forms of $\sin \theta$ and $\cos \theta$, show that $\sin (\pi/2 - \theta)$ $= \sin (\theta + \pi/2) = \cos \theta$ and $\cos (\theta + \pi) = -\cos \theta$.

4 Given that $\sin (\frac{1}{2}\theta) \neq 0$, prove that
   (a) $\cos \theta + \cos 2\theta + \ldots + \cos n\theta = [\sin (\frac{1}{2}n\theta) \cos \frac{1}{2}(n + 1)\theta]/\sin (\frac{1}{2}\theta)$,
   (b) $\sin \theta + \sin 2\theta + \ldots + \sin n\theta = [\sin (\frac{1}{2}n\theta) \sin \frac{1}{2}(n + 1)\theta]/\sin (\frac{1}{2}\theta)$.

5 Show that

$$32 \sin^6 \theta = 10 - 15 \cos 2\theta + 6 \cos 4\theta - \cos 6\theta.$$

6 Show that

$$256 \sin^9 \theta = \sin 9\theta - 9 \sin 7\theta + 36 \sin 5\theta - 84 \sin 3\theta + 126 \sin \theta.$$

7 Use the exponential forms of $\sin x$ and $\cos x$ to show that
   (a) $\sin 2x = 2 \sin x \cos x$, (b) $\cos^2 x + \sin^2 x = 1$.

2.7 Straight lines in the Argand diagram

The complex number z can be represented in an Argand diagram by the point $P(x, y)$, where $z = x + iy$. Alternatively, z can be represented by the vector $\overrightarrow{OP}$, where O is the origin, or by any vector equivalent to $\overrightarrow{OP}$.

The straight line through two given points

Let the vectors $\overrightarrow{OA}$ and $\overrightarrow{OB}$ in Fig. 2.7 represent the complex numbers z_1 and z_2 respectively. Then the vector $\overrightarrow{AB}$ represents $z_2 - z_1$.

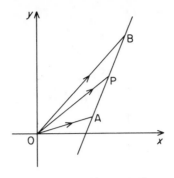

Fig. 2.7 $\overrightarrow{AP} = t\,\overrightarrow{AB}$

Let P be any point on the line through the points A and B, and let the vector $\overrightarrow{OP}$ represent the complex number z.

For all points P on the line, $\overrightarrow{AP}$ is parallel to $\overrightarrow{AB}$, so that $\overrightarrow{AP}$ equals $\overrightarrow{AB}$ multiplied by a scalar, i.e.

$$\overrightarrow{AP} = t\overrightarrow{AB},$$

where t is a real number.

Since $\overrightarrow{AP} = z - z_1$ and $\overrightarrow{AB} = z_2 - z_1$, this gives

$$z - z_1 = t(z_2 - z_1)$$
$$\Rightarrow \qquad z = z_1 + t(z_2 - z_1).$$

This equation represents the straight line through the points A and B.

For values of the parameter t between 0 and 1, the point P lies between A and B. For $t > 1$, P lies on AB produced, and for $t < 0$, P lies on BA produced.

Example 1
Find the equation of the straight line through the points representing the complex numbers $2 - i$ and $6 + 3i$.

Let $z_1 = 2 - i$ and let $z_2 = 6 + 3i$.
Then $z_2 - z_1 = 4 + 4i$, and the equation of the line is

$$z = 2 - i + t(4 + 4i).$$

The perpendicular bisector of a line
In Fig. 2.8, the points P, A and B represent the complex numbers z, z_1 and z_2 respectively. M is the mid-point of AB and PM is at right angles to AB. For any point P on the perpendicular bisector of AB, the lengths of AP and BP are equal.

Since $AP = |z - z_1|$ and $BP = |z - z_2|$, this gives

$$|z - z_1| = |z - z_2|.$$

This is the equation of the line PM.

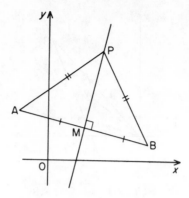

Fig. 2.8 $|z - z_2| = |z - z_2|$

Example 2

Write down the equation of the perpendicular bisector of the line joining the points representing $2 - 3i$ and $4 + i$.

The required equation is

$$|z - 2 + 3i| = |z - 4 - i|.$$

The half-line in a given direction

In Fig. 2.9, the points P and A represent z and z_1 respectively. Let the vector $\overrightarrow{AP}$ make a constant angle α with Ox for all positions of P, i.e.

$$\arg (z - z_1) = \alpha.$$

This is the equation of a half-line starting at the point A and drawn in the given direction.

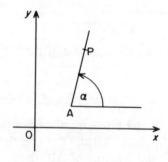

Fig. 2.9 $\arg (z - z_1) = \alpha$

Example 3

Use the Argand diagram to solve the simultaneous equations

$$\arg (z - i) = \pi/4, \qquad \arg (z - 3) = \pi/2.$$

In Fig. 2.10, the points A and B represent i and 3 respectively. The equation $\arg (z - i) = \pi/4$ gives a half-line drawn from A in the direction making an angle $\pi/4$ with the positive x-axis. The equation $\arg (z - 3) = \pi/2$ gives a half-line drawn from B in the direction shown, i.e. parallel to Oy.

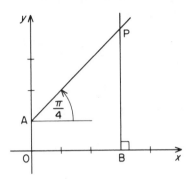

Fig. 2.10 $\arg (z - i) = \pi/4$
$\arg (z - 3) = \pi/2$

The two half-lines intersect at the point P, which represents the complex number $3 + 4i$. Hence the solution of the given equations is $z = 3 + 4i$.

Exercises 2.7

1   Find the equation of the straight line through the points representing
(a) $1 + 3i$ and $2 + i$,   (b) $2 - 5i$ and $5 + 2i$,
(c) $-2 + i$ and $4 - 5i$,   (d) $5 + i$ and i.

2   The points A, B, C and D represent the complex numbers $2 + i$, $1 + 2i$, $4 - 3i$ and $5 - 2i$, respectively. Find the equations of the perpendicular bisectors of AB and CD, and show that these bisectors meet at the point given by $z = 1 + i$.

3   Find the value of z at the point of intersection of the straight lines given by

$$z = 1 + i + t(6 - 4i),$$
$$z = -2i + s(4 + i).$$

4   Draw in an Argand diagram the half-lines

$$\arg (z + 2) = \pi/4, \qquad \arg (z - 4) = 3\pi/4 \qquad \text{and} \qquad \arg (z - 4i) = 0.$$

Find the area of the triangle formed by these lines.

5   Show that the straight line through the points representing $-2i$ and $2 - i$ passes through the point P given by $z = 6 + i$. Find the equation of the line through P at right angles to this line.

6 Show that the equations $|z - 1| = |z - i|$ and $z = t(1 + i)$ represent the same straight line.

7 Show that the points given by $z = -3 + 7i$, $1 + i$, $7 - 8i$ are collinear.

8 Find the distance from the origin to the line $z = 5i + t(2 - i)$.

9 Find the area of the triangle formed by the line $z = 2 + t(1 - i\sqrt{3})$ and the half-lines $\arg z = 0$ and $\arg z = \pi/3$.

10 Show that the lines $|z + 2 - 3i| = |z + i|$ and $z = 6 + t(1 - i)$ and the half-line $\arg z = \pi/4$ are concurrent.

2.8 Circles in the Argand diagram

When the points P and C represent the complex numbers z and c in the Argand diagram, the modulus $|z - c|$ equals the length of CP. If this length is constant and equal to R, the locus of P is a circle with radius R and centre C. Hence the equation of this circle is

$$|z - c| = R.$$

Example 1

Find the cartesian equation of the circle $|z - c| = R$ when $c = 3 + 2i$ and $R = 4$.

Let $z = x + iy$. Then

$$\begin{aligned} |z - c| &= |x + iy - 3 - 2i| \\ &= |(x - 3) + i(y - 2)|. \end{aligned}$$

When
$$|z - c| = 4,$$
$$|z - c|^2 = 16$$
$$\Rightarrow \qquad (x - 3)^2 + (y - 2)^2 = 16.$$

This is the corresponding cartesian equation.

Example 2

Express the equation $|z - c| = R$ in terms of z and z^*.

The conjugate of $z - c$ is $z^* - c^*$.

Hence
$$|z^* - c^*| = |z - c| = R.$$

The product of $z - c$ and $z^* - c^*$ is real and equal to R^2, i.e.

$$zz^* - zc^* - z^*c + cc^* = R^2.$$

Any equation of this form represents a circle.

The circle $|z - c| = k|z|$, where $k \neq 1$

This equation states that the distance of the point P representing z from the point C representing c is k times the distance of P from the origin. Note that if k were equal to 1 then the locus of P would be a straight line.

Let $c = a + ib$, where a and b are real.

Then
$$|z - c| = k|z|$$
$\Rightarrow$
$$|z - c|^2 = k^2|z|^2$$
$\Rightarrow$
$$(x - a)^2 + (y - b)^2 = k^2(x^2 + y^2)$$
$\Rightarrow$
$$(1 - k^2)(x^2 + y^2) - 2ax - 2by + a^2 + b^2 = 0.$$

This equation reduces to
$$\left[x - \frac{a}{1 - k^2} \right]^2 + \left[y - \frac{b}{1 - k^2} \right]^2 = \frac{k^2(a^2 + b^2)}{(1 - k^2)^2},$$

which is the cartesian equation of a circle. The centre of this circle in the Argand diagram is given by
$$z = \frac{a + ib}{1 - k^2} = \frac{c}{1 - k^2},$$

and, since $a^2 + b^2 = |c|^2$, the radius R of the circle is given by
$$R = \frac{k|c|}{|1 - k^2|}.$$

The case when $k = \frac{1}{2}$ is illustrated in Fig. 2.11. For each point P on the circle, $PC = \frac{1}{2}PO$. The centre of the circle is given by $z = 4c/3$ and its radius is $\frac{2}{3}|c|$.

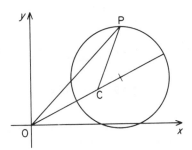

Fig. 2.11   CP = $\frac{1}{2}$OP

The equation
$$|z - p| = k|z - q|, \text{ where } k \neq 1,$$

represents a circle with radius $k|p - q|/|1 - k^2|$ and with its centre at the point given by $z = (p - qk^2)/(1 - k^2)$.

This can be shown by the method above, or by considering the transformation $Z = z - q$. This type of transformation is considered in the next section.

A circle defined in this way is known as a *circle of Apollonius*.

Example 3

Find the cartesian equation of the circle $|z - 3| = 2|z - i|$.

$$|z - 3| = 2|z - i|$$
$$\Rightarrow \qquad |z - 3|^2 = 4|z - i|^2$$
$$\Rightarrow \qquad (x - 3)^2 + y^2 = 4[x^2 + (y - 1)^2]$$
$$\Rightarrow \qquad x^2 - 6x + 9 + y^2 = 4x^2 + 4y^2 - 8y + 4$$
$$\Rightarrow \qquad 3x^2 + 3y^2 + 6x - 8y - 5 = 0$$
$$\Rightarrow \qquad (x + 1)^2 + (y - 4/3)^2 = 40/9.$$

Circular arcs

In Fig. 2.12, the point P lies on the arc of a circle standing on the chord AB. The magnitude of the angle APB is the same for all positions of P on this arc, i.e. if the point Q lies on the same arc, the angle AQB is equal to the angle APB.

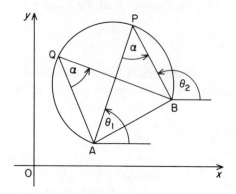

Fig. 2.12 $\theta_2 - \theta_1 = \alpha$

Let the points P, A and B represent z, z_1 and z_2 respectively.
Let arg $(z - z_1) = \theta_1$ and arg $(z - z_2) = \theta_2$.
If the angle APB equals α, $\theta_2 - \theta_1 = \alpha$

$$\Rightarrow \qquad \text{arg } (z - z_2) - \text{arg } (z - z_1) = \alpha.$$

This is the equation of the circular arc APB.

Example 4

Show that the equation arg $(z - 1) - $ arg $(z + 1) = \pi/2$ is satisfied at all points on the semicircle given by $|z| = 1, 0 < \theta < \pi$.

Find the equation which is satisfied on the semicircle given by $|z| = 1$, $-\pi < \theta < 0$.

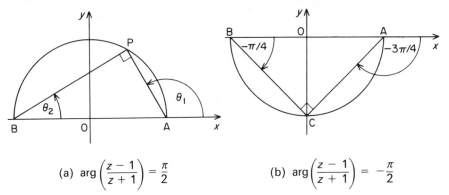

(a) $\arg\left(\dfrac{z-1}{z+1}\right) = \dfrac{\pi}{2}$ (b) $\arg\left(\dfrac{z-1}{z+1}\right) = -\dfrac{\pi}{2}$

Fig. 2.13

In Fig. 2.13(a) the points P, A and B represent z, 1 and -1, respectively. Let $\arg(z-1) = \theta_1$ and $\arg(z+1) = \theta_2$.

From the given equation, $\theta_1 - \theta_2 = \pi/2$, so that the angle BPA is a right angle.

Hence the locus of P is the semicircle given by $|z| = 1$, $0 < \theta < \pi$.

In Fig. 2.13(b), let the point C represent the complex number $-i$. At C, $\arg(z-1) = -3\pi/4$ and $\arg(z+1) = -\pi/4$. This gives

$$\arg(z+1) - \arg(z-1) = \pi/2.$$

Since the value of $\arg(z+1) - \arg(z-1)$ is constant on the semicircle given by $|z| = 1$, $-\pi < \theta < 0$, this is the required equation.

Example 5

Show that the equation $\arg(z-1) - \arg(z-i) = \pi/4$ is satisfied on an arc of the circle $|z - 1 - i| = 1$. Find the equation of the remaining arc of this circle.

Let $\theta_1 = \arg(z-1) = \arg(x-1+iy)$.

Then $\tan\theta_1 = y/(x-1)$.

Let $\theta_2 = \arg(z-i) = \arg(x+iy-i)$.

Then $\tan\theta_2 = (y-1)/x$.

From the given equation, $\theta_1 - \theta_2 = \pi/4$

$\Rightarrow$ $\tan(\theta_1 - \theta_2) = 1$

$\Rightarrow$ $\tan\theta_1 - \tan\theta_2 = 1 + \tan\theta_1 \tan\theta_2$.

This gives

$$y/(x-1) - (y-1)/x = 1 + (y^2 - y)/(x^2 - x)$$

$\Rightarrow$ $x^2 + y^2 - 2x - 2y + 1 = 0$

$\Rightarrow$ $(x-1)^2 + (y-1)^2 = 1$.

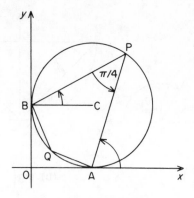

Fig. 2.14 $\angle xAP - \angle CBP = \pi/4$

This is the cartesian equation of the circle $|z - 1 - i| = 1$, with radius 1 and centre C at the point given by $z = 1 + i$. Fig. 2.14 shows this circle, with the points P, A, B and C representing the complex numbers z, $1 + 0i$, $0 + i$ and $1 + i$, respectively.

Let P move along the major arc AB from A to B.
Then arg $(z - 1)$, i.e. the angle xAP, increases from zero to $3\pi/4$.
Also, arg $(z - i)$, i.e. the angle CBP, increases from $-\pi/4$ to $\pi/2$.
At each point on the major arc,

$$\arg (z - 1) - \arg (z - i) = \pi/4.$$

Now consider a point Q on the minor arc BA of the circle.
Let Q move along the minor arc from B to A.
Then arg $(z - 1)$, i.e. the angle xAQ, increases from $3\pi/4$ to π.
Also arg $(z - i)$, i.e. the angle CBQ, increases from $-\pi/2$ to $-\pi/4$.
At each point on the minor arc,

$$\arg (z - 1) - \arg (z - i) = 5\pi/4.$$

This is the equation of the minor arc.
Since tan $(5\pi/4)$ = tan $(\pi/4)$ = 1, each of the statements

$$\arg (z - 1) - \arg (z - i) = \pi/4 \qquad \qquad \text{... (1)}$$
and $$\qquad \arg (z - 1) - \arg (z - i) = 5\pi/4 \qquad \qquad \text{... (2)}$$

implies the statement

$$(x - 1)^2 + (y - 1)^2 = 1, \qquad \qquad \text{... (3)}$$

but neither statement (1) nor statement (2) is equivalent to statement (3).

Example 6
Find the value of z at the point at which the circular arc

$$\arg\left(\frac{z - 2i}{z + 2}\right) = \frac{3\pi}{4} \text{ intersects the circle } \left|\frac{z - 2i}{z + 2}\right| = \sqrt{2}.$$

Let
$$w = \frac{z - 2i}{z + 2}.$$

At the point of intersection, $|w| = \sqrt{2}$ and $\arg w = 3\pi/4$.

Hence
$$w = \sqrt{2}[\cos(3\pi/4) + i\sin(3\pi/4)]$$
$$\Rightarrow \qquad w = -1 + i$$

$$\Rightarrow \qquad \frac{z - 2i}{z + 2} = -1 + i$$

$$\Rightarrow \qquad z = (-8 + 6i)/5.$$

Exercises 2.8

1 Find in the form $|z - c| = R$ the equation of the circle which passes through the origin and has its centre at the point representing
   (a) $1 + i$,   (b) $4 - 3i$,   (c) $3i$,   (d) $3 + 3i$.

2 Find in the form $|z - c| = R$ the equation of the circle on AB as diameter when the points A and B represent the complex numbers
   (a) $5 + 5i$ and $-1 - 3i$,   (b) $2 + 3i$ and $-2 + 3i$,
   (c) $4 + i$ and $4 - i$,   (d) $-1 + 3i$ and $3 - i$.

3 Obtain the cartesian equation of the circle
   (a) $|z - 1 - i| = \sqrt{2}$ (b) $|z - 3i| = 2|z|$
   (c) $|z + 3 + 4i| = \sqrt{2}|z|$ (d) $|z + 12 - 9i| = 4|z|$.

4 Show that the equations $|z - i| = 2|z + 2i|$ and $|z + 3i| = 2$ represent the same circle.

5 Draw in an Argand diagram the straight line $z = 3 + t(3 + 4i)$ and the circle $3|z + 4i| = 2|z - 3|$. Show that the radius of the circle is 6.

6 The points A and B represent the complex numbers $4i$ and $-2i$ respectively, and the point P is such that $PA = 2PB$. Show that the locus of P is a circle of radius 4.

7 Show that
$$\arg(z + 2) - \arg(z - 2i) = \pi/2$$

is the equation of a semicircular arc which passes through the origin.

8 Show that the equation $\arg\left[\dfrac{z - 2}{z + 2}\right] = \dfrac{\pi}{6}$ represents an arc of a circle of radius 4.

   Draw this arc in an Argand diagram, together with the half-lines $\arg(z - 2) = 2\pi/3$ and $\arg(z + 2) = \pi/2$.

9 The circle $|z - 4| = 5$ cuts the y-axis at the points A and B. Show that the equation of the major arc of this circle standing on **AB** is

$$\arg\left[\frac{z + 3i}{z - 3i}\right] = \tan^{-1}(3/4).$$

10 Shade in an Argand diagram the region given by

$$\pi/3 \leqslant \arg\left[\frac{z - i}{z + i}\right] \leqslant 2\pi/3.$$

Show that the area of this region is $(4\pi + 6\sqrt{3})/9$.

2.9 The transformation $w = az + b$

The equation $w = az + b$, where a and b are complex and $a \neq 0$, expresses w as a function of z. To each point in the z-plane there corresponds one point in the w-plane. Since $z = (w - b)/a$, to each point in the w-plane there corresponds one point in the z-plane.

Consider first three special cases.

1. Let $a = 1$, $b = p + iq$, where p and q are real.

Then $$w = z + p + iq.$$

If $z = x + iy$ and $w = u + iv$, this gives

$\Rightarrow \qquad \qquad u + iv = (x + iy) + (p + iq)$

$\Rightarrow \qquad \qquad u = x + p \quad \text{and} \quad v = y + q.$

Each point in the z-plane is moved through the same distance in the same direction. Such a transformation, known as a translation, leaves the size, the shape and the orientation of any figure unchanged.

Figure 2.15 shows the effect of the transformation $w = z + 2 + i$ on the unit square with vertices given by $z = 0, 1, 1 + i$ and i.

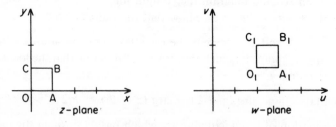

Fig 2.15 $\quad w = z + 2 + i$

2. Let $a = k$, where k is real and positive, and let $b = 0$.

Then $$w = kz$$

$\Rightarrow \qquad \qquad |w| = k|z| \quad \text{and} \quad \arg w = \arg z.$

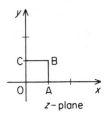
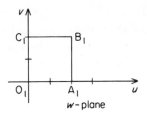

Fig. 2.16 $w = 2z$

This transformation is a magnification by the scale factor k, with centre the origin. The shape and the orientation of any figure will be unchanged.

Figure 2.16 illustrates the case when $w = 2z$.

3.   Let $b = 0$, so that $w = az$.

Then $$|w| = |a| \times |z|$$
and $$\arg w = \arg z + \arg a.$$

The distance of any point in the z-plane from the origin will be multiplied by $|a|$, and each point will be rotated through the angle $\arg a$ about the origin.

In the transformation $w = 2iz$, $|w| = 2|z|$.

Also $$\arg w = \arg z + \arg (2i)$$
$$= \arg z + \pi/2,$$

giving an anticlockwise rotation about O through a right angle. This is illustrated in Fig. 2.17.

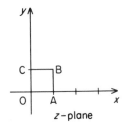
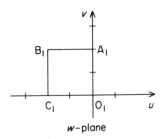

Fig. 2.17 $w = 2iz$

In each of the three basic cases considered above, the shape of any figure is left unchanged; a straight line in the z-plane is mapped to a straight line in the w-plane, a square is mapped to a square and a circle is mapped to a circle.

The same will be true of any combination of these three cases.

The general form of any such combination is given by

$$w = az + b,$$

where a and b are complex.

Example 1

The triangle OAB in the z-plane has its vertices at the points given by $z = 0$, $1 + i$ and $-1 + i$. Find the figure in the w-plane to which this triangle is mapped by the transformation

$$w = (1 + i)z + 2 + i.$$

When $z = 0$, $w = 2 + i$.
When $z = 1 + i$, $w = (1 + i)(1 + i) + 2 + i = 2 + 3i$.
When $z = -1 + i$, $w = (1 + i)(-1 + i) + 2 + i = i$.

Since the shape of any figure remains unchanged, the triangle OAB is mapped to the triangle in the w-plane with vertices at the points given by $w = 2 + i$, $2 + 3i$ and i.

This transformation can be made in two stages.

Let $Z = (1 + i)z$ and $w = Z + 2 + i$.

Since $1 + i = \sqrt{2}(\cos \pi/4 + i \sin \pi/4)$, the transformation from the z-plane to the Z-plane gives a magnification by a scale factor $\sqrt{2}$, centre O, followed by an anticlockwise rotation about the origin through $\pi/4$ radians.

The transformation from the Z-plane to the w-plane is a translation which moves the origin to the point $2 + i$.

This is illustrated in Fig. 2.18.

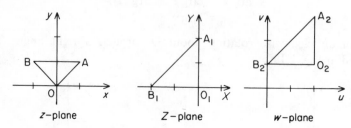

Fig. 2.18 $Z = (1 + i)z$, $w = Z + 2 + i$

Example 2

Express the transformation $w = 3iz - 6$ as the combination of a magnification, a rotation and translation in that order.

Let $Z = 3z$, a magnification by the scale factor 3, centre the origin,
         $W = iZ$, an anticlockwise rotation through a right angle,
and $w = W - 6$, a translation.
Then $w = W - 6$
$\Rightarrow$ $w = iZ - 6$
$\Rightarrow$ $w = 3iz - 6$.

Exercises 2.9

1 Find the transformation from the z-plane to the w-plane that gives a rotation about the origin

(a) anticlockwise through $3\pi/4$ radians,
(b) anticlockwise through $\pi/3$ radians,
(c) clockwise through $2\pi/3$ radians,
(d) clockwise through $\pi/6$ radians,

2 The vertices of the square ABCD in the z-plane are given by $z = 2 + 2i$, $2 - i, -1 - i$ and $-1 + 2i$. Show in a diagram the figure in the w-plane to which this square is mapped by the transformation
(a) $w = z + 2i$,   (b) $w = -z + 1 + i$,
(c) $w = 2z + 2$,   (d) $w = iz + 1$.

3 Express the transformation $w = 4z - 12i$
(a) as a magnification followed by a translation,
(b) as a translation followed by a magnification.
  Show that the circle $|z - 4i| = 1$ is mapped to the circle $|w - 4i| = 4$.

4 Find which of the following combinations of transformations are equivalent to the transformation $w = 2iz + 8$.
(a) $Z = iz, W = 2Z, w = W + 8$,
(b) $Z = iz, W = Z + 4, w = 2W$,
(c) $Z = 2z, W = iZ, w = W + 8$,
(d) $Z = 2z, W = Z - 8i, w = iW$,
(e) $Z = z - 4i, W = iZ, w = 2W$,
(f) $Z = z - 4i, W = 2Z, w = iW$.

5 Show that the transformation $w = 2iz + 2$ maps the circle $|z| = 2$ to the circle $|w - 2| = 4$, and the circle $|z - 4| = 2$ to the circle $|w - 2 - 8i| = 4$. Sketch these circles.

6 Express each of the following transformations as a magnification followed by a rotation and then a translation.
(a) $w = -2i(z + i)$,   (b) $w = (4 + 4i)z + i$,
(c) $w = (\sqrt{3} + i)(z - \sqrt{3} + i)$,   (d) $w = (1 + i)(3z - 1 + i)$.

2.10 The transformation $w = \dfrac{1}{z}, z \neq 0$

When $w = 1/z$, $|w| = 1/|z|$.
  This shows that points on the circle $|z| = 1$ will be mapped to points on the circle $|w| = 1$.
  Points for which $|z|$ is less than 1, i.e. points inside the unit circle in the z-plane, will be mapped to points for which $|w|$ is greater than 1, i.e. points outside the unit circle in the w-plane.
  Points for which $|z| > 1$ will be mapped to points for which $|w| < 1$.
  Also, when $w = \dfrac{1}{z}$, $\arg w = \arg\left(\dfrac{1}{z}\right) = -\arg z$.

This shows that points in the upper half of the z-plane will be mapped to points in the lower half of the w-plane. Points in the lower half of the z-plane will be mapped to points in the upper half of the w-plane.

Let $|z| = r$, and arg $z = \theta$, so that $z = r(\cos \theta + i \sin \theta)$.

Then, if $w = 1/z$,

$$z = r(\cos \theta + i \sin \theta) \quad \Rightarrow \quad w = \frac{1}{r}(\cos \theta - i \sin \theta).$$

For example,

$$z = 2(\cos \pi/3 + i \sin \pi/3) \Rightarrow w = \tfrac{1}{2}(\cos \pi/3 - i \sin \pi/3),$$
$$z = \tfrac{2}{3}(\cos \pi/4 - i \sin \pi/4) \Rightarrow w = \tfrac{3}{2}(\cos \pi/4 + i \sin \pi/4).$$
$$z = \cos (3\pi/4) + i \sin (3\pi/4) \Rightarrow w = \cos (3\pi/4) - i \sin (3\pi/4).$$

The points A, B, C given by these values of z and the corresponding points A_1, B_1, C_1 in the w-plane are shown in Fig. 2.19.

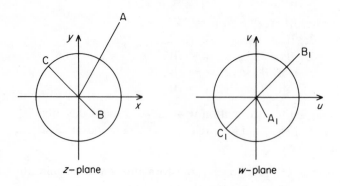

Fig. 2.19 $w = \dfrac{1}{z}$

Let $z = x + iy$ and $w = u + iv$, where u and v are real.

When $w = 1/z$,

$$u + iv = \frac{1}{x + iy} = \frac{x - iy}{x^2 + y^2}$$

$$\Rightarrow \qquad u = \frac{x}{x^2 + y^2}, \qquad v = \frac{-y}{x^2 + y^2}$$

$$\Rightarrow \qquad v/u = -y/x.$$

Let m be a real number. Then

$$y = mx \quad \Rightarrow \quad v = -mu.$$

This shows that the straight line $y = mx$, which passes through the origin in the z-plane, is mapped to the straight line $v = -mu$, which passes through the origin in the w-plane.

Also,
$$x + iy = \frac{1}{u + iv} = \frac{u - iv}{u^2 + v^2}$$

$\Rightarrow$
$$x = \frac{u}{u^2 + v^2}, \quad y = \frac{-v}{u^2 + v^2}.$$

The equation $y = mx + c$ becomes

$$\frac{-v}{u^2 + v^2} = \frac{mu}{u^2 + v^2} + c$$

$\Rightarrow$
$$c(u^2 + v^2) + mu + v = 0$$
$\Rightarrow$
$$u^2 + v^2 + mu/c + v/c = 0,$$

provided that $c \neq 0$.

This shows that the straight line $y = mx + c$, which does not pass through the origin, is mapped to a circle in the w-plane. Note that this circle passes through the origin.

Finally, consider the circle with equation

$$x^2 + y^2 + 2gx + 2fy + c = 0.$$

Since $x^2 + y^2 = 1/(u^2 + v^2)$, this becomes

$$\frac{1}{u^2 + v^2} + \frac{2gu}{u^2 + v^2} - \frac{2fv}{u^2 + v^2} + c = 0$$

$\Rightarrow$
$$cu^2 + cv^2 + 2gu - 2fv + 1 = 0.$$

When c is not zero, the circle in the z-plane is mapped to a circle in the w-plane, neither circle passing through the origin.

When c is zero, the circle in the z-plane passes through the origin and is mapped to a straight line in the w-plane.

Example 1

Given that $w = \dfrac{1}{z}$, find the loci in the w-plane to which the circles $|z - 1| = 1$ and $|z - 1| = 2$ are mapped. Find also the region in the w-plane corresponding to the region defined by $1 \leqslant |z - 1| \leqslant 2$.

When $w = 1/z$,
$$|z - 1| = 1$$

$\Rightarrow$
$$\left| \frac{1}{w} - 1 \right| = 1$$

$$\Rightarrow \qquad |1 - w| = |w|$$
$$\Rightarrow \qquad |w - 1| = |w|.$$

This is the equation of the perpendicular bisector of the line joining the points given by $w = 0$ and $w = 1$, i.e. the line $u = \frac{1}{2}$.

Also
$$|z - 1| = 2$$

$$\Rightarrow \qquad \left|\frac{1}{w} - 1\right| = 2$$

$$\Rightarrow \qquad |w - 1| = 2|w|$$
$$\Rightarrow \qquad |u + iv - 1|^2 = 4|u + iv|^2$$
$$\Rightarrow \qquad 3u^2 + 3v^2 + 2u = 1$$
$$\Rightarrow \qquad (u + \tfrac{1}{3})^2 + v^2 = \tfrac{4}{9}.$$

This is the equation of the circle with centre given by $w = -\frac{1}{3}$ and radius $\frac{2}{3}$. These loci are shown in Fig. 2.20.

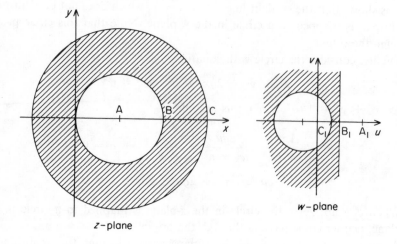

z-plane

w-plane

Fig. 2.20 $\quad w = \dfrac{1}{z}$

Since $w = 1/z$, w is real when z is real.

Also
$$z = 2 \Rightarrow w = \tfrac{1}{2} \quad \text{and} \quad z = 3 \Rightarrow w = \tfrac{1}{3}.$$

As z increases through real values from 2 to 3, w decreases through real values from $\frac{1}{2}$ to $\frac{1}{3}$.

This shows that points in the z-plane lying between the two circles are mapped to points in the w-plane that lie outside the circle $|w - 1| = 2|w|$ but to the left of the line $|w - 1| = |w|$.

The corresponding regions are shown shaded.

Example 2
The vertices of the square OABC in the z-plane are given by $z = 0, 1, 1 + i$ and i. Find the loci in the w-plane to which the sides of this square are mapped by the transformation $w = 1/z$.

(a) On OA, z is real and so w is real. As z increases from 0 to 1, w decreases from infinity to 1. This is illustrated in Fig. 2.21.

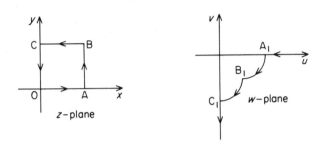

Fig. 2.21 $w = \dfrac{1}{z}$

(b) On AB, let $z = 1 + it$, where $0 \leqslant t \leqslant 1$. This gives

$$w = \frac{1}{1 + it} = \frac{1 - it}{1 + t^2}$$

$\Rightarrow$
$$u = \frac{1}{1 + t^2}, v = \frac{-t}{1 + t^2}$$

$\Rightarrow$
$$u^2 + v^2 = \frac{1 + t^2}{(1 + t^2)^2} = \frac{1}{1 + t^2}$$

$\Rightarrow$
$$u^2 + v^2 = u$$
$\Rightarrow$
$$(u - \tfrac{1}{2})^2 + v^2 = \tfrac{1}{4}.$$

This is the equation of the circle in the w-plane with centre given by $w = \tfrac{1}{2}$ and with radius $\tfrac{1}{2}$.

As t increases from 0 to 1, u decreases from 1 to $\tfrac{1}{2}$ and v decreases from 0 to $-\tfrac{1}{2}$.

Hence as z moves from A to B, w moves along the arc A_1B_1 shown in Fig. 2.21.

(c) On BC, let $z = i + t$, where $0 \leqslant t \leqslant 1$.

This gives
$$w = \frac{1}{i + t} = \frac{t - i}{1 + t^2}$$

$\Rightarrow$
$$u = \frac{t}{1 + t^2}, v = \frac{-1}{1 + t^2}$$

$$\Rightarrow \qquad u^2 + v^2 = \frac{1}{1 + t^2} = -v$$

$$\Rightarrow \qquad u^2 + (v + \tfrac{1}{2})^2 = \tfrac{1}{4}.$$

This is the equation of the circle in the w-plane with centre given by $w = -i/2$ and with radius $\tfrac{1}{2}$.

As t decreases from 1 to 0, u decreases from $\tfrac{1}{2}$ to 0 and v decreases from $-\tfrac{1}{2}$ to -1.

Hence in Fig. 2.21 w moves along the arc $B_1 C_1$.

(d) On CO, let $z = it$, where $0 \leqslant t \leqslant 1$.
Then $w = 1/(it) = -i/t$.

As t decreases from 1 to 0, w moves along the negative v-axis away from the origin.

Note that the centre of the square, given by $z = (1 + i)/2$, is mapped to the point given by $w = 1 - i$.

Exercises 2.10

1 A region is defined in the z-plane by $1 \leqslant |z| \leqslant 2, 0 \leqslant \arg z \leqslant \pi/4$. Find the region to which this is mapped by the transformation $w = 1/z$.

2 Show that the points given by $z = 1, 1 + i$ and i lie on the circle $|2z - 1 - i| = \sqrt{2}$. Show also that the transformation $w = 1/z$ maps these points to points on a straight line.

3 Given that $w = 1/z$, show that the circle $|z - 2| = 4$ is mapped to the circle $12u^2 + 12v^2 + 4u = 1$.

4 Given that $w = 1/z$, show that the straight line $|z - 2i| = |z|$ is mapped to a circle of radius $\tfrac{1}{2}$.

5 Show that the transformation $w = 1/z$ maps the line $y = x$ to the line $v = -u$ and the line $y = x + 1$ to the circle $u^2 + v^2 + u + v = 0$. Sketch these loci.

6 A sector of a circle is given by $\pi/6 \leqslant \arg z \leqslant \pi/3, 0 \leqslant |z| \leqslant 2$. Given that $w = 1/z$, find the corresponding region in the w-plane.

7 Find the cartesian equation of the circle in the w-plane to which the line $|z - 2| = |z - 4i|$ is mapped by the transformation $w = 1/z$.

8 Show that the transformation $w = 1/z$ maps the circle $|z - 3| = 2|z|$ to the circle $|3w - 1| = 2$.

9 The points given by $z = \pm 1 \pm i$ are the vertices of a square. Find the loci in the w-plane to which the transformation $w = 1/z$ maps the sides of this square.

10 When $w = 1/z$ and $z = x + iy$, $w = u + iv$, show that $x = u/(u^2 + v^2)$.

Find the area of the region in the w-plane to which the transformation $w = 1/z$ maps the region given by $1 \leqslant x \leqslant 2$.

2.11 The transformation $w = \dfrac{az + b}{cz + d}$

Let $w = \dfrac{az + b}{cz + d}$, where a, b, c, $d \in \mathbb{C}$, the set of complex numbers, and $z \neq -d/c$.

It will be assumed that $c \neq 0$, as the case when c is zero is the same as that considered in Section 2.9.

If $a/c = b/d$, the relation between w and z reduces to $w = a/c$ for all values of z. For this reason, the constants a, b, c and d must be such that $ad - bc$ is not zero.

In this transformation, each value of z gives one value of w.

Also $\qquad\qquad (cz + d)w = az + b$

$\Rightarrow \qquad\qquad (cw - a)z = -dw + b$

$\Rightarrow \qquad\qquad z = \dfrac{b - dw}{cw - a}$,

showing that each value of w, except a/c, gives one value of z.

This transformation can be expressed as a combination of transformations of the form $w = az + b$ and $w = 1/z$.

$$w = \frac{az + b}{cz + d}$$

$\Rightarrow \qquad\qquad w = \dfrac{b - (ad/c)}{cz + d} + \dfrac{a}{c}.$

Let $Z = z + d/c$. Then $\quad w = \dfrac{b - (ad/c)}{cZ} + \dfrac{a}{c}.$

Let $W = \dfrac{1}{Z}$. Then $\qquad w = \left[\dfrac{bc - ad}{c^2}\right]W + \dfrac{a}{c}$

$\Rightarrow \qquad\qquad w = kW + \dfrac{a}{c}$,

where $k = (bc - ad)/c^2$.

This shows that the transformation $w = \dfrac{az + b}{cz + d}$ is equivalent to the combination of the transformations

$$Z = z + d/c,$$
$$W = 1/Z,$$
$$w = kW + a/c.$$

Now the transformations $Z = z + d/c$ and $w = kW + a/c$ map a circle to a circle, while the transformation $W = 1/Z$ maps a circle either to a circle or to a straight line.

It follows that the transformation $w = \dfrac{az + b}{cz + d}$ maps a circle either to a circle or to a straight line.

Example 1

Show that the transformation $w = \dfrac{z + 1}{iz}$ maps the circle $|z| = 1$ to the circle $|w + i| = 1$.

Find the cartesian equation of the locus in the w-plane to which the circle $|z + 2| = 1$ is mapped.

Since
$$w = \frac{z + 1}{iz},$$

$$\Rightarrow \qquad izw = z + 1$$
$$z(iw - 1) = 1$$

$$\Rightarrow \qquad z = \frac{-i}{w + i}$$

$$\Rightarrow \qquad |z| = \frac{1}{|w + i|}.$$

Fig. 2.22 $w = \dfrac{z + 1}{iz}$

Hence when $|z| = 1$, $|w + i| = 1$, showing that the circle $|z| = 1$ is mapped to the circle $|w + i| = 1$.

The points A, B, C given by $z = 1, i, -1$ are mapped to the points A_1, B_1, C_1 given by $w = -2i, -1 - i, 0$ respectively.

Note that since $|z| = 1/|w + i|$, the region given by $|z| \geqslant 1$ is mapped to the region given by $|w + i| \leqslant 1$.

This means that the exterior of the circle $|z| = 1$ is mapped to the interior of the circle $|w + i| = 1$.

From
$$z = \frac{-i}{w + i},$$

$$z + 2 = \frac{2w + i}{w + i}.$$

When $|z + 2| = 1$, $|2w + i| = |w + i|$.

Let $w = u + iv$. Then

$$|2u + 2iv + i|^2 = |u + iv + i|^2$$
$$\Rightarrow \qquad 4u^2 + 4v^2 + 4v + 1 = u^2 + v^2 + 2v + 1$$
$$\Rightarrow \qquad 3u^2 + 3v^2 + 2v = 0$$
$$\Rightarrow \qquad u^2 + (v + \tfrac{1}{3})^2 = \tfrac{1}{9}.$$

This shows that the circle $|z + 2| = 1$ is mapped to a circle of radius $\tfrac{1}{3}$ with centre at the point given by $w = -i/3$. This circle passes through the origin in the w-plane.

Alternatively, the transformation can be put in the form

$$w = \frac{z + 1}{iz} = -\frac{i}{z} - i.$$

This shows that the transformation is equivalent to the combination of the three simple transformations

$$Z = \frac{1}{z}, \quad W = -iZ, \quad w = W - i.$$

The circle $|z| = 1$ is mapped to the circle $|Z| = 1$.

Since $|-iZ| = |Z|$, the transformation $W = -iZ$ maps the circle $|Z| = 1$ to the circle $|W| = 1$.

Since $W = w + i$, the circle $|W| = 1$ is mapped to the circle $|w + i| = 1$. This shows that the circle $|z| = 1$ is mapped to the circle $|w + i| = 1$.

Since $z + 2 = \dfrac{1}{Z} + 2 = \dfrac{2Z + 1}{Z}$, the transformation $Z = \dfrac{1}{z}$ maps the circle $|z + 2| = 1$ to the circle $|2Z + 1| = |Z|$.

This circle is mapped by the transformation $W = -iZ$ to the circle $|2W - i| = |W|$.

Finally, the transformation $W = w + i$ maps this circle to the circle $|2w + i| = |w + i|$, as before.

Example 2

A transformation from the z-plane to the w-plane is given by

$$w = \frac{2z - 1}{z - 2}.$$

(a) Show that the circle $|z| = 1$ is mapped to the circle $|w| = 1$.
(b) Find the locus in the w-plane which is the image of the circle $|4z - 3| = 5$.

Note that the coefficients in the relation between z and w are real, so that the real axis in the z-plane will be mapped to the real axis in the w-plane.

Since

$$\frac{2z - 1}{z - 2} = \frac{3}{z - 2} + 2,$$

the given transformation is equivalent to the combination of the three transformations

$$Z = z - 2, \quad W = \frac{1}{Z}, \quad w = 3W + 2.$$

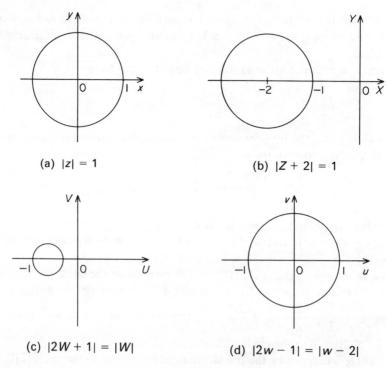

(a) $|z| = 1$

(b) $|Z + 2| = 1$

(c) $|2W + 1| = |W|$

(d) $|2w - 1| = |w - 2|$

Fig. 2.23

(a) The transformation $Z = z - 2$ maps the circle $|z| = 1$ to the circle $|Z + 2| = 1$.

This circle is mapped by the transformation $W = \dfrac{1}{Z}$ to the circle $|2W + 1| = |W|$.

This circle is symmetrical about the real axis, and passes through the points given by $W = -1$, $W = -\frac{1}{3}$ (Fig. 2.23).

Since $W = \frac{1}{3}(w - 2)$, this circle is mapped to the circle

$$|\tfrac{2}{3}(w - 2) + 1| = \tfrac{1}{3}|w - 2|$$

$\Rightarrow$
$$|2w - 1| = |w - 2|.$$

Let $w = u + iv$. Then the cartesian equation of this circle is

$$|2u + 2iv - 1|^2 = |u + iv - 2|^2$$
$\Rightarrow$
$$4u^2 - 4u + 1 + 4v^2 = u^2 - 4u + 4 + v^2$$
$\Rightarrow$
$$u^2 + v^2 = 1.$$

This shows that the circle $|z| = 1$ is mapped to the circle $|w| = 1$.

(b) The transformation $Z = z - 2$ maps the circle $|4z - 3| = 5$ to the circle $|4Z + 5| = 5$, with centre $Z = -5/4$ and radius $5/4$ (Fig. 2.24).

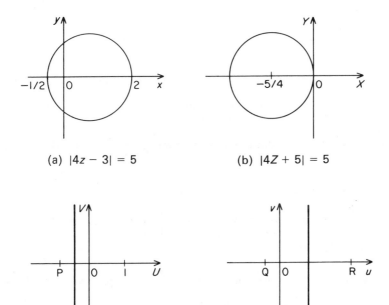

(a) $|4z - 3| = 5$

(b) $|4Z + 5| = 5$

(c) $|5W + 4| = 5|W|$

(d) $|w + 2/5| = |w - 2|$

Fig. 2.24

This circle is mapped by the transformation $W = \dfrac{1}{Z}$ to the straight line

$$|5W + 4| = |5W|$$
$$\Rightarrow \qquad |W + 4/5| = |W|.$$

This line is the perpendicular bisector of the line joining the point P given by $W = -4/5$ to the origin $W = 0$.

The transformation $w = 3W + 2$, or $W = \frac{1}{3}(w - 2)$, maps this line to the line

$$|\tfrac{1}{3}(w - 2) + 4/5| = \tfrac{1}{3}|w - 2|$$
$$\Rightarrow \qquad |w + 2/5| = |w - 2|.$$

This line is the perpendicular bisector of the line joining the points Q and R given by $w = -2/5$ and $w = 2$.

Let $w = u + iv$. Then the cartesian equation of this line is $u = 4/5$.

Hence the transformation maps the circle $|4z - 3| = 5$ to the straight line $u = 4/5$.

Example 3

Show that, if a is real and $a \neq 1$, the transformation $w = \dfrac{z - a}{az - 1}$ maps the circle $|z| = 1$ to the circle $|w| = 1$.

Let $z = \cos \theta + i \sin \theta = e^{i\theta}$. Then $|z| = 1$.

Also
$$w = \frac{e^{i\theta} - a}{ae^{i\theta} - 1} = \frac{-(a - e^{i\theta})}{e^{i\theta}(a - e^{-i\theta})}.$$

Since a is real, $a - e^{i\theta}$ and $a - e^{-i\theta}$ are conjugate complex numbers, so that

$$|a - e^{i\theta}| = |a - e^{-i\theta}|.$$

Also, $|e^{i\theta}| = 1$.

It follows that $|w| = 1$, showing that the circle $|z| = 1$ is mapped to the circle $|w| = 1$.

Exercises 2.11

1   Show that the transformation $w = \dfrac{z - i}{z + 1}$ maps the circle $|z| = 1$ to the line $|w - 1| = |w + i|$.

2   Show that the transformation $w = \dfrac{2z + 4}{iz + 1}$ maps the straight line $|z - i| = |z + 2|$ to the circle $|w| = 2$.

3   Show that the transformation $w = (z + 1)/z$ maps the circle $|z| = 1$ to the circle $|w - 1| = 1$, and maps the circle $|z + 1| = 1$ to the straight line $|w - 1| = |w|$.

4   Given that $w = (z + 1)/(z - 1)$, show that
(a) the real axis in the z-plane is mapped to the real axis in the w-plane,
(b) the y-axis in the z-plane is mapped to the circle $|w| = 1$.
   Show also that the left-hand half of the z-plane is mapped to the region $|w| < 1$.

5   Show that the transformation $w = \dfrac{z + 2}{2z + 1}$ maps the circle $|z| = 1$ to the circle $|w| = 1$. Show also that the region $|z| \leqslant 1$ is mapped to the region $|w| \geqslant 1$.

6   Given that $w \dfrac{1 + iz}{z + i}$, show that $z = \dfrac{1 - iw}{w - i}$. Deduce that the circle $|z| = 2$ is mapped to the circle $|w + i| = 2|w - i|$.

7   Show that the transformation $w = \dfrac{z - 2i}{2iz + 1}$ maps the region $|z| \leqslant 1$ to the region $|w| \geqslant 1$.

8   Show that the transformation $w = \dfrac{z - i}{z + i}$ is equivalent to the combination of transformations $Z = z + i$, $W = 1/Z$, $w = 1 - 2iW$.
   Show in diagrams the loci in the Z, W and w-planes corresponding to the circle $|z| = 1$.

9   Show that the transformation $w = i(1 - z)/(1 + z)$ maps the circle $|z| = 1$ to the real axis in the w-plane, and the region $|z| < 1$ to the upper half of the w-plane.

10   A transformation is given by $w = (i - z)/(i + z)$. If $z = t$, where t is real, and $w = u + iv$, show that

$$u = \frac{1 - t^2}{1 + t^2}, \quad v = \frac{2t}{1 + t^2}.$$

Deduce that when $z = \tan(\theta/2)$, $w = \cos\theta + i\sin\theta$. Show that the transformation maps the region Im $z > 0$ to the region $|w| < 1$.

Miscellaneous exercises 2

1   Express $1 + i\sqrt{3}$ and $1 - i\sqrt{3}$ in the form $re^{i\theta}$, where $r > 0$ and $-\pi < \theta \leqslant \pi$. Hence evaluate

$$\left(\frac{1 + i\sqrt{3}}{1 - i\sqrt{3}}\right)^{20},$$

expressing your result in the form $a + ib$, where a and b are real numbers.

[L]

2 Given that $(1 + i)^n = x + iy$, where x and y are real and n is an integer, prove that $x^2 + y^2 = 2^n$. [L]

3 Find the modulus and argument of

$$\frac{1 + \sin \theta + i \cos \theta}{1 + \sin \theta - i \cos \theta}.$$

Hence show that

$$\left[\frac{1 + \sin \pi/3 + i \cos \pi/3}{1 + \sin \pi/3 - i \cos \pi/3}\right]^3 = i. \qquad [L]$$

4 Given that $2 + 3i$ is a root of the equation

$$z^3 - 6z^2 + 21z - 26 = 0,$$

find the other two roots.

5 Represent on one Argand diagram
(a) the three roots of the equation $z^3 + 1 = 0$,
(b) the six roots of the equation $z^6 + 64 = 0$.

6 Given that $z = \cos \theta + i \sin \theta$, use de Moivre's theorem to show that $z^n + z^{-n} = 2 \cos n\theta$.
  Use this result to express $\cos^5 \theta$ in terms of cosines of multiples of θ. Show that

$$\int_0^{\pi/6} \cos^5 \theta \, d\theta = \frac{203}{480}.$$

7 (a) The complex numbers z and a are such that $z = x + iy$ and $a = 1 + i$. Sketch and obtain the cartesian equation of the locus of z in each of the following cases:
(i) $|z - a| = 3$, (ii) $\arg (z - a) = \pi/2$.
(b) When $z = 4\sqrt{3}e^{i\pi/3} - 4e^{i5\pi/6}$, express z in the form $re^{i\theta}$. Hence
(i) show that

$$\frac{z}{8} + i\left(\frac{z}{8}\right)^2 + \left(\frac{z}{8}\right)^3 = 2e^{i\pi/2},$$

(ii) find the cube roots of z in the form $re^{i\theta}$. [AEB]

8 Given that $z = (2 + 3t) + i(4 - t)$, where t is a real parameter, obtain the cartesian equation of the locus of z as t varies and sketch this locus in an Argand diagram.
  Find the minimum value of $|z|$ for real values of t. [L]

9 Prove that $\cos 5\theta = 16 \cos^5 \theta - 20 \cos^3 \theta + 5 \cos \theta$.
  By considering the equation $\cos 5\theta = 0$, prove that

$$\cos(\pi/10)\cos(3\pi/10) = \tfrac{1}{4}\sqrt{5}.$$ [L]

10 Given that $z = e^{i\theta}$ show that $z^n - z^{-n} = 2i\sin n\theta$.
Hence, or otherwise, prove that

$$16\sin^5\theta = \sin 5\theta - 5\sin 3\theta + 10\sin\theta.$$

Find the general solution of the equation $16\sin^5\theta = \sin 5\theta$. [L]

11 A rectangle R in the z-plane has vertices at the points corresponding to 0, 2, $2 + i$, i. Find the coordinates of the vertices of the rectangle in the w-plane into which R is mapped by the transformation $w = \sqrt{2}e^{i\pi/4}z$.

12 Express the roots of the equation $z^7 = 1$ in the form $re^{i\theta}$, where $r > 0$ and $-\pi < \theta \leqslant \pi$, and show the corresponding points on an Argand diagram.
Expand $(z - e^{i\theta})(z - e^{-i\theta})$ as a quadratic expression in z involving only trigonometrical functions of θ.
Hence, or otherwise, show that

$$z^6 + z^5 + z^4 + z^3 + z^2 + z + 1$$

$$\equiv \left(z^2 - 2z\cos\frac{2\pi}{7} + 1\right)\left(z^2 - 2z\cos\frac{4\pi}{7} + 1\right)\left(z^2 - 2z\cos\frac{6\pi}{7} + 1\right).$$
[L]

13 (a) Use de Moivre's theorem in the form

$$\cos 5\theta + i\sin 5\theta = (\cos\theta + i\sin\theta)^5$$

to find $\cos 5\theta$ and $\sin 5\theta$ each in terms of $\cos\theta$ and $\sin\theta$. Hence find $\tan 5\theta$ in terms of $\tan\theta$.
(b) Show that the equation $|z + 1| = 2|z - 1|$ represents a circle in the Argand diagram. Find the centre and radius of this circle. [L]

14 By expressing $\cos 4\theta$ and $\sin 4\theta$ in terms of $\sin\theta$ and $\cos\theta$, show that

$$\tan 4\theta = \frac{4t - 4t^3}{1 - 6t^2 + t^4},$$

where $t = \tan\theta$. Obtain each of the roots of the equation

$$t^4 + 4t^3 - 6t^2 - 4t + 1 = 0$$

in the form $\tan(p\pi)$.

15 (a) Find the complex number z such that $(2 - 3i)z = 4 + i$.
(b) In an Argand diagram shade the region in which the point representing the complex number z can lie if

$$|z + 2i - 1| < |z - i|.$$

(c) Use de Moivre's theorem to show that

$$\cos 4\theta = 8\cos^4\theta - 8\cos^2\theta + 1.$$ [L]

16 If $z = \cos \theta + i \sin \theta$, show that

$$z^n + 1/z^n = 2 \cos n\theta, \quad z^n - 1/z^n = 2i \sin n\theta.$$

By expanding $(z + 1/z)^4(z - 1/z)^4$, show that

$$128 \sin^4 \theta \cos^4 \theta = \cos 8\theta - 4 \cos 4\theta + 3.$$

17 Show that $\qquad 2i \sin n\theta = z^n - z^{-n}$,

where $z = \cos \theta + i \sin \theta$. Hence, or otherwise, prove that

$$\sum_{r=1}^{\infty} \frac{\sin r\theta}{2^r} = \frac{2 \sin \theta}{5 - 4 \cos \theta}. \qquad \text{[L]}$$

18 Write down the factors of $x^5 - 1$ in the field of complex numbers. Hence show that

$$x^2 - x + 1 + \frac{1}{x} + \frac{1}{x^2} = \left(x - 2 \cos \frac{2\pi}{5} + \frac{1}{x}\right)\left(x - 2 \cos \frac{4\pi}{5} + \frac{1}{x}\right).$$

By writing $y = x + \dfrac{1}{x}$ in this expression, or otherwise, obtain the value of $\cos (2\pi/5)$ in surd form. **[JMB]**

19 With the aid of a diagram, or otherwise, show that the argument of the complex number

$$z = 1 + \sin \phi + i \cos \phi \qquad (0 < \phi < \pi/2)$$

is $\dfrac{\pi}{2} - \dfrac{\phi}{2}$, and find its modulus.

Prove that

$$\frac{1 + \sin \phi + i \cos \phi}{1 + \sin \phi - i \cos \phi} = \sin \phi + i \cos \phi.$$

Deduce that

$$(1 + \sin \pi/5 + i \cos \pi/5)^5 + i(1 + \sin \pi/5 - i \cos \pi/5)^5 = 0. \quad \text{[JMB]}$$

20 If the complex numbers z_1 and z_2 are represented in the Argand diagram by the points P_1 and P_2 respectively, interpret geometrically the modulus and argument of $z_2 - z_1$.

Find (a) the general value of θ which satisfies the relation

$$\arg (1 + \cos \theta + i \sin \theta) = \pi/3,$$

(b) in cartesian form the equation of the locus of z if

$$\left| \frac{z + i}{z - 2 - i} \right| = 1. \qquad \text{[JMB]}$$

21 The sets A, B and C of points z in the complex plane are given by

$$A = \{z : |z - 2| = 2\},$$
$$B = \{z : |z - 2| = |z|\},$$
$$C = \{z : \arg (z - 2) = \arg z\}.$$

Determine the locus of z in each case, and illustrate the sets of points on one diagram.

Show that the points of the set $A \cap (B \cup C)$ are the vertices of an equilateral triangle. [JMB]

22 Given that $z = \cos \theta + i \sin \theta$ show that $z + z^{-1} = 2 \cos \theta$.

Use de Moivre's theorem to prove that

$$z^n + z^{-n} = 2 \cos n\theta.$$

Hence, or otherwise, solve the equation

$$2z^4 + 3z^3 + 5z^2 + 3z + 2 = 0. \qquad \text{[JMB]}$$

23 Use de Moivre's theorem to express $\cos 7\theta$ in terms of powers of $\cos \theta$. Express $\cos^7 \theta$ in the form

$$A \cos 7\theta + B \cos 5\theta + C \cos 3\theta + D \cos \theta,$$

where A, B, C and D are constants.

Use your expression for $\cos 7\theta$ in terms of powers of $\cos \theta$ to show that the largest root of the equation

$$64x^3 - 112x^2 + 56x - 7 = 0$$

is $\cos^2 (\pi/14)$. [L]

24 By shading in three separate Argand diagrams, show the regions in which the point z can lie when (a) $|z| > 3$, (b) $|z - 2| < |z - 4|$, (c) $0 < \arg (z + 3) < \pi/6$.

Shade in another sketch the region in which z can lie when all these inequalities apply. [L]

25 Describe in terms of a translation, a rotation and an enlargement, the transformation in the complex plane given by

$$w = 2i(z - 2).$$

Sketch on an Argand diagram the circle $|z| = 2$, and sketch on another Argand diagram the image of this circle under the above transformation, giving the equation of the image.

Find the three complex values of w that satisfy the equation

$$(w^2 - 48)w + 12w^2 i = 0.$$

Hence find the corresponding values of z given by the transformation $w = 2i(z - 2)$. [L]

26 In an Argand diagram, P represents a complex number z such that

$$2|z - 2| = |z - 6i|.$$

Show that P lies on a circle, and find
(a) the radius of this circle,
(b) the complex number represented by its centre.

If, further,

$$\arg\left(\frac{z - 2}{z - 6i}\right) = \pi/2,$$

find the complex number represented by P.   [L]

27 By expressing $-2^{12}i$ in the form $re^{i\theta}$, where $r > 0$ and $-\pi < \theta \leqslant \pi$, find the roots of the equation $z^4 = -2^{12}i$ and represent them by points A, B, C and D on an Argand diagram.

Represent on Argand diagrams the points into which A, B, C and D are mapped by the transformations from z to w defined by (a) $w = 2e^{i\pi/8}z$, (b) $w = 1/z$.

28 Show that in an Argand diagram the four points representing the roots of the equation $z^4 = -1$ lie at the corners of a square of side $\sqrt{2}$.

Obtain the image of this square under the transformation from z to w given by

$$w = (1 + i)z + \sqrt{2} + i\sqrt{2},$$

and sketch it on a diagram.

Specify the effect of the transformation in terms of translation, rotation and magnification.   [L]

29 Given that θ has a constant value between 0 and $\pi/2$, represent in the Argand diagram the roots of the equations
(a) $z^2 - 2z \cos \theta + 1 = 0$,   (b) $z^2 - 2z \sec \theta + 1 = 0$.

Show that the four points representing the roots lie on a circle, and find the equation of this circle.   [L]

30 (a) Simplify $(1 + \omega)(1 + \omega^2)$, where

$$\omega = \cos (2\pi/3) + i \sin (2\pi/3).$$

Express in the form $r(\cos \theta + i \sin \theta)$ each of the three roots of the equation $(z - \omega)^3 = 1$.
(b) Express in terms of θ the roots α and β of the equation

$$z + z^{-1} = 2 \cos \theta.$$

In the Argand diagram the points P and Q represent the numbers $(\alpha^n + \beta^n)$ and $(n^\alpha - \beta^n)$ respectively. Show that the length of PQ does not depend on the integer n.   [L]

31 (a) Show that $1 + i$ is a root of the equation

$$z^4 + 3z^2 - 6z + 10 = 0.$$

Find the other roots of this equation.
(b) Sketch the curve in the Argand diagram defined by

$$|z - 1| = 1, \quad \text{Im } z \geqslant 0.$$

Find the value of z at the point P in which this curve is cut by the line $|z - 1| = |z - 2|$.
Find also the values of arg z and arg $(z - 2)$ at P.　　　　　[L]

32 Show that the transformation $w = -2iz + 5i$ is equivalent to a rotation about the origin followed by a magnification and then by a translation.
　　Find the point in the complex plane which is mapped to itself by this transformation. Show that the transformation is equivalent to a rotation about this point followed by a magnification.　　　　　[L]

33 Solve the equation $z^3 + i = 0$, giving the roots in the form $r(\cos \theta + i \sin \theta)$, where $r > 0$ and $-\pi < \theta \leqslant \pi$.
　　The points A, B and C represent these roots in an Argand diagram. The image of the triangle ABC under the transformation from z to w given by

$$w = 2iz + 1 - i$$

is the triangle $A'B'C'$. Find the coordinates of the vertices of the triangle $A'B'C'$.
　　Show in an Argand diagram
(a) the set of points S, where $S = \{z : z \in \mathbb{C}, |z - 1| = |z - i|\}$,
(b) the image of S under the transformation from z to w given by $w = 2iz + 1 - i$.　　　　　[L]

34 Sketch on separate Argand diagrams the loci represented by
(a) $|z| = 4$,　　　　　　　　　　(b) $|z - i| = |z - 3i|$.
The complex variables w and z are related by the transformation $w = 2z + 5$. Sketch, also on separate Argand diagrams, the loci of w as z describes the loci (a) and (b) above.
　　Write down a complex equation of each w-locus.　　　　　[L]

35 (a) Two complex numbers z_1 and z_2 have arguments between 0 and π. The product $z_1 z_2$ is $(-\sqrt{3} + i)$ and the quotient z_1/z_2 is $2i$. Find the values of z_1 and z_2.
(b) If $w = 3(\cos \theta + i \sin \theta)$, find the moduli and arguments of (i) w^2, (ii) $1/w$, (iii) $(w^2 + 1)/w$. Show these four complex numbers on an Argand diagram when $\theta = \pi/6$.　　　　　[AEB]

36 (a) Express the complex number $(1 - i)$ in polar form. Deduce the three solutions z_1, z_2, z_3 of the equation $z^3 = 8 - 8i$.

Show that when represented on an Argand diagram z_1, z_2, z_3 are the vertices of an equilateral triangle.

(b) Expand $(z + 1/z)^4$ and $(z - 1/z)^4$. By substituting $z = \cos\theta + i\sin\theta$, deduce that

$$\cos^4\theta + \sin^4\theta = \tfrac{1}{4}(\cos 4\theta + 3).$$ [AEB]

37 (a) If $z = 5 - 12i$, express $1/z$ and $z^{1/2}$ in the form $a + ib$, where a and b are real. Illustrate z and $1/z$ on an Argand diagram.

(b) If P represents the complex number z on an Argand diagram, find the cartesian equation of the locus of P

(i) when $\dfrac{|z + i|}{|z + 1|} = 3$, (ii) when $\dfrac{z + i}{z + 1}$ is real. [AEB]

38 Given that $\left|\dfrac{z - 1}{z + 1}\right| = 2$, find the cartesian equation of the locus of z, and represent the locus by a sketch in the Argand diagram. Shade the region for which both the inequalities $\left|\dfrac{z - 1}{z + 1}\right| > 2$ and $0 < \arg z < 3\pi/4$ are satisfied. [L]

39 The vertices O, A, B of a triangle in the Argand diagram are the points corresponding to the numbers 0, 1 and $1 + i$ respectively. Show that, when the point z describes the perimeter of the triangle OAB, the locus of the point representing z^2 consists of part of the real axis, part of a parabola and part of the imaginary axis.

Sketch this locus. [L]

40 If z_1 and z_2 are any two complex numbers, show that
(a) $|z_1 + z_2| \leqslant |z_1| + |z_2|$, (b) $|z_1 + z_2| \geqslant |z_1| - |z_2|$.
Show that the equation $z^4 + z + 2 = 0$ cannot have a root z such that $|z| < 1$.
Show also that if α is real then both roots of the equation

$$z^2 \cos 2\alpha + z \cos \alpha = 1$$

lie outside the circle $2|z| = \sqrt{5} - 1$. [L]

41 Prove that if n is a positive integer

$$\sum_{r=0}^{n} \frac{n!}{(n - r)!r!} \cos 2r\theta = 2^n \cos^n\theta \cos n\theta,$$

by substituting $z = e^{2i\theta}$ in the expansion of $(1 + z)^n$ as a series of ascending powers of z, or otherwise. [L]

42 Show that the roots of the equation $(z + 1)^n = z^n$ can be expressed in the form

$$-\tfrac{1}{2}[1 + i \cot (k\pi/n)], \qquad k = 1, 2, \ldots, n - 1.$$

Hence, or otherwise, find the roots of the equation

$$(z + 1)^6 = z^6$$

in the form $a + ib$, where a and b are real. Plot the points representing these roots in an Argand diagram. [L]

3 Hyperbolic functions

3.1 Properties of functions

A function maps elements of a set A to elements of a set B, each element of the set A being mapped to a single element of the set B.

Let the function f map elements of the set $\mathbb{R}$ of real numbers to elements of the same set.

If $f(-x) = f(x)$ for all values of x, f is said to be an even function of x. In this case, the curve $y = f(x)$ is symmetrical about the y-axis, as in Fig. 3.1, for if the point (a, b) lies on the curve, so does the point $(-a, b)$.

If $f(-x) = -f(x)$ for all values of x, f is said to be an odd function of x. In this case, the curve $y = f(x)$ is symmetrical about the origin, as in Fig. 3.2. For if the point (a, b) lies on the curve, so does the point $(-a, -b)$.

When the function f is neither even nor odd, it can be expressed as the sum of an even function and an odd function.

Let $f_1(x) = \frac{1}{2}f(x) + \frac{1}{2}f(-x)$.

Since $f_1(-x) = f_1(x)$, f_1 is an even function.

Let $f_2(x) = \frac{1}{2}f(x) - \frac{1}{2}f(-x)$.

Since $f_2(-x) = \frac{1}{2}f(-x) - \frac{1}{2}f(x) = -f_2(x)$, f_2 is an odd function. Then the equation $f(x) = f_1(x) + f_2(x)$ expresses f as the sum of an even function and an odd function.

Note that a constant function is even.

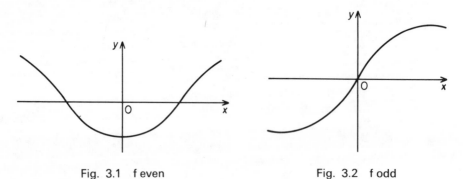

Fig. 3.1 f even   Fig. 3.2 f odd

Continuity

Let $f(x)$ be defined at each point of an interval $b < x < c$, and let a lie between b and c.

Then the function f is said to be continuous at $x = a$ if $\lim_{x \to a} f(x) = f(a)$.

This will be true if, as x approaches a from either side, f(x) tends to f(a).

Consider the function f defined for $x \in \mathbb{R}$ by

$$f(x) = x - 1 \qquad \text{for } x \leqslant 2,$$
$$f(x) = x \qquad \text{for } x > 2.$$

The graph of this function is shown in Fig. 3.3. As x increases towards the value 2, f(x) tends to 1, i.e. the limit of f(x) as x tends to 2 from the left is 1. This is expressed by

$$\lim_{x \to 2-} f(x) = 1.$$

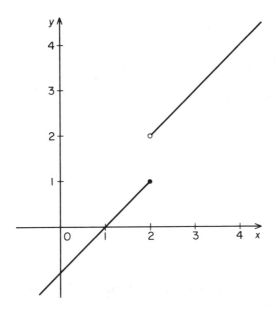

Fig. 3.3   Discontinuous function

As x decreases towards the value 2, f(x) tends to 2, i.e. the limit of f(x) as x tends to 2 from the right is 2. This is expressed as

$$\lim_{x \to 2+} f(x) = 2.$$

When x increases through the value 2, f(x) jumps in value from 1 to 2. Note that in the graph there is no vertical line drawn joining the points (2, 1) and (2, 2).

The graph of the function f is discontinuous at $x = 2$, and f is said to have a finite discontinuity at $x = 2$.

This is in contrast to a function such as tan x, which has an infinite discontinuity at $x = \pi/2$.

Periodic functions

Let the function f be defined for $x \in \mathbb{R}$, and let there be a constant p such that, for all values of x,

$$f(x + p) = f(x).$$

Then
$$f(x + 2p) = f(x + p) = f(x),$$

and
$$f(x + np) = f(x), \quad \text{where } n \in \mathbb{Z}.$$

The function f is said to be a periodic function, and the smallest positive value of p is known as the period of the function f.

Example

Find the period of the function $f(x)$, where

$$f(x) = \sin 8x + \cos^2 6x.$$

The period of $\sin x$ is 2π, so that the period of $\sin 8x$ is $\pi/4$. The period of $\cos^2 x$ is π, so that the period of $\cos^2 6x$ is $\pi/6$.

The least common multiple of $1/4$ and $1/6$ is $1/2$.

Hence the period of $f(x)$ is $\pi/2$.

Exercises 3.1

1   Find whether the function f is even, odd or neither when
    (a) $f(x) = (x - 1)^2$,   (b) $f(x) = x \cos x$,
    (c) $f(x) = \sin(2x + \pi/2)$,   (d) $f(x) = x^2 \tan x$.

2   Express each of the following as the sum of an even function and an odd function.
    (a) $(2x + 1)^3$,   (b) $(x + 1)/(x - 1)$,
    (c) $\sqrt{2} \cos(2x - \pi/4)$, (d) $\tan(x + \pi/4)$.

3   Given that $f(x) = \cos x$ and $g(x) = x^3$, find whether (a) fg, (b) gf, (c) gg are even, odd, or neither.

4   The function $f: \mathbb{R} \to \mathbb{R}$ is defined by

$$f(x) = 1 + x^2 \quad \text{for } x \geqslant 0, \ f(x) = 1 - x^2 \quad \text{for } x < 0.$$

Find whether the curves $y = f(x)$ and $y = f'(x)$ are continuous for all values of x.

5   Sketch the curve $y = f(x)$ in the following cases. State the value of x at any point of discontinuity.
    (a) $f(x) = |x|/x, \ x \neq 0; \ f(0) = 0$,
    (b) $f(x) = (x^2 - 4)/(x + 2), \ x \neq -2; \ f(-2) = 4$,
    (c) $f(x) = \sin x, \ -\pi \leqslant x \leqslant \pi, \ f(x) = x$ otherwise,
    (d) $f(x) = x - 2, \ x \geqslant 0; \ f(x) = x + 2, \ x < 0$.

6   Find the period of $f(x)$ when $f(x)$ is

(a) $\tan (2x - \pi/4)$,   (b) $\sin 4x + \tan (3x/2)$,

(c) $\sin x \cos 3x$,   (d) $\sin 4x \sec 12x$.

7   Given that $f(x)$ is periodic with period 2, and that $f(x) = 1 - x$ in the interval $0 < x \leqslant 1$, sketch the graph of $f(x)$ for $-2 \leqslant x \leqslant 2$ when (a) f is even, (b) f is odd.

8   Given that $f(x)$ is periodic with period π, and that $f(x) = 1 - \cos x$ in the interval $0 \leqslant x < \pi/2$, sketch the graph of $f(x)$ for $-\pi \leqslant x \leqslant \pi$ when (a) f is even, (b) f is odd.

9   The function $f : \mathbb{R} \to \mathbb{R}$ has period p. Sketch the graph of $f(x)$ for $0 \leqslant x < 3p$ given that

(a) $p = 2$, $f(x) = x$ for $0 \leqslant x < 2$,

(b) $p = \pi$, $f(x) = \pi - x$ for $0 \leqslant x < \pi$,

(c) $p = 1$, $f(x) = \sin \pi x$ for $0 \leqslant x < 1$,

(d) $p = 2$, $f(x) = 1 - x^2$ for $-1 \leqslant x < 1$.

10   The function f is defined for $x \in \mathbb{R}$ by $f(x) = 4e^{3x} + 2$. Show that the inverse function f^{-1} exists, and state its domain and its range.

     Sketch the curves $y = f(x)$ and $y = f^{-1}(x)$ on the same axes.

11   An even function $g(x)$ of period 2π is defined by

$$g(x) = x^2 \qquad \text{for } 0 \leqslant x \leqslant \pi/2,$$
$$g(x) = \pi^2/4 \qquad \text{for } \pi/2 < x \leqslant \pi.$$

Sketch the graph of $g(x)$ for $-2\pi \leqslant x \leqslant 2\pi$.   [L]

12   The function f is defined by

$$f(x) = \sin x \qquad \text{for } x \leqslant 0,$$
$$f(x) = x \qquad \text{for } x > 0.$$

Sketch the graphs of $f(x)$ and its derivative $f'(x)$ for $-\pi/2 < x < \pi/2$, and decide whether the functions f and f' are continuous at $x = 0$ or not. [L]

3.2   Cosh *x* and sinh *x*

The hyperbolic function $\cosh : \mathbb{R} \to \mathbb{R}$ is defined by

$$\cosh x = \tfrac{1}{2}(e^x + e^{-x}).$$

It is clear from the definition that $\cosh x$ is an even function of x, and that $\cosh 0 = 1$.

     The reciprocal of $\cosh x$ is known as $\text{sech } x$, i.e.

$$\text{sech } x = \frac{1}{\cosh x}.$$

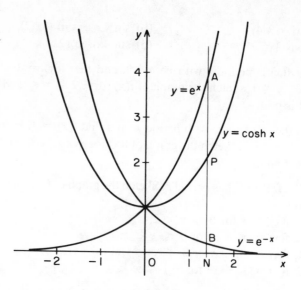

Fig. 3.4 $y = \cosh x$

The curve $y = \cosh x$ is shown in Fig. 3.4.

The line ABN is parallel to the y-axis, and cuts on the curve $y = e^x$ at A and the curve $y = e^{-x}$ at B. Then if $ON = x$, $AN = e^x$ and $BN = e^{-x}$. From the definition above,

$$\begin{aligned}
\cosh x &= \tfrac{1}{2}(e^x + e^{-x}) \\
&= \tfrac{1}{2}(AN + BN) \\
&= PN,
\end{aligned}$$

where P is the mid-point of AB.

As expected, the curve $y = \cosh x$ is symmetrical about the y-axis. Note that $\cosh x \geqslant 1$ for all values of x.

The hyperbolic function $\sinh : \mathbb{R} \to \mathbb{R}$ is defined by

$$\sinh x = \tfrac{1}{2}(e^x - e^{-x}).$$

Since
$$\begin{aligned}
\sinh(-x) &= \tfrac{1}{2}(e^{-x} - e^x) \\
&= -\sinh x,
\end{aligned}$$

$\sinh x$ is an odd function of x, with $\sinh 0 = 0$.

The reciprocal of $\sinh x$ is known as $\operatorname{cosech} x$, i.e.

$$\operatorname{cosech} x = \frac{1}{\sinh x}, \qquad x \neq 0.$$

The curve $y = \sinh x$ is shown in Fig. 3.5, with $y = \cosh x$ for comparison. As $\sinh x$ is an odd function, this curve is symmetrical about the origin.

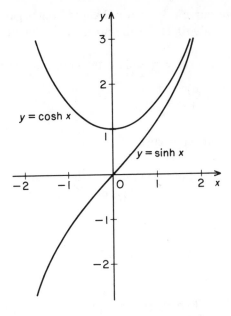

Fig. 3.5 $y = \sinh x$ and $y = \cosh x$

From the definitions of cosh x and sinh x,

$$\cosh x + \sinh x \equiv \tfrac{1}{2}(e^x + e^{-x}) + \tfrac{1}{2}(e^x - e^{-x})$$
$$\Rightarrow \qquad \cosh x + \sinh x \equiv e^x.$$
Also $\qquad \cosh x - \sinh x \equiv \tfrac{1}{2}(e^x + e^{-x}) - \tfrac{1}{2}(e^x - e^{-x})$
$$\Rightarrow \qquad \cosh x - \sinh x \equiv e^{-x}.$$

It follows that

$$\cosh^2 x - \sinh^2 x \equiv (\cosh x + \sinh x)(\cosh x - \sinh x)$$
$$\equiv e^x \times e^{-x}$$
$$\Rightarrow \qquad \cosh^2 x - \sinh^2 x \equiv 1.$$

This useful identity is obviously equivalent to

$$\cosh^2 x \equiv 1 + \sinh^2 x,$$
and to $\qquad \sinh^2 x \equiv \cosh^2 x - 1.$

Derived functions
The gradient at any point on the curve $y = \cosh x$ is given by

$$\frac{dy}{dx} = \frac{1}{2}\frac{d}{dx}(e^x + e^{-x}) = \frac{1}{2}(e^x - e^{-x})$$

$$\Rightarrow \qquad \frac{d}{dx}(\cosh x) = \sinh x.$$

The gradient of the curve $y = \cosh x$ changes from negative to positive as x increases through the value 0. Hence the curve has a minimum point at $(0, 1)$, as shown in Fig. 3.4.

At any point on the curve $y = \sinh x$, the gradient is given by

$$\frac{dy}{dx} = \frac{1}{2}\frac{d}{dx}(e^x - e^{-x}) = \frac{1}{2}(e^x + e^{-x})$$

$\Rightarrow \qquad \dfrac{d}{dx}(\sinh x) = \cosh x.$

It follows that the gradient of the curve $y = \sinh x$ is always positive and has a minimum value of 1 at the origin. At this point the line $y = x$ is the tangent to the curve.

Expansion in series
The Maclaurin series for $\cosh x$ and $\sinh x$ can be found directly from the exponential series:

$$e^x = 1 + x + \frac{1}{2!}x^2 + \frac{1}{3!}x^3 + \ldots + \frac{1}{n!}x^n + \ldots,$$

and $\qquad e^{-x} = 1 - x + \dfrac{1}{2!}x^2 - \dfrac{1}{3!}x^3 + \ldots + \dfrac{(-1)^n}{n!}x^n + \ldots.$

Hence $\qquad \cosh x = \dfrac{1}{2}(e^x + e^{-x})$

$$= 1 + \frac{1}{2!}x^2 + \frac{1}{4!}x^4 + \ldots + \frac{1}{(2n)!}x^{2n} + \ldots,$$

and $\qquad \sinh x = x + \dfrac{1}{3!}x^3 + \dfrac{1}{5!}x^5 + \ldots + \dfrac{1}{(2n-1)!}x^{2n-1} + \ldots.$

Each of these series converges for all values of x.

The expansion of $\cosh x$, an even function, contains only even powers of x, while the expansion of $\sinh x$, an odd function, contains only odd powers of x.

The series show that, when x is so small that terms in x^3 and higher powers of x can be neglected,

$$\sinh x \approx x,$$
$$\cosh x \approx 1 + \tfrac{1}{2}x^2.$$

Example 1
Show that (a) $\sinh 2x = 2 \sinh x \cosh x$,
(b) $\cosh 2x = \cosh^2 x + \sinh^2 x$.

An identity involving hyperbolic functions can always be verified by expressing the hyperbolic functions in terms of exponential functions.

(a) $2 \sinh x \cosh x \equiv 2[\frac{1}{2}(e^x - e^{-x})][\frac{1}{2}(e^x + e^{-x})]$

$\equiv \frac{1}{2}(e^{2x} - e^{-2x})$

$\equiv \sinh 2x.$

(b) $\cosh^2 x + \sinh^2 x \equiv \frac{1}{4}(e^x + e^{-x})^2 + \frac{1}{4}(e^x - e^{-x})^2$

$\equiv \frac{1}{4}(e^{2x} + 2 + e^{-2x}) + \frac{1}{4}(e^{2x} - 2 + e^{-2x})$

$\equiv \frac{1}{2}(e^{2x} + e^{-2x})$

$\equiv \cosh 2x.$

Example 2

Show that

$$\cosh (x + y) \equiv \cosh x \cosh y + \sinh x \sinh y,$$
$$\sinh (x + y) \equiv \sinh x \cosh y + \cosh x \sinh y.$$

For brevity, it is often convenient to denote cosh by ch and sinh by sh.

Since $e^x \equiv \text{ch } x + \text{sh } x$ and $e^y \equiv \text{ch } y + \text{sh } y,$

$e^{x+y} \equiv e^x \times e^y$

$\equiv (\text{ch } x + \text{sh } x)(\text{ch } y + \text{sh } y)$

$\equiv \text{ch } x \text{ ch } y + \text{ch } x \text{ sh } y + \text{sh } x \text{ ch } y + \text{sh } x \text{ sh } y.$

$e^{-x-y} \equiv e^{-x} \times e^{-y}$

$\equiv (\text{ch } x - \text{sh } x)(\text{ch } y - \text{sh } y)$

$\equiv \text{ch } x \text{ ch } y - \text{ch } x \text{ sh } y - \text{sh } x \text{ ch } y + \text{sh } x \text{ sh } y.$

$\cosh (x + y) \equiv \frac{1}{2}(e^{x+y} + e^{-x-y})$

$\equiv \text{ch } x \text{ ch } y + \text{sh } x \text{ sh } y,$

$\sinh (x + y) \equiv \frac{1}{2}(e^{x+y} - e^{-x-y})$

$\equiv \text{sh } x \text{ ch } y + \text{ch } x \text{ sh } y.$

Example 3

Given that $\sinh x = 3/4$, find the values of $\cosh x$ and x.

Since $\cosh^2 x - \sinh^2 x \equiv 1,$

$\cosh^2 x \equiv 1 + \sinh^2 x.$

When $\sinh x = 3/4, \cosh^2 x = 1 + 9/16 = 25/16.$

As cosh x is always positive, this gives $\cosh x = 5/4.$

Since $e^x \equiv \cosh x + \sinh x,$

$e^x = 5/4 + 3/4 = 2$

$\Rightarrow$ $x = \ln 2.$

Example 4

Solve the equation $6 \cosh x + 2 \sinh x = 9.$

Since $\cosh x \equiv \frac{1}{2}(e^x + e^{-x})$ and $\sinh x \equiv \frac{1}{2}(e^x - e^{-x})$, the given equation is equivalent to

$$3(e^x + e^{-x}) + (e^x - e^{-x}) = 9$$

$\Rightarrow$ $\qquad 4e^x + 2e^{-x} - 9 = 0$

$$\Rightarrow \qquad\qquad 4e^{2x} - 9e^x + 2 = 0$$
$$\Rightarrow \qquad\qquad (e^x - 2)(4e^x - 1) = 0$$

Hence $e^x = 2$, giving $x = \ln 2$, or $e^x = \frac{1}{4}$, giving $x = -\ln 4$.

Exercises 3.2

1 Express $\cosh^2 x$ and $\sinh^2 x$ in terms of $\cosh 2x$.

2 Evaluate $\sinh x$ and $\cosh x$ when $x = \ln(\sqrt{2} + 1)$.

3 Evaluate $\cosh x$ and x
   (a) when $\sinh x = -1$,   (b) when $\operatorname{cosech} x = 2\cdot4$.

4 Given that $\cosh x = 2$, find the possible values of $\sinh x$ and x.

5 Sketch the curves $y = \operatorname{sech} x$ and $y = \operatorname{cosech} x$.

6 Find the derivative of (a) $\cosh^2 x$, (b) $\sinh^2 x$.
   Sketch the curves $y = \cosh^2 x$ and $y = \sinh^2 x$ on the same axes.

7 Find the derivative of
   (a) $\cosh 3x$,   (b) $\sinh (\ln x)$,   (c) $\operatorname{sech} 2x$.

8 Show that
   (a) $\cosh 2x + \cosh 2y \equiv 2 \cosh (x + y) \cosh (x - y)$,
   (b) $\cosh 2x - \cosh 2y \equiv 2 \sinh (x + y) \sinh (x - y)$.

9 Solve the equations
   (a) $3 \cosh x - 2 \sinh x = 3$,
   (b) $4 \cosh x + 8 \sinh x = 11$,
   (c) $9 \cosh x \sinh x = 20$.

10 Show that
   (a) $\cosh 3x \equiv 4 \cosh^3 x - 3 \cosh x$,
   (b) $\sinh 3x \equiv 4 \sinh^3 x + 3 \sinh x$.

3.3 Tanh x

The hyperbolic function $\tanh : \mathbb{R} \to \mathbb{R}$ is defined by

$$\tanh x = \frac{\sinh x}{\cosh x} = \frac{e^x - e^{-x}}{e^x + e^{-x}}.$$

From the definition, it can be seen that $\tanh x$ is an odd function of x and that $\tanh 0 = 0$.

The reciprocal of $\tanh x$ is known as $\coth x$, i.e.

$$\coth x = \frac{1}{\tanh x}, \qquad x \neq 0.$$

Since
$$\tanh x \equiv \frac{e^{2x} - 1}{e^{2x} + 1} \equiv \frac{1 - e^{-2x}}{1 + e^{-2x}},$$

the value of tanh x lies between 0 and 1 when x is positive, and tends to the limit 1 as x tends to infinity through positive values.

When x tends to infinity through negative values, tanh x tends to the limit -1.

The curve $y = \tanh x$ is shown in Fig. 3.6. As tanh x is an odd function, this curve is symmetrical about the origin. The two lines $y = 1$ and $y = -1$ are asymptotes.

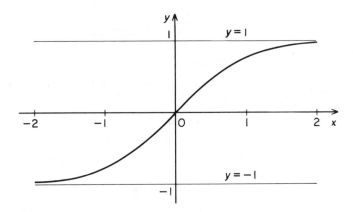

Fig. 3.6 $y = \tanh x$

The gradient of the curve $y = \tanh x$ is given by

$$\frac{dy}{dx} = \frac{d}{dx}(\tanh x)$$

$$= \frac{d}{dx}\left(\frac{\sinh x}{\cosh x}\right)$$

$$= \frac{\cosh^2 x - \sinh^2 x}{\cosh^2 x}$$

$$= \frac{1}{\cosh^2 x}$$

$$= \operatorname{sech}^2 x.$$

Hence the gradient of the curve is always positive. Since cosh x has its minimum value 1 when $x = 0$, the gradient of the curve $y = \tanh x$ has its maximum value 1 at the origin.

Exercises 3.3

1   Find the value of tanh x (a) when $x = \ln 10$, (b) when $x = -\ln 5$.

2   Simplify (a) $(1 + \tanh x)/(1 - \tanh x)$,
        (b) $(\cosh^2 x - 1)(1 - \coth^2 x)$.

3 Solve the equations (a) $\coth x + \operatorname{cosech} x = 3$,

                 (b) $3 \sinh^2 x - 2 \cosh x = 2$.

4 Find the derivatives of

  (a) $\tanh 2x$,   (b) $\ln \cosh x$,   (c) $\coth x$,

  (d) $\ln \sinh x$,   (e) $\operatorname{sech} x$,   (f) $\ln \tanh x$.

5 Show that (a) $1 - \tanh^2 x \equiv \operatorname{sech}^2 x$,

         (b) $\tanh (x + y) \equiv \dfrac{\tanh x + \tanh y}{1 + \tanh x \tanh y}$.

3.4 Summary of properties

There is a strong similarity between the properties of the hyperbolic functions and those of the circular functions.

Circular functions

$$\cos^2 x + \sin^2 x \equiv 1$$

$$\cos^2 x - \sin^2 x \equiv \cos 2x$$

$$2 \sin x \cos x \equiv \sin 2x$$

$$\sin (x + y) \equiv \sin x \cos y + \cos x \sin y$$

$$\cos (x + y) \equiv \cos x \cos y - \sin x \sin y$$

$$\sin x = x - \frac{1}{3!}x^3 + \frac{1}{5!}x^5 - \cdots$$

$$\cos x = 1 - \frac{1}{2!}x^2 + \frac{1}{4!}x^4 - \cdots$$

$$\frac{d}{dx}(\sin x) = \cos x$$

$$\frac{d}{dx}(\cos x) = -\sin x$$

$$\frac{d}{dx}(\tan x) = \sec^2 x$$

$$\cos x + i \sin x \equiv e^{ix}$$

Hyperbolic functions

$$\cosh^2 x - \sinh^2 x \equiv 1$$

$$\cosh^2 x + \sinh^2 x \equiv \cosh 2x$$

$$2 \sinh x \cosh x \equiv \sinh 2x$$

$$\sinh (x + y) \equiv \sinh x \cosh y + \cosh x \sinh y$$

$$\cosh (x + y) \equiv \cosh x \cosh y + \sinh x \sinh y$$

$$\sinh x = x + \frac{1}{3!}x^3 + \frac{1}{5!}x^5 + \cdots$$

$$\cosh x = 1 + \frac{1}{2!}x^2 + \frac{1}{4!}x^4 + \cdots$$

$$\frac{d}{dx}(\sinh x) = \cosh x$$

$$\frac{d}{dx}(\cosh x) = \sinh x$$

$$\frac{d}{dx}(\tanh x) = \operatorname{sech}^2 x$$

$$\cosh x + \sinh x \equiv e^x$$

An identity involving circular functions can be converted into an identity involving hyperbolic functions by replacing cos by cosh and sin by sinh, and changing the sign in front of any product of two sines.

Since $\cos^2 t + \sin^2 t = 1$, the coordinates of a point on the circle $x^2 + y^2 = a^2$ can be given in parametric form as $x = a \cos t$, $y = a \sin t$.

Since $\cosh^2 t - \sinh^2 t = 1$, the coordinates of a point on one branch of the hyperbola $x^2 - y^2 = a^2$ can be given in parametric form as $x = a \cosh t$, $y = a \sinh t$. This is the origin of the name hyperbolic functions.

3.5 Integration of hyperbolic functions

Since
$$\frac{d}{dx}(\sinh x) = \cosh x,$$

$$\int \cosh x \, dx = \sinh x + c,$$

where c is a constant.

Since
$$\frac{d}{dx}(\cosh x) = \sinh x,$$

$$\int \sinh x \, dx = \cosh x + c.$$

Since
$$\frac{d}{dx}(\tanh x) = \operatorname{sech}^2 x,$$

$$\int \operatorname{sech}^2 x \, dx = \tanh x + c.$$

Also
$$\int \tanh x \, dx = \int \frac{\sinh x}{\cosh x} \, dx.$$

Since $\dfrac{d}{dx}(\cosh x) = \sinh x$, this integral is of the form

$$\int \frac{f'(x)}{f(x)} \, dx = \ln f(x) + c.$$

Hence
$$\int \tanh x \, dx = \ln (\cosh x) + c.$$

Similarly
$$\int \coth x \, dx = \int \frac{\cosh x}{\sinh x} \, dx$$
$$= \ln (\sinh x) + c.$$

Example 1
Find $\int \sinh^2 x \, dx$.

From the identity

$$\cosh 2x \equiv \cosh^2 x + \sinh^2 x \equiv 1 + 2 \sinh^2 x,$$
$$\sinh^2 x \equiv \tfrac{1}{2}(\cosh 2x - 1).$$
$$\int \sinh^2 x \, dx = \int \tfrac{1}{2}(\cosh 2x - 1) \, dx$$
$$= \tfrac{1}{4} \sinh 2x - \tfrac{1}{2}x + c.$$

Example 2
Find $\int \operatorname{sech} x \, dx$.

$$\text{sech } x = \frac{1}{\cosh x} = \frac{2}{e^x + e^{-x}} = \frac{2e^x}{e^{2x} + 1}.$$

Let $e^x = t$. Then $e^x \dfrac{dx}{dt} = 1.$

$$\int \text{sech } x \, dx = \int \frac{2e^x}{e^{2x} + 1} \frac{dx}{dt} \, dt$$

$$= \int \frac{2}{t^2 + 1} \, dt$$

$$= 2 \tan^{-1} t + c$$

$\Rightarrow \qquad \int \text{sech } x \, dx = 2 \tan^{-1} (e^x) + c.$

Exercises 3.5

1 Find the integrals
    (a) $\int \sinh 2x \, dx$,   (b) $\int \tanh (x/2) \, dx$,
    (c) $\int \cosh^2 x \, dx$,   (d) $\int \coth 2x \, dx$.

2 Use integration by parts to find the integrals
    (a) $\int x \sinh x \, dx$,   (b) $\int x \cosh x \, dx$,
    (c) $\int x \text{ sech}^2 x \, dx$.

3 Show that $\int \text{cosech } 2x \, dx = \frac{1}{2}\ln (\tanh x) + c$.

4 Evaluate to three decimal places
    (a) $\displaystyle\int_0^1 \cosh^2 x \, dx$,   (b) $\displaystyle\int_0^1 x \sinh x \, dx$,

    (c) $\displaystyle\int_0^1 \tanh^2 x \, dx$,   (d) $\displaystyle\int_0^1 \text{sech } x \, dx$.

5 Show that the area of the region bounded by the x-axis, the lines $x = 0$, $x = 1$ and the arc of the curve $y = \text{sech } x$ for which $0 \leqslant x \leqslant 1$ equals $2 \tan^{-1} (e) - \pi/2$.

6 Calculate the area of the region defined by the inequalities $0 \leqslant x \leqslant 1$, $\sinh x \leqslant y \leqslant \cosh x$.

7 The finite region bounded by the y-axis, the line $y = \sinh 1$ and the arc of the curve $y = \sinh x$ for which $0 \leqslant x \leqslant 1$ is rotated completely about the y-axis. Find the volume generated.

8 Show that the point P ($a \cosh t, a \sinh t$), where a is a positive constant, lies on the hyperbola $x^2 - y^2 = a^2$.
    Given that A is the point $(a, 0)$ and that O is the origin, show that the area of the region bounded by the x-axis, the arc AP of the hyperbola and the line OP is $\frac{1}{2}a^2 t$.

3.6 Inverse hyperbolic functions

The function sinh⁻¹ or arsinh

The function $\sinh : \mathbb{R} \to \mathbb{R}$ maps each element of its domain $\mathbb{R}$ to one and only one element of its range $\mathbb{R}$. No two elements of the domain are mapped to the same element of the range. The mapping is one-one, and the function sinh possesses an inverse. This inverse function is denoted by $\sinh^{-1}$ or arsinh.

The curves $y = \sinh x$ and $y = \sinh^{-1} x$ are shown in Fig. 3.7. Each curve is the reflection of the other in the line $y = x$.

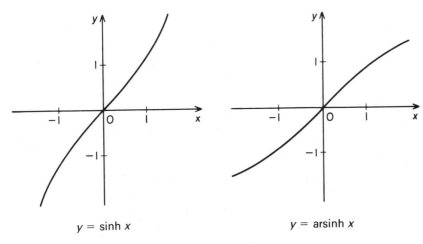

$y = \sinh x$ $y = \text{arsinh } x$

Fig. 3.7

When
$$y = \sinh^{-1} x,$$
$$x = \sinh y$$

$\Rightarrow$
$$\frac{dx}{dy} = \cosh y$$

$\Rightarrow$
$$\frac{dy}{dx} = \text{sech } y.$$

Since $\text{sech } y$ is always positive, the gradient of the curve $y = \sinh^{-1} x$ is always positive.

Also
$$\cosh^2 y = 1 + \sinh^2 y = 1 + x^2$$

$\Rightarrow$
$$\text{sech } y = \frac{1}{\cosh y} = \frac{1}{\sqrt{(1 + x^2)}}.$$

Note that the positive square root must be taken, since $\cosh y$ is always positive.

It follows that

$$\frac{d}{dx}(\sinh^{-1} x) = \frac{1}{\sqrt{(x^2 + 1)}}.$$

Since sinh $y = x$ and cosh $y = \sqrt{(x^2 + 1)}$,

$$e^y = \sinh y + \cosh y$$
$\Rightarrow \qquad e^y = x + \sqrt{(x^2 + 1)}$
$\Rightarrow \qquad y = \ln [x + \sqrt{(x^2 + 1)}].$

This shows that $\sinh^{-1} x$ is expressed in logarithmic form by

$$\sinh^{-1} x = \ln [x + \sqrt{(x^2 + 1)}].$$

This relation holds good for all real values of x.

The function cosh^{-1} or arcosh

The function cosh$: \mathbb{R} \rightarrow \mathbb{R}$ maps each element of its domain $\mathbb{R}$ to one and only one element of its range, which is the set $\{x : x \geq 1\}$. However, two elements of the domain are mapped to each element of the range, with the exception of the element 1. In consequence, the function cosh with domain $\mathbb{R}$ does not possess an inverse.

When the domain of the function cosh is restricted to the set $\{x : x \geq 0\}$, the mapping will be one-one and there will be an inverse function.

The inverse of the function cosh with domain the set $\{x : x \geq 0\}$ is denoted by cosh^{-1} or arcosh.

The domain of the function cosh^{-1} is the set $\{x : x \geq 1\}$.

The curves $y = \cosh x$, where $x \geq 0$, and $y = \cosh^{-1} x$ are shown in Fig. 3.8. Each curve is the reflection of the other in the line $y = x$.

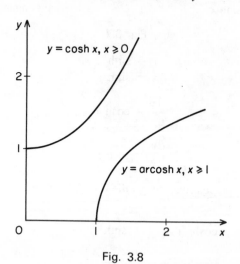

Fig. 3.8

When $\qquad\qquad\qquad y = \cosh^{-1} x,$

$$x = \cosh y$$

$\Rightarrow \qquad\qquad\qquad \dfrac{\mathrm{d}x}{\mathrm{d}y} = \sinh y$

$$\Rightarrow \qquad \frac{dy}{dx} = \text{cosech } y.$$

Now the range of the function $\cosh^{-1}$ is the set $\{x : x \geq 0\}$. This means that y and cosech y cannot be negative.

Since
$$\sinh^2 y = \cosh^2 y - 1 = x^2 - 1,$$
$$\text{cosech}^2 y = 1/(x^2 - 1), \qquad x > 1.$$

Hence
$$\frac{dy}{dx} = \frac{1}{\sqrt{(x^2 - 1)}},$$

the positive square root being taken as the gradient is positive.

This shows that
$$\frac{d}{dx}(\cosh^{-1} x) = \frac{1}{\sqrt{(x^2 - 1)}}$$

For a given value of x greater than 1, the equation $\cosh y = x$ is satisfied by two values of y, one positive and one negative.

Since
$$\sinh^2 y = \cosh^2 y - 1,$$

these two values of y are given by
$$\sinh y = \pm\sqrt{(x^2 - 1)}, \qquad x \geq 1.$$

From the identity
$$e^y \equiv \cosh y + \sinh y,$$
$$e^y = x \pm \sqrt{(x^2 - 1)}, \qquad x \geq 1$$
$$\Rightarrow \qquad y = \ln[x \pm \sqrt{(x^2 - 1)}], \qquad x \geq 1$$
$$\Rightarrow \qquad y = \pm \ln[x + \sqrt{(x^2 - 1)}], \qquad x \geq 1,$$

since $x - \sqrt{(x^2 - 1)}$ is the reciprocal of $x + \sqrt{(x^2 - 1)}$.
  The positive value of y gives $\cosh^{-1} x$, i.e.
$$\cosh^{-1} x = \ln[x + \sqrt{(x^2 - 1)}], \qquad x \geq 1.$$

This expresses $\cosh^{-1} x$ in logarithmic form.

The function $\tanh^{-1}$ or artanh
The domain of the function tanh is the set $\mathbb{R}$ and its range is the set $\{x : -1 < x < 1\}$. The mapping is one–one, and so this function possesses an inverse, which is denoted by $\tanh^{-1}$ or artanh.
  The curves $y = \tanh x$ and $y = \text{artanh } x$ are shown in Fig. 3.9. Each curve is the reflection of the other in the line $y = x$.

When
$$y = \tanh^{-1} x,$$
$$x = \tanh y$$

$$\Rightarrow \qquad \frac{dx}{dy} = \text{sech}^2 y = 1 - \tanh^2 y = 1 - x^2$$

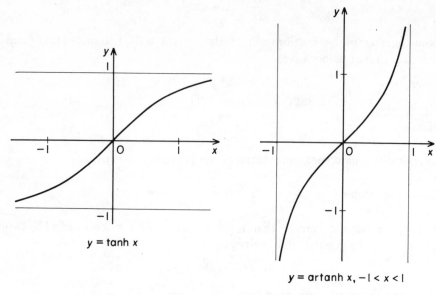

$y = \tanh x$

$y = \operatorname{artanh} x, -1 < x < 1$

Fig. 3.9

$\Rightarrow$
$$\frac{dy}{dx} = \frac{1}{1 - x^2}$$

$\Rightarrow$
$$\frac{d}{dx}(\tanh^{-1} x) = \frac{1}{1 - x^2}, \qquad -1 < x < 1.$$

This shows that the gradient of the curve $y = \tanh^{-1} x$ is never less than 1, and equals 1 at the origin.

When
$$y = \tanh^{-1} x,$$

$$x = \tanh y = \frac{e^{2y} - 1}{e^{2y} + 1}$$

$\Rightarrow$
$$(e^{2y} + 1)x = e^{2y} - 1$$

$\Rightarrow$
$$e^{2y} = \frac{1 + x}{1 - x}$$

$\Rightarrow$
$$y = \frac{1}{2}\ln\left(\frac{1 + x}{1 - x}\right), \qquad -1 < x < 1.$$

This expresses $\tanh^{-1} x$ as a logarithm.

Exercises 3.6

1 Show that (a) $\sinh(\operatorname{arcosh} x) = \sqrt{(x^2 - 1)}, \qquad x \geqslant 1,$
   (b) $\cosh(\operatorname{arsinh} x) = \sqrt{(x^2 + 1)}.$

2 Find the derivatives of

(a) $\ln [x + \sqrt{(x^2 + 1)}]$,

(b) $\ln [x + \sqrt{(x^2 - 1)}]$, $\quad x \geqslant 1$,

(c) $\ln \left(\dfrac{1 + x}{1 - x} \right)$, $\quad -1 < x < 1$,

(d) $\operatorname{artanh} \left(\dfrac{x^2 - 1}{x^2 + 1} \right)$.

3 Differentiate with respect to x

(a) $x\sqrt{(x^2 + 1)} + \sinh^{-1} x$,

(b) $x\sqrt{(x^2 - 1)} - \cosh^{-1} x$,

(c) $\ln \left(\dfrac{2 - x}{2 + x} \right) + 2 \tanh^{-1} (x/2)$, $\quad -2 < x < 2$,

(d) $\cosh^{-1} (x/2)$.

4 Evaluate to three decimal places

(a) arsinh 1,

(b) arcosh 2,

(c) artanh (0·9),

(d) arsinh $(-0·75)$,

(e) arcosh (5/4),

(f) artanh (0·5).

5 Show that the shortest distance between the curves $y = \cosh x$ and $y = \cosh^{-1} x$ is $2 - \sqrt{2} \ln (\sqrt{2} + 1)$.

6 Show that the area of the finite region bounded by the x-axis and the curves $y = \operatorname{arsinh} x$, $y = \operatorname{arcosh} (2x)$ is $1 - \frac{1}{2}\sqrt{3}$.

3.7 Applications to integration

The substitution $x = a \sinh t$ is suitable for the integral

$$\int \frac{1}{\sqrt{(a^2 + x^2)}} \, dx, \qquad a > 0,$$

because it enables the square root to be simplified.

When $x = a \sinh t$, $\dfrac{dx}{dt} = a \cosh t$ and $a^2 + x^2 = a^2 \cosh^2 t$.

$$\int \frac{1}{\sqrt{(a^2 + x^2)}} \, dx = \int \frac{1}{\sqrt{(a^2 + x^2)}} \frac{dx}{dt} \, dt$$

$$= \int \frac{1}{a \cosh t} (a \cosh t) \, dt$$

$$= \int dt = t + c.$$

This gives

$$\int \frac{1}{\sqrt{(a^2 + x^2)}} \, dx = \sinh^{-1} (x/a) + c$$

$$= \ln [x + \sqrt{(x^2 + a^2)}] + k, \qquad a > 0,$$

where $k = c - \ln a$.

The substitution $x = a \cosh t$ is suitable for the integral

$$\int \frac{1}{\sqrt{(x^2 - a^2)}} \, dx, \qquad x \geqslant a > 0.$$

When $x = a \cosh t$, $\dfrac{dx}{dt} = a \sinh t$ and $x^2 - a^2 = a^2 \sinh^2 t$.

$$\int \frac{1}{\sqrt{(x^2 - a^2)}} \, dx = \int \frac{1}{\sqrt{(x^2 - a^2)}} \frac{dx}{dt} \, dt$$

$$= \int \frac{1}{a \sinh t} \, (a \sinh t) \, dt$$

$$= \int 1 \, dt = t + c.$$

This gives

$$\int \frac{1}{\sqrt{(x^2 - a^2)}} \, dx = \cosh^{-1}(x/a) + c$$

$$= \ln [x + \sqrt{(x^2 - a^2)}] + k, \qquad x \geqslant a > 0,$$

where $k = c - \ln a$.

Example 1

Find $\displaystyle\int \frac{1}{\sqrt{(x^2 + 2x + 2)}} \, dx$.

Since $\qquad\qquad\qquad x^2 + 2x + 2 = (x + 1)^2 + 1,$

a suitable substitution is given by $x + 1 = \sinh t$.

Then $\qquad\qquad\qquad x^2 + 2x + 2 = \sinh^2 t + 1 = \cosh^2 t,$

and $\qquad\qquad\qquad\qquad \dfrac{dx}{dt} = \cosh t.$

$$\int \frac{1}{\sqrt{(x^2 + 2x + 2)}} \, dx = \int \frac{1}{\cosh t} \, (\cosh t) \, dt$$

$$= \int 1 \, dt = t + c$$

$$= \sinh^{-1}(x + 1) + c.$$

Example 2

Evaluate $\displaystyle\int_0^2 \sqrt{(x^2 + 4)} \, dx$.

The substitution $x = 2 \sinh t$ gives $\dfrac{dx}{dt} = 2 \cosh t$, and

$$\sqrt{(x^2 + 4)} = \sqrt{(4 \sinh^2 t + 4)} = 2 \cosh t.$$

$$\int \sqrt{(x^2 + 4)} \, dx = \int (2 \cosh t) \frac{dx}{dt} \, dt$$

$$= \int 4 \cosh^2 t \, dt$$
$$= \int 2 (\cosh 2t + 1) \, dt$$
$$= \sinh 2t + 2t$$
$$= 2 \sinh t \cosh t + 2t$$
$$= \tfrac{1}{2} x \sqrt{(x^2 + 4)} + 2 \sinh^{-1} (x/2)$$

$$\int_0^2 \sqrt{(x^2 + 4)} \, dx = \left[\tfrac{1}{2} x \sqrt{(x^2 + 4)} + 2 \sinh^{-1} (x/2) \right]_0^2$$

$$= 2\sqrt{2} + 2 \sinh^{-1} 1$$
$$= 2\sqrt{2} + 2 \ln (\sqrt{2} + 1).$$

Example 3

Find $\displaystyle\int \frac{\sqrt{(x^2 - 1)}}{x^2} \, dx$, where $x \geqslant 1$.

When $x = \cosh t$, $x^2 - 1 = \sinh^2 t$ and $\dfrac{dx}{dt} = \sinh t$.

$$\int \frac{\sqrt{(x^2 - 1)}}{x^2} \, dx = \int \frac{\sinh t}{\cosh^2 t} (\sinh t) \, dt$$

$$= \int \tanh^2 t \, dt$$
$$= \int (1 - \operatorname{sech}^2 t) \, dt$$
$$= t - \tanh t + c$$

$$= \cosh^{-1} x - \frac{\sqrt{(x^2 - 1)}}{x} + c, \qquad \text{where } x \geqslant 1.$$

Example 4

Evaluate to three decimal places $\displaystyle\int_0^1 \frac{x + 3}{\sqrt{(x^2 + 4x + 5)}} \, dx.$

This integral can be expressed as the sum of two integrals I_1 and I_2 of standard type, where

$$I_1 = \int_0^1 \frac{x + 2}{\sqrt{(x^2 + 4x + 5)}} \, dx, \qquad I_2 = \int_0^1 \frac{1}{\sqrt{(x^2 + 4x + 5)}} \, dx.$$

$$I_1 = \left[\sqrt{(x^2 + 4x + 5)} \right]_0^1 = \sqrt{10} - \sqrt{5}.$$

By the substitution $x + 2 = \sinh t$,

$$\int \frac{1}{\sqrt{(x^2 + 4x + 5)}} dx = \int \frac{1}{\sqrt{(\sinh^2 t + 1)}} (\cosh t) dt$$

$$= \int 1 \, dt = t + c.$$

This gives
$$I_2 = \left[\sinh^{-1}(x + 2) \right]_0^1$$

$$= \sinh^{-1} 3 - \sinh^{-1} 2$$
$$= \ln(3 + \sqrt{10}) - \ln(2 + \sqrt{5}).$$

Hence
$$I_1 + I_2 = \sqrt{10} - \sqrt{5} + \ln(3 + \sqrt{10}) - \ln(2 + \sqrt{5})$$
$$= 1 \cdot 301 \text{ to three decimal places.}$$

Exercises 3.7

1 Find the following integrals by means of the substitutions given.

(a) $\displaystyle\int \frac{1}{\sqrt{(x^2 + 4)}} dx, \quad x = 2 \sinh t;$

(b) $\displaystyle\int \frac{1}{\sqrt{(4x^2 + 9)}} dx, \quad 2x = 3 \sinh t;$

(c) $\displaystyle\int \frac{1}{\sqrt{(x^2 - 16)}} dx, \quad x > 4, x = 4 \cosh t;$

(d) $\displaystyle\int \frac{1}{\sqrt{(4x^2 - 1)}} dx, \quad 2x > 1, 2x = \cosh t.$

2 Evaluate the following integrals:

(a) $\displaystyle\int_2^3 \frac{1}{\sqrt{(x^2 - 4x + 5)}} dx$, by the substitution $x - 2 = \sinh t$,

(b) $\displaystyle\int_0^2 \frac{1}{\sqrt{(x^2 + 4x + 3)}} dx$, by the substitution $x + 2 = \cosh t$,

(c) $\displaystyle\int_2^4 \frac{1}{\sqrt{(x^2 + 4x)}} dx$, by the substitution $x + 2 = 2 \cosh t$,

(d) $\displaystyle\int_{-1}^2 \frac{1}{\sqrt{(x^2 + 2x + 10)}} dx$, by the substitution $x + 1 = 3 \sinh t$.

3 Find the following integrals:

(a) $\displaystyle\int \frac{1}{\sqrt{(x^2 - 4x + 3)}} dx, \quad x > 3,$

(b) $\displaystyle\int \frac{1}{\sqrt{(x^2 + 6x + 10)}} dx,$

(c) $\displaystyle\int \frac{1}{\sqrt{(x^2 + 8x + 20)}}dx,$

(d) $\displaystyle\int \frac{1}{\sqrt{(x^2 - 2x - 3)}}dx, \quad x > 3.$

4   Find the following integrals:

(a) $\displaystyle\int \frac{1}{\sqrt{(4x^2 + 4x - 3)}}dx, \quad 2x > 1,$

(b) $\displaystyle\int \frac{1}{\sqrt{(4x^2 + 4x + 10)}}dx,$

(c) $\displaystyle\int \frac{1}{\sqrt{(x^2 + 6x)}}dx, \quad x > 0,$

(d) $\displaystyle\int \frac{1}{\sqrt{(2x^2 - 2x + 1)}}dx.$

5   Find the following integrals:
(a) $\int \sqrt{(x^2 + 9)}\,dx,$ (b) $\int \sqrt{(x^2 + 25)}\,dx,$
(c) $\int \sqrt{(x^2 - 9)}\,dx, \quad x > 3,$ (d) $\int \sqrt{(4x^2 - 1)}\,dx, \quad 2x > 1.$

6   Evaluate to three decimal places

(a) $\displaystyle\int_0^1 \frac{2x + 1}{\sqrt{(x^2 + 1)}}dx,$ (b) $\displaystyle\int_0^1 \frac{\sqrt{x}}{\sqrt{(x + 2)}}dx.$

Miscellaneous exercises 3

1   The function f is such that $f(x + \pi) = f(x)$ for all values of x. In the interval $0 \leqslant x < \pi$, $f(x) = x - \sin x$. Sketch the curve $y = f(x)$ for $-2\pi \leqslant x \leqslant 2\pi$, and state all the values of x for which the function f is discontinuous. Evaluate the integrals

(a) $\displaystyle\int_{-\pi/2}^{\pi/2} f(x)\,dx,$ (b) $\displaystyle\int_0^{3\pi/2} f(x)\,dx.$ [L]

2   The periodic function $f(x)$ is given by

$$f(x) = \sin x \qquad \text{for } 0 \leqslant x < \pi,$$
$$f(x) = 0 \qquad \text{for } \pi \leqslant x < 2\pi,$$
$$f(x) = f(x + 2\pi).$$

Sketch the graph of $f(x)$ for $-2\pi \leqslant x \leqslant 4\pi$ and evaluate

$$\int_2^{2\pi+2} f(x)\,dx. \qquad\qquad [L]$$

3 The periodic function f(x) is given by

$$f(x) = \quad 1, \qquad\qquad 0 \leqslant x < 1,$$
$$f(x) = -1, \qquad\qquad 1 \leqslant x < 2,$$
$$f(x) \equiv f(x + 2).$$

Sketch the graph of f(x) for $-4 \leqslant x \leqslant 4$.

Explain why you consider this function to be continuous or discontinuous at $x = 3$, and write down the value of

$$\int_{-\pi}^{\pi} f(x)\,dx.$$
[L]

4 The functions f and g are defined for $x \in \mathbb{R}$ by

$$f(x) = x - \pi\cos(x/2),$$
$$g(x) = x + \pi\sin(x/3).$$

Determine whether f and g are even, odd or neither.

Calculate the derivative of g at the point where $x = 3\pi$.

Sketch the graph of the function g for $-3\pi \leqslant x \leqslant 3\pi$, showing the gradient at $x = 3\pi$ on your graph.

Find a constant k such that $f + kg$ is a periodic function and state its period.
[L]

5 Show that the derivative of an odd function is an even function, and that the derivative of an even function is an odd function. Illustrate these results with suitable diagrams.
[AEB]

6 Given that $a = \ln 2$, evaluate the integral

$$\int_{a}^{2a} \operatorname{cosech}^2 x \,dx$$

by the substitution $e^{2x} = u$, or otherwise.
[L]

7 Find (a) the minimum value of $5 \cosh x + 3 \sinh x$,
(b) the maximum value of $4 \sinh x - 5 \cosh x$.

8 Given that $\sinh x = \tan t$, where $-\pi/2 < t < \pi/2$, show that
(a) $\cosh x = \sec t$, \qquad (b) $\tanh x = \sin t$,
(c) $x = \ln(\sec t + \tan t)$.

9 Find the limit as x tends to zero of
(a) $\dfrac{\sinh x}{x}$, \quad (b) $\dfrac{\cosh x - 1}{x^2}$, \quad (c) $\dfrac{\cosh x - \cos x}{x^2}$.

10 Solve, for $x \in \mathbb{R}$, the equations
(a) $3 \cosh x + 5 \sinh x = 0$, \qquad (b) $2 \cosh x - \sinh x = 2$,
(c) $3 \sinh^2 x - 2 \cosh x = 2$.

11 Show that to three decimal places

(a) $\displaystyle\int_0^4 \frac{x+1}{\sqrt{(x^2+4)}}\,dx = 3\cdot916,$

(b) $\displaystyle\int_0^1 \frac{2x-1}{\sqrt{(x^2+3x+2)}}\,dx = -0\cdot048.$

12 Show that
(a) $\int \sinh^3 x\,dx = \frac{1}{3}\cosh^3 x - \cosh x + c,$
(b) $\int \tanh^3 x\,dx = \ln\cosh x - \frac{1}{2}\tanh^2 x + c.$

13 (a) Find $\int x\cosh 3x\,dx.$
(b) Solve the equations
  (i) $4\cosh x + 8\sinh x = 1,$
  (ii) $3\operatorname{sech}^2 x + 4\tanh x + 1 = 0,$
for real values of x.   [AEB]

14 (a) Use Maclaurin's theorem to find the first three non-zero terms in the expansion of $\cosh x$ in a series of ascending powers of x.
  If $y = c\cosh(x/c)$, show that, for small values of x/c, an approximation for y/c is $1 + x^2/(2c^2)$.
  Obtain the percentage error incurred in making this approximation when $x/c = 1/2$, given that $\cosh(1/2) = 1\cdot1276.$
(b) Define $\sinh y$ and $\cosh y$ in terms of exponential functions, and show that

$$2y = \ln\left(\frac{\cosh y + \sinh y}{\cosh y - \sinh y}\right).$$

By putting $\tanh y = 1/3$, deduce that $\tanh^{-1}(1/3) = \frac{1}{2}\ln 2.$ [AEB]

15 Show that $(1+x)\sinh(3x-2)$ has a stationary value which occurs at the intersection of $y = \tanh(3x-2)$ with the straight line $y + 3x + 3 = 0$.
  Sketch the curve $y = \tanh(3x-2)$ and the straight line, and show that the stationary value lies between $x = -1$ and $x = 0$.
  Use Newton's method once only, with starting value -1, to obtain an improved value, taking $\tanh(-5)$ to be -1 and $\operatorname{sech}(-5)$ to be 0.
                                                                    [AEB]

16 Solve the simultaneous equations

$$\cosh x - 3\sinh y = 0,$$
$$2\sinh x + 6\cosh y = 5,$$

leaving your answers in logarithmic form.   [AEB]

17 Sketch the curve $y = x \sinh^{-1} x$.

Show that to four decimal places $\displaystyle\int_1^2 x \sinh^{-1} x \, dx = 1 \cdot 8227$.

18 The finite region bounded by the axes, the line $x = 1$ and the curve $y = \cosh x$ is rotated through one revolution about Ox. Show that, to five decimal places,

(a) the volume swept out is $4 \cdot 419\,33$,

(b) the distance from the origin of the centroid of the solid of revolution formed is $0 \cdot 576\,83$.

19 (a) Solve the simultaneous equations

$$4 \cosh x + 3 \cosh y = 10,$$
$$4 \sinh x + 3 \sinh y = 7.$$

(b) Show that $\cosh x$ is an even function. Sketch the curve

$$\cosh x + \cosh y = 4.$$

20 Show that the curve $y = \cosh 2x - 4 \sinh x$ has just one stationary point, and find its coordinates, giving the x-coordinate in logarithmic form. Determine the nature of the stationary point.   [JMB]

21 Evaluate $\displaystyle\int_0^{\sinh 1} x^2 \sinh^{-1} x \, dx$, expressing your answer in terms of hyperbolic sine and cosine.   [JMB]

22 Use the definitions of $\sinh x$ and $\cosh x$ in terms of exponential functions to prove that $\sinh^2 x = \cosh^2 x - 1$. Hence show that $\tanh^2 x = 1 - \operatorname{sech}^2 x$. Find $\int \tanh^3 x \, dx$.   [JMB]

23 Using one pair of axes, indicate the finite regions R_1 and R_2, where R_1 is bounded by $y = \sinh x$, $y = 0$, $x = 1$, and R_2 is bounded by $y = \sinh^{-1} x$, $y = 0$ and $x = \sinh 1$.

By using your diagram, find the sum of the areas of R_1 and R_2. Calculate each of these areas.   [JMB]

24 Given that the domain is the set $\mathbb{R}$ of real numbers, state the range of (a) $\cosh x$, (b) $\sinh x$.

The domain of the function f given by

$$f(x) = a \cosh x - b \sinh x$$

is also $\mathbb{R}$, and a and b are positive. Determine the range of f in each of the cases (a) $b < a$, (b) $b > a$.   [JMB]

25 (a) Express $\tanh x$ in terms of e^x and e^{-x}, and sketch the graph of $\tanh x$.

Prove that

$$\tanh^{-1} x = \frac{1}{2} \ln \left(\frac{1+x}{1-x} \right).$$

Hence obtain a series expansion of $\tanh^{-1} x$ in powers of x, giving a formula for the general term of the series.

(b) Prove that

$$8 \sinh^4 x = \cosh 4x - 4 \cosh 2x + 3,$$

and evaluate

$$\int_0^1 \sinh^4 x \, dx,$$

giving your answer in terms of e. [JMB]

26 The normal at the point P ($a \cosh u$, $a \sinh u$) on the curve

$$x^2 - y^2 = a^2, \qquad (x > 0)$$

meets the x-axis at A and the y-axis at B.

(a) Find the area of the triangle OAB, where O is the origin.

(b) Show that $PA^2 = PB^2 = a^2 \cosh 2u$.

The tangent at P meets the x-axis at C and the y-axis at D. Show that the triangles APC and DOC are similar, and that the ratio of the area of triangle APC to the area of triangle DOC is $\cosh 2u \sinh^2 u$. [JMB]

27 Prove that $\sinh^{-1} x = \ln [x + \sqrt{(x^2 + 1)}]$.

Show that $\dfrac{d}{dx} (\sinh^{-1} x) = \dfrac{1}{\sqrt{(x^2 + 1)}}$.

Evaluate $\displaystyle\int_1^8 \frac{1}{\sqrt{(x^2 - 2x + 2)}} dx$,

expressing your answer as a natural logarithm.

Show that $\displaystyle\int_1^2 \frac{3}{\sqrt{(x^2 - 2x + 2)}} dx = \int_1^8 \frac{1}{\sqrt{(x^2 - 2x + 2)}} dx$. [JMB]

28 Show that, for all real values of x, $1 + \frac{1}{2}x^2 > x$.

Deduce that $\cosh x > x$.

Prove that the point P on the curve $y = \cosh x$ which is closest to the line $y = x$ has coordinates $\ln (1 + \sqrt{2})$, $\sqrt{2}$.

The region bounded by the curve $y = \cosh x$, the x-axis, the y-axis and the ordinate through P is rotated through an angle of 2π about the x-axis. Show that the volume swept out is

$$\tfrac{1}{2}\pi[\ln (1 + \sqrt{2}) + \sqrt{2}].$$ [JMB]

29 The curve C has parametric equations $x = \cosh \theta$, $y = \sinh \theta$. Show that the perpendicular distance p of the origin from the normal to C at the point P $(\cosh \theta, \sinh \theta)$ is given by

$$p^2 = \sinh 2\theta \tanh 2\theta.$$ [JMB]

30 Obtain expressions for the real values x which satisfy $\cosh x = u$, where $u > 1$. Evaluate $\tanh x$ when u takes the value $5/4$.

Find the real solutions of the simultaneous equations

$$\cos x = 0, \qquad \sin x \cosh y = 3,$$

expressing y in logarithmic form. [JMB]

31 Show that
(a) $\sinh^{-1} 1 = \ln (1 + \sqrt{2})$,
(b) $16 \sinh^4 \theta \cosh^2 \theta = (2 \sinh^2 2\theta \cosh 2\theta - \cosh 4\theta + 1)$.
Hence, or otherwise, show that

$$\int_0^1 x^4(1 + x^2)^{1/2}\,dx = (7/48)\sqrt{2} + (1/16) \ln (1 + \sqrt{2}).$$ [JMB]

32 Sketch the curve $y = \operatorname{arsinh} x$.

Find the angle between the x-axis and the tangent to this curve at the origin, and draw this tangent on your sketch.

Find the volume of the solid generated when the finite region bounded by the curve and the lines $y = 0$ and $x = 1$ is rotated through one complete revolution about the y-axis. [L]

33 Prove that (a) for $x > 0$, $x \cosh x + 2x > 3 \sinh x$,
  (b) for all x, $x \sinh x + 2 \geqslant 2 \cosh x$.

34 (a) Express $\sinh x$ and $\cosh x$ in terms of $\tanh \frac{1}{2}x$.

By means of the substitution $t = \tanh \frac{1}{2}x$, or otherwise, evaluate the integral

$$\int_0^a \frac{1 + \sinh x}{1 + \cosh x}\,dx,$$

where $a = \ln 3$.
(b) Sketch the curve $y = \operatorname{artanh} x$.

Prove that, for $|x| < 1$, $\operatorname{artanh} x = \dfrac{1}{2} \ln \dfrac{1 + x}{1 - x}$. [L]

Express $\operatorname{arcoth} x$ as a logarithm.

35 (a) Sketch the graphs of $\sinh x$ and $\cosh x$.
  Prove that (i) $\cosh^2 x - \sinh^2 x \equiv 1$, (ii) $\cosh^2 x + \sinh^2 x \equiv \cosh 2x$.
  Find the real roots of the equation $2 \cosh x - \sinh x = 2$.

(b) Evaluate the integral $\displaystyle\int_0^1 \frac{1}{\sqrt{(x^2 + 2x + 2)}}\,dx$,

giving the answer to three decimal places. [L]

36 (a) Using the definitions of $\cosh x$ and $\sinh x$ in terms of e^x, express e^x and e^{-x} in terms of $\cosh x$ and $\sinh x$.

  Show that $\cosh 5x = 16\cosh^5 x - 20\cosh^3 x + 5\cosh x$.

  (b) Given that $\tanh 2x = -4/5$ and that x is real, show that $x = -\frac{1}{2}\ln 3$ and find the value of $\tanh x$. [L]

37 (a) Solve for real x and y the simultaneous equations

$$\cosh x = 3\sinh y,$$
$$2\sinh x = 5 - 6\cosh y,$$

expressing your answers in terms of natural logarithms.

  (b) Show that $\displaystyle\int_{3/4}^{4/3}\operatorname{arsinh}\left(\frac{1}{x}\right)dx = \frac{1}{12}\ln 432.$ [L]

38 (a) Show that when x is small

$$\ln(\cosh x) \approx x^2/2 - x^4/12,$$

and that when x is large and positive

$$\ln(\cosh x) \approx x - \ln 2.$$

  Sketch the graph of $\ln(\cosh x)$.

  (b) Find the condition that must be satisfied by the constant λ if $(\cosh x + \lambda\sin x)$ is to have a minimum value.

  Find the minimum value of $(3\cosh x + 2\sinh x)$. [L]

39 Show that the equation

$$2\sinh x + 3\cosh x = 3 + k,$$

where k is small and positive, has two real roots. Show also that, if k^3 is neglected, the positive root is approximately

$$\tfrac{1}{2}k - 3k^2/16.$$ [L]

40 (a) Show graphically that the equation $\sinh^{-1} x = \operatorname{sech}^{-1} x$ has only one real root. Prove that this root is $[(\sqrt{5} - 1)/2]^{1/2}$.

  (b) Prove that

$$\int_{4/5}^1 \operatorname{sech}^{-1} x \, dx = 2\tan^{-1} 2 - \frac{\pi}{2} - \frac{4}{5}\ln 2.$$ [L]

4 Differential equations

4.1 Introduction

Any equation between the independent variable x and the dependent variable y which involves the derivatives of y with respect to x is called an *ordinary differential* equation. The *order* of the differential equation is the order of the highest derivative occurring in that equation. In *Advanced Mathematics 1* we considered the solutions of some simple separable differential equations. In this book we explain some methods of solving some other types occurring in elementary mathematics.

We consider first the solution of ordinary differential equations of the first order. Any such equation is of the form

$$\frac{dy}{dx} = f(x, y),$$

where x is the independent variable and y is the dependent variable. This equation is of the 'first order' since it involves only a derivative with respect to x of the first order. It is an 'ordinary' differential equation since partial derivatives do not appear. The 'solution' involves finding, where possible, a relation linking y and x. The determination of such a solution depends on the nature of $f(x, y)$. For example, the equation is easy to solve when $f(x, y) = x^2y^2$, but when $f(x, y) = 1 + x^2y^2$ the solution is difficult and beyond the scope of this book.

4.2 Separable differential equations

We first solve some separable differential equations considered in *Advanced Mathematics 1*, page 301, in order to illustrate some of the basic features which must be borne in mind. Any differential equation of the form

$$\frac{dy}{dx} = F(x)\,G(y)$$

is said to be separable since it can be written in the form

$$\frac{1}{G(y)}\frac{dy}{dx} = F(x),$$

and integration with respect to x gives

$$\int \frac{1}{G(y)}\frac{dy}{dx}\,dx = \int \frac{1}{G(y)}\,dy = \int F(x)\,dx,$$

i.e. the variables have been separated and the solution is obtained if the respective indefinite integrals can be found. It should be noted that, although we have to carry out two integrations, we need introduce only one arbitrary constant. (Although *each* indefinite integral gives rise to a constant of integration, these two constants can be combined and regarded as a single constant.)

Example 1
Find the general solution of the differential equation

$$y\frac{dy}{dx} = x(1 + y^2).$$

This equation can be written (in separated form)

$$\frac{y}{1 + y^2}\frac{dy}{dx} = x.$$

Integrating with respect to x gives

$$\int \frac{y}{1 + y^2}\frac{dy}{dx}dx = \int x\,dx$$

$\Rightarrow$
$$\int \frac{y}{1 + y^2}dy = \int x\,dx$$

$\Rightarrow$
$$\frac{1}{2}\ln(1 + y^2) = \frac{x^2}{2} + C,$$

where C is any constant (usually called an arbitrary constant),

$\Rightarrow$
$$\ln(1 + y^2) = x^2 + K,$$

where K is another arbitrary constant ($= 2C$), since it does not matter how we 'name' an arbitrary constant,

$\Rightarrow$
$$1 + y^2 = e^{x^2 + K} = e^K e^{x^2} = Le^{x^2},$$

where $L = e^K$ is yet another arbitrary constant, which is necessarily positive.
 The general solution is, therefore,

$$y^2 = Le^{x^2} - 1.$$

Note that, for real y, L is not completely arbitrary, since, if the solution is to be valid for all x, $L \geqslant 1$.

Example 2
Given that

$$x(x + 1)\frac{dy}{dx} = y(y + 1)$$

and $y = 1$ when $x = 2$, find the value of y when $x = 1$.

Upon separation of the variables the equation becomes

$$\int \frac{1}{y(y + 1)} \frac{dy}{dx} dx = \int \frac{1}{x(x + 1)} dx$$

$$\Rightarrow \quad \int \left(\frac{1}{y} - \frac{1}{y + 1} \right) dy = \int \left(\frac{1}{x} - \frac{1}{x + 1} \right) dx$$

$$\Rightarrow \quad \ln \left(\frac{y}{y + 1} \right) = \ln \left(\frac{x}{x + 1} \right) + C$$

$$= \ln \left(\frac{x}{x + 1} \right) + \ln A, \qquad \text{where } C = \ln A,$$

$$= \ln \left(\frac{Ax}{x + 1} \right)$$

$$\Rightarrow \quad \frac{y}{y + 1} = A \frac{x}{x + 1}. \qquad \qquad \dots (1)$$

(Note that here we have assumed that x and y are both positive when evaluating the integrals.)

The *general solution*, obtained by solving for y in terms of x and A, is therefore

$$y = \frac{Ax}{(1 - A)x + 1}$$

and involves the one arbitrary constant A.

We are given that $y = 1$ when $x = 2$. Therefore, from (1), $A = \frac{3}{4}$ and

$$y = \frac{3x}{x + 4}.$$

This is the particular solution satisfying the condition $y = 1$ when $x = 2$. In this case, when $x = 1$, $y = \frac{3}{5}$.

Example 3
Given that $x > 0$,

$$\frac{dy}{dx} = xy \ln x,$$

and $y = 1$ when $x = 1$, express $\ln y$ in terms of x.

This equation can be written

$$\frac{1}{y} \frac{dy}{dx} = x \ln x.$$

Integrating with respect to x,

$$\int \frac{1}{y}\frac{dy}{dx}dx = \int x \ln x\, dx$$

$\Rightarrow$
$$\int \frac{1}{y}dy = \int x \ln x\, dx$$

$\Rightarrow$
$$\ln y = \tfrac{1}{2}x^2 \ln x - \tfrac{1}{4}x^2 + C.$$

$$y = 1 \text{ when } x = 1 \text{ gives } C = \tfrac{1}{4}.$$

$\therefore$
$$\ln y = \tfrac{1}{2}x^2 \ln x + \tfrac{1}{4}(1 - x^2).$$

Example 4
Given that $x > 0$, $y > 0$, find that solution of the differential equation

$$(x^2 - x^2 y)\frac{dy}{dx} + xy^2 + y^2 = 0$$

for which $y = 1$ when $x = 1$.

At first sight, this differential equation does not look separable, but when it is written in the form

$$x^2(1 - y)\frac{dy}{dx} = -(x + 1)y^2$$

it is clearly separable. Then

$$\left(\frac{1}{y^2} - \frac{1}{y}\right)\frac{dy}{dx} = -\frac{1}{x} - \frac{1}{x^2}.$$

Integration gives
$$-\frac{1}{y} - \ln y = -\ln x + \frac{1}{x} + C.$$

This is the general solution.
   The condition $y = 1$ when $x = 1$ gives $C = -2$.

$\therefore$
$$\frac{1}{y} + \ln y = 2 - \frac{1}{x} + \ln x.$$

This is the particular solution corresponding to the condition that $y = 1$ when $x = 1$. Here we cannot solve for y explicitly in terms of x and so we must leave the solution in this form. Where feasible, y should be expressed explicitly in terms of x, but this may not always be possible, in which case we leave the general solution in the form $f(x, y, C) = 0$.

Example 5
Find the solution $y = f(x)$ of the differential equation

$$(1 + x^2)\frac{dy}{dx} = 4x(1 + y)$$

for which $y = 1$ when $x = 1$.

Show that, for $x > 0$, the gradient of the curve $y = f(x)$ is always positive.

[L]

Separating the variables, we have

$$\frac{1}{1 + y}\frac{dy}{dx} = \frac{4x}{1 + x^2}. \qquad \ldots (1)$$

In problems such as this, where a particular solution is required, we can proceed directly to this solution by definite integration between corresponding values of x and y:

$$\int_1^y \frac{1}{1 + y}\,dy = \int_1^x \frac{4x}{1 + x^2}\,dx$$

$\Rightarrow$
$$\left[\ln (1 + y)\right]_1^y = \left[2 \ln (1 + x^2)\right]_1^x$$

$\Rightarrow$
$$\ln\left(\frac{1 + y}{2}\right) = 2 \ln\left(\frac{1 + x^2}{2}\right) \qquad \ldots (2)$$

$\Rightarrow$
$$\frac{1 + y}{2} = \left(\frac{1 + x^2}{2}\right)^2$$

$\Rightarrow$
$$y = \tfrac{1}{2}(x^2 + 1)^2 - 1.$$

This is the particular solution required.

The reader should check that integrating equation (1) and using the condition $y = 1$ when $x = 1$ leads to equation (2).

It is now evident that

$$\frac{dy}{dx} = 2x(x^2 + 1) > 0, \qquad \text{for all } x > 0.$$

The gradient of the curve $y = f(x)$ is, therefore, always positive for $x > 0$.

Exercises 4.2
In 1–7 find the general solution of the differential equations.

1 $\dfrac{dy}{dx} = \dfrac{y}{x} \ln y$.

2 $\dfrac{dy}{dx} + \dfrac{1}{x} = \dfrac{e^y}{x}$.

3 $\sec x \dfrac{dy}{dx} = y$.

4 $\dfrac{dy}{dx} + \dfrac{x^2}{y} = x^2$.

5 $\dfrac{dy}{dx} + e^x = y^2 e^x$.

6 $xy(1 - x^2)\dfrac{dy}{dx} = y^2 - 1$.

7 $\cos y \dfrac{dy}{dx} + \cos x = \cos x \sin y.$

In 8–12 find the particular solution of the differential equation satisfying the given conditions.

8 $xy(1 + x^2)\dfrac{dy}{dx} = 1 + y^2 \qquad$ given that $y = 1$ when $x = 1$.

9 $(x - 1)\dfrac{dy}{dx} = y - 4 \qquad$ given that $y = 0$ when $x = 0$.

10 $y^2 e^x + \dfrac{dy}{dx} = y^2 \qquad$ given that $y = 1$ when $x = 0$.

11 $\cosh x \dfrac{dy}{dx} = \cos y \qquad$ given that $y = 0$ when $x = 0$.

12 $(\sin x + \cos x)\dfrac{dy}{dx} = (\cos x - \sin x)\tan y \quad$ given that $y = \frac{1}{2}\pi$ when $x = \frac{1}{2}\pi$.

13 Solve the differential equation

$$\cos y \frac{dy}{dx} = \cot x(1 + \sin y),$$

given that $y = \pi/2$ when $x = \pi/2$. $\hfill$ [AEB]

14 Solve the differential equation

$$\cos x \frac{dy}{dx} - (\sqrt{y}) \sin x = y \sin x,$$

given that $y = 4$ when $x = 0$. $\hfill$ [AEB]

15 Find y in terms of x when

$$(1 + x^2)\frac{dy}{dx} = x(1 - y^2),$$

given that $y = 0$ when $x = 1$. $\hfill$ [L]

16 Find the general solution, valid for $x > 0$, of the differential equation

$$y \ln y + x\frac{dy}{dx} = 0. \hfill \text{[L]}$$

17 Find the general solution of the equation

$$\frac{dy}{dx} = y \ln x. \hfill \text{[OC]}$$

18 Find y in terms of x when

$$\frac{dy}{dx} = \frac{(x-1)\sqrt{(y^2+1)}}{xy},$$

given that $y = -\sqrt{(e^2 - 2e)}$ when $x = e$. [L]

4.3 First-order linear differential equations

The differential equation

$$\frac{dy}{dx} + P(x)y = Q(x), \qquad \ldots (1)$$

where $P(x)$ and $Q(x)$ are given expressions in x, is a *linear differential* equation of the first order. We use the term linear because it contains no powers or products of y and $\dfrac{dy}{dx}$. If this equation, when multiplied by the expression $\phi(x)$, becomes of the form

$$\frac{d}{dx}[\phi(x)y] = \phi(x)\,Q(x), \qquad \ldots (2)$$

then it can be integrated at once to give

$$\phi(x)y = \int \phi(x)\,Q(x)\,dx + C,$$

which is the general solution. Here $\phi(x)$ is called an *integrating factor* of the differential equation. Sometimes an integrating factor can be seen 'by inspection'.

Example
Solve the differential equation

$$\frac{dy}{dx} + \frac{y}{x} = e^x.$$

Writing the equation in the form

$$x\frac{dy}{dx} + y = xe^x,$$

we see that it can be expressed in the form

$$\frac{d}{dx}(xy) = xe^x.$$

(In fact, an integrating factor of the equation is x.)

$$\Rightarrow \qquad xy = \int xe^x\,dx + C$$
$$\Rightarrow \qquad xy = (x-1)e^x + C$$
$$\Rightarrow \qquad y = (x-1)e^x/x + C/x.$$

This is the general solution of the equation.

Equation (2) is equivalent to

$$\phi(x)\frac{dy}{dx} + \phi'(x)y = \phi(x)\,Q(x)$$

$$\Rightarrow \qquad \frac{dy}{dx} + \frac{\phi'(x)}{\phi(x)}y = Q(x),$$

assuming that $\phi(x) \neq 0$. This equation is identical with equation (1) if

$$\frac{\phi'(x)}{\phi(x)} = P(x)$$

$$\Rightarrow \qquad \ln \phi(x) = \int P(x)\,dx$$
$$\Rightarrow \qquad \phi(x) = e^{\int P(x)\,dx}. \qquad \ldots (3)$$

Therefore an integrating factor for equation (1) is $e^{\int P(x)\,dx}$.

Notes

(1) It is not necessary to introduce an arbitrary constant when evaluating $\int P(x)\,dx$ for an integrating factor. (This would merely multiply the equation by a constant factor.)

(2) Frequently, the integrating factor takes the form $e^{\ln[f(x)]}$, which is the same as $f(x)$. (This follows because

$$e^{\ln[f(x)]} = f(x),$$

since the exponential and logarithmic functions are inverse functions.)

(3) When equation (1) has been multiplied by the integrating factor $\phi(x)$ it *must* be of the form

$$\frac{d}{dx}[\phi(x)y] = \phi(x)\,Q(x),$$

i.e. $$\phi(x)\frac{dy}{dx} + \phi'(x)y = \phi(x)\,Q(x).$$

This forms a useful check on the correctness of the integrating factor found.

Example 1

We reconsider the example on page 106. The equation is in standard form, with $P(x) = 1/x$. So

$$\int P(x)\,dx = \int \frac{1}{x}dx = \ln x,$$

and the integrating factor is $e^{\ln x} = x$.

Therefore the equation becomes

$$x\frac{\mathrm{d}y}{\mathrm{d}x} + y = xe^x$$

$$\Rightarrow \qquad \frac{\mathrm{d}}{\mathrm{d}x}(xy) = xe^x.$$

The solution proceeds as before.

Example 2

Solve the differential equation

$$\frac{\mathrm{d}y}{\mathrm{d}x} - y \tan x = \sec x.$$

This equation is in 'standard form' with $P(x) = -\tan x$.

The integrating factor is

$$\exp(-\textstyle\int \tan x\,\mathrm{d}x) = \exp(\ln \cos x) = \cos x.$$

(Note that an integrating factor is necessarily positive because $e^{f(x)} > 0$ for all $f(x)$; also that the integrating factor is $\exp(-\int \tan x\,\mathrm{d}x)$ *not* $\exp(\int \tan x\,\mathrm{d}x)$.)

When multiplied by this factor the equation becomes

$$(\cos x)\frac{\mathrm{d}y}{\mathrm{d}x} - y \sin x = 1$$

$$\Leftrightarrow \qquad \frac{\mathrm{d}}{\mathrm{d}x}(y \cos x) = 1$$

$$\Leftrightarrow \qquad y \cos x = x + C$$
$$\Leftrightarrow \qquad y = x \sec x + C \sec x,$$

since $\cos x \neq 0$. Note that, in this case, an integrating factor could be 'seen by inspection'.

Example 3

Solve the differential equation

$$x\frac{\mathrm{d}y}{\mathrm{d}x} + 3y = x^2,$$

given that $y = \tfrac{2}{3}$ when $x = -1$.

Note that this equation is not in 'standard' form. However, division by x brings the equation to 'standard' form with $P(x) = 3/x$.

An integrating factor is

$$e^{\int 3/x\,\mathrm{d}x} = e^{3\ln x} = e^{\ln(x^3)} = x^3.$$

The equation now becomes

$$x^3 \frac{dy}{dx} + 3x^2 y = x^4$$

$\Leftrightarrow$
$$\frac{d}{dx}(x^3 y) = x^4.$$

$\Leftrightarrow$
$$x^3 y = \frac{x^5}{5} + C$$

$\Leftrightarrow$
$$y = \frac{x^2}{5} + \frac{C}{x^3}.$$

This is the general solution.

$y = \frac{2}{5}$ when $x = -1$ gives $\frac{2}{5} = \frac{1}{5} - C \Rightarrow C = -\frac{1}{5}.$

The particular solution is

$$y = (x^5 - 1)/(5x^3).$$

Example 4
Solve the differential equation

$$(x - 1)\frac{dy}{dx} + xy = (x - 1)e^{-x},$$

given that $y = 0$ when $x = 0$.

Writing the equation in 'standard' form gives $P(x) = x/(x - 1)$ and so

$$\int P(x)\,dx = \int \frac{x}{(x - 1)}\,dx = \int\left(1 + \frac{1}{x - 1}\right)dx = x + \ln(x - 1).$$

An integrating factor is $(x - 1)e^x$.
    The differential equation now becomes

$$(x - 1)e^x \frac{dy}{dx} + xe^x y = (x - 1),$$

which must be of the form

$$\frac{d}{dx}[(x - 1)e^x y] = (x - 1)$$

$\Rightarrow$
$$(x - 1)e^x y = \int(x - 1)\,dx + C$$
$$= \frac{1}{2}(x - 1)^2 + C.$$

$\therefore$
$$y = \frac{1}{2}(x - 1)e^{-x} + \frac{Ce^{-x}}{(x - 1)}.$$

This is the general solution.

The condition $y = 0$ when $x = 0$ gives $C = -\frac{1}{2}$ and the particular solution satisfying these conditions is

$$y = \tfrac{1}{2}(x - 1)e^{-x} - \tfrac{1}{2}e^{-x}/(x - 1).$$

Example 5
Find the general solution of the differential equation

$$\frac{dy}{dx} + 6y = 3x. \qquad \ldots (1)$$

Find also the particular solution for which $y = 0$ when $x = 0$.

An integrating factor is $e^{\int 6dx} = e^{6x}$ and so, writing the differential equation in the form

$$e^{6x}\frac{dy}{dx} + 6\,e^{6x}y = 3xe^{6x}$$

$$\frac{d}{dx}(e^{6x}y) = 3xe^{6x}$$

$\Leftrightarrow$
$$e^{6x}y = \frac{1}{2}\left(x - \frac{1}{6}\right)e^{6x} + C$$

$\Leftrightarrow$
$$y = \frac{1}{2}\left(x - \frac{1}{6}\right) + Ce^{-6x}. \qquad \ldots (2)$$

This is the general solution of equation (1).

The condition $y = 0$ when $x = 0$ gives $C = 1/12$ and the particular solution

$$y = (6x - 1 + e^{-6x})/12.$$

Note here that the general solution (2) of equation (1) consists of two parts (added together):
(a) the part Ce^{-6x}, involving the arbitrary constant C, is the solution of the equation

$$\frac{dy}{dx} + 6y = 0$$

which is equation (1) with zero on the right-hand side;
(b) the part $\frac{1}{2}(x - \frac{1}{6})$, with no arbitrary constant, is a special solution of (1). We return to differential equations of this form in Section 4.5.

Exercises 4.3
Solve the differential equations:

1 $\dfrac{dy}{dx} + y \cot x = 2 \sin x.$

2 $\dfrac{dy}{dx}\operatorname{cosec} x + y \sec x = 2 \cos x$, given that $y = 0$ for $x = 0$.

3 $\dfrac{dy}{dx} = 2x - y$.

4 $\sin x\dfrac{dy}{dx} - 2y \cos x = 3 \sin x$.

5 $(x - 1)\dfrac{dy}{dx} + xy = x^2$.

6 $\dfrac{dy}{dx}\cos x + y \sin x = x \sin 2x + x^2$.

7 $x\dfrac{dy}{dx} - y = 3x^4$, given that $y = -1$ when $x = 1$.

8 $\dfrac{dy}{dx} - \sin 2x = y \cot x$.

9 $x\dfrac{dy}{dx} + 2y = x^3$, given that $y = 0$ when $x = 1$.

10 $\dfrac{dy}{dx} + y \tanh x = \sinh 2x$.

11 $x^2\dfrac{dy}{dx} + 1 + (1 - 2x)y = 0$.

12 $\dfrac{dy}{dx} + \dfrac{y}{x} = 2x^2 e^{x^2}$.   [L]

13 $\dfrac{dy}{dx} + y \tan x = \cos^2 x$ for $-\pi/2 < x < \pi/2$, given that $y = \frac{1}{2}$ when $x = \pi/4$.

        [L]

14 $x\dfrac{dy}{dx} - y = x^3 \ln x$, given that $y = \frac{3}{4}$ when $x = 1$.   [L]

15 $\dfrac{dx}{dt} + 2x = t$, given that $\dfrac{dx}{dt} = 0$ when $t = 0$.   [L]

4.4 First-order differential equations reducible to separable and linear types

In this section we illustrate some techniques which can be applied in solving a variety of differential equations. In each case we reduce the problem to one of the forms considered in Section 4.2 or 4.3.

Homogeneous differential equations

A differential equation of the form

$$\frac{dy}{dx} = f\left(\frac{y}{x}\right), \qquad\qquad \ldots (1)$$

is said to be *homogeneous* since, if each of x and y is regarded as having the dimensions of a length, then all the terms of the equation are zero-dimensional in length. (The expression $g(x, y)$ is said to be homogeneous in x and y and of degree n if $g(\lambda x, \lambda y) \equiv \lambda^n g(x, y)$.)

Such equations are solved using the obvious substitution $y/x = v$, i.e. $y = xv$, where v is an expression in x only. Equation (1) then becomes

$$v + x\frac{dv}{dx} = f(v)$$

and this is clearly a 'variables-separable' differential equation for v in terms of x. As with variables-separable equations, rearrangement is frequently required prior to solution.

Note that we solve the separable differential equation for v but, since the original equation (1) involved x and y, we must write $v = y/x$ and give our final result as an equation relating x and y.

Example 1

Solve, for $x > 0$, the differential equation

$$xy\frac{dy}{dx} = x^2 + y^2.$$

Rearranging the equation in the form

$$\frac{dy}{dx} = \frac{x^2 + y^2}{xy} = \frac{x}{y} + \frac{y}{x}$$

shows it to be homogeneous and so, putting $y = xv$, we find

$$v + x\frac{dv}{dx} = \frac{1}{v} + v$$

$$\Leftrightarrow \qquad v\frac{dv}{dx} = \frac{1}{x}$$

$$\Leftrightarrow \qquad \int v\,dv = \int \frac{1}{x}dx$$

$$\Leftrightarrow \qquad v^2/2 = \ln x + C, \qquad \text{since } x > 0,$$
$$\Leftrightarrow \qquad\quad v^2 = \ln (Bx^2), \qquad \text{where } \ln B = 2C,$$
$$\Leftrightarrow \qquad\quad y^2 = x^2 \ln (Bx^2),$$

which is the general solution.

Example 2
Find the general solution of the differential equation

$$(x - y)\frac{dy}{dx} = x + y.$$

$$\frac{dy}{dx} = \frac{x + y}{x - y} = \frac{1 + y/x}{1 - y/x}.$$

The substitution $y = xv$ gives

$$v + x\frac{dv}{dx} = \frac{1 + v}{1 - v}$$

⇔ $$x\frac{dv}{dx} = \frac{1 + v}{1 - v} - v = \frac{1 + v^2}{1 - v}$$

⇔ $$\frac{1 - v}{1 + v^2}\frac{dv}{dx} = \frac{1}{x}$$

⇔ $$\int \frac{1 - v}{1 + v^2}\,dv = \int \frac{1}{x}\,dx + C$$

⇔ $$\tan^{-1} v - \tfrac{1}{2}\ln (1 + v^2) = \ln x + C$$
⇔ $$\tan^{-1} v = \tfrac{1}{2}\ln [x^2(1 + v^2)] + C$$

⇔ $$\tan^{-1}\left(\frac{y}{x}\right) = \tfrac{1}{2}\ln (x^2 + y^2) + C.$$

Some differential equations can be made homogeneous by a simple substitution.

Example 3
Find the general solution of the differential equation

$$\frac{dy}{dx} = \frac{3x + y - 3}{x + 3y - 1}. \qquad \ldots (1)$$

As it stands, this equation is clearly not homogeneous. However, if we change the origin of coordinates by making the substitution $x = X + h$, $y = Y + k$, where h, k are as yet undetermined constants, we have

$$\frac{dY}{dX} = \frac{3X + Y + (3h + k - 3)}{X + 3Y + (h + 3k - 1)}. \qquad \ldots (2)$$

Clearly we can make (2) homogeneous by choosing h, k so that

$$3h + k - 3 = 0, \qquad h + 3k - 1 = 0,$$

i.e., $h = 1$, $k = 0$. (We have transferred the origin to the point of intersection of the lines $3x + y - 3 = 0$, $x + 3y - 1 = 0$.) Equation (2) now becomes

$$\frac{dY}{dX} = \frac{3X + Y}{X + 3Y},$$

which is clearly homogeneous, and the substitution $Y = XV$ gives

$$X\frac{dV}{dX} + V = \frac{3 + V}{1 + 3V}$$

$\Leftrightarrow \qquad\qquad X\frac{dV}{dX} = \frac{3(1 - V^2)}{1 + 3V}.$

Separation and integration with respect to X gives

$$\int \frac{1 + 3V}{1 - V^2} dV = 3 \int \frac{1}{X} dX$$

$\Leftrightarrow \qquad \int \left(\frac{2}{1 - V} - \frac{1}{1 + V} \right) dV = 3 \int \frac{1}{X} dX$

$\Leftrightarrow \qquad -2 \ln (1 - V) - \ln (1 + V) = 3 \ln X + C$

$\Leftrightarrow \qquad \ln [X^3 (1 - V)^2 (1 + V] = A$

$\Leftrightarrow \qquad X^3 (1 - V)^2 (1 + V) = B$

$\Leftrightarrow \qquad (X - Y)^2 (X + Y) = B$

$\Leftrightarrow \qquad (x - y + 1)^2 (x + y - 1) = B.$

This is the general solution.

Example 4
Find the general solution of the differential equation

$$\frac{dy}{dx} = \frac{y - x - 1}{x - y - 1}.$$

The method used in Example 3 above is not applicable in this case since the lines $y - x - 1 = 0$ and $x - y - 1 = 0$ are parallel.

However, if we write $z = x - y$, then

$$\frac{dz}{dx} = 1 - \frac{dy}{dx}$$

and the equation can be written

$$1 - \frac{dz}{dx} = \frac{-z - 1}{z - 1}$$

which is a separable equation and reduces to

$$\frac{dz}{dx} = \frac{2z}{z - 1}$$

$$\Leftrightarrow \qquad \left(1 - \frac{1}{z}\right)\frac{dz}{dx} = 2.$$

Integration gives

$$z - \ln z = 2x + C$$
$$\Leftrightarrow \qquad x + y + \ln (x - y) + C = 0$$
$$\Leftrightarrow \qquad (x - y)e^{x+y} = A,$$

which is the general solution.

Bernoulli's equation

The differential equation

$$\frac{dy}{dx} + P(x)y = Q(x)y^n, \qquad \qquad \dots (1)$$

where $n \neq 0$ or 1, is called a *Bernoulli* equation. Note that this equation is separable when $n = 1$, and is of the linear form (equation (1) on page 106) when $n = 0$.

A Bernoulli equation is clearly *not* a linear equation. However, it may be written as

$$\frac{1}{y^n}\frac{dy}{dx} + P(x)\frac{1}{y^{n-1}} = Q(x).$$

The left-hand side suggests, by comparison with the standard form of a linear equation, that we write $z = 1/y^{n-1}$, whence

$$\frac{dz}{dx} = -(n - 1)\frac{1}{y^n}\frac{dy}{dx},$$

and so the equation becomes a linear equation in z.

Example

Find the general solution of the differential equation

$$\frac{dy}{dx} + \frac{y}{x} = x^2 y^3.$$

Dividing by y^3 gives

$$\frac{1}{y^3}\frac{dy}{dx} + \frac{1}{x}\frac{1}{y^2} = x^2 \qquad \qquad \dots (1)$$

Putting $z = \dfrac{1}{y^2}$ gives $\dfrac{dz}{dx} = -\dfrac{2}{y^3}\dfrac{dy}{dx}$ and so (1) becomes

$$-\frac{1}{2}\frac{dz}{dx} + \frac{z}{x} = x^2$$

$$\Leftrightarrow \qquad \frac{dz}{dx} - \frac{2}{x}z = -2x^2. \qquad \qquad \dots (2)$$

Equation (2) is a linear differential equation for z in terms of x, which is in 'standard' form with $P(x) = -2/x$, giving an integrating factor of $1/x^2$. Then

$$\frac{1}{x^2}\frac{dz}{dx} - \frac{2}{x^3}z = -2$$

$$\Leftrightarrow \qquad \frac{d}{dx}\left(\frac{z}{x^2}\right) = -2$$

$$\Leftrightarrow \qquad \frac{z}{x^2} = C - 2x$$

$$\Leftrightarrow \qquad z = Cx^2 - 2x^3$$

$$\Leftrightarrow \qquad \frac{1}{y^2} = Cx^2 - 2x^3$$

$$\Leftrightarrow \qquad y^2 = \frac{1}{Cx^2 - 2x^3}$$

is the general solution.

Exercises 4.4
As indicated in some of the exercises below, a differential equation can sometimes be converted to integrable form by a special substitution, appropriate to the relevant differential equation.

1 By means of the substitution $y = v - x^3$, or otherwise, find the solution of

$$(x - 1)\frac{dy}{dx} - 3y = 3x^2,$$

for which $y = 0$ when $x = 2$. [JMB]

2 By making the substitution $y = zx$, show that the differential equation

$$x^2\frac{dy}{dx} = y(x + y)$$

becomes

$$\frac{dz}{dx} = \frac{z^2}{x}.$$

Hence find y in terms of x for $x > 0$, given that $y = -1$ when $x = 1$. [L]

3 Using the substitution $y = xv$, transform the differential equation

$$x^2\frac{dy}{dx} = x^2 + xy + y^2,$$

where $x > 0$, into a differential equation relating v and x.

Hence find y in terms of x, given that $y = 0$ when $x = 1$. [L]

4 Use the substitution $z = y^2$ to transform the differential equation

$$2y\frac{dy}{dx} + y^2 = 2 \sin x$$

into a differential equation relating z and x.

Hence find y^2 in terms of x, given that $y = 0$ when $x = 0$. [L]

5 By writing $z = dy/dx$, or otherwise, find the general solution of the differential equation

$$x\frac{d^2y}{dx^2} + 2\frac{dy}{dx} = 0.$$ [L]

6 By making the substitution $z = 1/y^2$, transform the differential equation

$$(1 + x^2)\frac{dy}{dx} + xy = x^3y^3$$

into a first-order linear differential equation.

Hence obtain the solution of the original differential equation which satisfies the condition $y = 1$ when $x = 0$. [L]

7 Find by means of the substitution $u = \cos y$, or otherwise, the solution of the differential equation

$$\sin y\frac{dy}{dx} = \cos x(2 \cos y - \sin x)$$

such that $y = \pi/2$ when $x = 0$. [L]

8 By using the substitution $x + y = v$, obtain the particular solution of the differential equation

$$\frac{dy}{dx} = (x + y)^2,$$

for which $y = 1$ when $x = 0$.

9 By the substitution $z = x + y$, transform the differential equation

$$\frac{dy}{dx} = \frac{1 + 2x + 2y}{1 - 2x - 2y}$$

into an equation involving z and x only. Hence, or otherwise, solve the given differential equation.

Solve the homogeneous differential equations in **10–14**.

10 $\dfrac{dy}{dx} + \dfrac{2x(x + y)}{x^2 + y^2} = 0.$ **11** $(5x - 2y)\dfrac{dy}{dx} - y = 2x.$

12 $\dfrac{dy}{dx} = e^{-y/x} + y/x.$ **13** $x\dfrac{dy}{dx} = y + \sqrt{(x^2 + y^2)}.$

14 $\dfrac{dy}{dx} = \dfrac{x + 3y}{3x + y}$, given that $y = 0$ when $x = 1$.

Solve the Bernoulli type differential equations in **15–20**.

15 $\dfrac{dy}{dx} = y \tan x + y^3 \tan^3 x.$

16 $x(1 - x^2)\dfrac{dy}{dx} + (2x^2 - 1)y = x^3 y^3.$

17 $\dfrac{dy}{dx}\cos x - y \sin x = y^3 \cos^2 x.$

18 $\dfrac{dy}{dx} = 2y - xy^2$, given that $y = 1$ at $x = 0$.

19 $\dfrac{dy}{dx} = xy^3 - y$, given that $y = 1$ at $x = 0$.

20 $\dfrac{dy}{dx} + \dfrac{y}{x} = y^2.$

4.5 Linear differential equations with constant coefficients

When, in the linear first-order differential equation

$$\frac{dy}{dx} + P(x)y = Q(x),$$

the expression $P(x)$ is a constant, the equation is said to have *constant coefficients*, i.e. the coefficients of y and dy/dx are constants. When P is constant, the method of solution using an integrating factor is unnecessarily tedious and the following approach is to be preferred. We will first consider equations for which $Q(x) = 0$, and, as a simple example, the differential equation

$$\frac{dy}{dx} + 2y = 0,$$

which is separable with general solution $y = Ce^{-2x}$, but we now give an alternative method of solution. Let us *try* as a solution e^{mx}, where m is a constant to be determined. Then

$$me^{mx} + 2e^{mx} = 0$$

$\Leftrightarrow$

$$(m + 2)e^{mx} = 0$$

$\Leftrightarrow$

$$m = -2,$$

since e^{mx} never vanishes. Therefore, $y = e^{-2x}$ is a solution of the given differential equation and clearly $y = Ce^{-2x}$, where C is an arbitrary constant, is also a solution. This is the general solution of the differential equation.

(Note that, knowing the solution $y = e^{-2x}$, we can assert that $y = Ce^{-2x}$ is also a solution only because the differential equation is linear in y. This does not, in general, hold for nonlinear differential equations. For example, $y = e^{-2x}$ satisfies $y\dfrac{dy}{dx} + 2e^{-4x} = 0$ but $y = 2e^{-2x}$ does *not* satisfy this differential equation.)

Later, we shall extend this technique to differential equations of higher order, but first we consider the formation of differential equations from known solutions.

Example 1
Find a differential equation whose general solution is

$$y = x^3 + Ce^{-2x}, \qquad \ldots (1)$$

where C is an arbitrary constant.

$$\frac{dy}{dx} = 3x^2 - 2Ce^{-2x}. \qquad \ldots (2)$$

We 'find' the differential equation by eliminating the constant C between (1) and (2), giving

$$\frac{dy}{dx} + 2y = 2x^3 + 3x^2,$$

which is the required differential equation.

Example 2
Find a differential equation whose general solution is

$$y = C_1 e^x + C_2 e^{-2x}. \qquad \ldots (1)$$

$$\frac{dy}{dx} = C_1 e^x - 2C_2 e^{-2x}. \qquad \ldots (2)$$

$$\frac{d^2 y}{dx^2} = C_1 e^x + 4C_2 e^{-2x}. \qquad \ldots (3)$$

(We need three equations in order to eliminate the two constants C_1 and C_2.)

From (1) and (2),

$$y - \frac{dy}{dx} = 3C_2 e^{-2x}. \qquad \ldots (4)$$

From (2) and (3),

$$\frac{d^2 y}{dx^2} - \frac{dy}{dx} = 6C_2 e^{-2x}. \qquad \ldots (5)$$

From (4) and (5),

$$\frac{d^2 y}{dx^2} + \frac{dy}{dx} - 2y = 0,$$

which is the required differential equation.

Note that, in Example 2, we obtain a second-order differential equation by eliminating two arbitrary constants. In general, if the expression for y contains n arbitrary constants then the result of eliminating them by the process illustrated in Examples 1 and 2 above leads to a differential equation of the nth order.

Conversely, it can be proved that the general solution of an nth-order differential equation contains n arbitrary constants. In particular, the general solution of a second-order linear equation contains two arbitrary constants.

Let us now consider the second-order linear equation

$$\frac{d^2 y}{dx^2} - 4\frac{dy}{dx} + 3y = 0$$

and again try a solution $y = e^{mx}$. Then

$$(m^2 - 4m + 3)e^{mx} = 0$$
$$\Leftrightarrow \quad m^2 - 4m + 3 = 0$$
$$\Leftrightarrow \quad (m - 3)(m - 1) = 0$$
$$\Leftrightarrow \quad m = 3 \quad \text{or} \quad 1.$$

Hence $y = e^{3x}$ and $y = e^x$ are solutions of the given differential equation. Also, clearly, $y = C_1 e^{3x}$ and $y = C_2 e^x$, where C_1 and C_2 are distinct arbitrary constants, are also solutions.

Finally, suppose that $y = C_1 e^{3x} + C_2 e^x$. Then

$$\frac{d^2 y}{dx^2} - 4\frac{dy}{dx} + 3y = 9C_1 e^{3x} + C_2 e^x - 4(3C_1 e^{3x} + C_2 e^x) + 3(C_1 e^{3x} + C_2 e^x)$$

$$= 0.$$

Therefore, $y = C_1 e^{3x} + C_2 e^x$ is a solution of the given differential equation and, since it contains two arbitrary constants, it must be the unique general solution. (We *assume* that a differential equation of the nth order has a unique general solution containing n arbitrary constants. We can add the solutions $C_1 e^{3x}$ and $C_2 e^x$ only because the given differential equation is linear.)

We consider now the second-order differential equation

$$a\frac{d^2y}{dx^2} + 2h\frac{dy}{dx} + by = 0, \qquad \dots (1)$$

where a, h, b are constants. Suppose that $y = f(x)$ and $y = g(x)$ are solutions of equation (1) and that $f(x)$ is *not* a constant multiple of $g(x)$, i.e. we cannot find a constant λ so that $f(x) + \lambda g(x) \equiv 0$. (When we cannot find this λ we say that $f(x)$ and $g(x)$ are *linearly independent*.) Since $y = f(x)$ and $y = g(x)$ are solutions of equation (1),

then
$$a\frac{d^2}{dx^2}[C_1 f(x)] + 2h\frac{d}{dx}[C_1 f(x)] + bC_1 f(x) = 0,$$

$$a\frac{d^2}{dx^2}[C_2 g(x)] + 2h\frac{d}{dx}[C_2 g(x)] + bC_2 g(x) = 0,$$

where C_1, C_2 are arbitrary constants.

By addition,

$$a\frac{d^2}{dx^2}[C_1 f(x) + C_2 g(x)] + 2h\frac{d}{dx}[C_1 f(x) + C_2 g(x)] + b[C_1 f(x) + C_2 g(x)] = 0$$

i.e., $y = C_1 f(x) + C_2 g(x)$ is also a solution of equation (1).

In fact, this is the general solution of equation (1).

It follows that, if we can find two linearly independent solutions $f(x)$, $g(x)$ of equation (1) above, then the general solution, involving two arbitrary constants, is $y = C_1 f(x) + C_2 g(x)$.

But if we *try* $y = e^{mx}$ as a solution, we find

$$am^2 + 2hm + b = 0, \qquad \dots (2)$$

a quadratic in m, with roots m_1, m_2 say, which will give two linearly independent solutions $e^{m_1 x}$, $e^{m_2 x}$ provided $m_1 \neq m_2$.

Equation (2) is called the *auxiliary* equation.

The general solution is then

$$y = C_1 e^{m_1 x} + C_2 e^{m_2 x}. \qquad \dots (3)$$

Three separate cases need consideration, depending on the nature of the roots of equation (2).

Case 1 the auxiliary equation has real and distinct roots

Example
Find the general solution of the differential equation

$$2\frac{d^2y}{dx^2} + 5\frac{dy}{dx} - 3y = 0.$$

The auxiliary equation is

$$2m^2 + 5m - 3 = 0$$
$$\Leftrightarrow \qquad (2m - 1)(m + 3) = 0$$
$$\Leftrightarrow \qquad m_1 = \tfrac{1}{2} \ \text{or} \ -3.$$
$$\therefore \qquad y = C_1 e^{x/2} + C_2 e^{-3x}$$

is the general solution.

Exercises 4.5(i)
Find the general solution of each of the following differential equations:

1 $y' - 3y = 0$.

2 $y' + 2y = 0$.

3 $y'' - 5y' + 4y = 0$.

4 $y'' + y' - 12y = 0$.

5 $y'' - 4y = 0$.

6 $y'' + 3y' + 2y = 0$.

7 $y'' + 4y' - 5y = 0$.

8 $y''' + 2y'' - y' - 2y = 0$.

Case 2   the auxiliary equation has complex roots

Suppose now that m_1, m_2 are complex conjugates, say $m_1 = \alpha + i\beta$, $m_2 = \alpha - i\beta$, where $\alpha, \beta \in \mathbb{R}$, $\beta \neq 0$. Then the general solution of equation (1) on page 121 may be written

$$y = C_1 e^{(\alpha + i\beta)x} + C_2 e^{(\alpha - i\beta)x}$$
$$\Leftrightarrow \qquad y = C_1 e^{\alpha x} \cdot e^{i\beta x} + C_2 e^{\alpha x} \cdot e^{-i\beta x}$$
$$\Leftrightarrow \qquad y = e^{\alpha x}(C_1 e^{i\beta x} + C_2 e^{-i\beta x}).$$
$$\therefore \qquad y = e^{\alpha x}[C_1(\cos \beta x + i \sin \beta x) + C_2(\cos \beta x - i \sin \beta x)]$$
$$\Leftrightarrow \qquad y = e^{\alpha x}[(C_1 + C_2) \cos \beta x + i(C_1 - C_2) \sin \beta x]$$
$$\Rightarrow \qquad y = e^{\alpha x}[A \cos \beta x + B \sin \beta x], \qquad \qquad \dots (1)$$

where A, B, defined by

$$A = (C_1 + C_2), \quad B = i(C_1 - C_2)$$

are new arbitrary constants.

The solution should always be given in this (real) form. (Do not be worried by the fact that B appears to be an imaginary number. In fact, C_1 and C_2 are conjugate complex constants, so that $C_1 + C_2$ is real and $C_1 - C_2$ is purely imaginary and therefore $i(C_1 - C_2)$ is real.)

Example 1
Find the general solution of the differential equation

$$\frac{d^2y}{dx^2} - 6\frac{dy}{dx} + 13y = 0.$$

The auxiliary equation is

$$m^2 - 6m + 13 = 0$$
$$\Leftrightarrow \qquad m = 3 \pm 2i.$$

The general solution is

$$y = e^{3x}[C_1 \cos 2x + C_2 \sin 2x].$$

Example 2
Find y in terms of k and x, given that

$$\frac{d^2y}{dx^2} + k^2y = 0,$$

where k is a non-zero, real constant, and $y = 1$ and $dy/dx = 1$ when $x = 0$.
The auxiliary equation is

$$m^2 + k^2 = 0 \quad \Leftrightarrow \quad m = \pm ik.$$

The general solution is, therefore

$$y = A_1 e^{ikx} + A_2 e^{-ikx}$$
i.e., $$y = C_1 \cos kx + C_2 \sin kx.$$

$y = 1 \quad$ when $\quad x = 0 \quad \Rightarrow \quad C_1 = 1.$

$\dfrac{dy}{dx} = 1 \quad$ when $\quad x = 0 \quad \Rightarrow \quad kC_2 = 1.$

The required solution is therefore

$$y = \cos kx + \frac{1}{k} \sin kx.$$

Exercises 4.5(ii)
Find the general solution of each of the following differential equations.

1 $y'' + 2y' + 5y = 0.$

2 $y'' + 9y = 0.$

3 $y'' + 4y' + 5y = 0.$

4 $y'' - 8y' + 25y = 0.$

5 $y'' + 16y = 0.$

6 $y'' + 2ky + (n^2 + k^2)y = 0,$ where k, n are real non-zero constants.

Case 3 the auxiliary equation has real equal roots

If, in equation (1), page 121, $h^2 = ab$, then the auxiliary equation has equal roots and the equation can be expressed in the form

$$\frac{d^2 y}{dx^2} - 2n\frac{dy}{dx} + n^2 y = 0. \qquad \ldots (1)$$

The auxiliary equation is $(m - n)^2 = 0$ giving only $m = n$. Equation (1) can be written as

$$\frac{d}{dx}\left(\frac{dy}{dx} - ny\right) - n\left(\frac{dy}{dx} - ny\right) = 0, \qquad \ldots (2)$$

which suggests writing $u = \dfrac{dy}{dx} - ny$. The equation (2) now becomes

$$\frac{du}{dx} - nu = 0. \qquad \ldots (3)$$

(The second-order differential equation has been reduced to two equivalent linear equations. This reduction can always be effected, even in the general case of a second-order differential equation with constant coefficients, but is particularly helpful here.)

The solution of (3) is

$$u = A\mathrm{e}^{nx}$$

so that

$$\frac{dy}{dx} - ny = A\mathrm{e}^{nx},$$

which is first-order linear with an integrating factor e^{-nx}. Then

$$\frac{d}{dx}(y\mathrm{e}^{-nx}) = A$$

$$\Leftrightarrow \qquad\qquad y\mathrm{e}^{-nx} = Ax + B$$
$$\Leftrightarrow \qquad\qquad y = (Ax + B)\mathrm{e}^{nx} \qquad \ldots (4)$$

where A, B are arbitrary constants.

Example
Solve the differential equation

$$\frac{d^2 y}{dx^2} - 2\frac{dy}{dx} + y = 0,$$

given that, when $x = 0$, $y = 1$ and $\dfrac{dy}{dx} = -2$.

The auxiliary equation is

$$m^2 - 2m + 1 = 0$$
$$\Leftrightarrow \qquad (m - 1)^2 = 0$$
$$\Leftrightarrow \qquad m = 1, 1$$
$$\Rightarrow \qquad y = (C_1 x + C_2)e^x.$$

This is the general solution.

$$y = 1 \quad \text{when} \quad x = 0 \quad \Rightarrow \quad C_2 = 1$$

$$\frac{dy}{dx} = -2 \quad \text{when} \quad x = 0 \quad \Rightarrow \quad C_1 + C_2 = -2 \quad \Rightarrow \quad C_1 = -3.$$

The required solution is, therefore,

$$y = (1 - 3x)e^x.$$

Exercises 4.5(iii)

Find the general solution of each of the following differential equations.

1 $y'' + 4y' + 4y = 0.$

2 $y'' - 10y' + 25y = 0.$

3 $4y'' + 4y' + y = 0.$

4 $9y'' + 24y' + 16y = 0.$

We now consider some miscellaneous examples.

Example 1

Solve the differential equation

$$\frac{d^2 y}{dx^2} + 3\frac{dy}{dx} = 0$$

given that $y \to 2$ as $x \to \infty$ and that $\dfrac{dy}{dx} = -3$ when $x = 0.$

The auxiliary equation is

$$m^2 + 3m = 0$$
$$\Leftrightarrow \qquad m = 0, -3$$
$$\Leftrightarrow \qquad y = C_1 e^{0x} + C_2 e^{-3x}$$
$$\Leftrightarrow \qquad y = C_1 + C_2 e^{-3x}$$

is the general solution.

$$y \to 2 \quad \text{as} \quad x \to \infty \quad \Rightarrow \quad C_1 = 2.$$

$$\frac{dy}{dx} = -3 \quad \text{when} \quad x = 0 \quad \Rightarrow \quad C_2 = 1.$$

The required solution is, therefore, $y = 2 + e^{-3x}$.

Example 2
Find the general solution of the differential equation

$$y''' - 5y'' + 8y' - 4y = 0.$$

The auxiliary equation is

$$m^3 - 5m^2 + 8m - 4 = 0$$
$$\Leftrightarrow \qquad (m - 1)(m^2 - 4m + 4) = 0$$
$$\Leftrightarrow \qquad (m - 1)(m - 2)^2 = 0$$
$$\Leftrightarrow \qquad m = 1, 2, 2.$$

The general solution is, therefore,

$$y = C_1 e^x + (C_2 + C_3 x)e^{2x}.$$

Example 3
Find the general solution of the differential equation

$$\frac{d^3 y}{dx^3} + 8y = 0.$$

For equations of higher order than the second, we follow the procedure outlined above. In this case the auxiliary equation is

$$m^3 + 8 = 0$$
$$\Leftrightarrow \qquad (m + 2)(m^2 - 2m + 4) = 0$$
$$\Leftrightarrow \qquad m = -2, 1 \pm i\sqrt{3}.$$

The general solution is, therefore,

$$y = C_1 e^{-2x} + e^x[C_2 \cos (x\sqrt{3}) + C_3 \sin (x\sqrt{3})].$$

Example 4
Solve the differential equation

$$\frac{d^2 y}{dx^2} - n^2 y = 0.$$

where $n \neq 0$, given that, when $x = 0$, $y = 0$ and $\frac{dy}{dx} = n$.

The auxiliary equation is

$$m^2 - n^2 = 0$$
$$\Leftrightarrow \qquad m = \pm n$$
$$\Rightarrow \qquad y = C_1 e^{nx} + C_2 e^{-nx}.$$

Note that this can be written in the alternative form

$$y = A \cosh nx + B \sinh nx$$

$(A = C_1 + C_2, B = C_1 - C_2)$, which is more useful when applying given conditions at $x = 0$.

$$y = 0 \quad \text{at} \quad x = 0 \quad \Rightarrow \quad A = 0.$$

$$\frac{dy}{dx} = n \quad \text{at} \quad x = 0 \quad \Rightarrow \quad B = 1.$$

The required solution is therefore $y = \sinh nx$.

Exercises 4.5(iv)
Find the general solution of the following differential equations.

1 $y''' - 2y'' = 0$.

2 $y''' + 64y = 0$.

3 $y''' + y'' + 3y' + 3y = 0$.

4 $y^{(4)} - n^4 y = 0$, where n is a non-zero real constant.

5 $y''' - 3y'' + 4y' - 12y = 0$.

Suppose now that we have to solve the differential equation

$$a\frac{d^2 y}{dx^2} + 2h\frac{dy}{dx} + by = \phi(x), \qquad \qquad \dots (1)$$

where, as before, a, h and b are constants, and $\phi(x)$ is a given function of x.

Let $y = F(x)$ be any solution of equation (1), involving *no* arbitrary constants. Then $F(x)$ is called a *particular integral* of equation (1).

Let $y = G(x)$, involving two arbitrary constants, be the general solution of the equation

$$a\frac{d^2 y}{dx^2} + 2h\frac{dy}{dx} + by = 0,$$

i.e. equation (1) with zero on the right-hand side. Then $G(x)$ is known as the *complementary function* of equation (1), and the general solution of the equation is

$$y = G(x) + F(x),$$

i.e. the general solution is the sum of the complementary function and a particular integral.

At this level we determine particular integrals by trial, using intuitive assumptions based on the form of $\phi(x)$. The following examples illustrate some of the cases which arise.

Example 1

Find the general solution of the differential equation

$$\frac{d^2y}{dx^2} - 3\frac{dy}{dx} + 2y = 4. \qquad \ldots (1)$$

The auxiliary equation is

$$m^2 - 3m + 2 = 0$$
$$\Leftrightarrow \qquad (m - 1)(m - 2) = 0$$
$$\Leftrightarrow \qquad m = 1 \text{ or } 2.$$

The complementary function $G(x)$ is, therefore, given by

$$G(x) = C_1e^x + C_2e^{2x}.$$

In this case the $\phi(x)$ of equation (1) on p. 127 is 4, a constant, and we *try* as the particular integral $F(x)$ a constant, k say. Substituting in (1) we find $2k = 4 \Leftrightarrow k = 2$, so that a particular integral is given by $F(x) = 2$. The general solution is, therefore,

$$y = G(x) + F(x) = C_1e^x + C_2e^{2x} + 2.$$

Example 2

We reconsider the differential equation

$$\frac{dy}{dx} + 6y = 3x$$

of Example 5, page 110.

The auxiliary equation is

$$m + 6 = 0 \quad \Leftrightarrow \quad m = -6.$$

Then $G(x) = Ce^{-6x}$.

For the particular integral $F(x)$, we *try px + q*, where p and q are constants. Then, substituting $px + q$ for y, we require

$$p + 6(px + q) \equiv 3x \qquad \text{(to be true for *all* x)}$$
$$\Leftrightarrow \qquad 6px + (p + 6q) \equiv 3x$$
$$\Leftrightarrow \qquad 6p = 3, \qquad p + 6q = 0.$$
$$\therefore \qquad p = \tfrac{1}{2}, \qquad q = -\tfrac{1}{12}.$$
$$\therefore \qquad F(x) = \tfrac{1}{2}(x - \tfrac{1}{6}).$$

The general solution is, therefore,

$$y = G(x) + F(x) = Ce^{-6x} + \tfrac{1}{2}(x - \tfrac{1}{6})$$

as before.

Example 3

Find the general solution of the differential equation

$$\frac{d^2y}{dx^2} - 4\frac{dy}{dx} + 3y = 3x - 10.$$

The complementary function is given by $G(x) = C_1e^x + C_2e^{3x}$.
   As a particular integral $F(x)$ let us try

$$F(x) = px + q,$$

where p, q are constants. Then, substituting as in Example 2,

$$-4p + 3(px + q) \equiv 3x - 10$$
$$\Leftrightarrow \qquad 3p = 3, 3q - 4p = -10 \quad \Leftrightarrow \quad p = 1, q = -2.$$

   A particular integral is, therefore, $x - 2$.
   The general solution is

$$y = C_1e^x + C_2e^{3x} + x - 2.$$

   Note that if the right-hand side of the given equation had, for example, been $6x^2 + 2x + 1$, then an appropriate trial particular integral would have been $px^2 + qx + r$ leading to $p = 2, q = 6, r = 7$. We choose as our trial function the simplest possible function and are guided by experience.

Example 4
Find the value of the constant p for which pe^{2x} is a solution of the differential equation

$$\frac{d^2y}{dx^2} - 6\frac{dy}{dx} + 13y = 10e^{2x}.$$

   Solve this differential equation completely given that $y = 0$ and $\dfrac{dy}{dx} = 0$ when $x = 0$.

The equation for p is

$$p[4e^{2x} - 6(2e^{2x}) + 13e^{2x}] = 10e^{2x}$$
$$\Rightarrow \qquad 5p = 10 \quad \Leftrightarrow \quad p = 2.$$

   A particular integral is, therefore, $2e^{2x}$.
   To find the complementary function $G(x)$ the auxiliary equation is

$$m^2 - 6m + 13 = 0$$
$$\Leftrightarrow \qquad m = 3 \pm 2i.$$
$$\therefore \qquad G(x) = e^{3x}[C \cos 2x + D \sin 2x].$$

The general solution is

$$y = e^{3x}[C \cos 2x + D \sin 2x] + 2e^{2x}.$$

$y = 0$ when $x = 0$ gives $C + 2 = 0 \Leftrightarrow C = -2$.
$\dfrac{dy}{dx} = 0$ when $x = 0$ gives $3C + 2D + 4 = 0 \Rightarrow D = 1$.

The required solution is, therefore,

$$y = e^{3x}(\sin 2x - 2 \cos 2x) + 2e^{2x}.$$

Note that this example illustrates the fact that, if the right-hand side of the given differential equation is of the form ke^{ax}, then an appropriate trial particular integral is λe^{ax}, provided e^{ax} is not part of the complementary function. (See Example 5.)

Example 5
Find the general solution of the differential equation

$$\frac{d^2y}{dx^2} - 6\frac{dy}{dx} + 5y = e^{5x}. \qquad \ldots (1)$$

The complementary function $G(x)$ is given by

$$G(x) = C_1 e^x + C_2 e^{5x}.$$

Therefore, it is pointless to try λe^{5x} as the particular integral since the left-hand side of (1) vanishes when $y = \lambda e^{5x}$. Instead, experience tells us to try $\lambda x e^{5x}$.

When
$$y = \lambda x e^{5x},$$

$$\frac{dy}{dx} = 5\lambda x e^{5x} + \lambda e^{5x},$$

$$\frac{d^2y}{dx^2} = 25\lambda x e^{5x} + 10\lambda e^{5x},$$

and substitution in (1) gives

$$\lambda(25xe^{5x} + 10e^{5x}) - 6\lambda(5xe^{5x} + e^{5x}) + 5\lambda x e^{5x} \equiv e^{5x}$$
$$\Rightarrow \qquad 4\lambda e^{5x} \equiv e^{5x}$$

(the term in xe^{5x} vanishing indentically),

$$\Leftrightarrow \qquad \lambda = \tfrac{1}{4}.$$

A particular integral is, therefore, $\tfrac{1}{4}xe^{5x}$ and the general solution of (1) is

$$y = C_1 e^x + (C_2 + \tfrac{1}{4}x)e^{5x}.$$

Example 6
Find the values of the constants p and q if $y = p \cos x + q \sin x$ satisfies the differential equation

$$\frac{d^2y}{dx^2} + 8\frac{dy}{dx} + 25y = 48 \cos x - 16 \sin x, \qquad \ldots (1)$$

and hence find the general solution of this differential equation.
Find also the solution of this differential equation for which $y = 8$ and

$\dfrac{dy}{dx} = 3$ when $x = 0$.

When $y = p \cos x + q \sin x$,

$$\frac{dy}{dx} = -p \sin x + q \cos x,$$

$$\frac{d^2y}{dx^2} = -p \cos x - q \sin x.$$

$\therefore \quad \dfrac{d^2y}{dx^2} + 8\dfrac{dy}{dx} + 25y = (24p + 8q) \cos x + (-8p + 24q) \sin x.$

Substitution in the differential equation (1) gives

$$(24p + 8q) = 48, \qquad (-8p + 24q) = -16$$
$$\Leftrightarrow \qquad\qquad p = 2, \qquad q = 0.$$

A particular integral is, therefore, $2 \cos x$.

The auxiliary equation (for the complementary function) is

$$m^2 + 8m + 25 = 0$$
$$\Leftrightarrow \qquad\qquad m = -4 \pm 3i.$$

The complementary function is, therefore, $e^{-4x}(C_1 \cos 3x + C_2 \sin 3x)$ and so the general solution is

$$y = e^{-4x}(C_1 \cos 3x + C_2 \sin 3x) + 2 \cos x.$$

$y = 8$ when $x = 0$ gives $C_1 + 2 = 8$.

$\dfrac{dy}{dx} = 3$ when $x = 0$ gives $-4C_1 + 3C_2 = 3$.

Therefore, $C_1 = 6$, $C_2 = 9$ and the required solution is

$$y = 3e^{-4x}(2 \cos 3x + 3 \sin 3x) + 2 \cos x.$$

Example 7

(a) Obtain the general solution of the differential equation

$$\frac{d^2y}{dx^2} + k^2y = 5 \sin 2x,$$

where k is real and non-zero, and $|k| \neq 2$.

(b) Find the value of p for which $px \cos 2x$ is a particular integral of the differential equation

$$\frac{d^2y}{dx^2} + 4y = 5 \sin 2x.$$

Hence find the solution of this equation which satisfies the conditions $y = 2$ when $x = 0$ and $y = 0$ when $x = \pi/4$.

(a) The complementary function is $C_1 \cos kx + C_2 \sin kx$. For a particular integral we try $p \sin 2x + q \cos 2x$. Then

$$(-4p \sin 2x - 4q \cos 2x) + k^2(p \sin 2x + q \cos 2x) \equiv 5 \sin 2x$$
$$\Leftrightarrow \qquad p(k^2 - 4) \sin 2x + q(k^2 - 4) \cos 2x \equiv 5 \sin 2x$$
$$\Leftrightarrow \qquad p = 5/(k^2 - 4), \qquad q = 0.$$

The general solution is, therefore,

$$y = C_1 \cos kx + C_2 \sin kx + (5 \sin 2x)/(k^2 - 4).$$

Note that this solution does not hold when $|k| = 2$, i.e. $k^2 = 4$, since the particular integral is then undefined.

(b) When $k = 2$, the complementary function is

$$C_1 \cos 2x + C_2 \sin 2x.$$

With the suggested particular integral $px \cos 2x$, then

$$p(-4x \cos 2x - 4 \sin 2x) + 4px \cos 2x \equiv 5 \sin 2x$$
$$\Leftrightarrow \qquad -4p \sin 2x \equiv 5 \sin 2x \text{ (the terms in } x \cos 2x \text{ vanishing)},$$
$$\Leftrightarrow \qquad p = -5/4.$$

The general solution is, therefore,

$$y = C_1 \cos 2x + C_2 \sin 2x - (5x \cos 2x)/4.$$
$$y = 2 \quad \text{when} \quad x = 0 \quad \text{gives } C_1 = 2.$$
$$y = 0 \quad \text{when} \quad x = \pi/4 \quad \text{gives } C_2 = 0.$$

The required solution is, therefore,

$$y = \tfrac{1}{4}(8 - 5x) \cos 2x.$$

Note that, in this case, the expression on the right-hand side of the differential equation, namely $\sin 2x$, occurs in the complementary function, and the assumption of a particular integral of the form $p \cos 2x + q \sin 2x$ would be of no avail.

Exercises 4.5(v)

1 Find the value of the constant k for which ke^{3x} is a particular integral of the differential equation

$$\frac{d^2 y}{dx^2} + 9y = 9e^{3x}.$$

Obtain the general solution of this differential equation.

Find y in terms of x given that $y = 1$ and $\dfrac{dy}{dx} = 0$ when $x = 0$. By expanding y in a series of ascending powers of x, or otherwise, determine whether y has a maximum value, a minimum value or neither when $x = 0$. [L]

2   Given that the differential equation

$$\frac{d^2y}{dx^2} - 4\frac{dy}{dx} + 13y = e^{2x}$$

has a particular integral of the form ke^{2x}, determine the value of the constant k. Find the general solution of the equation. [L]

3   Find the value of the constant c for which $y = x + c$ is a solution of the differential equation

$$\frac{d^2y}{dx^2} - 3\frac{dy}{dx} - 4y = -4x.$$

Solve this differential equation given that $y = 5$ and $\frac{dy}{dx} = 0$ when $x = 0$. [L]

4   Obtain the values of the constants p and q when $p + qx$ satisfies the differential equation

$$\frac{d^2y}{dx^2} + 4\frac{dy}{dx} + 4y = 24 + 28x.$$

Obtain the general solution of this differential equation and find that solution for which $y = 0$ and $\frac{dy}{dx} = 0$ when $x = 0$. [L]

5   Find the constant k such that ke^{-2x} is a particular integral of the differential equation

$$\frac{d^2y}{dx^2} + 4\frac{dy}{dx} + 29y = 75e^{-2x}.$$

Hence solve the differential equation given that $y = 0$ and $\frac{dy}{dx} = 0$ when $x = 0$. [L]

6   Find the values of the constants a and b if $y = ax + b$ is a solution of the differential equation

$$6\frac{d^2y}{dx^2} + 7\frac{dy}{dx} - 3y = -9x + 21.$$

Hence solve the differential equation completely given that $y = 0$ and $\frac{dy}{dx} = \frac{11}{6}$ when $x = 0$. [L]

7 A particle moves in a straight line so that its distance x from a fixed point in the line satisfies the differential equation

$$\frac{d^2x}{dt^2} + 4\frac{dx}{dt} + 4x = 0.$$

The particle starts from rest at time $t = 0$ when $x = a$. Prove that its greatest speed in the ensuing motion is $2ae^{-1}$. **[JMB]**

8 Find the value of p, given that $px \sin 3x$ is a particular solution of the differential equation

$$\frac{d^2y}{dx^2} + 9y = 6 \cos 3x.$$

Hence find the general solution of this differential equation.

Find, also, y in terms of x given that $y = \frac{1}{3}$ and $\frac{dy}{dx} = -3$ when $x = 0$.

Find the values of x for which this solution has stationary values. **[L]**

9 A particular integral of the differential equation

$$\frac{d^2y}{dx^2} + n^2y = \sin nx$$

has the form $x(a \sin nx + b \cos nx)$, where a, b and n are constants. Determine a and b in terms of n. Hence find y in terms of x given that $y = 0$ and $\frac{dy}{dx} = 0$ when $x = 0$. **[L]**

10 Obtain a function f such that the general solution of the differential equation

$$\frac{d^2y}{dt^2} + 4\frac{dy}{dt} + 4y = 2e^{-2t}$$

may be written in the form $y = e^{-2t}f(t)$.

Find the values of the arbitrary constants in f(t) which make $y = 1$ and $\frac{dy}{dt} = 0$ at $t = 0$, and show that y decreases monotonically for all $t > 0$.

[JMB]

11 A particular integral of the differential equation

$$2\frac{d^2y}{dx^2} + 2\frac{dy}{dx} + 5y = \cos x - 5 \sin x$$

is $p \cos x + q \sin x$, where p and q are constants. Determine p and q and obtain the general solution of this differential equation.

Find also the solution for which $y = 0$ and $\frac{dy}{dx} = 0$ when $x = 0$. In this case show that, for large values of x as x increases, y oscillates between $-\sqrt{2}$ and $+\sqrt{2}$ approximately. [L]

12 Find the value of the constant a such that $y = axe^{-x}$ is a solution of the differential equation

$$\frac{d^2y}{dx^2} + 3\frac{dy}{dx} + 2y = 2e^{-x}.$$

Find the solution of this differential equation for which $y = 1$ and $\frac{dy}{dx} = 3$ when $x = 0$. [L]

13 A particular integral of the differential equation

$$\frac{d^2y}{dx^2} + 2\frac{dy}{dx} + 5y = 26 + 15x$$

is given by $y = a + bx$, where a, b are constants. Find the values of a, b and determine the general solution of the differential equation. [L]

14 Show that the substitution $y = z\cos x$, where z is a function of x, reduces the differential equation

$$\cos^2 x\frac{d^2y}{dx^2} + 2\cos x\sin x\frac{dy}{dx} + 2y = x\cos^3 x$$

to the differential equation

$$\frac{d^2z}{dx^2} + z = x.$$

Hence find y given that $y = 0$ and $\frac{dy}{dx} = 1$ when $x = 0$. [L]

15 Given that $x = e^t$, show that

$$x\frac{dy}{dx} = \frac{dy}{dt}, \qquad x^2\frac{d^2y}{dx^2} = \frac{d^2y}{dt^2} - \frac{dy}{dt}.$$

Hence, or otherwise, solve the differential equation

$$x^2\frac{d^2y}{dx^2} - 2x\frac{dy}{dx} + 2y = x^3,$$

given that $y = 0$ and $\frac{dy}{dx} = 0$ when $x = 1$. [L]

4.6 Integral curves

The general solution of a first-order differential equation is of the form $f(x, y, C) = 0$, where C is an arbitrary constant. This solution is represented graphically by a *family of curves*. Each member of the family is obtained by giving a specific value to C and is called an *integral curve*.

Example 1
The differential equation

$$y\frac{dy}{dx} = 2a,$$

where a is positive constant, integrates to give

$$y^2 = 4ax + C.$$

This is a family of non-intersecting parabolas all with their axes along Ox. Three members of the family are shown sketched in Fig. 4.1.

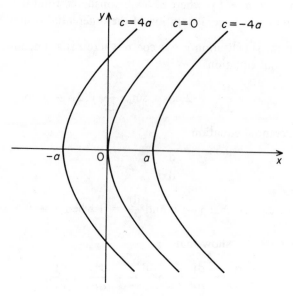

Fig. 4.1

Example 2
Sketch the integral curves of the differential equation

$$y\frac{dy}{dx} + x = 0.$$

Separating the variables and integrating we find

$$\int x \, dx + \int y \, dy = 0$$
$$\Leftrightarrow \qquad x^2 + y^2 = C.$$

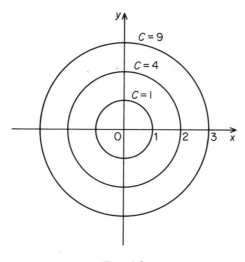

Fig. 4.2

For positive values of C the integral curves are concentric circles, all with centre O, and of radius $\sqrt{C}$. Three integral curves are shown sketched in Fig 4.2.

Example 3
Obtain the general solution of the differential equation

$$x\frac{dy}{dx} + (x - 1)y = 0.$$

Sketch the family of integral curves.

Separating the variables gives

$$\int \frac{1}{y}dy = \int \left(\frac{1}{x} - 1\right)dx$$

$\Leftrightarrow$ $\qquad\qquad \ln y = \ln x - x + A$

$\Leftrightarrow$ $\qquad\qquad\quad y = Cxe^{-x}.$

All the integral curves pass through the origin O.
  Since

$$\frac{dy}{dx} = C(1 - x)e^{-x},$$

when $C > 0$, each integral curve has a maximum when $x = 1$,
when $C < 0$, each integral curve has a minimum when $x = 1$.
  All the integral curves have the positive x-axis as an asymptote and, since $\frac{dy}{dx} = 0$ when $x = 1$, their stationary points lie on the line $x = 1$. Further,

since $\dfrac{d^2y}{dx^2} = C(x - 2)e^{-x}$, each of the curves has an inflexion on the line $x = 2$.

Some integral curves are shown sketched in Fig. 4.3.

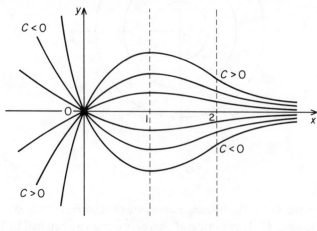

Fig. 4.3

Example 4

Obtain the general solution of the differential equation

$$2x^2 \frac{dy}{dx} + x^2y^2 = y^2.$$

Find the three particular solutions such that

(a) when $x = 1$, $y = 1$,
(b) when $x = 1$, $y = \frac{2}{3}$,
(c) when $x = 1$, $y = \frac{1}{2}$.

Show that each of the integral curves given by these particular solutions has a maximum point on the line $x = 1$. Sketch the three curves on the same axes.

The variables in the differential equation can be separated to give

$$\int \frac{2}{y^2} dy = \int \left(\frac{1}{x^2} - 1 \right) dx$$

$$\Leftrightarrow \qquad -\frac{2}{y} = -\frac{1}{x} - x - C$$

$$\Leftrightarrow \qquad y = \frac{2x}{x^2 + Cx + 1}.$$

(a) $x = 1$, $y = 1$ $\Rightarrow$ $C = 0$ $\Rightarrow$ $y = \dfrac{2x}{x^2 + 1}$.

(b) $x = 1$, $y = \dfrac{2}{3}$, $\Rightarrow$ $C = 1$ $\Rightarrow$ $y = \dfrac{2x}{x^2 + x + 1}$.

(c) $x = 1$, $y = \dfrac{1}{2}$ $\Rightarrow$ $C = 2$ $\Rightarrow$ $y = \dfrac{2x}{(x + 1)^2}$.

When $x \neq 0$, the differential equation gives

$$\frac{dy}{dx} = \frac{y^2(1 - x^2)}{2x^2}.$$

Since $\dfrac{dy}{dx}$ changes from positive through zero to negative as x increases through 1, each of the three integral curves has a maximum point on the line $x = 1$.

Curves (a) and (b) each have a minimum on the line $x = -1$.

At the origin all the curves have the same tangent $y = 2x$.

The x-axis is an asymptote to all three curves for $x > 0$ and for $x < 0$.

Curve (a) is the graph of an odd function and is symmetrical about 0.

Curve (b) lies below curve (a) for $x > 0$ and for $x < 0$.

Curve (c) has an asymptote $x = -1$.

The curves are shown sketched in Fig. 4.4.

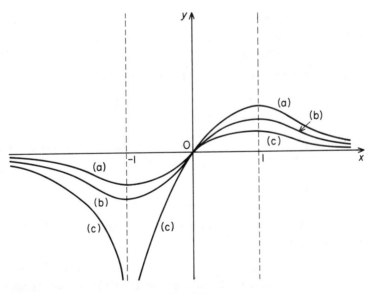

Fig. 4.4

Exercises 4.6

1 Find y in terms of x given that

$$\frac{dy}{dx} = 2y + 1,$$

and that $y = 0$ when $x = 0$.

Sketch the curve represented by this solution and find the area of the finite region enclosed by the curve, the line $x = 1$ and the x-axis. [L]

2 Solve the differential equation

$$(x + 2)\frac{dy}{dx} + (x + 1)y = 0,$$

(where $x > -2$), given that $y = 2$ when $x = 0$.

Show that the maximum value of y is e, and sketch the graph of y against x for $x > -2$.

Draw on your graph the tangent to the curve at its point of inflexion. Find the equation of this tangent. [L]

3 Find the solution of the differential equation

$$\frac{dy}{dx} = 2 - 2x$$

for which $y = 0$ when $x = 1$.

Find also the solution of the differential equation

$$\frac{dy}{dx} = \frac{1}{2 - x}.$$

for which $y = 0$ when $x = 1$.

Sketch the graphs of these solutions on the same diagram, indicating clearly where each graph meets the y-axis. [JMB]

4 The gradient of a curve is given by the differential equation

$$\frac{dy}{dx} + \frac{2xy}{x^2 - 1} = \frac{2x - 4}{x^2 - 1}$$

and the curve passes through the point $(2, 0)$. By solving this differential equation, show that the equation of the curve is

$$y = \frac{(x - 2)^2}{x^2 - 1}.$$

Find the equations of the asymptotes and the coordinates of the stationary points of the curve. Sketch the curve. [JMB]

5 The gradient at any point (x, y) on a certain curve is given by $(2y + x + 1)/x$. The curve passes through the point $(1, 0)$. Find its equation. [L]

6 Solve the differential equation

$$\left(\frac{dy}{dx} - y\right)e^x = 1$$

and show that if the graph corresponding to any particular solution has a stationary point, then this point must be a maximum point. [L]

7 It is required to find a curve, or curves, passing through the point $(0, c)$ and satisfying the differential equation

$$\left(\frac{dy}{dx}\right)^2 + y^2 = c^2.$$

By first differentiating both sides of this differential equation with respect to x, find two distinct solutions and sketch both their graphs in a diagram. [OC]

8 Find the most general solution of the differential equation

$$\frac{d^2y}{dx^2} + y = 5x,$$

for which $y = 0$ when $x = 0$.

Show that all the curves given by this solution pass through the point $(\pi, 5\pi)$, and that they all have the same gradient when $x = (2n + 1)\pi/2$, where n is an integer.

Find that solution of this differential equation for which $y = 0$ when $x = 0$ and when $x = \pi/2$.

Prove that, for $0 < x < \pi/2$, the minimum value of y for this solution is

$$5 \arccos (2/\pi) - \frac{5}{2}\sqrt{(\pi^2 - 4)}.$$ [L]

9 By differentiation, or otherwise, solve the differential equation

$$\left(\frac{dy}{dx}\right)^2 - x\frac{dy}{dx} + y = 0,$$

showing that the set of solution curves includes

(a) precisely one parabola, and
(b) any tangent to this parabola.

Is the curve defined by

$$y = \begin{cases} \frac{1}{4}x^2 & \text{if } x \leqslant 1, \\ \frac{1}{2}x - \frac{1}{4} & \text{if } x > 1, \end{cases}$$

a solution curve? Give your reasons. [OC]

4.7 Applications

Many problems in applied mathematics lead to differential equations. Here we give a few illustrative examples.

Example 1   The law of natural growth and decay
A body is placed in a room which is kept at a constant temperature. The temperature of the body falls at a rate $k\theta$ K per minute, where k is a positive constant and θ is the difference in K between the temperature of the body and that of the room at time t minutes. Express this information in the form of a differential equation and hence show that $\theta = \theta_0 e^{-kt}$, where θ_0 is the temperature difference in K at time $t = 0$.

The temperature of the body falls 5 K in the first minute and 4 K in the second minute. Show that the fall of temperature in the third minute is 3·2 K.

The rate of increase of the temperature, T K say, of the body is $\dfrac{dT}{dt}$, where t is the time in minutes. But we are given that the temperature *decreases* at the rate $k\theta$ K per minute and so

$$\frac{dT}{dt} = -k\theta. \qquad \qquad \ldots (1)$$

Also, if α K is the constant room temperature, $T = \alpha + \theta$ and so substitution in (1) gives

$$\frac{d\theta}{dt} = -k\theta. \qquad \qquad \ldots (2)$$

This is the required differential equation, which has the general solution

$$\theta = Ae^{-kt},$$

where A is constant. The initial condition $\theta = \theta_0$ at $t = 0$ gives $A = \theta_0$.

$$\therefore \qquad \theta = \theta_0 e^{-kt}. \qquad \qquad \ldots (3)$$

Since the temperature falls by 5 K in the first minute,

$$\theta_{t=0} - \theta_{t=1} = 5.$$
$$\therefore \qquad \theta_0(1 - e^{-k}) = 5. \qquad \qquad \ldots (4)$$

Similarly, the fall by 4 K in the second minute gives

$$\theta_0(e^{-k} - e^{-2k}) = 4. \qquad \qquad \ldots (5)$$

Dividing corresponding sides of (5) and (4) gives

$$e^{-k} = \tfrac{4}{5}. \qquad \qquad \ldots (6)$$

The fall in temperature in the third minute is

$$\theta_{t=2} - \theta_{t=3} = \theta_0(e^{-2k} - e^{-3k})$$
$$= \theta_0(e^{-k} - e^{-2k})e^{-k},$$

which from (5) and (6), by multiplication, gives

$$\theta_{t=2} - \theta_{t=3} = 4 \times \tfrac{4}{5} = 3\cdot2$$

as required.

Note that the successive values of θ_0 at equal intervals of time constitute the terms of a geometric progression (and hence the differences between successive terms are also terms of a geometric progression) so that, if x K is the required answer, $5x = 4^2 \Rightarrow x = 3\cdot2$.

The above example illustrates the (exponential) law of natural decay in that, if the rate of decrease of some quantity is proportional to that quantity, then that quantity decays exponentially. For example, if the rate of decay of a radioactive substance is kx, where x is the mass of the substance remaining, then

$$\frac{dx}{dt} = -kx \quad \Rightarrow \quad x = x_0 e^{-kt},$$

where x_0 is the initial mass (at time $t = 0$). The *half-life*, τ, is the time that elapses before the moment when just half the original mass remains and is given by

$$\tfrac{1}{2}x_0 = x_0 e^{-k\tau} \quad \Rightarrow \quad \tau = (\ln 2)/k.$$

Of course, if the quantity under consideration *increases* at a rate proportional to the quantity already present, then that quantity increases exponentially. (This is the exponential (natural) law of growth.) For example, suppose that, at time t, a large population of organisms has N members, and assuming that N can be regarded as a continuous variable, and the net rate of increase of N, due to reproduction less deaths, is proportional to N, i.e. is kN, where k is a positive constant, then

$$\frac{dN}{dt} = kN \quad \Rightarrow \quad N = N_0 e^{kt},$$

where N_0 is the initial population.

Example 2
In appropriate units, the relation between the concentration c and the time t in a chemical reaction is given by

$$\frac{dc}{dt} = -kc^2,$$

where k is a constant. Prove that

$$kt = \frac{1}{c} - A,$$

where A is a constant.
It is given that when $t = 0$, $c = 1$ and when $t = 30$, $c = 0\cdot5$. Find c when $t = 120$.   [L]

Separating the variables and integrating the given differential equation gives

$$\int k \, dt = -\int \frac{1}{c^2} dc$$

$\Leftrightarrow$
$$kt + A = \frac{1}{c}$$

$\Leftrightarrow$
$$kt = \frac{1}{c} - A.$$

$c = 1$ when $t = 0$ gives $A = 1$.
$c = 0.5$ when $t = 30$ gives $30k = 2 - 1 \Leftrightarrow k = \frac{1}{30}$.
When $t = 120$,

$$\frac{1}{c} = \frac{120}{30} + 1 = 5 \Leftrightarrow c = \frac{1}{5} = 0.2.$$

Example 3
A cyclindrical tank, of volume V and height h with its axis vertical, has water flowing in at a constant rate k and leaving through a hole in the base at a rate $c\sqrt{x}$, where x is the depth of water at any instant and c is a constant. Find the rate at which x changes.

Given that $k^2 < c^2 h$, show that the water will not overflow at the top of the cylinder.

The cross-section of the tank is V/h so that, when the depth of water is x, the volume of water in the tank is v, where $v = Vx/h$. The rate of change of v with respect to t is the difference between two expressions:
(a) the rate at which water flows into the tank, i.e. the constant rate k; and
(b) the rate at which the water leaves through the hole in the bottom, i.e. the rate $c\sqrt{x}$.

Therefore,

$$\frac{dv}{dt} = k - c\sqrt{x}$$

$\Rightarrow$
$$\frac{V}{h}\frac{dx}{dt} = k - c\sqrt{x}$$

$\Leftrightarrow$
$$\frac{dx}{dt} = \frac{h}{V}(k - c\sqrt{x}).$$

This differential equation shows that $\dfrac{dx}{dt} < 0$ if $k < c\sqrt{x}$, and so the water level in the tank falls when $x > k^2/c^2$. It follows at once that, if $h > k^2/c^2$, the water level can never reach the top of the cylinder and so the water will not overflow.

Note that, in this example, it has not been necessary to solve the differential equation (although this could be done by separating the variables and making the substitution $x = u^2$).

Example 4
In an influenza epidemic, the fraction of the population which has caught the disease is denoted by x, while y denotes the fraction of the population which has not caught the disease. The spreading rate of the epidemic is defined to be the rate of change of x with respect to t, the time, in days, measured from the instant when the disease was first noticed. In a mathematical model of the spreading of influenza, it is assumed that the spreading rate is directly proportional to the product of x and y, but depends on no other factors. A particular epidemic was noticed when one quarter of the population had caught influenza. The spreading rate at that time was such that, if the rate had remained constant for 20 days, the remainder of the population would have caught influenza in that time. Show that

$$5\frac{dx}{dt} = x(1 - x).$$

Show also that, according to this model, 10 days after the disease was first noticed, approximately 71 per cent of the population would have caught influenza.

Since $y = 1 - x$ and the spreading rate $\dfrac{dx}{dt}$ is proportional to xy, it follows that

$$\frac{dx}{dt} = kx(1 - x),$$

where k is a constant.

We take time $t = 0$ as the instant when 25 per cent of the population had caught influenza, so that $x = \frac{1}{4}$ when $t = 0$.

At this instant $\dfrac{dx}{dt} = k \times \dfrac{1}{4}\dfrac{3}{4} \times = \dfrac{3k}{16}$.

If the spreading rate had remained constant at this value, the rest of the population would have caught influenza in another 20 days. Hence

$$\frac{3k}{16} \times 20 = \frac{3}{4} \quad \Rightarrow \quad k = \frac{1}{5}.$$

$$\therefore \qquad 5\frac{dx}{dt} = x(1 - x).$$

Separating the variables and integrating gives

$$5 \int_{1/4}^{x} \frac{1}{x(1 - x)} dx = \int_{0}^{t} 1 \, dt.$$

(Notice that here we avoid the introduction of an arbitrary constant by using definite integrals with the limits as corresponding values of x and t.)

Hence (using partial fractions),

$$5 \int_{1/4}^{x} \left(\frac{1}{x} + \frac{1}{1-x} \right) dx = t$$

$\Rightarrow \qquad 5[\ln x - \ln (1/4) - \ln (1-x) + \ln (3/4)] = t$

$$\Rightarrow \qquad 5 \ln \left(\frac{3x}{1-x} \right) = t$$

$$\Leftrightarrow \qquad \frac{3x}{1-x} = e^{t/5}$$

$$\Leftrightarrow \qquad x = \frac{e^{t/5}}{3 + e^{t/5}}.$$

When $t = 10$, $x = e^2/(3 + e^2) \approx 7 \cdot 389/10 \cdot 389 \approx 0 \cdot 71$.

Hence, after 10 days approximately 71 per cent of the population would have caught influenza.

Example 5

A particle is moving in a straight line so that at time t s its displacement x m from a fixed point O of the line satisfies the differential equation

$$\frac{d^2 x}{dt^2} - 4\frac{dx}{dt} + 5x = -3e^{2t}.$$

When $t = 0$, the particle is at the point where $x = -1$, moving away from O with speed 2 m s^{-1}. Find x in terms of t and show that $x < 0$ for all $t > 0$.

The auxiliary equation is

$$m^2 - 4m + 5 = 0 \quad \Leftrightarrow \quad m = 2 \pm i.$$

Hence the complementary function is

$$e^{2t}(C \cos t + D \sin t).$$

For the particular integral we try λe^{2t}. Then

$$\lambda(4 - 8 + 5)e^{2t} \equiv -3e^{2t} \quad \Rightarrow \quad \lambda = -3.$$

The general solution is therefore

$$x = e^{2t}(C \cos t + D \sin t) - 3e^{2t}.$$

$x = -1$ when $t = 0$ gives $C = 2$.

$\dfrac{dx}{dt} = -2$ when $t = 0$ gives $2C + D - 6 = -2 \quad \Rightarrow \quad D = 0$.

Therefore

$$x = e^{2t}(2 \cos t - 3).$$

Since $e^{2t} > 0$ and $2 \cos t < 3$ for all t, it follows that $x < 0$ for all $t > 0$.

Exercises 4.7

1   In a chemical reaction in which a compound X is formed from a compound Y and other substances, the masses of X and Y present at time t are x and y respectively. The sum of the two masses is constant, and at any time the rate at which x is increasing is proportional to the product of the two masses at that time. Show that the equation governing the reaction is of the form

$$\frac{dx}{dt} = kx(a - x)$$

and interpret the constant a. Given that $k = a/10$ at time $t = 0$, find, in terms of k and a, the time at which $y = a/10$.   [L]

2   In established forest fires, the proportion of the total area of the forest which has been destroyed is denoted by x, and the rate of change of x with respect to time, t hours, is called the destruction rate. Investigations show that the destruction rate is directly proportional to the product of x and $(1 - x)$. A particular fire is initially noticed when one half of the forest is destroyed, and it is found that the destruction rate at this time is such that, if it remained constant thereafter, the forest would be completely destroyed in a further 24 hours. Show that

$$12\frac{dx}{dt} = x(1 - x)$$

and deduce that approximately 73 per cent of the forest is destroyed 12 hours after it is first noticed.   [L]

3   At time $t = 0$ the number of bacteria in a certain culture is N_0. At time t the birth-rate of the bacteria is numerically equal to half of the number N of living bacteria and there is a constant death-rate of $\frac{1}{4}N_0$. Assuming N to be a continuous variable, form a differential equation for N and obtain N in terms of t. Find the time that elapses before the number of living bacteria doubles.

4   In the manufacture of wine, a jar initially contains 20 litres of a mixture, which consists of a liquid into which 5 kg of sugar have been dissolved. At any given time the sugar is converted into alcohol at a rate of $4c$ kg/day, where c kg/litre is the concentration of unconverted sugar in the mixture. Assuming that there is no change in the volume of the mixture, show that, if y kg of sugar remain after t days, then

$$5\frac{dy}{dt} = -y.$$

Hence find the time taken for the quantity of sugar in the mixture to fall to 0·005 kg.

A second jar also initially contains 5 kg of sugar in 20 litres of mixture but, as it is not sealed correctly, the liquid evaporates at a constant rate of 100 ml/day. Show that, if x kg of sugar remain after t days, then

$$\frac{dx}{dt} = -\frac{40x}{200 - t}.$$

Hence find, to the nearest day, the time taken for the sugar content to fall to 0·005 kg in the second jar. [L]

5   At any instant, a spherical meteorite is gaining mass because of two effects: (a) mass is condensing onto it at a rate which is proportional to the surface area of the meteorite at that instant; (b) the gravitational field of the meteorite attracts mass onto itself, the rate being proportional to the meteorite mass at that instant. Assuming that the two effects can be added together and that the meteorite remains spherical and of constant density, show that its radius r at time t satisfies the differential equation

$$\frac{dr}{dt} = A + Br,$$

where A and B are constants.
If $r = r_0$ at $t = 0$, show that

$$r = r_0 e^{Bt} + \frac{A}{B}(e^{Bt} - 1).$$ [L]

6   The population x of a colony of bacteria increases at a rate equal to the product of the number x of bacteria present at time t and the capability C of the environment to support the number present at time t. The capability C is measured by the excess of the maximum number b of bacteria that can be supported by the environment over the number of bacteria actually present. Write down the differential equation governing the growth of bacteria.

Solve this differential equation, expressing x in terms of t, given that the population at time $t = 0$ is p, where $p < b$. State what happens to the population after the passage of a large interval of time. [L]

7   A chemical reaction involves the substances X and Y, which are present at time t s in amounts x g and y g respectively. The rate of increase in the amount of X present is λ times the amount of Y present, and the rate of increase in the total amount of X and Y present is μ times the amount of X present. Express these statements as first-order differential equations, and hence obtain a second-order differential equation for x.

Given that the solution of this differential equation is $x = 5e^t - e^{-3t}$, find the constants λ and μ, and express y in terms of t. [L]

8 A tank being cleaned initially contains 3900 litres of water and 100 kg of dissolved dye. The mixture is kept uniform by stirring. Pure water is run in at the rate of 130 litres per minute and the mixture is removed at the same rate. The mass of dissolved dye remaining after t minutes is x kg.

Show that

$$\frac{dx}{dt} = -\frac{x}{30}.$$

How long will it be before half the dye has been removed?

Determine the mass of dissolved dye remaining after 30 minutes.

9 The equation of motion of a particle moving on the x-axis is

$$\ddot{x} + 2k\dot{x} + n^2 x = 0.$$

Given that k and n are positive constants, with $k < n$, show that the time between two successive maxima of $|x|$ is constant. [L]

10 The temperature in the atmosphere falls, at a uniform rate with increasing height, from T_0 on the ground to T_1 at height h above the ground. Show that, for $0 \leqslant x \leqslant h$, the temperature T at height x above the ground is given by $T = T_0(1 - \beta x)$, where β is constant, and find β in terms of T_0, T_1 and h.

The air pressure p at height x above the ground satisfies the equation

$$\frac{dp}{dx} = -\frac{kp}{T},$$

where k is a constant. The pressure at ground level is p_0. Obtain a differential equation giving dp/dx in terms of p and x in the interval $0 \leqslant x \leqslant h$, and solve it to express p in terms of p_0, β, k, T_0 and x. Show that the pressure p_1 at height h is given by

$$p_1 = p_0 \left(\frac{T_1}{T_0}\right)^{k/(T_0\beta)}$$

For $x \geqslant h$ the temperature is constant. Write down an expression for dp/dx valid for $x \geqslant h$ and hence find p in terms of p_1, k, T_1, h and x for $x \geqslant h$. [JMB]

11 (a) In a chemical reaction the concentration x of a substance at time t satisfies the differential equation

$$\frac{dx}{dt} = a - bx,$$

where a and b are constants. Given that initially $x = c$, find x in terms of a, b, c and t. Find the condition that the concentration should decrease initially.

(b) A hemispherical bowl of radius a is filled with water which can escape through a hole of effective cross-sectional area A in the bottom of the bowl. The velocity of the escaping water at any instant is $\sqrt{(2gx)}$, where x is the depth of water remaining in the bowl. By considering the change in the volume of water remaining due to a small increase δx in x, show that

$$\frac{\mathrm{d}x}{\mathrm{d}t} = -\frac{A\sqrt{(2gx)}}{\pi(2ax - x^2)}.$$

Hence show that the time required for the bowl to empty is

$$14\pi a^3/[15A\sqrt{(2ga)}].$$

12   When a water tank is emptied through a hole in its base the rate of decrease of the volume of water in the tank at any time may be assumed to be directly proportional to the square root of the height, at that time, of the water level above the hole. The constant of proportionality may be assumed to depend only on the shape and size of the hole, and not on the shape or size of the tank.

Water is leaving a cylindrical tank, of radius a and height h, through a small hole in its horizontal circular base at a rate of $k\sqrt{y}$, where k is a positive constant and y is the depth of water in the tank at time t from the beginning of the outflow. Show that

$$\frac{\mathrm{d}y}{\mathrm{d}t} = -\frac{k\sqrt{y}}{\pi a^2},$$

and deduce that, if the tank is initially full and is emptied in time T, then

$$k = \frac{2\pi a^2\sqrt{h}}{T}.$$

Another tank is in the form of a right circular cone of height $3h$ and base radius a, and has its axis vertical and its vertex at its lowest point. There is a hole at the vertex identical to the hole in the base of the cylindrical tank described above. Show that, if V is the volume of water in the conical tank when the depth of water is y, then

$$\frac{\mathrm{d}V}{\mathrm{d}y} = \frac{\pi a^2 y^2}{9h^2}.$$

Show that the time T_1 required to empty the conical tank when initially full is given by

$$T_1 = \frac{\sqrt{3}}{5}T. \qquad\qquad\qquad\qquad [\text{JMB}]$$

4.8 Vector differential equations

In *Advanced Mathematics 1*, page 321, we defined the derivative of a vector **a**, whose components (a_1, a_2, a_3) parallel to fixed rectangular cartesian axes are functions of a single variable t, by the equation

$$\frac{d\mathbf{a}}{dt} = \frac{da_1}{dt}\mathbf{i} + \frac{da_2}{dt}\mathbf{j} + \frac{da_3}{dt}\mathbf{k}.$$

Similarly, if $\mathbf{b} = b_1(t)\mathbf{i} + b_2(t)\mathbf{j} + b_3(t)\mathbf{k}$, then we *define* $\displaystyle\int_{t_0}^{t_1} \mathbf{b}\,dt$ by the equation

$$\int_{t_0}^{t_1} \mathbf{b}\,dt = \int_{t_0}^{t_1} b_1\,dt\,\mathbf{i} + \int_{t_0}^{t_1} b_2\,dt\,\mathbf{j} + \int_{t_0}^{t_1} b_3\,dt\,\mathbf{k}.$$

In this way we can differentiate and integrate a vector $\mathbf{v}$ when its components parallel to a *fixed* set of cartesian axes are functions of a single variable.

It follows that linear vector differential equations can be solved by resolving into components parallel to fixed axes, as illustrated by the following examples.

Example 1

Given that $\dfrac{d\mathbf{v}}{dt} = -k\mathbf{v}$, where k is constant scalar and $\mathbf{v} = \mathbf{V}$ when $t = 0$, find $\mathbf{v}$ in terms of $\mathbf{V}$ and k.

Given also that $\mathbf{v} = \dfrac{d\mathbf{r}}{dt}$ and $\mathbf{r} = \mathbf{a}$ when $t = 0$, find $\mathbf{r}$ in terms of $\mathbf{a}$, $\mathbf{V}$ and k.

Let $\mathbf{v} = v_1\mathbf{i} + v_2\mathbf{j} + v_3\mathbf{k}$. Then the components of the given equation are

$$\frac{dv_1}{dt} = -kv_1, \qquad \frac{dv_2}{dt} = -kv_2, \qquad \frac{dv_3}{dt} = -kv_3,$$

which integrate to give

$$v_1 = A_1 e^{-kt}, \qquad v_2 = A_2 e^{-kt}, \qquad v_3 = A_3 e^{-kt},$$

where A_1, A_2, A_3 are arbitrary scalars. It follows that

$$\mathbf{v} = (A_1\mathbf{i} + A_2\mathbf{j} + A_3\mathbf{k})e^{-kt}$$

⇔ $$\mathbf{v} = \mathbf{A}e^{-kt},$$

where $\mathbf{A}$ is an arbitrary *vector*. Since $\mathbf{v} = \mathbf{V}$ when $t = 0$ it follows that $\mathbf{A} = \mathbf{V}$ and so

$$\mathbf{v} = \mathbf{V}e^{-kt}$$

⇒ $$\frac{d\mathbf{r}}{dt} = \mathbf{V}e^{-kt}. \qquad \ldots (1)$$

Writing $\mathbf{r} = x\mathbf{i} + y\mathbf{j} + z\mathbf{k}$ and $\mathbf{V} = V_1\mathbf{i} + V_2\mathbf{j} + V_3\mathbf{k}$, the components of equation (1) are

$$\frac{dx}{dt} = V_1 e^{-kt}, \qquad \frac{dy}{dt} = V_2 e^{-kt}, \qquad \frac{dz}{dt} = V_3 e^{-kt},$$

which integrate to give

$$x = -\frac{V_1}{k} e^{-kt} + B_1, \qquad y = -\frac{V_2}{k} e^{-kt} + B_2, \qquad z = -\frac{V_3}{k} e^{-kt} + B_3.$$

$$\therefore \quad \mathbf{r} = -\frac{\mathbf{V}}{k} e^{-kt} + \mathbf{B},$$

where $\mathbf{B}$ is another arbitrary vector.

The initial condition $\mathbf{r} = \mathbf{a}$ at $t = 0$ gives

$$\mathbf{a} = -\frac{\mathbf{V}}{k} + \mathbf{B} \qquad \Leftrightarrow \qquad \mathbf{B} = \mathbf{a} + \frac{\mathbf{V}}{k}.$$

The required solution is, therefore,

$$\mathbf{r} = \mathbf{a} + \frac{\mathbf{V}}{k}(1 - e^{-kt}).$$

Example 2

If $\dfrac{d^2\mathbf{r}}{dt^2} + 3n\dfrac{d\mathbf{r}}{dt} + 2n^2\mathbf{r} = \mathbf{0}$, find $\mathbf{r}$ in terms of t given that, when $t = 0$, $\mathbf{r} = \mathbf{0}$

and $\dfrac{d\mathbf{r}}{dt} = \mathbf{V}$.

If $\mathbf{r} = x\mathbf{i} + y\mathbf{j} + z\mathbf{k}$, then the x-component of the given equation is

$$\frac{d^2x}{dt^2} + 3n\frac{dx}{dt} + 2n^2x = 0$$

with general solution

$$x = A_1 e^{-nt} + B_1 e^{-2nt}.$$

Similar results hold for the y and z components and so

$$\mathbf{r} = \mathbf{A}e^{-nt} + \mathbf{B}e^{-2nt},$$

where $\mathbf{A}$ and $\mathbf{B}$ are arbitrary vectors.
$\mathbf{r} = \mathbf{0}$ when $t = 0$ gives $\mathbf{A} + \mathbf{B} = \mathbf{0}$.

$\dfrac{d\mathbf{r}}{dt} = \mathbf{V}$ when $t = 0$ gives $-n(\mathbf{A} + 2\mathbf{B}) = \mathbf{V}$.

$$\therefore \qquad \qquad \mathbf{A} = \mathbf{V}/n, \qquad \mathbf{B} = -\mathbf{V}/n.$$

$$\therefore \qquad \qquad \mathbf{r} = \frac{\mathbf{V}}{n}(e^{-nt} - e^{-2nt}).$$

Example 3
Solve the vector differential equation

$$\frac{d^2\mathbf{r}}{dt^2} + k\frac{d\mathbf{r}}{dt} = \mathbf{a} \cos pt, \qquad \dots (1)$$

given that $\mathbf{a}$ is a non-zero constant vector and k, p are constant non-zero scalars.

Again we write $\mathbf{r} = x\mathbf{i} + y\mathbf{j} + z\mathbf{k}$ and $\mathbf{a} = a_1\mathbf{i} + a_2\mathbf{j} + a_3\mathbf{k}$; the x-component of equation (1) is

$$\frac{d^2x}{dt^2} + k\frac{dx}{dt} = a_1 \cos pt.$$

The auxiliary equation is

$$m^2 + km = 0 \quad \Leftrightarrow \quad m = 0 \quad \text{or} \quad -k.$$

Therefore the x-component of the complementary function is

$$A_1 e^{0t} + B_1 e^{-kt} = A_1 + B_1 e^{-kt}.$$

Similarly for the y and z components, and so the complementary function for equation (1) is $\mathbf{A} + \mathbf{B}e^{-kt}$, $\mathbf{A}$ and $\mathbf{B}$ being arbitrary vectors.

To obtain the particular integral we try

$$\mathbf{a}(\lambda \cos pt + \mu \sin pt),$$

where λ, μ are scalars to be determined.

Then $\quad -p^2\mathbf{a}(\lambda \cos pt + \mu \sin pt) + pk\mathbf{a}(-\lambda \sin pt + \mu \cos pt) \equiv \mathbf{a} \cos pt$

$\Leftrightarrow \qquad (-\lambda p^2 + \mu pk - 1)\mathbf{a} \cos pt + (-\mu p^2 - \lambda pk)\mathbf{a} \sin pt = 0$

$\Leftrightarrow \qquad -\lambda p^2 + \mu pk - 1 = 0, \qquad -\mu p^2 - \lambda pk = 0$

$$\Leftrightarrow \qquad \lambda = \frac{-1}{p^2 + k^2}, \qquad \mu = \frac{k}{p(p^2 + k^2)}.$$

The general solution of equation (1) is therefore

$$\mathbf{r} = \mathbf{A} + \mathbf{B}e^{-kt} + \frac{\mathbf{a}}{p(p^2 + k^2)}(k \sin pt - p \cos pt).$$

Exercises 4.8

1   The motion of a particle satisfies the differential equation

$$\frac{d^2\mathbf{r}}{dt^2} + 5\lambda\frac{d\mathbf{r}}{dt} + 4\lambda^2\mathbf{r} = \mathbf{0}, \qquad \text{where } \lambda \neq 0.$$

Find $\mathbf{r}$ in terms of $\lambda, \mathbf{U}$ and t, given that $\mathbf{r} = \mathbf{0}$ and $\dfrac{d\mathbf{r}}{dt} = \mathbf{U}$ when $t = 0$.

[L]

2 Solve the vector differential equation

$$\frac{d^2\mathbf{r}}{dt^2} + \lambda\frac{d\mathbf{r}}{dt} = \mathbf{a},$$

where **a** is a constant vector and λ is a non-zero constant scalar.　　　[L]

3 At time t the position vector **r** of a point P satisfies the vector differential equation

$$\frac{d^2\mathbf{r}}{dt^2} + 2k\frac{d\mathbf{r}}{dt} + (k^2 + n^2)\mathbf{r} = \mathbf{g},$$

where k and n are positive constants and **g** is a constant vector. Solve this differential equation given that $\mathbf{r} = \mathbf{a}$ and $\dfrac{d\mathbf{r}}{dt} = \mathbf{V}$ when $t = 0$, **a** and **V** being constant vectors.

Show that P moves in a plane and write down a vector equation of this plane.

Describe the path of P as $t \to \infty$.　　　[L]

4 (a) At time t seconds the velocity $\mathbf{v}$ m s^{-1} of a particle P satisfies the equation $d\mathbf{v}/dt + 2\mathbf{v} = \mathbf{0}$. When $t = 0$ the position vector and velocity of P are $(\mathbf{i} + \mathbf{j})$ m and $4(-\mathbf{i} + \mathbf{j})$ m s^{-1} respectively.

Show that at time t seconds the position vector of P is **r** m, where

$$\mathbf{r} = (-\mathbf{i} + 3\mathbf{j}) + 2(\mathbf{i} - \mathbf{j})e^{-2t}.$$

(b) A particle moves in the x–y plane so that its position vector **r** m at time t seconds satisfies the differential equation

$$\frac{d^2\mathbf{r}}{dt^2} + 4\mathbf{r} = \mathbf{0}.$$

Given that $\mathbf{r} = 3\mathbf{i}$ when $t = 0$, and $\mathbf{r} = \mathbf{j}$ when $t = \pi/4$, show that the particle describes the ellipse

$$x^2 + 9y^2 = 9.$$

5 (a) Solve the differential equation $d\mathbf{r}/dt = 4\mathbf{r}$, given that, when $t = 0$, $\mathbf{r} \cdot \mathbf{i} = 1$ and $\mathbf{r} \times \mathbf{i} = \mathbf{j} + \mathbf{k}$.

(b) Find the solution of the differential equation

$$\frac{d^2\mathbf{r}}{dt^2} + 2\frac{d\mathbf{r}}{dt} + 2\mathbf{r} = \mathbf{0}$$

such that $\mathbf{r} = \mathbf{i} + \mathbf{k}$ when $t = 0$, and $d\mathbf{r}/dt = \mathbf{0}$ when $t = \pi/2$.

(c) At time t seconds the position vector **r** m of the point P satisfies the differential equation

$$\frac{d^2\mathbf{r}}{dt^2} + 4\mathbf{r} = 3\mathbf{i} \sin t.$$

When $t = 0$, P passes through the point whose position vector is $\mathbf{j}$ m with velocity $\mathbf{i}$ m s^{-1}.

Find an equation of the locus of P. [L]

Miscellaneous exercises 4

1 Solve the differential equations

(a) $2\sqrt{(1 + x^2)}\dfrac{dy}{dx} = 3y$, given that $y = 1$ when $x = 0$;

(b) $\dfrac{1}{x}\dfrac{dy}{dx} - \dfrac{y}{x^2} = 3x^2$, given that $y = -1$ when $x = 1$ by writing the left-hand side as $\dfrac{d(y/x)}{dx}$;

(c) $x(x - y)\dfrac{dy}{dx} + y^2 = 0$ by using the substitution $y = vx$.

2 (a) The gradient at any point on a curve is given by

$$\frac{dy}{dx} = \frac{y - x}{x}.$$

Find the equation of the curve, given that it passes through the point $(1, 1)$.
(b) Find the value of the constant k for which $y = kx \sin 2x$ is a solution of the differential equation

$$\frac{d^2y}{dx^2} + 4y = \cos 2x.$$

Obtain the general solution of the differential equation. [L]

3 (a) Find the general solution of the differential equation

$$\frac{dy}{dx} + \frac{y}{x} = 2e^{x^2}.$$

(b) Find the value of the constants p and q for which $y = px + q$ is a solution of the differential equation

$$\frac{d^2y}{dx^2} - 3\frac{dy}{dx} - 4y = -4x.$$

Solve this differential equation, given that $y = 0$ and $\dfrac{dy}{dx} = \dfrac{3}{2}$ when $x = 0$.

4 Solve the differential equations

(a) $\dfrac{dy}{dx} + y \cot x = \cos 2x$, given that $y = 1$ when $x = \pi/2$;

(b) $\dfrac{d^2 y}{dx^2} - 4\dfrac{dy}{dx} + 4y = 8(x + 1)$, given that $y = 8$ and $\dfrac{dy}{dx} = 13$ when $x = 0$. [L]

5   By putting $z = y + 1$, or otherwise, obtain the general solution $y = f(x)$ of the differential equation

$$\frac{d^2 y}{dx^2} = 4(y + 1)$$

and show that when $f'(0) = 0$ and $f(0) = 1$ then $f(x) = 2 \cosh 2x - 1$. By expanding $\cosh 2x$ in powers of x as far as the term in x^4 find the value of $f(0.02)$ correct to seven decimal places. [JMB]

6   Solve the differential equations

(a) $\sqrt{(3x - x^2)}\dfrac{dy}{dx} = 1 + \cos 2y$,

(b) $(1 - x^2)\dfrac{dy}{dx} = x(1 - 2y + x)$. [AEB]

7   (a) Find the general solution of the differential equation

$$\frac{dy}{dx} + y \tan x = x \cos x.$$

(b) The differential equation

$$\frac{d^2 y}{dx^2} + 4\frac{dy}{dx} + 4y = e^{-2x}$$

has a particular integral of the form $kx^2 e^{-2x}$. Find the value of the constant k.

Solve the equation, given that $y = \dfrac{1}{2}, \dfrac{dy}{dx} = -\dfrac{1}{2}$ when $x = 0$. [L]

8   (a) Find the general solution of the differential equation

$$\frac{dy}{dx} + 2xy = 2x^3.$$

(b) Use the substitution $z = 1/y^2$ to transform the differential equation

$$2x\frac{dy}{dx} + y = 2x^2(x + 1)y^3$$

into a differential equation involving z and x. Hence find $1/y^2$ in terms of x, given that $y = 1$ when $x = 1$.

9   (a) Find the general solution of the differential equation

$$x\frac{dy}{dx} - 2y + mx = 0,$$

where m is constant.

(b) Find the solution of the differential equation

$$\frac{d^2y}{dx^2} + 4\frac{dy}{dx} + 5y = 2e^{-x}$$

for which $y = 0$ and $\frac{dy}{dx} = 1$ when $x = 0$.

10 (a) Show that, if $x = e^t$, then

$$x\frac{dy}{dx} = \frac{dy}{dt}.$$

Use the substitution $x = e^t$ to show that the differential equation

$$x^2\frac{d^2y}{dx^2} + x\frac{dy}{dx} + y = 0$$

becomes

$$\frac{d^2y}{dt^2} + y = 0.$$

Hence, or otherwise, find y in terms of x given that $y = 0$ when $x = 1$, and $\frac{dy}{dx} = 2$ when $x = e^\pi$.

(b) Find the general solution of the differential equation

$$x\frac{dy}{dx} + y = 3x^2.$$ [L]

11 (a) Find the general solution of the differential equation

$$x(x - 1)\frac{dy}{dx} + y = x^2(x - 1).$$

(b) Obtain constants a and b such that $a\cos 2x + b\sin 2x$ is a particular integral of the differential equation

$$\frac{d^2y}{dx^2} + 4\frac{dy}{dx} + 5y = 65\cos 2x,$$

and hence obtain the general solution of this differential equation.   [L]

12 (a) A particular integral of the differential equation

$$3\frac{d^2y}{dx^2} + 8\frac{dy}{dx} + 5y = 9\cos 2x - 23\sin 2x$$

is $P\cos 2x + Q\sin 2x$, where P and Q are constants. Find P and Q, and obtain the general solution of the differential equation.

(b) By differentiating twice, and eliminating the arbitrary constants A and B, obtain the differential equation of which the general solution is

$$y = A \cosh 3x + B \sinh 3x + x^2 + 2.$$ [L]

13 (a) Find the general solution of the differential equation

$$(x^2 - 1)\frac{dy}{dx} + 2y = (x^2 - 1)(x^2 + 2x + 1).$$

(b) Find the constant k so that $kx^2 e^x$ is a particular integral of the differential equation

$$\frac{d^2 y}{dx^2} - 2\frac{dy}{dx} + y = e^x.$$

Obtain the solution of this differential equation which satisfies the conditions $y = 1$ at $x = 0$ and $\frac{dy}{dx} = -1$ at $x = 0$. [L]

14 (a) Solve the differential equation

$$2xy\frac{dy}{dx} = x^2 + y^2$$

given that $y = 0$ when $x = 2$.

(b) Given that $y = Ae^{Bx}$ where A and B are constants, form a differential equation which does not contain these constants. Find the possible values of $\frac{dy}{dx}$ when $y = 4$ and $\frac{d^2 y}{dx^2} = 9$.

15 A compound C is formed by the composition of two substances A and B in such a way that one molecule of C arises from one molecule of A combining with **two** molecules of B. Let N be the number of molecules of C which have been formed at time t and let the rate of increase of N be proportional to the product of the number of molecules of A and the number of molecules of B present at time t. Suppose that at time $t = 0$ there are a molecules of A, $3a$ molecules of B and no molecules of C. Show that (regarding N as a continuous variable)

$$\frac{dN}{dt} = k(3a^2 - 5aN + 2N^2)$$

where k is a positive constant.
    Show also that

$$\ln\left(\frac{3a - 2N}{3a - 3N}\right) = kat,$$

and deduce that

$$N = \frac{3(1 - e^{kat})}{2 - 3e^{kat}} a.$$

[JMB]

16 An army of red ants encounters an army of black ants. Each red ant then proceeds to kill α black ants per unit time while each black ant kills 4α red ants per unit time, where α is a positive constant. The number of live red ants and the number of live black ants at time t are x and y respectively. Assuming that x and y may be considered as continuous variables, show that

$$\frac{d^2 x}{dt^2} = 4\alpha^2 x.$$

Solve this equation for x, and hence determine x and y, given that $x = 5N$ and $y = 2N$ when $t = 0$. Show that your solutions are not realistic for $\alpha t > \frac{1}{2} \ln 3$ and find the time at which the army of black ants is reduced to half of its original size. [L]

17 The temperature θ inside a box is maintained by a heater controlled by a thermostat. When the heater is switched off the interior temperature falls at a rate $\beta(\theta - \theta_0)$, where β is a positive constant and θ_0 is the constant exterior temperature. Show that, if the interior temperature is θ_1 at time t_1, and the heater remains off, then at time $t(>t_1)$ the temperature is

$$\theta_0 + (\theta_1 - \theta_0)e^{-\beta(t - t_1)}.$$

When the heater is switched on, the interior temperature increases at a constant rate α (the effect of cooling then being negligible). The thermostat switches the heater on when $\theta = \theta_2$, and off when $\theta = \theta_1$, where $\theta_0 < \theta_2 < \theta_1$.

Find the time that elapses between t_1 and the next occasion when the thermostat switches the heater off. [JMB]

18 (a) The amounts x, y and z of three substances at time t satisfy the differential equations

$$\frac{dx}{dt} = -2x; \quad \frac{dy}{dt} = 2x - 3y; \quad \frac{dz}{dt} = 3y.$$

Solve these equations with the initial conditions that when $t = 0$, $x = 1$ and $y = z = 0$.

(b) Find the solution of the differential equation

$$\frac{d^2 x}{dt^2} + \omega^2 x = E \sin pt,$$

given that $p^2 \neq \omega^2$ and that when $t = 0$, $x = dx/dt = 0$.

Verify that, when $p = \omega$, $-(E/2\omega)t \cos \omega t$ is a particular integral of the above equation and show that the solution satisfying the initial condition is then

$$x = (E/2\omega^2)(\sin \omega t - \omega t \cos \omega t).$$

19 Find the general solution of the differential equation

$$\frac{d^2u}{dx^2} + 4u = 8.$$

If $u = x\dfrac{dy}{dx} + 2y$, prove that

$$\frac{d^2u}{dx^2} + 4u \equiv x\frac{d^3y}{dx^3} + 4\frac{d^2y}{dx^2} + 4x\frac{dy}{dx} + 8y$$

and hence, or otherwise, obtain the solution of the differential equation

$$x\frac{d^3y}{dx^3} + 4\frac{d^2y}{dx^2} + 4x\frac{dy}{dx} + 8y = 8, \qquad \text{for } x > 0,$$

given that $y = \dfrac{dy}{dx} = \dfrac{d^2y}{dx^2} = 0$ when $x = \frac{1}{2}\pi$. [OC]

20 (a) Find the general solution of the differential equation

$$(x^2 + 1)\frac{dy}{dx} + 4xy = \frac{x^2 - 1}{x^2 + 1}.$$

(b) Find the value of the constant k such that kx^2e^{ax} is a particular integral of the differential equation

$$\frac{d^2y}{dx^2} - 2a\frac{dy}{dx} + a^2y = 4e^{ax},$$

where a is a positive constant.

Obtain the solution of this differential equation which satisfies the conditions $y = 1$ and $\dfrac{dy}{dx} = 1 + a$ at $x = 0$. [L]

21 (a) Find the general solution of the differential equation

$$\sin x\frac{dy}{dx} + (\sin x + \cos x)y = 2 + \sin x.$$

(b) Find the value of the constant k such that kx^2e^{3x} is a particular integral of the differential equation

$$\frac{d^2y}{dx^2} - 6\frac{dy}{dx} + 9y = 4e^{3x}.$$

Write down the general solution of the above differential equation and find the solution for which $y = 0$ when $x = 0$ and $y = 3e^3$ when $x = 1$. [L]

22 (a) The position vector **r** of a point P at time t satisfies the equation

$$\frac{d\mathbf{r}}{dt} = \boldsymbol{\omega} \times \mathbf{r},$$

where $\boldsymbol{\omega}$ is a constant non-zero vector. Show that $|\mathbf{r}|$ is constant and that P lies in a fixed plane.

Deduce that P describes a circle with constant speed.

(b) A particle Q moves so that its position vector $\mathbf{r}$ satisfies the equation

$$\frac{d^2\mathbf{r}}{dt^2} = -n^2\mathbf{r},$$

where $n \neq 0$. Show that Q describes a closed plane curve in time $2\pi/n$.

[L]

23 Given that $\mathbf{r}$ is a time-dependent vector, differentiate with respect to the time t:

(a) $\dot{\mathbf{r}} \cdot \dot{\mathbf{r}}$,   (b) $\mathbf{r} \times \mathbf{H}$,   (c) $\dot{\mathbf{r}} \cdot \mathbf{H}$,

where $\mathbf{H}$ is a constant vector.

The position vector $\mathbf{r}$ of a particle satisfies the equation

$$\ddot{\mathbf{r}} = \mathbf{E} + \dot{\mathbf{r}} \times \mathbf{H},$$

with $\mathbf{r} = \mathbf{0}, \dot{\mathbf{r}} = \lambda\mathbf{H}$ at $t = 0$, where $\mathbf{E}$ and $\mathbf{H}$ are constant vectors and λ is a constant scalar. Obtain

(a) an expression for $\dot{\mathbf{r}}^2$ in terms of λ, $\mathbf{E}$, $\mathbf{H}$ and $\mathbf{r}$,

(b) an expression for $\dot{\mathbf{r}}$ in terms of λ, $\mathbf{E}$, $\mathbf{H}$, $\mathbf{r}$ and t.

Further, if $\mathbf{E} \cdot \mathbf{H} = 0$, show that $\dot{\mathbf{r}} \cdot \mathbf{H} = \lambda\mathbf{H}^2$ and deduce that

$$\dot{\mathbf{r}} = \mathbf{r} \times \mathbf{H} + \frac{(\mathbf{r} \cdot \mathbf{H})}{\lambda\mathbf{H}^2}\mathbf{E} + \lambda\mathbf{H}.$$

[L]

24 Given that $\dfrac{dy}{dx} - 3z = 6, \dfrac{dz}{dx} - 3y = 12$, and $y = z = 0$ when $x = 0$, find y and z in terms of x, and show that

$$y^2 - z^2 + 8y - 4z = 0.$$

5 Polar coordinates

5.1 The straight line

The polar coordinates (r, θ) of a point P give the distance r of P from the pole O and the angle θ, measured anticlockwise, between OP and the initial line. Since the distance OP is a positive number, the convention that r takes positive values only will be adopted in this chapter.

Consider a straight line AB passing through the pole O. The equation $\theta = \alpha$ is satisfied at all points on the half-line OA, where α is the angle between the initial line $\theta = 0$ and OA measured anticlockwise (Fig. 5.1).

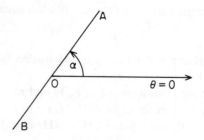

Fig. 5.1

The equation $\theta = \alpha + \pi$ is satisfied at all points on the half-line OB. Since tan $(\alpha + \pi) = \tan \alpha$, the equation $\tan \theta = \tan \alpha$ gives the complete line through the points A and B.

In Fig. 5.2, ON is the perpendicular from the pole O to the straight line LM, and α is the angle measured anticlockwise between the initial line and ON. The length of ON is denoted by p.

Let the polar coordinates of any point P on the line LM be (r, θ). Then the diagram shows that θ lies between $\alpha - \pi/2$ and $\alpha + \pi/2$. In the triangle OPN,

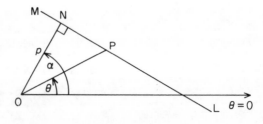

Fig. 5.2

$$ON = OP \cos PON$$
$$\Rightarrow \qquad p = r \cos(\alpha - \theta).$$
$$\Rightarrow \qquad r = p \sec(\alpha - \theta).$$

This is the polar equation of the line LM.

As θ increases from $\alpha - \pi/2$, the point P moves along the line in the direction from L towards M. When $\theta = \alpha$, $\cos(\alpha - \theta)$ will have its maximum value 1, $\sec(\alpha - \theta)$ will have its minimum value 1 and P will coincide with N.

The two lines given by the equations

$$r = p_1 \sec(\alpha - \theta), \qquad r = p_2 \sec(\alpha - \theta)$$

are parallel, since they are both perpendicular to ON. The distance from O to the first line is p_1 and the distance from O to the second line is p_2. Hence the distance between the two lines is $|p_1 - p_2|$.

The two lines given by the equations

$$r = p_1 \sec(\alpha - \theta), \qquad r = p_2 \sec(\beta - \theta),$$

where $\alpha > \beta$, intersect at an angle $\alpha - \beta$, since this is the angle between the normals ON_1, ON_2 from O to these lines. In Fig. 5.3, the point M is the point of intersection of the two lines, and the angle $N_2 M N_1$ equals $\alpha - \beta$.

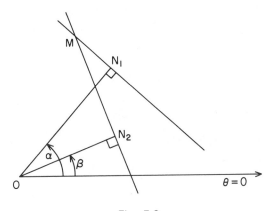

Fig. 5.3

Example 1
Calculate the length of the line AB, where the point A is given by $r = 2\sqrt{3}$, $\theta = \pi/6$ and the point B is given by $r = 6$, $\theta = 2\pi/3$.
Find the polar equation of the line AB.

In Fig. 5.4 let ON be the perpendicular from the pole O to the line AB. The angle AOB is a right angle, since $2\pi/3 - \pi/6 = \pi/2$.
Hence

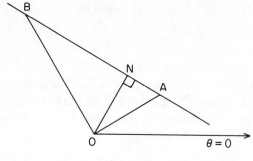

Fig. 5.4

$$AB^2 = OA^2 + OB^2 = 12 + 36 = 48$$
$$\Rightarrow \qquad AB = 4\sqrt{3}.$$

Area of the triangle AOB $= \frac{1}{2}AB \times ON = 2\sqrt{3}\ ON$.
This area is also given by $\frac{1}{2}OA \times OB$, which equals $6\sqrt{3}$.
It follows that $ON = 3$.

The polar equation of the line AB will be

$$r = 3 \sec(\alpha - \theta),$$

where α is the angle between ON and the initial line.

This equation will be satisfied by the polar coordinates of A, i.e by $r = 2\sqrt{3}$, $\theta = \pi/6$. Therefore

$$2\sqrt{3} = 3 \sec(\alpha - \pi/6)$$
$$\Rightarrow \qquad \cos(\alpha - \pi/6) = \sqrt{3}/2$$
$$\Rightarrow \qquad \alpha - \pi/6 = \pi/6.$$
$$\Rightarrow \qquad \alpha = \pi/3.$$

The polar equation of the line AB is therefore

$$r = 3 \sec(\pi/3 - \theta).$$

Example 2
Find the distance from the point P, given by $r = 4$, $\theta = 2\pi/3$, to the line $r = 3 \sec(\pi/4 - \theta)$.

In Fig. 5.5, AB represents the given line.
The perpendicular distance ON from the pole O to AB equals 3. Let the line CD drawn through P parallel to AB meet ON at M. The equation of CD will be of the form

$$r = p \sec(\alpha - \theta).$$

Since CD is parallel to AB, $\alpha = \pi/4$.
Since CD passes through the point P, at which $r = 4$ and $\theta = 2\pi/3$,

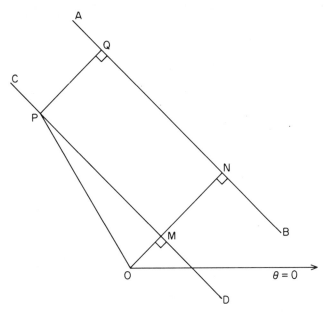

Fig. 5.5

$$4 = p \sec (\pi/4 - 2\pi/3)$$
$$\Rightarrow \qquad p = 4 \cos (5\pi/12).$$

This is the length of OM.

Let PQ be the perpendicular drawn from P to AB.

$$PQ = MN = ON - OM$$
$$\Rightarrow \qquad PQ = 3 - 4 \cos (5\pi/12).$$

Example 3

Express the polar equations

(a) $r = p \sec (\alpha - \theta)$,   (b) $\dfrac{c}{r} = a \cos \theta + b \sin \theta, \, c \neq 0,$

in terms of cartesian coordinates.

In Fig. 5.6, the origin for cartesian coordinates coincides with the pole O for polar coordinates, and the positive x-axis coincides with the initial line.

Let the point P have cartesian coordinates (x, y) and polar coordinates (r, θ). Then

$$x = OM = r \cos \theta,$$
$$y = MP = r \sin \theta.$$

(a) The equation $r = p \sec (\alpha - \theta)$ gives

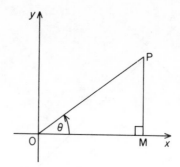

Fig. 5.6

$$r \cos (\alpha - \theta) = p$$
$$\Rightarrow \qquad r(\cos \alpha \cos \theta + \sin \alpha \sin \theta) = p$$
$$\Rightarrow \qquad (r \cos \theta) \cos \alpha + (r \sin \theta) \sin \alpha = p$$
$$\Rightarrow \qquad x \cos \alpha + y \sin \alpha = p.$$

This form of the cartesian equation for a straight line is known as the perpendicular form. The perpendicular distance of the line from the origin equals p, and the angle between the x-axis and this perpendicular is α.

(b)
$$\frac{c}{r} = a \cos \theta + b \sin \theta$$

$$\Rightarrow \qquad c = a(r \cos \theta) + b(r \sin \theta)$$
$$\Rightarrow \qquad ax + by = c.$$

Since c is not zero, this is the equation of a straight line which does not pass through the origin.

Exercises 5.1

1 Draw a diagram showing the lines given by the polar equations $r = 4 \sec (\pi/2 - \theta)$ and $r = 2 \sec (\pi/6 - \theta)$.

    Find the acute angle between the lines and the polar coordinates of the point at which they meet.

2 Express the equation $1/r = \cos \theta + \sin \theta$ in the form $r = p \sec (\alpha - \theta)$.

3 Use the relations $x = r \cos \theta$, $y = r \sin \theta$ to convert to cartesian coordinates the polar equations
    (a) $r = 5 \sec (\pi/4 - \theta)$,   (c) $r = 3 \sec (2\pi/3 - \theta)$,
    (b) $\tan \theta = 1$,   (d) $r = 2 \sec \theta$.

4 Find the polar equation of the straight line drawn through the point $(2\sqrt{2}, \pi/2)$ parallel to the line $r = 4 \sec (\pi/4 - \theta)$.

5 Find the polar equations of the two lines which pass through the point $A(4, \pi/4)$ and which make an angle $\pi/4$ with OA, where O is the pole.

6 Find the distance from the point $(2, \pi/3)$ to the line $r = 3 \sec (2\pi/3 - \theta)$.

7 Find the polar equation of the straight line through the point $(2, \pi/2)$ which is at right angles to the line $r = 3 \sec (\pi/4 - \theta)$.

8 Find the polar equation of the line through the points $(1, \pi/6)$, $(2, -\pi/6)$.

9 Find the polar coordinates of the point of intersection of the lines $r = 2 \sec (\pi - \theta)$ and $r = 2 \sec (\pi/3 - \theta)$.

10 Draw a diagram showing the lines $r = \sec (\pi/4 - \theta)$, $r = \sec (3\pi/4 - \theta)$ and $r = \sec (\pi/2 + \theta)$. Calculate the area of the triangle formed by these lines.

11 Show that the polar equation $\dfrac{2a}{r} = 1 - \cos \theta$ and the cartesian equation $y^2 = 4a(x + a)$ represent the same curve.

12 The polar coordinates of the points A and B are $(a, 0)$ and (a, π) respectively, where $a > 0$. Given that $PA \times PB = a^2$, show that the polar equation of the locus of the point P is $r^2 = 2a^2 \cos 2\theta$.
  Show also that the cartesian equation of this locus is

$$(x^2 + y^2)^2 = 2a^2(x^2 - y^2).$$

5.2 The circle

The circle of radius a with centre at the pole
Each point on the circle is at a distance a from the pole O, so that the polar equation of the circle is $r = a$.

Example 1
Find the equation of the tangent to the circle $r = a$ at the point (a, α).

Let N be the point (a, α) in Fig. 5.7. Then ON is the perpendicular from the pole to the tangent at N. Since a is the length of ON and α is the angle between the initial line and ON, the equation of the tangent at N is $r = a \sec (\alpha - \theta)$.

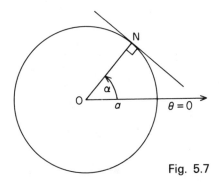

Fig. 5.7

Example 2

Find the polar coordinates of the points in which the line $r = 2 \sec (2\pi/3 - \theta)$ meets the circle $r = 4$ (Fig. 5.8).

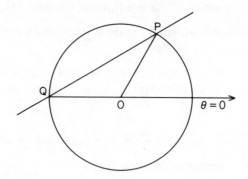

Fig. 5.8

The line meets the circle where

$$2 \sec (2\pi/3 - \theta) = 4$$
$$\Rightarrow \qquad \sec (2\pi/3 - \theta) = 2$$
$$\Rightarrow \qquad \cos (2\pi/3 - \theta) = \tfrac{1}{2} = \cos (\pi/3)$$
$$\Rightarrow \qquad 2\pi/3 - \theta = \pm \pi/3$$
$$\Rightarrow \qquad \theta = \pi/3 \text{ or } \pi.$$

Hence the points of intersection are P(4, $\pi/3$) and Q(4, π).

The circle of radius a and centre C $(a, 0)$

The centre C is on the initial line, and, since OC $= a$, the circle passes through the pole O. In Fig. 5.9, OA is a diameter of the circle, and P(r, θ) is any point on the circle. Since the angle OPA is a right angle,

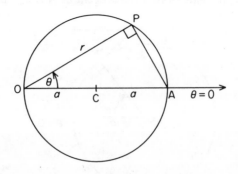

Fig. 5.9

$$\text{OP} = \text{OA cos AOP}$$
$$\Rightarrow \qquad r = 2a \cos \theta,$$

where $-\pi/2 < \theta < \pi/2$. This is the polar equation of the circle. If this equation is multiplied by r, it becomes $r^2 = 2ar \cos \theta$. This is equivalent to the cartesian equation $x^2 + y^2 = 2ax$.

Example 3
Find the equation of the tangent to the circle $r = 2a \cos \theta$ at the point T $(2a \cos \beta, \beta)$.

In Fig. 5.10,　OT $= 2a \cos \beta$,
$$\text{ON} = \text{OT} \cos \beta = 2a \cos^2 \beta.$$

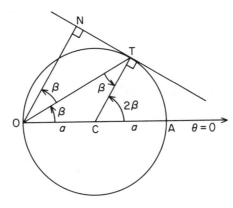

Fig. 5.10

Let the equation of the tangent at T be $r = p \sec (\alpha - \theta)$.

Then
$$p = \text{ON} = 2a \cos^2 \beta.$$
$$\alpha = \text{angle AON} = 2\beta.$$

Hence the equation of the tangent is

$$r = (2a \cos^2 \beta) \sec (2\beta - \theta).$$

The circle of radius a and centre C $(a, \pi/2)$
In Fig. 5.11, the centre C is the point $(a, \pi/2)$. Since OC $= a$, the circle passes through the pole O. The diameter OA is at right angles to the initial line.

Let P(r, θ) be a point on the circle.
Since the angle OPA is a right angle,

$$\text{OP} = \text{OA cos POA}$$
$$\Rightarrow \qquad r = 2a \cos (\pi/2 - \theta)$$
$$\Rightarrow \qquad r = 2a \sin \theta,$$

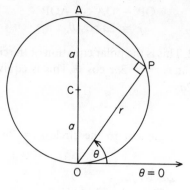

Fig. 5.11

where $0 < \theta < \pi$. This is the polar equation of the circle.

If this equation is multiplied by r, then it becomes $r^2 = 2ar \sin \theta$, which is equivalent to the cartesian equation $x^2 + y^2 = 2ay$.

The circle of radius a and centre C (c, α)

In Fig. 5.12, the point P(r, θ) lies on the circle of radius a and centre C (c, α). From the triangle OCP,

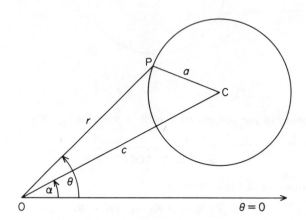

Fig. 5.12

$$CP^2 = OP^2 + OC^2 - 2OP \cdot OC \cos COP$$
$$\Rightarrow \quad a^2 = r^2 + c^2 - 2rc \cos (\theta - \alpha)$$
$$\Rightarrow \quad r^2 - 2rc \cos (\theta - \alpha) + c^2 - a^2 = 0.$$

This is the polar equation of the circle.

If $r \cos \theta$ is replaced by x and $r \sin \theta$ by y then the polar equation is converted to the cartesian equation

$$x^2 + y^2 - 2cx \cos \alpha - 2cy \sin \alpha + c^2 = a^2$$
$$\Rightarrow \qquad (x - c \cos \alpha)^2 + (y - c \sin \alpha)^2 = a^2.$$

Example 4

Find the polar equation of the circle with radius 2 and centre at the point $(3, \pi/4)$.

In Fig. 5.12, let $OC = 3$, $a = 2$, $\alpha = \pi/4$ and $OP = r$.
From the triangle OCP,

$$CP^2 = OP^2 + OC^2 - 2OP \cdot OC \cos COP$$
$$\Rightarrow \qquad 4 = r^2 + 9 - 6r \cos (\theta - \pi/4)$$
$$\Rightarrow \qquad r^2 - 6r \cos (\theta - \pi/4) + 5 = 0.$$

Exercises 5.2

1 Find the polar equation of the tangent
   (a) to the circle $r = 2$ at the point $(2, \pi/3)$,
   (b) to the circle $r = 5$ at the point $(5, 3\pi/4)$,
   (c) to the circle $r = 4$ at the point $(4, -\pi/4)$.

2 Find the polar coordinates of the points in which
   (a) the line $r = 3 \sec \theta$ meets the circle $r = 6$,
   (b) the line $r = 2 \sec (\pi/4 - \theta)$ meets the circle $r = 2\sqrt{2}$,
   (c) the line $r = 2 \sec (\pi/6 - \theta)$ meets the circle $r = 4$.

3 Find the polar equation of the tangent
   (a) to the circle $r = 4 \cos \theta$ at the point at which $\theta = \pi/3$,
   (b) to the circle $r = 6 \cos \theta$ at the point at which $\theta = \pi/4$,
   (c) to the circle $r = 2 \sin \theta$ at the point at which $\theta = \pi/6$.

4 Find the polar equation of the circle
   (a) with centre C $(5, \pi/4)$ and radius 4,
   (b) with centre C $(3, \pi/3)$ and radius 5,
   (c) with centre C $(2, \pi/6)$ and radius 2.

5 Show that the circles $r = 2a \cos \theta$ and $r = 2a \sin \theta$ cut at right angles.

6 Find the polar coordinates of the points in which the circle $r = 2$ meets the circle $r = 4 \cos \theta$.

7 Find the centre and the radius of each of the circles
   (a) $r^2 - 2r \cos \theta - 2 = 0$,
   (b) $r = 4 \cos (\theta - 2\pi/3)$,
   (c) $r = 2 \cos \theta + 2 \sin \theta$.

8 Calculate the area of the triangle OAB, where O is the pole, and A and B are the points $(6, \pi/6)$ and $(6, -\pi/6)$ respectively. Find the polar equation of the inscribed circle of this triangle.

9 Find the polar equation of the circle with centre $(2, \pi/2)$ and radius 1. Find also the polar equations of the tangents to this circle from the pole.

10 Show that the circle with polar equation $r^2 - 8r \cos(\theta - \pi/6) + 12 = 0$ touches the initial line.

Find the polar equation of the other tangent to this circle from the pole.

5.3 The angle ϕ between tangent and radius vector

Let the tangent be drawn to the curve $r = f(\theta)$ at the point (r, θ). Let $\overrightarrow{PT}$ be a vector along the tangent in the direction of the velocity of P when the radius OP is rotating anticlockwise about O. Then the angle ϕ at P is defined to be the angle between the vectors $\overrightarrow{OP}$ and $\overrightarrow{PT}$, i.e. in Fig. 5.13 ϕ equals the angle SPT.

Consider the points $P(r, \theta)$ and $Q(r + \delta r, \theta + \delta\theta)$ on the curve $r = f(\theta)$, δr and $\delta\theta$ being small (Fig. 5.14).

As $\delta\theta$ tends to zero, the angle between the chord PQ and the radius OQ tends to the angle ϕ, so that

$$\tan PQN \to \tan \phi$$

$$\Rightarrow \qquad \lim_{\delta\theta\to 0} \frac{PN}{NQ} = \tan \phi.$$

In Fig. 5.14, $PN = OP \sin \delta\theta = r \sin \delta\theta,$

$$\lim_{\delta\theta\to 0} \frac{PN}{\delta\theta} = \lim_{\delta\theta\to 0} \frac{r \sin \delta\theta}{\delta\theta} = r.$$

Since $NQ = OQ - ON = r + \delta r - r \cos \delta\theta,$

$$\frac{NQ}{\delta\theta} = \frac{\delta r}{\delta\theta} + \frac{r(1 - \cos \delta\theta)}{\delta\theta}.$$

Now $\lim\limits_{\delta\theta\to 0} \dfrac{1 - \cos \delta\theta}{\delta\theta} = 0$, and so $\lim\limits_{\delta\theta\to 0} \dfrac{NQ}{\delta\theta} = \dfrac{dr}{d\theta}.$

From the two results

$$\lim_{\delta\theta\to 0} \frac{PN}{\delta\theta} = r, \qquad \lim_{\delta\theta\to 0} \frac{NQ}{\delta\theta} = \frac{dr}{d\theta},$$

it follows that $\lim\limits_{\delta\theta\to 0} \dfrac{PN}{NQ} = r\dfrac{d\theta}{dr}$

$$\Rightarrow \qquad \tan \phi = r\frac{d\theta}{dr}, \quad \text{or} \quad r \left/ \frac{dr}{d\theta} \right.$$

Example 1

Find ϕ in terms of θ at any point $P(r, \theta)$ on the circle $r = 2a \cos \theta$ where $a > 0$. Find also the angle between the initial line and the tangent at P.

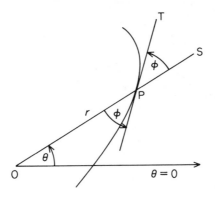

Fig. 5.13

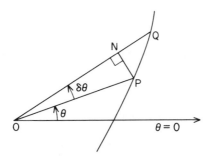

Fig. 5.14

From the equation $r = 2a \cos \theta$,

$$\frac{dr}{d\theta} = -2a \sin \theta$$

$\Rightarrow$

$$r\frac{d\theta}{dr} = \frac{2a \cos \theta}{-2a \sin \theta} = -\cot \theta$$

$\Rightarrow \qquad \tan \phi = -\cot \theta$

$\Rightarrow \qquad \tan \phi = \tan (\pi/2 + \theta).$

The general solution of this equation is $\phi = \pi/2 + \theta + n\pi$, where n is any integer or zero.

The value of n is found by observing the value of ϕ at a particular point on the curve.

In Fig. 5.15, the tangent at the point A(2a, 0) is at right angles to OA, so that when $\theta = 0$, ϕ is equal to $\pi/2$.

This gives $n = 0$, showing that

$$\phi = \pi/2 + \theta.$$

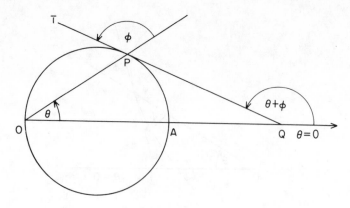

Fig. 5.15

Let the tangent at P meet the initial line at Q.
From Fig. 5.15, the angle between the initial line and QP equals $\theta + \phi$, i.e. $2\theta + \pi/2$.

Example 2
Express the angle ϕ in terms of θ for the curve given by

$$r = a(1 + \cos \theta),$$

where $a > 0$ and $-\pi < \theta \leqslant \pi$.
Sketch the curve, and find the points on the curve at which the tangent is parallel to the initial line.

This curve is known as a *cardioid*, a heart-shaped curve.

From the equation $\qquad\qquad r = a(1 + \cos \theta)$

$$\frac{dr}{d\theta} = -a \sin \theta$$

$\Rightarrow \qquad\qquad\qquad r\dfrac{d\theta}{dr} = \dfrac{1 + \cos \theta}{-\sin \theta}.$

Since $1 + \cos \theta = 2 \cos^2 (\theta/2)$ and $\sin \theta = 2 \sin (\theta/2) \cos (\theta/2)$, this gives

$$\tan \phi = -\cot (\theta/2) = \tan (\theta/2 + \pi/2)$$
$\Rightarrow \qquad\qquad \phi = \theta/2 + \pi/2 + n\pi,$

where n is an integer or zero.
When $\theta = 0$, r has its maximum value. The tangent to the curve is then at right angles to the radius vector, i.e. $\phi = \pi/2$ and hence $n = 0$. This gives

$$\phi = \theta/2 + \pi/2.$$

Since cos θ is an even function of θ, the curve $r = a(1 + \cos \theta)$ is symmetrical about the initial line.

It is sufficient to find the values of r for values of θ between 0° and 180° as shown in the table:

θ	0°	30°	60°	90°	120°	150°	180°
r/a	2	1·87	1·5	1	0·5	0·13	0

The curve is shown in Fig. 5.16. Note that the tangent to the curve at the pole is horizontal. As θ increases towards π, ϕ increases towards π, and the angle $(\theta + \phi)$ between the tangent and the initial line tends to 2π. As θ decreases towards $-\pi$, ϕ decreases towards 0, and so $(\theta + \phi)$ tends to $-\pi$.

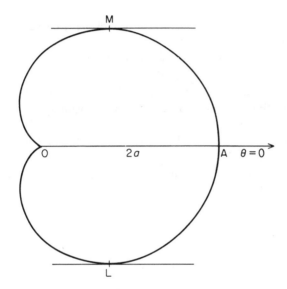

Fig. 5.16 $r = a(1 + \cos \theta)$

The tangent at the point (r, θ) will be parallel to the initial line when the angle $(\theta + \phi)$ between the tangent and the initial line is zero or a multiple of π.

Now $\theta + \phi = 3\theta/2 + \pi/2$. For values of θ in the interval $-\pi < \theta \leqslant \pi$ this gives

$$\theta + \phi = 0 \text{ when } \theta = -\pi/3,$$
$$\theta + \phi = \pi \text{ when } \theta = \pi/3,$$
$$\theta + \phi = 2\pi \text{ when } \theta = \pi.$$

Hence the tangent is parallel to the initial line at the points L($3a/2$, $-\pi/3$), M($3a/2$, $\pi/3$) and at the cusp at the pole O. Alternatively, the tangents parallel to the initial line can be found by considering the stationary values of $r \sin \theta$. When $r = a(1 + \cos \theta)$, the stationary values of $r \sin \theta$ will occur when

$$\frac{d}{d\theta}\Big[(1 + \cos\theta)\sin\theta\Big] = 0$$

$\Rightarrow$ $-\sin^2\theta + \cos\theta + \cos^2\theta = 0$

$\Rightarrow$ $2\cos^2\theta + \cos\theta - 1 = 0$

$\Rightarrow$ $(2\cos\theta - 1)(\cos\theta + 1) = 0.$

This gives $\cos\theta = \frac{1}{2}$ or -1, so that the relevant values of θ are $\pm\pi/3$ and π, as before.

The points at which the tangents to the curve $r = f(\theta)$ are at right angles to the initial line can be found by considering the stationary values of $r\cos\theta$.

Curve sketching

The main points to be borne in mind in sketching a curve given by a polar equation $r = f(\theta)$ are as follows:

(a) When f is an even function, i.e. when $f(-\theta) = f(\theta)$ for all values of θ, the curve is symmetrical about the initial line.

(b) When f is a periodic function with period 2π, i.e. when $f(\theta + 2\pi) = f(\theta)$ for all values of θ, the complete curve is given by values of θ in the interval $-\pi < \theta \leqslant \pi$.

(c) Any stationary value of r will be given by a value of θ which satisfies the equation $f'(\theta) = 0$.

(d) The half-lines which are tangential to the curve $r = f(\theta)$ at the pole can be found by solving the equation $f(\theta) = 0$.

(e) At the point (r, θ) the angle ϕ between the radius vector and the tangent to the curve is given by $\tan\phi = r\dfrac{d\theta}{dr}$.

Exercises 5.3

1   Sketch the curve $r = 2\theta$ for $0 \leqslant \theta \leqslant \pi$. Show that the tangent to this curve at the point at which $r = 2$ makes an angle $(\pi/4 + 1)$ with the initial line.

2   Show that if $f(\pi - \theta) = f(\theta)$ the curve $r = f(\theta)$ is symmetrical about the half-line $\theta = \pi/2$.

     Show that, for the circle $r = 2a\sin\theta$, the angle ϕ equals θ.

3   Sketch the cardioid $r = 2(1 + \sin\theta)$.

     Show that for this curve $\phi = \theta/2 + \pi/4$.

4   Sketch the curve $r = \tan(\theta/2)$ for $0 \leqslant \theta < \pi$.

     Show that the tangent to the curve at the point given by $r = 1$ makes an angle of $3\pi/4$ with the initial line.

5   Show that at any point on the spiral $r = e^{\theta}$ the tangent makes an angle of $\pi/4$ with the radius vector.

     Sketch the curve for $0 \leqslant \theta \leqslant 2\pi$.

6 Show that for the curve $r^2 = a^2 \cos 2\theta$ the angle ϕ equals $2\theta + \pi/2$. Sketch the curve for $-\pi/4 < \theta \leqslant \pi/4$.

7 Sketch the curve $2/r = 1 + \cos \theta$ for $-\pi < \theta < \pi$, and show that $\phi = \pi/2 - \theta/2$.
Show also that the cartesian equation of this curve can be put in the form $y^2 = 4 - 4x$.

8 Show that the perpendicular distance from the pole to the tangent to the curve $r = a(1 + \cos \theta)$ at the point given by $\theta = \pi/6$ is $2a \cos^3 (\pi/12)$.

9 Find the polar coordinates of the points other than the pole at which the two cardioids $r = 2 + 2 \cos \theta$ and $r = 6 - 6 \cos \theta$ intersect.
Show that at each of these points the tangents to the two curves are at right angles.

10 By considering the stationary values of $r \cos \theta$, find the polar coordinates of the points on the cardioid $r = 2(1 - \cos \theta)$ at which the tangents are perpendicular to the initial line.

5.4 The area of a sector

Consider the sector bounded by the radii OL, OM and the arc LM of the curve $r = f(\theta)$ shown in Fig. 5.17.

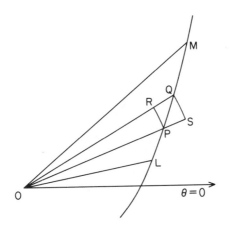

Fig. 5.17

Let OL be the half-line $\theta = \alpha$ and OM the half-line $\theta = \beta$.
Let $P(r, \theta)$ and $Q(r + \delta r, \theta + \delta\theta)$ be points on the arc LM.
The circle with centre O and radius OP meets OQ at R, and the circle with centre O and radius OQ meets OP produced at S. Then, in Fig. 5.17, the area δA of the sector OPQ is greater than the area of the sector OPR and less than the area of the sector OSQ, i.e.

$$\tfrac{1}{2}OP^2 \ \delta\theta < \delta A < \tfrac{1}{2}OQ^2 \ \delta\theta$$

$\Rightarrow$
$$\tfrac{1}{2}r^2 \ \delta\theta < \delta A < \tfrac{1}{2}(r + \delta r)^2 \ \delta\theta$$

$\Rightarrow$
$$\tfrac{1}{2}r^2 < \frac{\delta A}{\delta\theta} < \tfrac{1}{2}(r + \delta r)^2.$$

When $\delta\theta$ tends to zero, δr tends to zero. It follows that

$$\lim_{\delta\theta \to 0} \frac{\delta A}{\delta\theta} = \tfrac{1}{2}r^2$$

$\Rightarrow$
$$\frac{dA}{d\theta} = \tfrac{1}{2}r^2.$$

By integration with respect to θ from $\theta = \alpha$ to $\theta = \beta$, the area of the sector OLM is given by

$$A = \int_{\alpha}^{\beta} \tfrac{1}{2}r^2 \ d\theta.$$

It was assumed in Fig. 5.17 that r increases as θ increases. When r decreases as θ increases, the inequalities are reversed, but the same result is obtained.

Example 1
Calculate the area of the finite sector bounded by the initial line, the half-line $\theta = 2\pi/3$ and an arc of the parabola $r = 2a/(1 + \cos\theta)$.

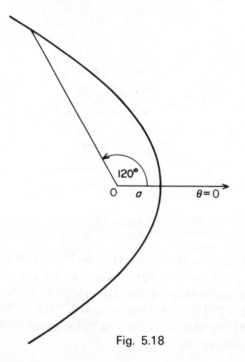

Fig. 5.18

The sector is shown in Fig. 5.18.

Since $1 + \cos \theta = 2 \cos^2 (\theta/2)$, $r = 2a/(2 \cos^2 \theta/2) = a \sec^2 (\theta/2)$.

The area A of the sector is given by

$$A = \int_0^{2\pi/3} \tfrac{1}{2} r^2 \, d\theta$$

$$= \int_0^{2\pi/3} \tfrac{1}{2} a^2 \sec^4 (\theta/2) \, d\theta$$

$$= \int_0^{2\pi/3} \tfrac{1}{2} a^2 [1 + \tan^2 (\theta/2)] \sec^2 (\theta/2) \, d\theta$$

$$\Rightarrow \qquad A = a^2 \left[\tan (\theta/2) + \tfrac{1}{3} \tan^3 (\theta/2) \right]_0^{2\pi/3}$$

$$= (2\sqrt{3}) a^2.$$

Example 2

Draw the curve $r^2 = a^2 \cos 2\theta$.

The half-line $\theta = \alpha$, where $0 < \alpha < \pi/4$, divides one loop of the curve into two regions. Calculate the area of each region.

Given that the ratio of these areas is $3:1$, find α.

Since the cosine function is an even function, the curve is symmetrical about the initial line.

Also $\cos 2\theta$ is periodic with period π, and so the curve is symmetrical about the pole.

For values of θ between 45° and 90°, $\cos 2\theta$ is negative and there is no real value for r.

For these reasons, when the arc of the curve for $0° \leqslant \theta \leqslant 45°$ has been drawn, the remainder of the curve can be completed by symmetry.

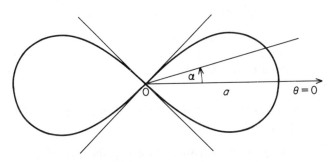

Fig. 5.19 $r^2 = a^2 \cos 2\theta$

The table gives values of r for the interval $0° \leqslant \theta \leqslant 45°$:

θ	0°	7·5°	15°	22·5°	30°	37·5°	45°
r	0	0·98a	0·93a	0·84a	0·71a	0·51a	0

The curve is shown in Fig 5.19. Note that, since $\cos 2\theta = 0$ when $\theta = \pm \pi/4$, $\pm 3\pi/4$, these are the values of θ giving the half-lines which are tangential to the curve at the pole.

The areas of the two regions are given by

(a) $\displaystyle\int_{-\pi/4}^{\alpha} \tfrac{1}{2}a^2 \cos 2\theta \, d\theta = \left[\tfrac{1}{4}a^2 \sin 2\theta \right]_{-\pi/4}^{\alpha} = \tfrac{1}{4}a^2 (\sin 2\alpha + 1),$

(b) $\displaystyle\int_{\alpha}^{\pi/4} \tfrac{1}{2}a^2 \cos 2\theta \, d\theta = \left[\tfrac{1}{4}a^2 \sin 2\theta \right]_{\alpha}^{\pi/4} = \tfrac{1}{4}a^2 (1 - \sin 2\alpha).$

When these two areas are in the ratio 3:1,

$$\sin 2\alpha + 1 = 3(1 - \sin 2\alpha)$$
$$\Rightarrow \qquad \sin 2\alpha = \tfrac{1}{2}$$
$$\Rightarrow \qquad 2\alpha = \pi/6$$
$$\Rightarrow \qquad \alpha = \pi/12, \text{ i.e. } 15°.$$

Exercises 5.4

1 Sketch the curve $r = e^{\theta}$ for $0 \leqslant \theta \leqslant 1$.
   Show that, to two decimal places, the area of the sector bounded by this curve, the initial line and the half-line $\theta = 1$ is 1·60.

2 Sketch the spiral $r = 2\theta$ for $0 \leqslant \theta \leqslant \pi$.
   Show that the area of the region enclosed by this curve and the half-line $\theta = \pi$ is $2\pi^3/3$.

3 Show that the area of the region enclosed by the curve

$$r^2 = a^2 \sin^2 \theta + b^2 \cos^2 \theta$$

is $\pi(a^2 + b^2)/2$.

4 Show that the area of each loop of the curve $r = a \cos^2 \theta$, where $a > 0$, is $3\pi a^2/16$.

5 Sketch the curve $r = a \sin 3\theta$, where $a > 0$.
   Show that the area of the loop of the curve for which $0 \leqslant \theta \leqslant \pi/3$ is $\pi a^2/12$.

6 Show that the area of the loop of the curve $r^2 = a^2 \sin^2 \theta \cos^3 \theta$ for which $0 \leqslant \theta \leqslant \pi/2$ is $a^2/15$.

7 Sketch the curve $r^2 = a^2 \sin 2\theta$ and show that the area of each loop of the curve is $a^2/2$.

8 Show that the area of the finite region enclosed by the curve $r = 4 \tan \theta$ and the half-line $\theta = \pi/4$ is $8 - 2\pi$.

9 Sketch the curve $r = 3 + 2 \cos \theta$.
   Show that the area of the region enclosed by the curve is 11π.

10 Show that the area of the region enclosed by the cardioid $r = 2a(1 + \cos \theta)$ is $6\pi a^2$.

Miscellaneous exercises 5

1 The point $P(r, \theta)$ lies on the line joining the points $A(r_1, \theta_1)$ and $B(r_2, \theta_2)$, where $0 < \theta_1 < \theta_2 < \pi$. By considering the areas of the triangles OAP, OPB, OAB, where O is the pole, show that the equation of the line AB is

$$rr_1 \sin (\theta - \theta_1) - rr_2 \sin (\theta - \theta_2) = r_1 r_2 \sin (\theta_2 - \theta_1).$$

Find the equation of the line through the points $(2, \pi/4)$ and $(4, 3\pi/4)$.

2 Verify that the equation $r = 2a \cos \alpha \cos \beta \sec (\alpha + \beta - \theta)$ represents the straight line through the points on the circle $r = 2a \cos \theta$ at which $\theta = \alpha$ and $\theta = \beta$.
   By letting β tend to α, obtain the equation of the tangent to the circle at the point $(2a \cos \alpha, \alpha)$.

3 Find the polar equation of the tangent to the circle $r = 4a \sin \theta$ at the point $(2a, \pi/6)$.

4 Show that the distance between the centres of the circles

$$r = 4 \cos (\theta - \pi/3), \quad r^2 - 6r \cos \theta + 6 = 0$$

is $\sqrt{7}$. Show also that the circles cut at right angles.

5 Find the centre and the radius of the circle

$$r^2 - 2r (\cos \theta + \sin \theta) + 1 = 0.$$

Show that this circle touches the initial line, the half-line $\theta = \pi/2$ and the line $r = (\sqrt{2} + 1) \sec (\pi/4 - \theta)$.

6 Sketch the curve $r^2 = a^2 \sec 2\theta$ for $0 \leqslant \theta < \pi/4$.
   Show that at any point on this curve $\phi = \pi/2 - 2\theta$.

7 Show that at any point (r, θ) on the circle $r = a(\cos \theta + \sin \theta)$ the tangent makes an angle $2\theta + \pi/4$ with the initial line.

8 Sketch in the same diagram the circles given by the polar equations $r = 2 \cos \theta$, where $-\pi/2 < \theta \leqslant \pi/2$, and $r = 2 \sin \theta$, where $0 \leqslant \theta < \pi$.
   Show that the circles intersect at the pole O, and find the polar coordinates of the other point P at which the circles intersect.
   Find the polar equations of the tangents to the circles at P.   [L]

9   Find the polar equation of the circle which has its centre at the point $(2, \pi/4)$ and which passes through the pole.

Find also the polar equations of the tangents to this circle which are parallel to the initial line.

10 Sketch the curve $r = 2a(1 - \cos \theta)$, where $a > 0$.

Find the polar coordinates of the points in which the curve meets the line $2r = a \sec \theta$.

11 Prove that the angle between the tangent to the curve $r = a(1 + \cos \theta)$ at the point for which $\theta = \alpha$ and the radius vector through the point is $\frac{1}{2}(\pi + \alpha)$.

A circle touches this curve at a point P and passes through the pole O. Prove that the centre of this circle lies on the circle $r = a \cos \theta$.   [L]

12 Sketch on the same diagram the curves whose equations in polar coordinates are $r = 2a \cos 2\theta$, $-\pi/4 \leqslant \theta \leqslant \pi/4$, and $r = a$, where $a > 0$. Show that the area of the finite region lying within both curves is $a^2(4\pi - 3\sqrt{3})/12$.   [L]

13 Find the area of the region enclosed by the curve $r = a(1 + \cos \theta)$, and show that the area of the part of this region for which $r > a$ is $a^2(\pi + 8)/4$.   [L]

14 Sketch the curve with polar equation $r = 3 + 2 \cos \theta$. Find the polar coordinates of the points in which this curve meets the line $r = 2 \sec \theta$.

Show that the area of the region defined by the inequalities

$$2 \sec \theta \leqslant r \leqslant 3 + 2 \cos \theta$$

is $11\pi/3 + (5\sqrt{3})/2$.   [L]

15 Sketch the curve given by $r = a \sin 2\theta$ and determine the area of one of its loops.   [AEB]

16 Sketch the curve $r = \sec^2 \theta$, where $-\pi/2 < \theta < \pi/2$. Find the area of the region contained between the curve and the straight lines $\theta = \pm \pi/4$.   [AEB]

17 Find the area of the sector enclosed by the curve whose equation in polar coordinates is $r = a \sec^2 (\theta/2)$ and the radii $\theta = 0$, $\theta = \alpha$ where $\alpha < \pi$.   [AEB]

18 Sketch the curve whose polar equation is $r = \sin 3\theta$, and find the area of one loop.   [AEB]

19 Sketch on one diagram the curves with polar equations $r = a(1 + \cos \theta)$ and $r = a(2 + \cos \theta)$, where $a > 0$.

Show that the area of the region lying between these curves is $3\pi a^2$.

20 Sketch the curve whose polar equation is $r^2 = a^2 \cos 2\theta$, where $a > 0, r > 0$, indicating the coordinates of any points where the curve cuts the initial line.

At the point P of the curve, $\theta = \pi/6$. Prove that the tangent at P is parallel to the initial line.

Find the area of the finite region enclosed by the initial line, the curve and the straight line joining the origin to P. [L]

21 A curve has polar equation $r = 2 \cos (2\theta/3)$, where $-3\pi/4 \leqslant \theta \leqslant 3\pi/4$. Show that the area of the region enclosed by this curve is $3\pi/2$. [L]

22 Sketch the curve with polar equation $r = a \cos 3\theta$, where $a > 0$, showing clearly the tangents at the pole.

Find the area of the finite region enclosed by one loop of the curve. [L]

23 (a) Sketch the curves C_1 and C_2 whose polar equations are

$$C_1 : r = 2\theta/\pi, \qquad \text{for } 0 \leqslant \theta \leqslant \pi/2,$$
$$C_2 : r = \sin \theta, \qquad \text{for } 0 \leqslant \theta \leqslant \pi/2.$$

The half-line $\theta = \alpha$ meets the curve C_1 at the pole O and at P, and it meets C_2 at O and at Q. Find the limit as $\alpha \to 0$ of PQ/OP.

(b) Solve the differential equation

$$r \sin \theta \frac{d\theta}{dr} = 1 - \cos \theta,$$

given that $r = 2a$ when $\theta = \pi$. [L]

24 A straight line through the pole cuts the circle $r = a \cos \theta$ at P. The points A and B on this line are such that $AP = PB = a$.

Sketch the locus of the points A and B, and show that its polar equation is $r = a(1 + \cos \theta)$.

Calculate the area of the finite region bounded by the arc of the locus for which $0 \leqslant \theta \leqslant \pi/4$ and by the lines $\theta = 0$, $\theta = \pi/4$. [L]

25 Sketch the curve with polar equation

$$\frac{1}{r} = 1 + \tfrac{1}{2} \cos \theta,$$

and show that it is an ellipse with one directrix $r \cos \theta = 2$.

Find the lengths of the major and minor axes, and hence obtain the area of the region enclosed by this curve, and the polar coordinates of the points on the curve where the tangent is parallel to the initial line. [L]

26 Sketch the curve $r = 2a(1 - \sin \theta)$, where $a > 0$. Find the polar coordinates of the points P and Q in which the curve intersects the circle $r = a$.

Show that the area of the finite region bounded by the minor arc PQ of the circle and the arcs OP, OQ of the curve, where O is the pole, is $(21\sqrt{3} - 10\pi)a^2/6$. [L]

27 Sketch on the same diagram the curves
(a) $r = 2 + \cos\theta$, for $0 \leqslant \theta \leqslant \pi$,
(b) $r^2 - 2r\cos\theta = 3$, for $0 \leqslant \theta \leqslant \pi$.
Find the area of the finite region between them. [L]

28 Show that the tangent at the point P (r, θ) on the curve C whose polar equation is $r^2 = a^2 \cos 2\theta$ makes an angle $\pi/2 + 2\theta$ with the radius vector OP. Find the four points on the curve at which the tangent is parallel to the line $\theta = 0$, and sketch the curve.
   On the same diagram sketch the curve whose equation is $r^2 = a^2 \sin 2\theta$. Calculate the area of the whole region within which both the relations $r^2 < a^2 \cos 2\theta$, $r^2 < a^2 \sin 2\theta$ hold. [JMB]

29 The angle between the radius vector OP to a point P on a curve C and the tangent at P to the curve is a given constant $\alpha\,(0 < \alpha < \pi/2)$. Show that the polar coordinates (r, θ) of P satisfy the equation $r = be^{\theta\cot\alpha}$, where b is constant.
   Find the area swept out by the radius vector as θ increases from 0 to 2π.
   Give the polar-coordinate equation of the circle with centre $r = c$, $\theta = \pi/2$ and radius c. Show that this circle touches the curve $r = e^{\theta\cot\alpha}$ provided that

$$2c \sin\alpha = e^{(\alpha + m\pi)\cot\alpha},$$

where m is any even integer. Explain why m must be even. [JMB]

30 Sketch the curve whose polar equation is $r = 2a(1 - \sin\theta)$. Find the values of θ at the points at which the curve intersects the circle $r = a$. At each point of intersection, determine the angle, lying in the range 0 to π, between the tangent to the curve and the radius vector to the point.
   Prove that the area swept out by the radius vector of the curve as θ increases from $\pi/6$ to $\pi/2$ is $[\pi - \frac{7}{4}\sqrt{3}]a^2$.
   Hence determine the area of the region common to the interiors of the curve and the circle $r = a$. [JMB]

31 Two curves have the equations in polar coordinates

$$r = 2\sin\theta \quad \text{and} \quad r = 4\sin^2\theta\,(0 \leqslant \theta \leqslant \pi).$$

Show that the curves intersect at the pole O and at two other points A and B, giving the polar coordinates of A and B.
   Sketch the two curves on the same diagram.
   Find the tangents of the angles of intersection of the two curves at A and B. Find also the area of the region defined by $r \leqslant 2\sin\theta$, $r \geqslant 4\sin^2\theta$. [JMB]

32 The straight line whose polar equation is $r \cos \theta = 3$ meets the curve whose polar equation is $r = 5 + 2 \cos \theta$ at the points Q and R. Show that the area of the triangle OQR (where O is the origin) is $9\sqrt{3}$.

Find the areas of the two parts into which QR divides the whole of the region defined by $r \leqslant 5 + 2 \cos \theta$. [JMB]

33 Referred to polar coordinates with pole O, the equation of a curve is

$$r = 16/(5 + 3 \cos \theta),$$

and P is a point (r, θ) on C. Find in terms of θ the tangent of the angle between OP and the tangent to C at P.

Show that the tangents to C at the points $(2, 0)$ and $(8, \pi)$ are each perpendicular to the initial line, and find the polar coordinates of the points Q and R at which the tangents are parallel to the initial line. Sketch the curve C.

The polar equation of another curve K is

$$r^2 + 12r \cos \theta + 11 = 0.$$

Express this equation in terms of cartesian coordinates and hence, or otherwise, sketch the curve K on the same diagram as the curve C.

Verify that K and C intersect at the points Q and R, and find the acute angle between their tangents at each of these points. [JMB]

34 Sketch the curves with polar equations
(a) $r = a \sin^2 \theta \sec^3 \theta$,   for $-\pi/2 < \theta < \pi/2$,
(b) $r^2 - 2ar \sec \theta + a^2 = 0$,   for $-\pi/2 < \theta < \pi/2$.

35 Show that the angle between the tangent at the point $P(r, \theta)$ on the curve $r = a(1 + \cos \theta)$ and the radius vector OP is $(\pi/2 + \theta/2)$.

The point N is the foot of the perpendicular to this tangent from the pole O. Show that $ON^2 = OP^3/2a$, and deduce that the polar equation of the locus of N is

$$r = 2a \cos^3 (\theta/3).$$

For $0 < \theta < \pi/3$, the points P_1 and N_1 are the feet of the perpendiculars drawn from P and N respectively to the initial line. Find the limit of the ratio PP_1/NN_1 as PP_1 tends to zero. [L]

6 Sequences and series: convergence

6.1 Convergence of sequences

A sequence $u_0, u_1, u_2, \ldots, u_n, \ldots$ is a succession of expressions, each formed uniquely by some given procedure, e.g. $1^1, 2^2, 3^3, \ldots, r^r, \ldots$ is a sequence defined by $u_r = r^r$.

$$2, 2, 4, 6, 10, \ldots, u_r, \ldots$$

is a sequence defined by $u_1 = u_2 = 2$, $u_r = u_{r-1} + u_{r-2}$, when $r > 2$.

A sequence which has a last term is called a *finite* sequence. When a sequence has no last term, it is called an *infinite* sequence.

With infinite sequences we are often concerned with the behaviour of u_n for large values of n.

Consider the sequence

$$1, \frac{1}{2}, \frac{1}{3}, \ldots, \frac{1}{n}, \ldots.$$

Let h be a small positive number. Then $1/n$ is less than h for all integers n greater than $1/h$, i.e.

$$n > \frac{1}{h} \quad \Rightarrow \quad \frac{1}{n} < h.$$

This statement is true for any choice of h, i.e. for any arbitrary small positive value of h. We express this result by saying that $1/n$ tends to zero as n tends to infinity, i.e.

$$\lim_{n \to \infty} \left(\frac{1}{n} \right) = 0.$$

The sequence

$$\frac{1}{2^2}, \frac{1}{3^2}, \frac{1}{4^2}, \ldots, \frac{1}{(n+1)^2}, \ldots$$

also tends to zero, because, for any positive value of h,

$$n + 1 > \frac{1}{\sqrt{h}} \quad \Rightarrow \quad \frac{1}{(n+1)^2} < h.$$

This shows that, for all sufficiently large values of n, the difference between $1/(n+1)^2$ and zero is less than h.

Consider next the sequence

$$1, \frac{3}{2}, \frac{5}{3}, \frac{7}{4}, \ldots, \frac{2n-1}{n}, \ldots$$

This sequence converges to the limit 2, since

$$\frac{2n-1}{n} - 2 = -\frac{1}{n}$$

$$\Rightarrow \qquad \lim_{n \to \infty} \left(\frac{2n-1}{n} - 2 \right) = \lim_{n \to \infty} \left(-\frac{1}{n} \right) = 0$$

Formally, the sequence $u_1, u_2, u_3, \ldots, u_n, \ldots$ is said to converge to the limit l if, for any small positive value of h, an integer N can be found such that

$$|u_n - l| < h \qquad \text{for every } n > N.$$

Thus, when $u_n = (2n-1)/n$ and $l = 2$,

$$|u_n - l| = \frac{1}{n} < h, \qquad \text{for every } n > \frac{1}{h}.$$

If a sequence does not converge it is said to *diverge* or to *be divergent* (or, in some cases, *oscillate*). (See Example 1 below.)

The sequence $2, 4, 8, \ldots, 2^n, \ldots$ is divergent because 2^n does not tend to a finite limit as $n \to \infty$.

The sequence $1, -1, 1, \ldots, (-1)^{n-1}, \ldots$ is not convergent because $(-1)^{n-1}$ does not tend to a unique limit as $n \to \infty$. This is an example of a finitely oscillating sequence.

The sequence $\{u_n\}$ is said to be monotonic increasing if $u_{n+1} \geqslant u_n$ for all n and to be monotonic decreasing if $u_{n+1} \leqslant u_n$ for all n.

Example 1

Examine the convergence of the sequence $\{x^n\}$, where $x \in \mathbb{R}$.

(a) Suppose first that $x > 1$ so that $x = 1 + k$, where $k > 0$. Then, for $n > 1$,

$$x^n = (1 + k)^n = 1 + nk + \ldots > 1 + nk.$$

Since k is positive, $nk \to \infty$ as $n \to \infty$ and so the sequence $\{x^n\}$ diverges.

(b) Suppose now that $0 < x < 1$, so that $x = \frac{1}{1 + t}$, where $t > 0$,

Then
$$x^n = \frac{1}{(1 + t)^n} < \frac{1}{1 + nt} \to 0, \qquad \text{as } n \to \infty.$$

(c) When $x = 1$, $x^n = 1$, for all $n \in \mathbb{N}$.

(d) When $x = 0$, $x^n = 0$, for all n.

(e) When $-1 < x < 0$, $x = -y$, where $0 < y < 1$, and $x^n = (-1)^n y^n$. It follows from the result of case (b) that $\lim_{n \to \infty} x^n = 0$ in this case also.

(f) When $x = -1$, $x^n = (-1)^n$ and the sequence $\{x^n\}$ oscillates boundedly between 1 and -1, and so does not converge.

(g) When $x < -1$, $x = -w$, where $w > 1$, and $x^n = (-1)^n w^n$.

It follows from the result of case (a) that, in this case, the sequence does not converge.

Conclusion   The sequence $\{x^n\}$ converges to 0 when $|x| < 1$, and converges to 1 when $x = 1$. For $|x| > 1$ and $x = -1$ the sequence does not converge.

Example 2

Find (a) $\displaystyle\lim_{n\to\infty} \left(\frac{2n + 3}{5n + 1}\right)$,   (b) $\displaystyle\lim_{n\to\infty} \left(\frac{4n^2 + 3n + 7}{6n^2 - 2n + 8}\right)$,

(c) $\displaystyle\lim_{n\to\infty} \left(\frac{n^2 + 7n + 6}{6n^3 + 3n^2 + 2n + 1}\right)$.

(a) $\displaystyle\lim_{n\to\infty} \left(\frac{2n + 3}{5n + 1}\right) = \lim_{n\to\infty} \left(\frac{2 + 3/n}{5 + 1/n}\right) = \frac{2}{5}$,   since $\displaystyle\lim_{n\to\infty} \left(\frac{1}{n}\right) = 0$.

(b) $\displaystyle\lim_{n\to\infty} \left(\frac{4n^2 + 3n + 7}{6n^2 - 2n + 8}\right) = \lim_{n\to\infty} \left(\frac{4 + 3/n + 7/n^2}{6 - 2/n + 8/n^2}\right) = \frac{4}{6} = \frac{2}{3}$,

since $\displaystyle\lim_{n\to\infty} \left(\frac{1}{n}\right) = 0$, $\displaystyle\lim_{n\to\infty} \left(\frac{1}{n^2}\right) = 0$.

(c) $\displaystyle\lim_{n\to\infty} \left(\frac{n^2 + 7n + 6}{6n^3 + 3n^2 + 2n + 1}\right) = \lim_{n\to\infty} \left(\frac{1/n + 7/n^2 + 6/n^3}{6 + 3/n + 2/n^2 + 1/n^3}\right) = 0$.

Example 3
Show that the sequence

$$1, \frac{3}{\sqrt{2}}, \frac{1}{\sqrt{3}}, \frac{3}{\sqrt{4}}, \cdots, \frac{[2 + (-1)^n]}{\sqrt{n}}, \cdots$$

converges to zero.

Denoting the nth term in the sequence by u_n, we have

$$u_{2n} = \frac{3}{\sqrt{(2n)}} \quad \text{and} \quad u_{2n+1} = \frac{1}{\sqrt{(2n + 1)}}.$$

Given a small positive number h, then

$$u_{2n} < h, \quad \text{for } n > 9/(2h^2),$$
$$u_{2n+1} < h, \quad \text{for } n > (1 - h^2)/(2h^2).$$

Therefore, both u_{2n} and u_{2n+1} are less than h (however small) for $n > 9/(2h^2)$ and hence the sequence converges to zero.

Note that here, since the sequence is not monotonic, we have considered the odd and even terms separately.

Example 4

A sequence is defined by the recurrence relation

$$u_{n+1} = \frac{5u_n}{u_n + 1}. \qquad \ldots (1)$$

By expressing $(u_{n+1} - 4)$ in terms of u_n, show that
(a) if $u_n > 4$, then $u_n > u_{n+1} > 4$; (b) if $0 < u_n < 4$, then $u_n < u_{n+1} < 4$.
Given that $u_1 > 0$, show that $|u_{n+1} - 4| \leqslant |u_n - 4|$ and prove that $u_n \to 4$ as $n \to \infty$.

From equation (1),

$$u_{n+1} - 4 = (u_n - 4)/(u_n + 1) \qquad \ldots (2)$$

(a) When $u_n > 4$, the right-hand side of equation (2) is positive and therefore $u_{n+1} > 4$.

Also, $\qquad (u_{n+1} - 4)/(u_n - 4) = 1/(u_n + 1) < 1$ since $u_n > 0$.
$\therefore \qquad\qquad u_{n+1} - 4 < u_n - 4 \qquad \Leftrightarrow \qquad u_{n+1} < u_n.$

(b) When $0 < u_n < 4$ the right-hand side of equation (2) is negative and therefore $u_{n+1} < 4$.
Also $(u_{n+1} - 4)/(u_n - 4) < 1$ but, because $u_n < 4$, it follows that

$$u_{n+1} - 4 > u_n - 4 \qquad \Leftrightarrow \qquad u_{n+1} > u_n.$$

When $u_1 > 0$, then $u_n > 0$. This can be formally proved by induction using equation (1).

Also, equation (2) gives $\qquad |u_{n+1} - 4| = \dfrac{|u_n - 4|}{|u_n + 1|} \qquad \ldots (3)$

$\Rightarrow \quad |u_{n+1} - 4| < |u_n - 4| \qquad$ since $\qquad |u_n + 1| = u_n + 1 > 1.$

From equation (3) by induction

$$|u_{n+1} - 4| = \frac{|u_1 - 4|}{|u_n + 1| \cdot |u_{n-1} + 1| \ldots |u_1 + 1|}$$

and, as $n \to \infty$, the denominator on the right-hand side of this equation tends to infinity and therefore

$$\lim_{n \to \infty} |u_{n+1} - 4| = 0$$

$\Rightarrow$
$$\lim_{n \to \infty} u_n = 4.$$

We could also obtain this result by assuming that $u_n \to$ the limit l as $n \to \infty$ and so, from equation (1),

$$\lim_{n\to\infty} u_{n+1} = \lim_{n\to\infty}\left(\frac{5u_n}{u_n + 1}\right)$$

$$\Rightarrow \qquad l = \frac{5l}{l + 1}$$

$$\Rightarrow \qquad l^2 - 4l = 0$$

$$\Rightarrow \qquad l = 0 \text{ or } 4.$$

But from (a) and (b) it follows that l cannot be zero and so $l = 4$.

Example 5
Show that

$$\lim_{n\to\infty}\left(1 + \frac{x}{n}\right)^n = e^x.$$

From the graphs of $\ln(1 + x)$, $x/(x + 1)$ and x, or using the methods of Section 8.4, it can be seen that

$$\frac{x}{x + 1} < \ln(1 + x) < x, \qquad \text{for } x > 0.$$

Writing x/n for x gives

$$\frac{nx}{n + x} < n \ln\left(1 + \frac{x}{n}\right) < x, \qquad \text{for } n > 0$$

$$\Rightarrow \qquad \lim_{n\to\infty}\left[n \ln\left(1 + \frac{x}{n}\right)\right] = x$$

$$\lim_{n\to\infty}\left(1 + \frac{x}{n}\right)^n = e^x.$$

assuming that the limit of the logarithm equals the logarithm of the limit.
This result is of great value in some aspects of probability theory. In particular

(a) $\lim_{n\to\infty}\left(1 + \frac{1}{n}\right)^n = e,$ \qquad (b) $\lim_{n\to\infty}\left(1 + \frac{1}{n}\right)^{nx} = e^x.$

Example 6
Show that, when n is a large positive number so that $\dfrac{1}{n^3}$ and higher powers of $\dfrac{1}{n}$ can be neglected, then

$$\left(1 + \frac{1}{n}\right)^n = e\left(1 - \frac{1}{2n} + \frac{11}{24n^2}\right).$$

$$\ln\left(1 + \frac{1}{n}\right)^n = n \ln\left(1 + \frac{1}{n}\right)$$

$$= n\left(\frac{1}{n} - \frac{1}{2n^2} + \frac{1}{3n^3} \cdots\right)$$

$$= 1 - \frac{1}{2n} + \frac{1}{3n^2} - \cdots$$

$$\Rightarrow \qquad \left(1 + \frac{1}{n}\right)^n = \exp\left(1 - \frac{1}{2n} + \frac{1}{3n^2} - \cdots\right)$$

$$= e \exp\left(-\frac{1}{2n}\right) \exp\left(\frac{1}{3n^2}\right) \cdots$$

$$= e\left(1 - \frac{1}{2n} + \frac{1}{8n^2}\right)\left(1 + \frac{1}{3n^2}\right)$$

$$= e\left(1 - \frac{1}{2n} + \frac{11}{24n^2}\right)$$

neglecting terms involving $1/n^3$ and higher powers of $1/n$.

This is a very good approximation to $\left(1 + \frac{1}{n}\right)^n$ when $n > 10$. When $n = 10$, $\left(1 + \frac{1}{n}\right)^n \approx 2\cdot594$, and the approximation gives the value as $2\cdot595$.

Exercises 6.1
Find the limit as $n \to \infty$ of the sequences whose nth terms are as follows:

1 $\dfrac{1}{2n - 1}$.

2 $\dfrac{3n}{4n + 7}$.

3 $\dfrac{6n}{2n^2 + 7}$.

4 $\dfrac{3n^2 - 6n + 2}{7n^2 - 2n + 9}$.

5 $\dfrac{\sin^2(n\pi/4)}{n}$.

6 $\dfrac{\cos(n\pi/3)}{n}$.

7 $\cosh nx - \sinh nx$, $x > 0$.

8 $\tanh nx$, $x > 0$.

9 $e^{-nx^2} \sin ax$, $x > 0$, $a \neq 0$.

10 $\left(\dfrac{2}{3}\right)^n$.

11 $\dfrac{n}{2^n}$.

12 $\dfrac{2 + n}{2^{n+1}}$.

13 $\dfrac{x^n - 1}{x^n + 1}$, $x > 0$.

14 $\dfrac{1}{2n + (-1)^n}$.

Investigate whether the sequences whose nth terms are given below converge, diverge or oscillate finitely.

15 $\dfrac{2^n}{n}$.

16 $n \sin(n\pi/6)$.

17 $\dfrac{1}{\ln(1 + n)}$.

18 $\dfrac{n^2}{2^{n/2}}.$ **19** $\sin^n (\pi/4).$ **20** $\sqrt{n} \sin^n (\pi/4).$

21 $\dfrac{\sin^n (\pi/4)}{\sqrt{n}}.$ **22** $\operatorname{sech} n.$ **23** $e^{-n} \cosh n.$

Investigate whether the sequences whose nth terms are given below converge, diverge or oscillate finitely, considering all values of x.

24 $\dfrac{\sin nx}{n}.$ **25** $\dfrac{x^n}{n}.$ **26** $nx^n.$

6.2 Convergence of series

Consider the infinite series

$$u_1 + u_2 + u_3 + \ldots + u_n + \ldots.$$

Let S_n denote the sum to n terms of the series, i.e.

$$S_n = u_1 + u_2 + \ldots + u_n = \sum_{r=1}^{n} u_r.$$

The terms S_n form a sequence of *partial sums*. If the partial sum S_n converges to a limit S as n tends to infinity, the infinite series is said to *converge* to the sum (or limit) S. If S_n does not tend to a limit as n tends to infinity, we say that the series *diverges*.

Note that we use the notation $\sum_{r=1}^{\infty} u_r$ to denote the limit, when it exists, of $\sum_{r=1}^{n} u_n$. When the series $u_1 + u_2 + \ldots + u_n + \ldots$ does not converge, then $\sum_{r=1}^{\infty} u_r$ does not exist. In such cases, it is convenient to use the abbreviation Σu_r to denote the *series*.

There are three important properties of convergent series.

(a) If the series Σu_n converges to the sum S, then the series Σku_n will converge to the sum kS, for any constant k.

For let $S_n = u_1 + u_2 + \ldots + u_n,$
and let $T_n = ku_1 + ku_2 + \ldots ku_n.$

Then as n tends to infinity S_n tends to the limit S, while T_n, which equals kS_n, tends to the limit kS.

This shows that the series Σku_n converges to the sum kS. A similar argument shows that, if the series Σu_n diverges, the series Σku_n will diverge (for $k \neq 0$).

(b) A necessary condition for the series

$$u_1 + u_2 + \ldots + u_n$$

to converge is that $u_n \to 0$ as n tends to infinity.

For when the partial sum S_n tends to the limit S as n tends to infinity, the partial sum S_{n-1} will also tend to S. Since $u_n = S_n - S_{n-1}$, this shows that u_n tends to zero as n tends to infinity, provided that the series converges. When u_n does not tend to zero as n tends to infinity, S_n cannot tend to a limit and the series cannot converge.

Note that the condition 'u_n tends to 0 as n tends to infinity' is not a sufficient condition for convergence. This is illustrated by example 2 below, in which $u_n = 1/n$.

Although $1/n$ tends to 0 as n tends to infinity the series $\Sigma\,(1/n)$ diverges.

(c) When a finite number of terms are omitted from a convergent series, the remaining series will be convergent.

Let the first m terms of a convergent series $\Sigma\,u_n$ be omitted. The remaining series will be

$$u_{m+1} + u_{m+2} + \ldots + u_{m+n} + \ldots.$$

The sum of the first n terms of the remaining series is

$$\sum_{r=1}^{m+n} u_r - \sum_{r=1}^{m} u_r.$$

As n tends to infinity, this tends to the limit $S - S_m$, where S is the sum of the series $\Sigma\,u_n$, and S_m is the sum of of the terms omitted. This shows that the remaining series is convergent.

If only some of the first m terms were omitted, the new series would converge to the sum $(S - \text{sum of terms omitted})$.

Example 1

By expressing $\dfrac{1}{r(r + 1)}$ in partial fractions, or otherwise, show that

$$\sum_{r=1}^{n} \frac{1}{r(r + 1)} = 1 - \frac{1}{n + 1}.$$

Deduce that the series $\displaystyle\sum_{r=1}^{n} \frac{1}{r(r + 1)}$ is convergent and find its sum.

By the usual methods for partial fractions

$$\frac{1}{r(r + 1)} \equiv \frac{1}{r} - \frac{1}{r + 1}.$$

Then
$$S_n = \sum_{r=1}^{n} \frac{1}{r(r+1)} = \frac{1}{1} - \frac{1}{2}$$
$$+ \frac{1}{2} - \frac{1}{3}$$
$$+ \ldots$$
$$+ \frac{1}{n} - \frac{1}{n+1}.$$

$$\therefore \qquad \sum_{r=1}^{n} \frac{1}{r(r+1)} = 1 - \frac{1}{n+1} \qquad \text{(by addition)}.$$

$$\therefore \qquad S_n = 1 - \frac{1}{n+1}$$

$$\Rightarrow \qquad \lim_{n \to \infty} S_n = 1.$$

Therefore the given series converges to unity.

Example 2
Show that the infinite series
$$1 + \frac{1}{2} + \frac{1}{3} + \ldots + \frac{1}{n} + \ldots$$
diverges.

$$S_{2^n} = \sum_{r=1}^{2^n} \frac{1}{r} = 1 + \frac{1}{2} + \left(\frac{1}{3} + \frac{1}{4}\right) + \left(\frac{1}{5} + \frac{1}{6} + \frac{1}{7} + \frac{1}{8}\right) + \ldots$$
$$+ \left(\frac{1}{2^{n-1}+1} + \ldots + \frac{1}{2^n}\right)$$
$$> 1 + \left(\tfrac{1}{2} + \tfrac{1}{2} + \ldots + \tfrac{1}{2}\right)$$
$$\Rightarrow \qquad S_{2^n} > 1 + n(\tfrac{1}{2}) = (n+2)/2.$$

Therefore, as $n \to \infty$, S_{2^n}, the sum of the first 2^n terms of the series, tends to ∞ and the series diverges.

Example 3
Simplify
$$[\sqrt{(x+1)} + \sqrt{x}][\sqrt{(x+1)} - \sqrt{x}]$$

and hence find S_n, where $S_n = \sum_{r=1}^{n} u_r$ and
$$u_r = \frac{1}{\sqrt{(r+1)} + \sqrt{r}}.$$

Show that $\Sigma\, u_r$ diverges.

$$[\sqrt{(x + 1)} + \sqrt{x}][\sqrt{(x + 1)} - \sqrt{x}] = (x + 1) - x = 1.$$

So $\quad u_r = \dfrac{1}{\sqrt{(r + 1)} + \sqrt{r}} = \sqrt{(r + 1)} - \sqrt{r}.$

$\therefore \qquad\qquad S_n \sum\limits_{r=1}^{n} u_r = \sqrt{(n + 1)} - 1$ (by addition as in Example 1).

As $n \to \infty$, $S_n \to \infty$, i.e. the series is divergent.

Example 4
Find the set of values of x for which the infinite geometric series

$$1 + \left(\frac{1 - 2x}{1 + x}\right) + \left(\frac{1 - 2x}{1 + x}\right)^2 + \ldots$$

is convergent.
   If, for suitable x, the sum to infinity of the series is denoted by $f(x)$, find an explicit expression for $f(x)$.

It was shown in *Advanced Mathematics 1* that, for $r \neq 1$,

$$\sum\limits_{k=1}^{n} ar^{k-1} = \frac{a}{1 - r} - \frac{ar^n}{1 - r},$$

and so the geometric series $\sum\limits_{k=1}^{\infty} ar^{k-1}$ converges to the sum $\dfrac{a}{1 - r}$ when $|r| < 1$.
For $|r| \geqslant 1$, the series does not converge.
   In the given example, the series converges when

$$\left|\frac{1 - 2x}{1 + x}\right| < 1$$

$\Rightarrow \qquad\qquad \left(\dfrac{1 - 2x}{1 + x}\right)^2 < 1$

$\Rightarrow \qquad\qquad \dfrac{(1 - 2x)^2 - (1 + x)^2}{(1 + x)^2} < 0$

$\Rightarrow \qquad\qquad \dfrac{3x(x - 2)}{(x + 1)^2} < 0$

$\Rightarrow \qquad\qquad 0 < x < 2.$

When x belongs to the set $\{x : 0 < x < 2\}$, the series converges to

$$\frac{1}{1 - \dfrac{1 - 2x}{1 + x}} = \frac{1 + x}{3x}$$

i.e. $\qquad\qquad f(x) = \dfrac{1 + x}{3x}.$

So, for example, when $x = 1$, the series is convergent with sum $\frac{2}{3}$. The series is then,

$$1 - \tfrac{1}{2} + \tfrac{1}{4} - \tfrac{1}{8} + \ldots,$$

i.e. a geometric progression with $a = 1$, $r = -\tfrac{1}{2}$ and sum $\dfrac{1}{1 + \frac{1}{2}} = \frac{2}{3}$.

Exercises 6.2

1 Show that

$$\sum_{r=1}^{n} \frac{1}{r(r + 2)} = \frac{3}{4} - \frac{2n + 3}{2(n + 1)(n + 2)},$$

and deduce that $\displaystyle\sum_{r=1}^{\infty} \frac{1}{r(r + 2)} = \frac{3}{4}$. [L]

2 Find constants A and B such that

$$\frac{Ar}{(r + 1)^2} + \frac{B(r + 1)}{(r + 2)^2} \equiv \frac{r^2 + r - 1}{(r + 1)^2(r + 2)^2}.$$

Hence find

$$\sum_{r=1}^{n} \frac{r^2 + r - 1}{(r + 1)^2(r + 2)^2} \text{ and } \sum_{r=1}^{\infty} \frac{r^2 + r - 1}{(r + 1)^2(r + 2)^2}. \qquad \text{[L]}$$

3 Defining arctan x to lie between $-\pi/2$ and $+\pi/2$, show that, for $r > 0$,

$$\arctan(2r + 1) - \arctan(2r - 1) = \arctan\left(\frac{1}{2r^2}\right).$$

Hence, or otherwise, find the value of

$$\sum_{r=1}^{n} \arctan\left(\frac{1}{2r^2}\right),$$

and deduce the value of

$$\sum_{r=1}^{\infty} \arctan\left(\frac{1}{2r^2}\right). \qquad \text{[L]}$$

4 If $r \neq 0$ or 1, show that

$$\sum_{k=1}^{n} ar^{k-1} = \frac{a(1 - r^n)}{1 - r}.$$

Deduce that, if $a \neq 0$, the series whose nth term is ar^{n-1} converges if and only if $|r| < 1$.

Investigate, for real x, the convergence of the series whose nth terms are
(a) $(1 + x)^n$, (b) $(1 + \ln x)^n$. [L]

6.3 Tests for convergence of series of positive terms

The comparison test

Consider the two series

$$1 + \frac{1}{2} + \frac{1}{2^2} + \ldots + \frac{1}{2^n} + \ldots$$

and $\quad 1 + \frac{1}{2} \cdot \left(\frac{1}{2}\right) + \frac{1}{3} \cdot \left(\frac{1}{2}\right)^2 + \left(\frac{1}{4}\right)\left(\frac{1}{2}\right)^3 + \ldots + \frac{1}{n+1}\left(\frac{1}{2}\right)^n + \ldots$

Let S_n be the sum of the first n terms of the first (geometric) series. Then S_n tends to the limit 2 as $n \to \infty$.

Let T_n be the sum of the first n terms of the second series. Each term of the second series (other than the first) is less than the corresponding term of the first series. Therefore $T_n < S_n$. Also $S_n < 2$.

$$\therefore \qquad\qquad 0 < T_n < S_n < 2.$$

As n increases, T_n increases since each term is positive. As $T_n < 2$, it can be proved that T_n tends to a limit, i.e. that the second series converges. (Here we have used a theorem of advanced mathematics that, if the terms of the sequence $\{u_n\}$ are such that $u_{n+1} > u_n$ and $u_n <$ a finite number for all n, then $u_n \to$ a finite limit as $n \to \infty$.)

(a) The *comparison test* for the *convergence* of a series of positive terms can be stated as follows:

Let $\qquad\qquad\qquad u_1 + u_2 + \ldots + u_n + \ldots$
and $\qquad\qquad\qquad v_1 + v_2 + \ldots + v_n + \ldots$

be two series of positive terms such that $v_n \leqslant u_n$ for all $n \in \mathbb{N}$. Then, if the series Σu_n converges, so does the series Σv_n.

Consider next the two series

$$1 + \frac{1}{2} + \frac{1}{3} + \frac{1}{4} + \ldots + \frac{1}{n} + \ldots$$

and $\qquad 1 + \frac{1}{\sqrt{2}} + \frac{1}{\sqrt{3}} + \frac{1}{\sqrt{4}} + \ldots + \frac{1}{\sqrt{n}} + \ldots$

Let S_n be the sum of the first n terms of the first series. Then S_n does not converge to a finite limit as $n \to \infty$. (See Example 2, page 194, where it was shown that this series diverges.)

Let T_n be the sum of the first n terms of the second series. Then, clearly, $T_n > S_n$ for $n > 1$. Therefore the second series diverges also.

(b) The *comparison test* for the *divergence* of a series of positive terms can be stated as follows:

If the series $u_1 + u_2 + u_3 + \ldots u_n + \ldots$ of positive terms is divergent and if $v_n \geqslant u_n$ for all $n \in \mathbb{N}$, then the series

$$v_1 + v_2 + v_3 + \ldots v_n + \ldots$$

is also divergent.

There is a useful alternative form of the comparison test for series of positive terms:

(c) If u_n/v_n tends to a finite non-zero limit as $n \to \infty$, then the series $\Sigma\, u_n$ and $\Sigma\, v_n$ are both convergent or both divergent.

A result frequently used in conjunction with the comparison test is:

The series $\Sigma \dfrac{1}{n^k}$ is convergent if $k > 1$ and divergent if $k \leqslant 1$.

The case $k = 1$ is discussed in Example 2 on page 194 and the result for $k < 1$ follows since $\dfrac{1}{n^k} > \dfrac{1}{n}$ if $k < 1$.

Consider $\displaystyle\sum_{n=1}^{N} \frac{1}{n^\alpha}$, where $\alpha > 1$ and $N < 2^n$.

$$S_N = \sum_{n=1}^{N} \frac{1}{n^\alpha} \leqslant 1 + (2^{-\alpha} + 3^{-\alpha}) + (4^{-\alpha} + \ldots + 7^{-\alpha}) + \ldots$$

$$+ \,[2^{(n-1)(-\alpha)} + \ldots + (2^n - 1)^{-\alpha}]$$

$$< 1 + 2(2^{-\alpha}) + 4(4^{-\alpha}) + \ldots + 2^{n-1}(2^{-(n-1)\alpha})$$

$$\Rightarrow \quad S_N = 1 + 2^{1-\alpha} + 4^{1-\alpha} + \ldots + (2^{n-1})^{1-\alpha}.$$

This last expression is a geometrical progression with common ratio r where, since $\alpha > 1$, $0 < r < 1$. So the given series has a sum lying between 0 and $\dfrac{1}{1 - 2^{1-\alpha}}$, and is thus convergent.

Example 1

Show that the series whose nth term is $\dfrac{n}{\sqrt{(n^3 + 1)}}$ diverges.

$$\frac{n}{\sqrt{(n^3 + 1)}} = \frac{1}{\sqrt{\left(1 + \dfrac{1}{n^3}\right)}} \cdot \frac{1}{\sqrt{n}} > \frac{1}{2\sqrt{n}}, \qquad \text{for } n > 1.$$

But the series with nth term $1/(2\sqrt{n})$ diverges.
Therefore the given series diverges also.

Example 2

Show that the series whose nth term is $\dfrac{1}{\sqrt{(n^3 + 1)}}$ converges.

Let $u_n = 1/\sqrt{(n^3 + 1)}$, $v_n = 1/n^{3/2}$.

Then $\qquad \dfrac{u_n}{v_n} = \dfrac{n^{3/2}}{\sqrt{(n^3 + 1)}} = \dfrac{n^{3/2}}{n^{3/2}\sqrt{(1 + n^{-3})}} = \dfrac{1}{\sqrt{(1 + n^{-3})}}.$

$$\therefore \qquad\qquad \lim_{n \to \infty} \left(\frac{u_n}{v_n}\right) = 1.$$

Therefore the series $\Sigma\, u_n$ and $\Sigma\, v_n$ are either both convergent or both divergent. But $\Sigma\, v_n$ converges and therefore $\Sigma\, u_n$ converges also.

The ratio test for series of positive terms
Let k be a (positive) number such that $0 < k < 1$. If $u_{n+1}/u_n < k$ for *all sufficiently large* n, then the series of positive terms $u_1 + u_2 + \ldots + u_n + \ldots$ is convergent.

Proof
Let $u_{n+1}/u_n < k$ for $n \geqslant N$, where N is constant, i.e. $u_{n+1} < ku_n$ for $n \geqslant N$. Then

$$u_N + u_{N+1} + u_{N+2} + \ldots + u_{N+r} < u_N (1 + k + k^2 + \ldots + k^r).$$

Since the geometric series $1 + k + k^2 + \ldots$ converges, so does the series $u_N + u_{N+1} + u_{N+2} + \ldots$.

Since the sum $u_1 + u_2 + \ldots + u_{N-1}$ is finite, it follows that the series

$$\sum_{n=1}^{\infty} u_n \text{ is convergent.}$$

(Note that the convergence or divergence of a series is unaffected by the removal of a *finite* number of terms.)

Conversely, if $\lim_{n \to \infty} (u_{n+1}/u_n) = p$ where $p > 1$, then the series $\Sigma\, u_n$ diverges.

It is important to observe that no conclusion can be reached yet on the case when $u_{n+1}/u_n < 1$ for all finite values of n. This condition holds for each of the series

$$1 + \frac{1}{2} + \frac{1}{3} + \frac{1}{4} + \ldots + \frac{1}{n} + \ldots,$$

$$1 + \frac{1}{2^2} + \frac{1}{3^2} + \frac{1}{4^2} + \ldots + \frac{1}{n^2} + \ldots,$$

yet the first diverges and the second converges.

Example 1

Show that the series

$$\frac{3}{1} + \frac{3^2}{2!} + \frac{3^3}{3!} + \frac{3^4}{4!} + \ldots + \frac{3^n}{n!} + \ldots$$

converges.

Here $u_n = 3^n/n!$, $u_{n+1} = 3^{n+1}/(n + 1)!$ and so

$$\frac{u_{n+1}}{u_n} = \frac{3}{n + 1} \quad \Rightarrow \quad \lim_{n \to \infty} \left(\frac{u_{n+1}}{u_n}\right) = 0.$$

Therefore, by the ratio test, the series $\Sigma\, u_n$ converges.

Example 2

Show that the series

$$\frac{2}{3} + 2^3\left(\frac{2}{3}\right)^2 + 3^3\left(\frac{2}{3}\right)^3 + 4^3\left(\frac{2}{3}\right)^4 + \ldots + n^3\left(\frac{2}{3}\right)^n + \ldots$$

converges.

Taking $u_n = n^3\left(\dfrac{2}{3}\right)^n$,

$$\lim_{n \to \infty} \left(\frac{u_{n+1}}{u_n}\right) = \lim_{n \to \infty} \left(\frac{(n + 1)^3\left(\dfrac{2}{3}\right)^{n+1}}{n^3\left(\dfrac{2}{3}\right)^n}\right)$$

$$= \lim_{n \to \infty} \left[\frac{2}{3}\left(1 + \frac{1}{n}\right)^3\right] = \frac{2}{3} < 1.$$

Therefore, by the ratio test, the series converges.

When the terms of a series involve powers of x, the ratio test is very useful.

Example 3

Find the set of values of x for which the series with nth term x^{2n}/n converges.

Here $u_n = x^{2n}/n$ and

$$\lim_{n \to \infty} \left(\frac{u_{n+1}}{u_n}\right) = \lim_{n \to \infty} \left(\frac{x^{2n+2}/(n + 1)}{x^{2n}/n}\right)$$

$$= \lim_{n \to \infty} [x^2/(1 + 1/n)] = x^2.$$

Hence, by the ratio test, if $x^2 < 1$, i.e. $|x| < 1$, the series converges. Conversely, if $x^2 > 1$, i.e. $|x| > 1$, the series diverges.

Finally, when $x^2 = 1$, i.e. $x = \pm 1$, the series becomes $1 + \tfrac{1}{2} + \tfrac{1}{3} + \ldots + \dfrac{1}{n} + \ldots$, which diverges.

Example 4

Given that $x \geqslant 0$, test for divergence or convergence the series whose nth terms are

(a) $\dfrac{x^n}{n^2}$;   (b) $n(n + 1)\left(\dfrac{x}{a}\right)^n$, where $a > 0$;   (c) $\dfrac{x^n}{n!}$;   (d) $n!x^n$.

(a) When $u_n = x^n/n^2$, then

$$\lim_{n \to \infty} \left(\frac{u_{n+1}}{u_n}\right) = \lim_{n \to \infty} \left(\frac{n^2 x^{n+1}}{(n + 1)^2 x^n}\right)$$

$$= \lim_{n \to \infty} \left(\frac{x}{(1 + 1/n)^2}\right) = x.$$

Therefore, by the ratio test, the series converges when $0 \leqslant x < 1$, and diverges when $x > 1$.

When $x = 1$ the series becomes $\sum \dfrac{1}{n^2}$, which converges.

(b) When $u_n = n(n + 1)(x/a)^n$, then

$$\lim_{n \to \infty} \left(\frac{u_{n+1}}{u_n}\right) = \lim_{n \to \infty} \left(\frac{(n + 1)(n + 2)(x/a)^{n+1}}{n(n + 1)(x/a)^n}\right)$$

$$= \lim_{n \to \infty} \left[\left(1 + \frac{2}{n}\right)\left(\frac{x}{a}\right)\right] = \frac{x}{a}.$$

Hence, by the ratio test, the series converges when $0 \leqslant x < a$ and diverges when $x > a$.

When $x = a$, $u_n = n(n + 1)$ and the series diverges (by comparison with $\sum n^2$).

(c) When $u_n = x^n/n!$, then

$$\lim_{n \to \infty} \left(\frac{u_{n+1}}{u_n}\right) = \lim_{n \to \infty} \left(\frac{n! x^{n+1}}{(n + 1)! x^n}\right)$$

$$= \lim_{n \to \infty} \left(\frac{x}{n + 1}\right) = 0 \text{ for all finite } x.$$

The series therefore converges for all $x \geqslant 0$.

(d) When $u_n = n!x^n$, then

$$\lim_{n \to \infty} \left(\frac{u_{n+1}}{u_n} \right) = \lim_{n \to \infty} \left(\frac{(n+1)!x^{n+1}}{n!x^n} \right)$$

$$= \lim_{n \to \infty} (n+1)x.$$

Unless $x = 0$, this limit does not exist and so the series diverges for all positive x. When $x = 0$, every term of the series is zero.

Exercises 6.3

1 Test whether the series $\Sigma\, u_n$ converges when u_n equals

(a) $3^n/2^{2n}$; (b) $2^n/n^2$; (c) $2^n/n^{10}$; (d) $n^{10}/10^n$;

(e) $n^n/n!$; (f) $(n!)^2/(2n)!$; (g) $2^n(n!)/n^n$; (h) $3^n(n!)/n^n$;

(i) $n/(n^2 + 1)$; (j) ne^{-n}; (k) $n/(2n + 1)^3$; (l) $(n/3)^n/n!$

2 Given that $x > 0$, test for convergence or divergence the series whose nth terms are

(a) $n(2x)^n$; (b) $(1 + \ln x)^n$; (c) $nx^n/[(n + 1)(n + 2)]$;

(d) $(\cos nx)/2^n$; (e) $(\cosh nx)/2^n$; (f) n/x^n;

(g) $x^n e^{-nx}$; (h) $x^n/(1 + x^{2n})$; (i) $(\sin^{2n} x)/n$.

6.4 Convergence of integrals

Thus far, when integrating $\int_a^b f(x)\, dx$ we have assumed that $f(x)$ remains finite in the closed interval $[a, b]$ and that a, b are finite.

We now define the *improper integral*

$$\int_a^\infty f(x)\, dx = \lim_{X \to \infty} \int_a^X f(x)\, dx,$$

and we say that, if this limit exists, then the integral $\int_a^\infty f(x)\, dx$ *converges*.

Example 1

Investigate the convergence of $\int_0^\infty \dfrac{1}{(1 + x)^{4/3}}\, dx$.

$$\int_0^\infty \frac{1}{(1 + x)^{4/3}}\, dx = \lim_{X \to \infty} \int_0^X \frac{1}{(1 + x)^{4/3}}\, dx$$

$$= \lim_{X \to \infty} \left[\frac{-3}{(1 + x)^{1/3}} \right]_0^X$$

$$= \lim_{X \to \infty} \left[3 - \frac{3}{(1 + X)^{1/3}} \right] = 3.$$

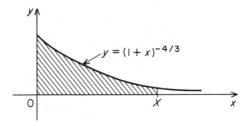

Fig. 6.1

The given integral therefore converges to the value 3. This result is illustrated in Fig. 6.1, in which the area of the shaded region tends to the limit 3 as $X \to \infty$.

Example 2

Investigate the convergence of $\displaystyle\int_1^\infty \frac{1}{4 + x}\,\mathrm{d}x$.

We first consider

$$\int_1^X \frac{1}{4 + x}\,\mathrm{d}x = \left[\; \ln(4 + x) \;\right]_1^X$$

$$= \ln(4 + X) - \ln 5.$$

But, as $X \to \infty$, $\ln(4 + X) \to \infty$ and so the integral is not convergent.

In this case the area of the shaded region shown in Fig. 6.2 does not tend to a limit as $X \to \infty$.

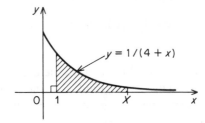

Fig. 6.2

Example 3

Show that the integral $\displaystyle\int_0^\infty \frac{1}{(a^2 + x^2)}\,\mathrm{d}x$, where $a > 0$, converges to the value $\pi/(2a)$.

By definition,

$$\int_0^\infty \frac{1}{a^2 + x^2} \, dx = \lim_{X \to \infty} \int_0^X \frac{1}{a^2 + x^2} \, dx$$

$$= \lim_{X \to \infty} \left[\frac{1}{a} \tan^{-1} \left(\frac{x}{a} \right) \right]_0^X$$

$$= \lim_{X \to \infty} \left(\frac{1}{a} \tan^{-1} \left(\frac{X}{a} \right) - 0 \right)$$

$$= \pi/(2a).$$

Example 4

Show that the integral $I = \displaystyle\int_0^\infty \frac{x}{(x^2 + 4)(x^2 + 9)} \, dx$, converges and find its value.

Since, by partial fractions,

$$\frac{x}{(x^2 + 4)(x^2 + 9)} \equiv \frac{1}{5} \left(\frac{x}{x^2 + 4} - \frac{x}{x^2 + 9} \right),$$

$$I = \lim_{X \to \infty} \frac{1}{5} \int_0^X \left(\frac{x}{x^2 + 4} - \frac{x}{x^2 + 9} \right) dx$$

$$= \lim_{X \to \infty} \frac{1}{10} \left[\ln (x^2 + 4) - \ln (x^2 + 9) \right]_0^X$$

$$= \lim_{X \to \infty} \frac{1}{10} \left[\ln \left(\frac{X^2 + 4}{X^2 + 9} \right) + \ln (9/4) \right].$$

But
$$\lim_{X \to \infty} \ln \left(\frac{X^2 + 4}{X^2 + 9} \right) = \lim_{X \to \infty} \ln \left(\frac{1 + 4/X^2}{1 + 9/X^2} \right) = \ln 1 = 0.$$

Therefore I converges to the value $\frac{1}{10} \ln (9/4)$.

Example 5

Find the area of the finite region bounded by the curve $y = x/(1 + x^2)$, **the** coordinate axes and the line $x = a$. Find also the volume of the solid of revolution generated when this region is rotated through 2π about Ox. Investigate what happens when $a \to \infty$.

The area is A, where

$$A = \int_0^a \frac{x}{1 + x^2} \, dx$$

$$= \left[\tfrac{1}{2} \ln (1 + x^2) \right]_0^a,$$

i.e.,
$$A = \tfrac{1}{2} \ln (1 + a^2).$$

The volume of the solid of revolution is V, where

$$V = \pi \int_0^a \frac{x^2}{(1 + x^2)^2} \, dx.$$

Making the substitution $x = \tan \theta$,

$$V = \pi \int_0^{\tan^{-1} a} \frac{\tan^2 \theta}{\sec^4 \theta} \sec^2 \theta \, d\theta$$

$$= \pi \int_0^{\tan^{-1} a} \sin^2 \theta \, d\theta$$

$$= \frac{\pi}{2} \int_0^{\tan^{-1} a} (1 - \cos 2\theta) \, d\theta$$

$$= \frac{\pi}{2} \left[\theta - \sin \theta \cos \theta \right]_0^{\tan^{-1} a}$$

$$\Rightarrow \qquad V = \frac{\pi}{2} \left(\tan^{-1} a - \frac{a}{1 + a^2} \right).$$

As $a \to \infty$, $A \to \infty$ and so the area is not finite. However, as $a \to \infty$, $V \to \frac{\pi}{2} \cdot \frac{\pi}{2} = \frac{\pi^2}{4}$ and so the volume of the solid of revolution is finite.

In some problems it is necessary to use the results:

(a) $\lim_{x \to \infty} (x^n \, e^{-x}) = 0,$ for all n;

(b) $\lim_{x \to \infty} \left(\frac{\ln x}{x^\alpha} \right) = 0,$ for all $\alpha > 0$;

(c) $\lim_{x \to 0} (x^\beta \ln x) = 0,$ for all $\beta > 0$ and $x > 0$.

The proofs of these results are left as an exercise for the reader. (*Hint.* In (a) use

$$e^{-x} = \frac{1}{e^x} = \frac{1}{1 + x + \ldots + x^n/n! + x^{n+1}/(n + 1)! + \ldots}.$$

Results (b) and (c) follow from (a) by a substitution.

Exercises 6.4(i)
Investigate the convergence of each of the integrals and, where possible, evaluate the integral.

1   $\int_1^\infty \frac{1}{x^2} \, dx.$ 2 $\int_1^\infty \frac{1}{x} \, dx.$

3   $\int_1^\infty \frac{1}{\sqrt{x}} \, dx.$ 4 $\int_0^\infty \frac{x}{(1 + x)(1 + x^2)} \, dx.$

5 $\displaystyle\int_0^\infty xe^{-x}\,dx.$ **6** $\displaystyle\int_1^\infty \frac{1}{x\sqrt{(1+x^2)}}\,dx.$

7 $\displaystyle\int_0^\infty \operatorname{sech} x\,dx.$ **8** $\displaystyle\int_0^\infty \operatorname{sech}^2 x\,dx.$

9 $\displaystyle\int_1^\infty \ln x\,dx.$ **10** $\displaystyle\int_1^\infty \frac{\ln x}{x}\,dx.$

11 $\displaystyle\int_1^\infty \frac{\ln x}{x^2}\,dx.$ **12** $\displaystyle\int_0^\infty e^{-x}\cos x\,dx.$

In some cases the integrand f(x) in $\displaystyle\int_a^b f(x)\,dx$ has a singularity, i.e. is not defined, at a or b. For example, $\displaystyle\int_0^1 \frac{1}{\sqrt{x}}\,dx$, where the integrand $1/\sqrt{x}$ tends to infinity as $x \to 0$.

When f(x) has a singularity at $x = a$ we define

$$\int_a^b f(x)\,dx = \lim_{\varepsilon \to 0+} \int_{a+\varepsilon}^b f(x)\,dx,$$

where $\varepsilon \to 0+$, means that ε tends to zero through positive values. If this limit exists then we say that the integral converges. Then

$$\int_0^1 \frac{1}{\sqrt{x}}\,dx = \lim_{\varepsilon \to 0+} \int_\varepsilon^1 \frac{1}{\sqrt{x}}\,dx$$

$$= \lim_{\varepsilon \to 0+} \left[2\sqrt{x}\right]_\varepsilon^1 = \lim_{\varepsilon \to 0+} (2 - 2\sqrt{\varepsilon}) = 2.$$

The integral therefore converges to the value 2. In this case the area of the shaded region shown in Fig. 6.3 tends to the limit 2 as $\varepsilon \to 0$.

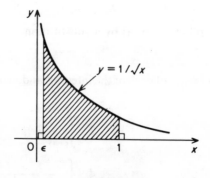

Fig. 6.3

Example 1

Show that the integral $\displaystyle\int_0^1 x \ln x \, dx$ converges to the limit $-\tfrac{1}{4}$.

By definition,

$$\int_0^1 x \ln x \, dx = \lim_{\varepsilon \to 0+} \int_\varepsilon^1 x \ln x \, dx$$

$$= \lim_{\varepsilon \to 0+} \left[\frac{x^2}{2} \ln x - \frac{x^2}{4} \right]_\varepsilon^1$$

$$= \lim_{\varepsilon \to 0+} \left(-\frac{1}{4} - \frac{\varepsilon^2}{2} \ln \varepsilon + \frac{\varepsilon^2}{4} \right)$$

$$= -\tfrac{1}{4} \quad \text{since} \quad \varepsilon^2 \to 0 \quad \text{and} \quad \varepsilon^2 \ln \varepsilon \to 0 \quad \text{as} \quad \varepsilon \to 0+.$$

Example 2

Show that the integral $\displaystyle\int_0^2 \sqrt{\left(\frac{x}{2-x}\right)} \, dx$ converges to the limit π.

In this case the integrand has a singularity at the upper limit and we define

$$\int_0^2 \sqrt{\left(\frac{x}{2-x}\right)} \, dx = \lim_{\varepsilon \to 0+} \int_0^{2-\varepsilon} \sqrt{\left(\frac{x}{2-x}\right)} \, dx.$$

The substitution $x = 2 \sin^2 \theta$ gives

$$\int \sqrt{\left(\frac{x}{2-x}\right)} \, dx = 4 \int \sin^2 \theta \, d\theta$$

$$= 2 \int (1 - \cos 2\theta) \, d\theta$$
$$= 2(\theta - \sin \theta \cos \theta)$$

$$= \left[2 \sin^{-1} \sqrt{\left(\frac{x}{2}\right)} - \sqrt{[x(2-x)]} \right].$$

$$\therefore \quad \lim_{\varepsilon \to 0+} \int_0^{2-\varepsilon} \sqrt{\left(\frac{x}{2-x}\right)} \, dx$$

$$= \lim_{\varepsilon \to 0+} \left[2 \sin^{-1} \sqrt{\left(\frac{2-\varepsilon}{2}\right)} - \sqrt{[\varepsilon(2-\varepsilon)]} \right]$$

$$= 2\pi/2 = \pi.$$

Example 3
Investigate the convergence of

$$\int_1^4 \frac{1}{4-x} \, dx.$$

We first consider

$$\lim_{\varepsilon \to 0} \int_1^{4-\varepsilon} \frac{1}{4-x}\,dx = \lim_{\varepsilon \to 0+} \left[-\ln(4-x) \right]_1^{4-\varepsilon}$$

$$= \lim_{\varepsilon \to 0+} (\ln 3 - \ln \varepsilon).$$

But as $\varepsilon \to 0+$, $\ln \varepsilon \to -\infty$ and so the integral does not converge.

Exercises 6.4(ii)
Investigate the convergence of each of the following integrals and, where possible, evaluate the integral.

1 $\displaystyle\int_0^1 \frac{1}{x}\,dx.$ **2** $\displaystyle\int_0^2 \frac{1}{\sqrt{(2-x)}}\,dx.$ **3** $\displaystyle\int_0^1 \frac{e^x}{e^x - 1}\,dx.$

4 $\displaystyle\int_0^1 \frac{1}{\sqrt{(1-x^2)}}\,dx.$ **5** $\displaystyle\int_1^2 \frac{1}{\sqrt{(x^2-1)}}\,dx.$ **6** $\displaystyle\int_0^1 \ln x\,dx.$

7 $\displaystyle\int_2^3 \frac{x}{\sqrt{(3-x)}}\,dx.$ **8** $\displaystyle\int_2^3 \frac{x}{3-x}\,dx.$ **9** $\displaystyle\int_{-1}^2 \sqrt{\left(\frac{1+x}{2-x}\right)}\,dx.$

10 $\displaystyle\int_0^1 \frac{1}{(1+x)\sqrt{(1-x^2)}}\,dx.$

When $f(x) \to \infty$ both as $x \to a+0$ and as $x \to b-0$ then $\displaystyle\int_a^b f(x)\,dx$ converges to the limit L if

$$\lim_{\substack{\varepsilon \to 0 \\ \varepsilon' \to 0}} \int_{a+\varepsilon'}^{b-\varepsilon} f(x)\,dx = L,$$

where ε and ε' are both positive and tend to zero independently.

Example 3

$$\int_0^1 \frac{1}{\sqrt{(x-x^2)}}\,dx = \lim_{\substack{\varepsilon \to 0 \\ \varepsilon' \to 0}} \int_{\varepsilon'}^{1-\varepsilon} \frac{1}{\sqrt{[\frac14 - (x - \frac12)^2]}}\,dx$$

$$= \lim_{\substack{\varepsilon \to 0 \\ \varepsilon' \to 0}} \left[\sin^{-1}(2x - 1) \right]_{\varepsilon'}^{1-\varepsilon}$$

$$= \lim_{\substack{\varepsilon \to 0 \\ \varepsilon' \to 0}} [\sin^{-1}(1 - 2\varepsilon) - \sin^{-1}(2\varepsilon' - 1)]$$

$$= \pi/2 - (-\pi/2) = \pi.$$

The integral, therefore, converges to π.

Suppose now that $f(x) \to \infty$ as $x \to c$, where $a < c < b$. Then $\displaystyle\int_a^b f(x)\,dx$ is said to converge to the limit L if

$$\lim_{\varepsilon \to 0} \int_a^{c-\varepsilon} f(x)\,dx + \lim_{\varepsilon' \to 0} \int_{c+\varepsilon'}^b f(x)\,dx = L,$$

where ε and ε' are positive and tend to zero independently.

Example 4

$$
\int_{-1}^2 \frac{1}{x^{1/5}}\,dx = \lim_{\varepsilon \to 0} \int_{-1}^{-\varepsilon} \frac{1}{x^{1/5}}\,dx + \lim_{\varepsilon' \to 0} \int_{\varepsilon'}^2 \frac{1}{x^{1/5}}\,dx
$$

$$
= \lim_{\varepsilon \to 0} \left[\tfrac{5}{4} x^{4/5}\right]_{-1}^{-\varepsilon} + \lim_{\varepsilon' \to 0} \left[\tfrac{5}{4} x^{4/5}\right]_{\varepsilon'}^2
$$

$$
= \lim_{\varepsilon \to 0} \left[\tfrac{5}{4}(\varepsilon^{4/5} - 1)\right] + \lim_{\varepsilon' \to 0} \left[\tfrac{5}{4}(2^{4/5} - \varepsilon'^{4/5})\right]
$$

$$
= \tfrac{5}{4}(2^{4/5} - 1).
$$

The integral therefore converges.

Exercises 6.4(iii)

For each of **1–12**, determine whether the integral exists, and, if it exists, obtain its value.

1 $\displaystyle\int_1^\infty \frac{1}{(x-1)}\,dx.$ **2** $\displaystyle\int_1^\infty \frac{1}{(x-1)^2}\,dx.$ **3** $\displaystyle\int_1^2 \frac{1}{x\sqrt{(x-1)}}\,dx.$

4 $\displaystyle\int_1^\infty e^{-ax} \sin bx\,dx,$ where $a > 0.$

5 $\displaystyle\int_0^1 x^2 \ln x\,dx.$ **6** $\displaystyle\int_0^1 \sqrt{\left(\frac{1+x}{1-x}\right)}\,dx.$

7 $\displaystyle\int_0^1 \frac{1}{(1-x)\sqrt{(1-x^2)}}\,dx.$ **8** $\displaystyle\int_2^3 \sqrt{\left(\frac{x-2}{3-x}\right)}\,dx.$

9 $\displaystyle\int_{-1}^1 \frac{1}{\sqrt{|x|}}\,dx.$ **10** $\displaystyle\int_{-1}^1 \frac{1}{x^2}\,dx.$

11 $\displaystyle\int_1^\infty \frac{x \tan^{-1} x}{(1+x^2)^2}\,dx.$ **12** $\displaystyle\int_1^\infty \frac{1}{\sinh x}\,dx.$

13 (a) Show that $\displaystyle\int_a^b \frac{x}{\sqrt{[(b-x)(x-a)]}}\,dx = \frac{\pi(a+b)}{2},$ where $a < b.$

(b) Prove that each of the integrals

(i) $\displaystyle\int_1^\infty \frac{1}{x(x+1)}\,dx,$ (ii) $\displaystyle\int_{-1}^{31} \frac{1}{(x+1)^{1/5}}\,dx$

is convergent, and find their values.

14 Evaluate the integral

$$\int_{-1}^1 \frac{1}{(2-x)\sqrt{(1-x^2)}}\,dx.$$

6.5 Maclaurin's and Taylor's series

Maclaurin's series

In *Advanced Mathematics 1* we showed that, subject to certain conditions, a Maclaurin series could be found for f(x) in the form

$$f(x) = f(0) + xf'(0) + \frac{x^2}{2!}f''(0) + \ldots + \frac{x^r}{r!}f^{(r)}(0) + \ldots.$$

Particular series obtainable by this method are:

(1) *The binomial series*

$$(1+x)^n = 1 + \binom{n}{1}x + \binom{n}{2}x^2 + \ldots + \binom{n}{r}x^r + \ldots,$$

where $\displaystyle\binom{n}{r} = \frac{n(n-1)\ldots(n-r+1)}{r!}$. It can be shown that this series is convergent for all values of n provided that $|x| < 1$.

(2) *The exponential series*

$$e^x = \exp x = 1 + x + \frac{x^2}{2!} + \ldots + \frac{x^n}{n!} + \ldots.$$

This series is convergent for all x.

(3) *The logarithmic series*

$$\ln(1+x) = x - \frac{x^2}{2} + \frac{x^3}{3} - \ldots + \frac{(-1)^{n-1}x^n}{n} + \ldots.$$

This series is convergent for $-1 < x \leqslant 1$.

(4) *The sine and cosine series*

$$\sin x = x - \frac{x^3}{3!} + \frac{x^5}{5!} - \ldots + \frac{(-1)^{n-1}x^{2n-1}}{(2n-1)!} + \ldots,$$

$$\cos x = 1 - \frac{x^2}{2!} + \frac{x^4}{4!} - \ldots + \frac{(-1)^{n-1}x^{2n-2}}{(2n-2)!} + \ldots.$$

Both of these series are convergent for all x.

(5)   *The sinh and cosh series*

$$\sinh x = x + \frac{x^3}{3!} + \frac{x^5}{5!} + \ldots + \frac{x^{2r-1}}{(2r-2)!} + \ldots$$

$$\cosh x = 1 + \frac{x^2}{2!} + \frac{x^4}{4!} + \ldots + \frac{x^{2r-2}}{(2r-2)!} + \ldots.$$

Both of these series are convergent for all x.
These series can be used in evaluating certain limits as $x \to 0$.

Example 1

Show that $\lim\limits_{x \to 0} \left(\dfrac{e^x - \cos 2x - x}{x^2} \right) = \dfrac{5}{2}$.

Using the series for e^x and $\cos 2x$, we have

$$\lim_{x \to 0} \left(\frac{\left(1 + x + \frac{x^2}{2!} + \frac{x^3}{3!} + \ldots \right) - \left(1 - \frac{(2x)^2}{2!} + \ldots \right) - x}{x^2} \right)$$

$$= \lim_{x \to 0} \left(\frac{\frac{5x^2}{2} + \text{terms involving higher powers of } x}{x^2} \right)$$

$$= \lim_{x \to 0} \left(\tfrac{5}{2} + \text{terms involving powers of } x \right)$$

$$= \tfrac{5}{2}.$$

Example 2

$$\lim_{x \to 0} \left(\frac{2 \cosh x + 2 \cos x - 4}{x^4} \right)$$

$$= \lim_{x \to 0} \left(\frac{2 \left(1 + \frac{x^2}{2!} + \frac{x^4}{4!} + \ldots \right) + 2 \left(1 - \frac{x^2}{2!} + \frac{x^4}{4!} - \ldots \right) - 4}{x^4} \right)$$

$$= \lim_{x \to 0} \left(\frac{x^4}{6} + \text{terms involving higher powers} \right) \bigg/ x^4$$

$$= \tfrac{1}{6}.$$

Example 3
The first three terms in the expansion of

$$\frac{\sqrt{(x^2 + 4x^3)} - x}{1 - \cos 2x}$$

are $A + Bx + Cx^2$. Find the constants A, B, C.

We can find the constants by equating coefficients in the identity

$$\sqrt{(x^2 + 4x^3)} - x \equiv (1 - \cos 2x)(A + Bx + Cx^2).$$

Left-hand side $= x\sqrt{(1 + 4x)} - x$
$= x(1 + 2x - 2x^2 + 4x^3 + \ldots) - x$
$= 2x^2 - 2x^3 + 4x^4 + \ldots.$

Right-hand side $= (2x^2 - \frac{2}{3}x^4 + \ldots)(A + Bx + Cx^2)$
$= 2Ax^2 + 2Bx^3 + 2Cx^4 - \frac{2}{3}Ax^4 + \ldots.$

By equating coefficients,

$$A = 1, B = -1, \quad \text{and} \quad 2C - \tfrac{2}{3}A = 4 \quad \Rightarrow \quad c = 7/3.$$

Example 4

Given that x is so small that its fourth and higher powers may be neglected compared with unity, show that

$$\frac{\cosh x - \cos 3x}{x \sin x} = 5\left(1 - \frac{x^2}{2}\right).$$

Using the standard series, we have

$$\frac{\cosh x - \cos 3x}{x \sin x} = \frac{\left(1 + \dfrac{x^2}{2} + \dfrac{x^4}{24} + \ldots\right) - \left(1 - \dfrac{9x^2}{2} + \dfrac{27x^4}{8} - \ldots\right)}{x\left(x - \dfrac{x^3}{6} + \ldots\right)}$$

$$= \frac{5x^2 - \dfrac{10x^4}{3} + \text{terms in } x^6 \text{ etc.}}{x^2\left(1 - \dfrac{x^2}{6}\right) + \text{terms in } x^6}$$

$$= \left(5 - \frac{10}{3}x^2\right)\left(1 + \frac{x^2}{6}\right) + \text{terms in } x^4$$

$$= 5 - \tfrac{5}{2}x^2 + \text{terms in } x^4,$$

which is the required result.

Notes

(1) Since the given expression is of the form f(x), where f(x) is even, then its expansion will contain only even powers of x. (Remember that a constant is an even function.) This is a useful check on power series expansions.

(2) Since the denominator $x \sin x$ is approximately x^2 when x is small, we must retain terms involving x^4 in the early part of the calculation, i.e. we have to

expand both the numerator and denominator up to and including the term in x^4.

Example 5
Given that x is so small that its fifth and higher powers may be neglected, show that

$$\ln (1 + \sin x) = x - \frac{x^2}{2} + \frac{x^3}{6} - \frac{x^4}{12}.$$

Find also the corresponding approximation for $\ln (1 - \sin x)$ and hence, or otherwise, show that

$$\ln (\cos x) \approx -\frac{x^2}{2} - \frac{x^4}{12}.$$

You are advised to try this problem using Maclaurin's series, but the differentiation can be very tedious. However, here we shall use the series for $\sin x$, neglecting powers of x higher than the fourth whenever they occur. (Check that this has been done.) Once a term in x^5 has been neglected, all terms involving x^5 and higher powers should be neglected,

$$\sin x = x - \frac{x^3}{6} + \text{higher powers}$$

$$\Rightarrow \quad \ln(1 + \sin x) = \ln\left[1 + \left(x - \frac{x^3}{6} \right) + \text{higher powers} \right]$$

$$= \left(x - \frac{x^3}{6} \right) - \frac{1}{2}\left(x - \frac{x^3}{6} \right)^2 + \frac{1}{3}\left(x - \frac{x^3}{6} \right)^3 - \frac{1}{4}\left(x - \frac{x^3}{6} \right)^4$$
$$+ \text{higher powers}$$

$$= x - \frac{x^3}{6} - \frac{1}{2}\left(x^2 - \frac{x^4}{3} \right) + \frac{x^3}{3} - \frac{x^4}{4} + \text{higher powers}$$

$$= x - \frac{x^2}{2} + \frac{x^3}{6} - \frac{x^4}{12} + \text{higher powers}.$$

The corresponding approximation for $\ln (1 - \sin x)$, obtained by writing $-x$ for x in the above, is

$$\ln (1 - \sin x) = -x - \frac{x^2}{2} - \frac{x^3}{6} - \frac{x^4}{12} + \text{higher powers}.$$

Then $\quad \ln (\cos x) = \frac{1}{2} \ln (\cos^2 x) = \frac{1}{2} \ln [(1 + \sin x)(1 - \sin x)]$
$$= \tfrac{1}{2}\ln (1 + \sin x) + \tfrac{1}{2} \ln (1 - \sin x)$$

$$= -\frac{x^2}{2} - \frac{x^4}{12} \text{ neglecting terms involving higher powers of } x.$$

An 'or otherwise' approach is

$$\ln(\cos x) = \ln\left(1 - \frac{x^2}{2} + \frac{x^4}{24} - \cdots\right)$$

$$= -\left(\frac{x^2}{2} - \frac{x^4}{24}\right) - \frac{1}{2}\left(\frac{x^2}{2} - \frac{x^4}{24}\right)^2 - \cdots$$

$$= -\frac{x^2}{2} + \frac{x^4}{24} - \frac{x^4}{8} + \text{higher powers of } x$$

$$= -\frac{x^2}{2} - \frac{x^4}{12} + \text{higher powers of } x.$$

Another 'or otherwise' approach is to integrate both sides of the equation

$$\tan x = x + \frac{x^3}{3} + \cdots$$

with respect to x and determine the constant of integration by putting $x = 0$.

Example 6
Use Maclaurin's series to express $\tan x$ as a series of ascending powers of x up to and including the term in x^5.

Let $f(x) = \tan x$, and let s and t denote $\sec x$ and $\tan x$ respectively.

$$
\begin{array}{llll}
f(x) &= t & & \Rightarrow f(0) = 0. \\
f'(x) &= s^2 & = 1 + t^2 & \Rightarrow f'(0) = 1. \\
f''(x) &= 2ts^2 & = 2t + 2t^3 & = f''(0) = 0. \\
f'''(x) &= 2s^2 + 6t^2s^2 & = 2 + 8t^2 + 6t^4 & \Rightarrow f'''(0) = 2. \\
f^{(4)}(x) &= 16ts^2 + 24t^3s^2 & = 16t + 40t^3 + 24t^5 & \Rightarrow f^{(4)}(0) = 0. \\
f^{(5)}(x) & & = 16 + \text{terms involving } t & \Rightarrow f^{(5)}(0) = 16.
\end{array}
$$

Maclaurin's series now gives

$$\tan x = x + \frac{2x^3}{3!} + \frac{16x^5}{5!} + \text{terms in } x^7$$

$$\Rightarrow \qquad \tan x = x + \frac{x^3}{3} + \frac{2x^5}{15} + \text{terms in } x^7.$$

(Note that since $\tan x$ is an odd function of x there can be no even powers of x in its Maclaurin expansion and so the coefficient of x^6 must be zero.)

The result of this example can be found in a variety of other ways.

Thus neglecting x^6 and higher powers,

$$\tan x = \frac{\sin x}{\cos x} = \frac{x - \dfrac{x^3}{6} + \dfrac{x^5}{120}}{1 - \dfrac{x^2}{2} + \dfrac{x^4}{24}}.$$

Then the required result can be obtained by long division or expansion of $\left(1 - \dfrac{x^2}{2} + \dfrac{x^4}{24}\right)^{-1}$ in ascending powers of $\left(\dfrac{x^2}{2} - \dfrac{x^4}{24}\right)$. In both cases terms involving x^6 and higher powers should be neglected whenever they arise.

Example 7
By expanding $(1 + x^2)^{-1/2}$ in ascending powers of x and then integrating, show that, when $|x| < 1$,

$$\sinh^{-1} x = \frac{x}{1} - \frac{1}{2}\frac{x^3}{3} + \frac{1 \cdot 3}{2 \cdot 4}\frac{x^5}{5} - \cdots$$

and state the general term of the series.

$$(1 + x^2)^{-1/2} = 1 + (-\tfrac{1}{2})x^2 + \frac{(-\tfrac{1}{2})(-\tfrac{1}{2} - 1)}{2!}(x^2)^2 + \cdots$$

$$+ \frac{(-\tfrac{1}{2})(-\tfrac{1}{2} - 1) \cdots (-\tfrac{1}{2} - n + 1)}{n!}(x^2)^n + \cdots$$

$$\Rightarrow \quad (1 + x^2)^{-1/2} = 1 - \frac{1}{2}x^2 + \frac{1 \cdot 3}{2 \cdot 4}x^4 - \cdots$$

$$+ \frac{(-1)^n 1 \cdot 3 \cdot 5 \cdots (2n - 1)}{2 \cdot 4 \cdot 6 \cdots (2n)}x^{2n} + \cdots \text{ for } |x| < 1.$$

Assuming that term by term integration is valid, integrating with respect to x we find

$$\sinh^{-1} x = x - \frac{1}{2}\frac{x^3}{3} + \frac{1 \cdot 3}{2 \cdot 4}\frac{x^5}{5} - \cdots$$

$$+ \frac{(-1)^n 1 \cdot 3 \cdot 5 \cdots (2n - 1)}{2 \cdot 4 \cdot 6 \cdots (2n)} \frac{x^{2n+1}}{2n + 1} + \cdots + C.$$

Putting $x = 0$ gives $C = 0$, and the required result follows.

Taylor's series
Suppose now that in Maclaurin's series

$$g(x) = g(0) + xg'(0) + \frac{x^2}{2!}g''(0) + \cdots \frac{x^n}{n!}g^{(n)}(0) + \cdots$$

we put $g(x) = f(a + x)$, where a is constant.
Then $g(0) = f(a)$, $g'(0) = f'(a)$, $\ldots$, $g^{(n)}(0) = f^{(n)}(a)$, and we obtain Taylor's series

$$f(a + x) = f(a) + xf'(a) + \frac{x^2}{2!}f''(a) + \cdots + \frac{x^n}{n!}f^{(n)}(a) + \cdots.$$

Taylor's series may be expressed in the form

$$f(x) = f(a) + (x - a)f'(a) + \frac{(x - a)^2}{2!}f''(a) + \ldots + \frac{(x - a)^n}{n!}f^{(n)}(a) + \ldots$$

This form is often referred to as the Taylor series expansion for $f(x)$ about the point $x = a$. (Maclaurin's series gives the expansion of $f(x)$ about $x = 0$.)

Example 1
Show that, when powers of h greater than the second can be neglected,

(a) $\sin (a + h) = \sin a + h \cos a - \dfrac{h^2}{2} \sin a,$

(b) $\cos (a + h) = \cos a - h \sin a - \dfrac{h^2}{2} \cos a,$

(c) $\tan (a + h) = \tan a + h \sec^2 a + h^2 \sec^2 a \tan a.$

These results follow directly from Taylor's series in the form

$$f(a + h) = f(a) + hf'(a) + \frac{h^2}{2!}f''(a) + \ldots$$

by taking $f(x)$ as $\sin x$, $\cos x$ and $\tan x$ respectively.

Approximations derived from Taylor's series are very useful in the numerical solution of differential equations (see Section 9.3). The approximations derived in Example 2 are particularly useful.

Example 2
Use Taylor's series to show that, when h is small, in general
(a) $f'(a) \approx [f(a + h) - f(a)]/h,$
(b) $f'(a) \approx [f(a + h) - f(a - h)]/(2h),$
(c) $f''(a) \approx [f(a + h) - 2f(a) + f(a - h)]/h^2.$

$$f(a + h) = f(a) + hf'(a) + \frac{h^2}{2!}f''(a) + \ldots$$

$\Rightarrow \quad [f(a + h) - f(a)]/h = f'(a) +$ terms in h and higher powers.

The result (a) follows at once.

Similarly, results (b) and (c) follow by using (1) and

$$f(a - h) = f(a) - hf'(a) + \frac{h^2}{2!}f''(a) + \ldots.$$

Exercises 6.5

1   State the first three terms and the nth term in the expansion of $\sin x$ in ascending powers of x.

Given that x is sufficiently small for x^6 and higher powers of x to be neglected show that

$$\frac{3 \sin x}{2 + \cos x} = x - \frac{x^5}{180}.$$ [JMB]

2 Show that, when x is small enough for powers of x higher than the fourth to be neglected,

$$\ln \left(\frac{1 + \sinh x}{1 + x} \right) = \frac{x^3}{6} - \frac{x^4}{6}.$$ [JMB]

3 Expand $(1 + x - x^2) \ln (1 - 2x)$ as a series of ascending powers of x as far as and including the term in x^2. [L]

4 Find the following limits:

(a) $\lim\limits_{x \to 0} \left[\dfrac{a \sin bx - b \sin ax}{x^3} \right]$, where a and b are unequal constants,

(b) $\lim\limits_{x \to 0} \left[\dfrac{x^2 + 2 \cos x - 2}{x^4} \right]$. [L]

5 Given that $\quad y = \left(\dfrac{1 + 2px}{1 + 2qx} \right)^{1/2}, \quad z = \left(\dfrac{1 - px}{1 - qx} \right)^{-1/2}, \quad (p \neq q),$

expand y and z in ascending powers of x as far as the terms in x^2.

Given that both px and qx are small
(a) show that if the terms in x^2 and higher powers of x are neglected, the expressions obtained for y and z satisfy the relation $1 + y = 2z$;
(b) find the ratio $p:q$ for which the relation $1 + y = 2z$ still holds if the terms in x^2 are retained but the terms in x^3 and higher powers of x are neglected.

6 Differentiate with respect to x

$$\ln (\sec x + \tan x).$$

Express $\sec x$ as a series in ascending powers of x up to and including the term in x^4. Hence, or otherwise, show that, as far as the term in x^5, the series for $\ln (\sec x + \tan x)$ in ascending powers of x is

$$x + \frac{x^3}{6} + \frac{x^5}{24}.$$ [L]

7 Give expansions in powers of x of

$$\cosh x + \cos x \quad \text{and} \quad \frac{\sinh x + \sin x}{x}.$$

Find the sum of the infinite series

$$\frac{x^4}{5!} + \frac{2x^8}{9!} + \ldots + \frac{nx^{4n}}{(4n + 1)!} + \ldots. \qquad \text{[JMB]}$$

8 (a) Expand

$$\frac{e^x + e^{-x}}{e^{2x}}$$

in ascending powers of x, as far as the term in x^3. Give the general term in the expansion.

(b) Expand

$$\ln\left[\frac{(1 + x)^2}{1 + 2x}\right]$$

in ascending powers of x as far as the term in x^4. Give the general term and the set of values of x for which the expansion is valid. [L]

9 Show that $e^x + e^{-x}$ is never less than 2 and that $(2^x - 2^{-x})/x$ is never less than $2 \ln 2$.

Obtain the first four terms in the Maclaurin series for

$$\ln (2 + 4x) + \ln (1 + x)^2 + \ln (1 - 3x + 2x^2)$$

and state the set of values of x for which the expansion is valid. [L]

10 In the triangle PQR, the angle QPR is $3\pi/4$, the angle PQR is θ and the length of the side QR is p. Show that the length of the side PQ is $p(\cos \theta - \sin \theta)$. Given that θ is small enough for its fourth and higher powers to be negligible, show that the area of the triangle is approximately equal to $\frac{1}{2}p^2(\theta - \theta^2 - \frac{2}{3}\theta^3)$. [JMB]

11 Find the coefficient of x^n, where $n > 1$, in the expansion of

$$(1 + 2x) \ln (1 + 2x)$$

as a series of ascending powers of x, where $|x| < \frac{1}{2}$. [L]

12 Assuming the expansion of $\ln (1 + x)$, express

$$\ln\left[\frac{1 + x}{1 - x}\right]$$

in a series of ascending powers of x, giving the set of values of x for which the series is convergent.

Deduce an expansion of $\ln\left(1 + \frac{1}{y}\right)$ in ascending powers of $\frac{1}{2y + 1}$.

Show that the geometric series $\sum\limits_{n=1}^{\infty} \frac{2}{(2y + 1)^n}$ is convergent when $y > 0$ and find its sum. Deduce that, if $y > 0$, then

(a) the expansion of $\ln\left(1 + \dfrac{1}{y}\right)$ in ascending powers of $\dfrac{1}{2y + 1}$ is convergent,

(b) $\dfrac{2}{2y + 1} < \ln\left(1 + \dfrac{1}{y}\right) < \dfrac{1}{y}.$ [L]

13 Given that $t = \tan x$, express $\dfrac{dt}{dx}, \dfrac{d^2t}{dx^2}, \dfrac{d^3t}{dx^3}, \dfrac{d^4t}{dx^4}$ in terms of t only.

Hence, or otherwise, express
(a) $\tan x$ as a Maclaurin series of ascending powers of x as far as and including the term in x^5,

(b) $\tan\left(\dfrac{\pi}{4} + h\right)$ as a Taylor series of ascending powers of h as far as and including the term in h^4. [L]

14 (a) Obtain the Maclaurin series for y as a series of ascending powers of x, given that $y = 1$ and that $\dfrac{d^n y}{dx^n} = (-1)^n$, $n \geqslant 1$, at $x = 0$.

Find the sum of the series at $x = 2$.
(b) Obtain the first three non-zero terms in
(i) the expansion of $\sin^2 2x$ in powers of x,
(ii) the expansion of $(\cos x - \sin x)$ in powers of $(x - \pi/4)$. [L]

15 Defining x and y by $x = \cos\theta$, $y = \cos 9\theta$, where $\theta \neq n\pi$, $n \in \mathbb{Z}$, show that

$$(1 - x^2)\frac{d^2y}{dx^2} - x\frac{dy}{dx} + 81y = 0.$$

By induction, or otherwise, prove that

$$(1 - x^2)\frac{d^{n+2}y}{dx^{n+2}} - x(2n + 1)\frac{d^{n+1}y}{dx^{n+1}} - (n^2 - 81)\frac{d^n y}{dx^n} = 0.$$

Denoting $\dfrac{d^n y}{dx^n}$ by $f^{(n)}$, deduce that

$$f^{(n+2)}(0) = (n^2 - 81)f^{(n)}(0).$$

Use Maclaurin's series to obtain y as a polynomial in x.

6.6 Repeated differentiation: Leibnitz's theorem and applications

We first consider some examples involving the determination of nth derivatives of simple expressions.

Example 1

Given that $y = x^m$, where $m \in \mathbb{Z}^+$, find $\dfrac{d^n y}{dx^n}$.

$$\frac{d^n y}{dx^n} = m(m-1) \dots (m-n+1)x^{m-n}$$

$$= \frac{m! x^{m-n}}{(m-n)!}, \qquad \text{for } n < m,$$

$$\frac{d^n y}{dx^n} = m!, \qquad \text{for } n = m,$$

$$\frac{d^n y}{dx^n} = 0, \qquad \text{for } n > m.$$

Example 2

Given that $y = e^{ax}$, where a is constant, find $\dfrac{d^n y}{dx^n}$.

$$\frac{d^n y}{dx^n} = a^n e^{ax}.$$

This result can be proved formally by induction.

Example 3

Given that $y = e^{ax} \sin bx$, where a and b are constants, show that

$$\frac{d^n y}{dx^n} = r^n e^{ax} \sin(bx + n\phi), \qquad \dots (1)$$

where $r = \sqrt{(a^2 + b^2)}$, $\sin \phi = b/r$ and $\cos \phi = a/r$.

$$\frac{dy}{dx} = ae^{ax} \sin bx + be^{ax} \cos bx$$

$$= e^{ax} \sqrt{(a^2 + b^2)} \left[\frac{a}{\sqrt{(a^2 + b^2)}} \sin bx + \frac{b}{\sqrt{(a^2 + b^2)}} \cos bx \right]$$

$$= re^{ax}[\cos \phi \sin bx + \sin \phi \cos bx]$$
$$= re^{ax} \sin(bx + \phi).$$

The result (1) therefore holds when $n = 1$.

Suppose now that it holds when $n = k$.

Then $\quad \dfrac{d^{k+1} y}{dx^k} = r^k \dfrac{d}{dx}[e^{ax} \sin(bx + k\phi)]$

$$= r^k e^{ax}[a \sin(bx + k\phi) + b \cos(bx + k\phi)]$$
$$= r^{k+1} e^{ax}[\cos \phi \sin(bx + k\phi) + \sin \phi \cos(bx + k\phi)]$$
$$= r^{k+1} e^{ax} \sin[bx + (k+1)\phi].$$

Therefore, if (1) holds for $n = k$, it holds for $n = k + 1$ also. But (1) is true when $n = 1$ and therefore by induction it holds for all $n \in \mathbb{Z}^+$.

[Note the special case (when $a = 0$)

$$\frac{d^n}{dx^n}(\sin bx) = b^n \sin (bx + n\pi/2).]$$

The effect of differentiating $e^{ax} \sin bx$ is to multiply the expression by r and add ϕ to the angle bx; so differentiation n times multiplies the expression by r^n and adds $n\phi$ to the angle. We can thus deduce that

$$\int e^{ax} \sin bx \, dx = \frac{1}{r} e^{ax} \sin (bx - \phi) + C.$$

In particular, differentiation of $\sin bx$ multiplies the expression by b and adds $\pi/2$ to the angle.

Thus

$$\int \sin bx \, dx = \frac{1}{b} \sin \left(bx - \frac{\pi}{2}\right) + C.$$

Leibnitz's theorem on repeated differentiation of a product may be stated as follows.

Given that u and v are functions of x and $\dfrac{d^r u}{dx^r}, \dfrac{d^s v}{dx^s}$ are denoted by u_r, v_s respectively, then

$$\frac{d^n}{dx^n}(uv) = u_0 v_n + \binom{n}{1} u_1 v_{n-1} + \binom{n}{2} u_2 v_{n-2} + \cdots$$

$$+ \binom{n}{r} u_r v_{n-r} + \cdots + u_n v_0.$$

Proof

Assume that the theorem is true when $n = k$. Then, differentiating (using the product formula),

$$\frac{d^{k+1}}{dx^{k+1}}(uv) = (u_0 v_{k+1} + u_1 v_k) + \binom{k}{1}(u_1 v_k + u_2 v_{k-1}) + \cdots$$

$$+ \binom{k}{r}(u_r v_{k+1-r} + u_{r+1} v_{k-r}) + \cdots + \cdots$$

$$+ (u_k v_1 + u_{k+1} v_0).$$

The coefficient of the term involving $u_r v_{k+1-r}$ is

$$\binom{k}{r-1} + \binom{k}{r} = \frac{k!}{(r-1)!(k+1-r)!} + \frac{k!}{r!(k-r)!}$$

$$= \frac{k!(k+1-r+r)}{r!(k+1-r)!} = \frac{(k+1)!}{r!(k+1-r)!}$$

$$= \binom{k+1}{r}.$$

$$\therefore \quad \frac{d^{k+1}}{dx^{k+1}}(uv) = u_0 v_{k+1} + \binom{k+1}{1} u_1 v_k + \binom{k+1}{2} u_2 v_{k-1} + \dots$$

$$+ \binom{k+1}{r} u_r v_{k+1-r} + \dots + u_{k+1} v_0.$$

This is merely the statement of the theorem with $k+1$ in the place of k. But the theorem is true for $k = 1$ (it then reduces to the product rule for differentiation). Hence by induction the theorem is true for $n \in \mathbb{Z}^+$.

Example 1
Given that $f(x) = x^3 e^{2x}$, find $f^{(6)}(x)$.

$$\frac{d^6(x^3 e^{2x})}{dx^6} = x^3 \frac{d^6}{dx^6}(e^{2x}) + \binom{6}{1}(3x^2)\frac{d^5}{dx^5}(e^{2x}) + \binom{6}{2}(6x)\frac{d^4}{dx^4}(e^{2x})$$

$$+ \binom{6}{3} 6 \frac{d^3}{dx^3}(e^{2x})$$

$$= x^3 . 2^6 e^{2x} + 6 . 3x^2 . 2^5 e^{2x} + 15 . 6x . 2^4 e^{2x} + 20 . 6 . 2^3 e^{2x}$$
$$= (64x^3 + 576x^2 + 1440x + 960)e^{2x}.$$

Example 2
Given that $f(x) = x^2 \sin x$, find $f^{(5)}(x)$.

$$\frac{d^5(x^2 \sin x)}{dx^5} = (x^2)\frac{d^5(\sin x)}{dx^5} + \binom{5}{1}(2x)\frac{d^4(\sin x)}{dx^4} + \binom{5}{2}(2)\frac{d^3(\sin x)}{dx^3}$$

$$= x^2 \cos x + 10x \sin x - 20 \cos x.$$

Leibnitz's theorem may be used in conjunction with Maclaurin's series to obtain expansions of certain expressions in ascending powers of x.

Example 1
Given that $y = (\arcsin x)^2$, show that, for $-1 < x < 1$,

(a) $(1 - x^2)\left(\frac{dy}{dx}\right)^2 = 4y$,   (b) $(1 - x^2)\frac{d^2 y}{dx^2} - x\frac{dy}{dx} = 2$.

Deduce that $y_{n+2}(0) = n^2 y_n(0)$ for $n \geqslant 1$, where $y_r(0)$ denotes the value, when $x = 0$, of the rth derivative of y with respect to x.

Write down the Maclaurin series for $(\arcsin x)^2$ up to and including the term in x^6.

Show also that, for $n \geqslant 1$,

$$y_{2n-1}(0) = 0, \qquad y_{2n}(0) = 2^{2n-1}[(n-1)!]^2$$

and hence find the general term in the Maclaurin series for $(\arcsin x)^2$.

(a)
$$\frac{dy}{dx} = \frac{2 \arcsin x}{\sqrt{(1-x^2)}} \qquad \ldots (1)$$

$\Rightarrow$
$$\sqrt{(1-x^2)} \frac{dy}{dx} = 2 \arcsin x$$

$\Rightarrow$
$$(1-x^2)\left(\frac{dy}{dx}\right)^2 = 4(\arcsin x)^2$$

$\Rightarrow$
$$(1-x^2)\left(\frac{dy}{dx}\right)^2 = 4y.$$

(b) Differentiating gives

$$(1-x^2)(2)\left(\frac{dy}{dx}\right)\left(\frac{d^2 y}{dx^2}\right) - 2x\left(\frac{dy}{dx}\right)^2 = 4\frac{dy}{dx}.$$

Division by $2\dfrac{dy}{dx}$, which in general is non-zero, gives

$$(1-x^2)\frac{d^2 y}{dx^2} - x\frac{dy}{dx} = 2. \qquad \ldots (2)$$

Differentiating equation (2) n times by Leibnitz's theorem and writing $\dfrac{d^r y}{dx^r}$ as y_r gives

$$(1-x^2)y_{n+2}(x) - 2nxy_{n+1}(x) - n(n-1)y_n(x) - xy_{n+1}(x) - ny_n(x) = 0,$$
for $n \geqslant 1$.

Putting $x = 0$ we find

$$y_{n+2}(0) = n^2 y_n(0), \qquad \text{for } n \geqslant 1. \qquad \ldots (3)$$

Clearly $y(0) = 0$; equation (1) gives $y_1(0) = 0$, and equation (2) gives $y_2(0) = 2$.

Then, from (3),

$$y_3(0) = 0, \qquad y_4(0) = 2^2 y_2(0) = 8, \qquad y_5(0) = 0,$$
$$y_6(0) = 4^2 y_4(0) = 128.$$

The Maclaurin series for $y(x)$ is

$$y(x) = y(0) + xy_1(0) + \frac{x^2}{2}y_2(0) + \frac{x^3}{3!}y_3(0) + \dots$$

$$\Rightarrow \qquad y(x) = x^2 + \frac{x^4}{3} + \frac{8x^6}{45}$$

as far as the first three non-zero terms.

Clearly, y is an even function and so the Maclaurin series will contain only even powers of x.

We have found already that

$$y_2(0) = 2,$$
$$y_4(0) = 2^2 2,$$
$$y_6(0) = 4^2 2^2 2,$$
$$y_8(0) = 6^2 4^2 2^2 2,$$
$$\dots \qquad \dots$$

So

$$\begin{aligned} y_{2n}(0) &= (2n-2)^2(2n-4)^2 \dots 6^2 4^2 2^2 2 \\ &= [(n-1)!]^2((2^2)^{n-1})2 \\ &= 2^{2n-1}[(n-1)!]^2. \end{aligned}$$

So the general term in the Maclaurin series for $(\arcsin x)^2$ is

$$\frac{2^{2n-1}[(n-1)!]^2 x^{2n}}{(2n)!}.$$

Example 2

Given that $x = \sin\theta$, $y = \sin p\theta$, where $-\pi/2 < \theta < \pi/2$ and p is a positive integer, prove that

$$(1 - x^2)\frac{d^2y}{dx^2} - x\frac{dy}{dx} + p^2y = 0.$$

Differentiate this equation n times using Leibnitz's theorem, and hence prove that

$$y_{n+2}(0) = (n^2 - p^2)y_n(0),$$

where $y_n(0)$ denotes the value of $\dfrac{d^n y}{dx^n}$ when $x = 0$.

Find the first three non-zero terms in the expansion of y as a series of ascending powers of x.

$$\frac{dy}{dx} = \frac{dy}{d\theta}\bigg/\frac{dx}{d\theta} = \frac{p\cos p\theta}{\cos\theta} \qquad \dots (1)$$

$$\Rightarrow \qquad \left(\frac{dy}{dx}\right)^2 = \frac{p^2\cos^2 p\theta}{\cos^2\theta} = \frac{p^2(1 - \sin^2 p\theta)}{1 - \sin^2\theta}$$

$$\Rightarrow \qquad (1 - x^2)\left(\frac{dy}{dx}\right)^2 = p^2(1 - y^2).$$

Differentiation with respect to x and division by $2\dfrac{dy}{dx}$ (not, in general, zero), gives

$$(1 - x^2)\frac{d^2y}{dx^2} - x\frac{dy}{dx} + p^2y = 0. \qquad\qquad \dots (2)$$

Differentiation n times by Leibnitz's theorem gives

$$(1 - x^2)y_{n+2}(x) - 2nxy_{n+1}(x) - n(n - 1)y_n(x) - xy_{n+1}(x) - ny_n(x)$$
$$+ p^2y_n(x) = 0, \qquad \text{for } n \geqslant 1$$
$$\Rightarrow \quad (1 - x^2)y_{n+2}(x) - (2n + 1)xy_{n+1}(x) - (n^2 - p^2)y_n(x) = 0.$$

Putting $x = 0$ gives

$$y_{n+2}(0) = (n^2 - p^2)y_n(0), \qquad \text{for } n \geqslant 0.$$

But $y(0) = 0$, since $\theta = 0$ gives $x = 0$, $y = 0$, and from equation (1) $y_1(0) = p$. The given relation is that

$$y = \sin (p \arcsin x)$$

and so y is an odd function, i.e. the Maclaurin series for y will contain only odd powers.

We then have
$$y_1(0) = p,$$
$$y_3(0) = (1^2 - p^2)p,$$
$$y_5(0) = (3^2 - p^2)(1^2 - p^2)p,$$

whence

$$\sin (p \arcsin x) = px + (1^2 - p^2)\frac{px^3}{3!} + (1^2 - p^2)(3^2 - p^2)\frac{px^5}{5!} + \dots.$$

Note that the series terminates when p is an odd integer.

Exercises 6.6

1 Given that $y = e^{-x} \sin x$, prove that $\dfrac{d^4y}{dx^4} + 4y = 0$.

2 Given that $y = (x^3 - 3x^2)e^{2x}$, express $\dfrac{d^6y}{dx^6}$ in terms of x.

3 Given that $y = (e^x - 1)^2$ show that

$$\frac{d^2y}{dx^2} - 3\frac{dy}{dx} + 2y = 2.$$

By repeated differentiation of this equation find the Maclaurin expansion of y in terms of x as far as the term in x^5.

Check your result by writing y as $e^{2x} - 2e^x + 1$ and using the exponential series.

4 The differential equation

$$(1 - x^2)\frac{d^2y}{dx^2} - 2x\frac{dy}{dx} + 20y = 0$$

has a solution $y = f(x)$. Obtain the equation connecting

$$\frac{d^{n+2}y}{dx^{n+2}}, \qquad \frac{d^{n+1}y}{dx^{n+1}} \quad \text{and} \quad \frac{d^ny}{dx^n}$$

and hence show that $f^{(n+2)}(0) = (n + 5)(n - 4)f^{(n)}(0)$ where $f^{(n)}(x)$ denotes the nth derivative of $f(x)$.

Given that $f(0) = 1$ and $f^{(1)}(0) = 0$, find $f(x)$ as a polynomial in x.

[JMB]

5 Given that

$$y'' - x^2y = 0$$

and $y(0) = 0$, $y'(0) = 1$, express y as a series of ascending powers of x as far as the term in x^9.

Miscellaneous exercises 6

1 (a) Evaluate $\lim\limits_{x\to 0} \left(\dfrac{\sqrt{(27 + x)} - 3\sqrt{3}}{\sqrt{(9 + x)} - 3} \right)$.

(b) Find the sum to infinity of the series whose nth term is

$$(n + 1)^2/n!.$$

(c) Find the set of values of x for which the series whose nth term is

$$\left(\frac{2x}{1 + x^2} \right)^n$$

is convergent.

[L]

2 Given that x is so small that its sixth and higher powers may be neglected, show that

$$\frac{\sin x}{\cos^4 x - \sin^4 x} = ax + bx^3 + cx^5,$$

where a, b, c are constants whose values are to be found.

[JMB]

3 (a) Examine for convergence the series whose nth terms are

(i) $\dfrac{n}{2^n}$, (ii) $\dfrac{1}{\sqrt{(n^3 + 1)}}$.

(b) Prove that

$$\arctan(r + 1) - \arctan r = \arctan[1/(1 + r + r^2)],$$

where $r > 0$.

Hence, or otherwise, obtain the sum of the series

$$\sum_{r=1}^{n} \arctan [1/(1 + r + r^2)].$$

Deduce that

$$\sum_{r=1}^{\infty} \arctan [1/(1 + r + r^2)]$$

converges, and find its sum. [L]

4 (a) Find the limit a of the sequence $\{a_n\}$ as n tends to infinity, where

$$a_n = (n + 10)/(10n + 1).$$

Find the smallest positive integer n for which $a_n - a < 10^{-3}$.

(b) The arc AB of a circle subtends an acute angle θ at the centre O. The tangent to the circle at B meets OA produced at T, and the perpendicular from A to OB meets OB at N. Show that the ratio of the areas of the two parts into which the arc AB divides the quadrilateral ANBT may be written as

$$(\tan \theta - \theta):(\theta - \sin \theta \cos \theta).$$

Find the limit of this ratio as θ tends to zero. [L]

5 (a) Find the following limits:

(i) $\lim\limits_{x \to 0} \left[\dfrac{\ln (1 + ax)}{\ln (1 + bx)} \right]$,   (ii) $\lim\limits_{x \to 0} \left[\dfrac{\sin x - x}{x \cos x - x} \right]$.

(b) Show that, for $a > 0$, the series whose nth term is

$$\left(\frac{1}{n} - \frac{1}{n + a} \right)$$

is convergent.

(c) Investigate the convergence or divergence of the series whose nth term is

$$\frac{1}{n}\left(\frac{2}{3}\right)^n.$$

6 (a) Obtain the first four terms in the expansion of

$$\left(\frac{1 + 2x}{1 - 2x}\right)^{1/2}$$

in a series of ascending powers of x.

(b) Use the comparison test to establish the convergence of the infinite series in which the nth term is

(i) $\dfrac{1}{n}\left(\dfrac{3}{4}\right)^n,$ (ii) $\dfrac{1}{n}\left(\dfrac{1}{2}\right)^{n-1}.$

Show that the two series converge to the same value. [L]

7 By using the power series expansions of $\sinh x$, $\cosh x$, $\sin x$ and $\cos x$ prove that, if x is so small that x^8 can be neglected,

$$\frac{\sinh x - \sin x}{\cosh x - \cos x} = \frac{x}{3}\left(1 - \frac{x^4}{630}\right).$$ [JMB]

8 Given that

$$x\frac{d^2y}{dx^2} + \frac{dy}{dx} + xy = 0,$$

and $\dfrac{dy}{dx} = 0$ when $x = 0$, show that, when $x = 0$,

$$\frac{d^{n+1}y}{dx^{n+1}} = \frac{-n}{n+1}\frac{d^{n-1}y}{dx^{n-1}}, \qquad n \geqslant 1.$$

Given also that $y = 1$ when $x = 0$, obtain the Maclaurin series for y, giving the first four non-zero terms and the general term. [L]

9 Examine the convergence or non-convergence of the integrals

(a) $\displaystyle\int_1^\infty \frac{\sin^2 x}{x^{3/2}}\,dx,$

(b) $\displaystyle\int_0^\infty \frac{e^{-x}}{x^{1/2}}\,dx,$

(c) $\displaystyle\int_1^\infty e^{-x^2}\,dx,$

(d) $\displaystyle\int_1^\infty \frac{1}{\sqrt{(1 + x^3)}}\,dx,$

(e) $\displaystyle\int_{-\infty}^\infty \frac{x}{\cosh x}\,dx,$

(f) $\displaystyle\int_0^\infty \sin x\,dx.$

10 Test for convergence the series whose nth term is

(a) $\dfrac{2n - 1}{n(n + 1)(n + 2)},$ (b) $\dfrac{1 + n}{2 + n^2},$ (c) $ne^{-n^2}.$ [L]

11 Show that, for all integers n and r greater than unity,

$$\frac{(n + 1)!}{(n + r)!} < \frac{1}{(n + 1)^{r-1}}$$

and deduce that

$$\frac{1}{(n + 1)!} + \frac{1}{(n + 2)!} + \ldots + \frac{1}{(n + r)!} + \ldots < \frac{1}{n \cdot n!}.$$

Hence show that the error in taking e as

$$1 + \frac{1}{1!} + \frac{1}{2!} + \ldots + \frac{1}{8!}$$

is of magnitude less than 4×10^{-6}. [L]

12 Investigate the convergence or divergence of the series whose nth term is

(a) $\dfrac{1}{\sqrt{(n + 1)}}$,　(b) $\dfrac{1}{n \cdot 2^n}$,　(c) $\dfrac{1 + n}{n^2}$,　(d) $\dfrac{1}{n + 3^n}$.

13 (a) Prove that the series whose rth term is

$$\frac{(2r^2 + 1)x^r}{r}$$

is convergent for $|x| < 1$.

(b) Find the limit of (i) $\dfrac{e^x - 1}{x}$ as $x \to 0$, (ii) $\dfrac{(\sinh^2 a - \sinh^2 x)}{(a^2 - x^2)}$ as

$x \to a$, where $a \neq 0$.

14 (a) Investigate the convergence of the series whose rth term is

$$\frac{1}{r}\left(\frac{3}{4}\right)^r.$$

(b) Show that the series whose rth term is $\dfrac{3^r + 2}{4^r}$ is convergent and find

its sum.

15 Evaluate the following:

(a) $\displaystyle\sum_{n=1}^{\infty} \frac{1}{2^{2n-1}}$,　(b) $\displaystyle\sum_{n=1}^{\infty} \frac{1}{2^n n!}$,　(c) $\displaystyle\sum_{n=1}^{\infty} \frac{1}{n2^n}$,　(d) $\displaystyle\sum_{n=1}^{\infty} \frac{n}{2^{3n-2}}$. [L]

16 Show that, for a certain value of k, the integral

$$\int_0^{\infty} \left[\frac{1}{\sqrt{(1 + x^2)}} - \frac{k}{1 + x}\right] dx$$

converges. Find this value of k and evaluate the integral in this case. [L]

17 Determine whether or not each of the following integrals is convergent. Evaluate each integral which is convergent.

(a) $\displaystyle\int_0^{\infty} \frac{1}{1 + x^2} dx$,　(b) $\displaystyle\int_0^{\infty} \frac{x^2}{1 + x^2} dx$. [L]

18 Test for convergence or divergence the series whose rth term is

(a) $\dfrac{1}{\sqrt{r}}$,

(b) $\dfrac{r + 1}{r\sqrt{r}}$,

(c) $\dfrac{1}{\sqrt{(r^2 + 1)}}$,

(d) $\dfrac{r!}{1 \cdot 3 \cdot 5 \cdot 7 \cdot \ldots \cdot (2r - 1)}$.

(e) $\dfrac{x^{2r}}{r\sqrt{r}}$,

(f) $\dfrac{\cos^2 rx}{r^2}$.

19 For the series whose nth terms are given in (a) and (b) determine the set of positive values of x for which the series will converge.

(a) $\dfrac{x^n}{\sqrt{n}}$,

(b) $\dfrac{1}{(1 + 2x^2)^n}$.

20 *The integral test for convergence.* Given that $f(x) > 0$ and $f'(x) < 0$ for $x \in \mathbb{R}^+$, show that, for $r \in \mathbb{Z}^+$,

$$f(r) > \int_r^{r+1} f(x)\, dx > f(r + 1).$$

Deduce that

$$\sum_{n=1}^{n} f(r) > \int_1^{n+1} f(x)\, dx > \sum_{r=2}^{n+1} f(r),$$

and illustrate your result graphically.

Hence show that the series $\Sigma\, f(r)$ and the integral $\displaystyle\int_1^{\infty} f(x)\, dx$ will both converge or both diverge.

Use this result to show that the series $\Sigma\, \dfrac{1}{r^\alpha}$ converges if $\alpha > 1$ and diverges if $\alpha < 1$.

Use the integral test to examine for convergence the series whose rth terms are as follows:

(a) $\dfrac{1}{r^2 + 4}$,

(b) $\dfrac{r}{r^2 + 3}$,

(c) $\dfrac{1}{(r + 1) \ln (r + 1)}$,

(d) $\dfrac{\ln r}{r}$,

(e) $\dfrac{\ln r}{r^2}$,

(f) $\dfrac{r}{e^r}$,

(g) $\dfrac{r}{(r + 1)(r^2 + 4)}$.

21 Prove that

$$\int_a^{\infty} \dfrac{1}{x^4 \sqrt{(a^2 + x^2)}}\, dx = \dfrac{2 - \sqrt{2}}{3a^4}.$$

22 (a) Use the integral test to show that the series whose rth term is

$$\dfrac{1}{(r + 1) \ln (r + 1) \ln [\ln (r + 1)]}$$

is divergent.

(b) Show that

$$\int_2^\infty \frac{1}{x\,(\ln x)^2}\,dx = \frac{1}{\ln 2},$$

and hence, or otherwise, examine the convergence of the series whose $(n - 1)$th term is

$$\frac{1}{n(\ln n)^2}.$$

23 Examine for convergence the integrals

(a) $\displaystyle\int_0^2 \frac{1}{\sqrt{[x(2 - x)]}}\,dx,$

(b) $\displaystyle\int_0^\infty \frac{e^x}{e^{2x} - 1}\,dx,$

(c) $\displaystyle\int_0^\infty x^2 e^{-x}\,dx,$

(d) $\displaystyle\int_0^\infty \frac{1}{1 + e^{-2x}}\,dx,$

(e) $\displaystyle\int_0^\infty \frac{1}{\cosh^3 x}\,dx,$

(f) $\displaystyle\int_1^\infty \frac{x^2 + 2}{x^2(x^2 + 1)}\,dx.$

7 Further integration and applications

7.1 Some properties of definite integrals

In *Advanced Mathematics 1* we discussed the techniques for the direct evaluation of integrals. We now establish a number of properties of definite integrals which may enable us to simplify the working. First, we define the definite integral by the relation

$$\int_a^b f(x)\,dx = \left[F(x) \right]_a^b = F(b) - F(a), \qquad \ldots (1)$$

where $F'(x) = f(x)$. Then

$$\int_a^b f(x)\,dx = \int_a^b f(y)\,dy = \int_a^b f(u)\,du \qquad (\text{etc.}) \qquad \ldots (2)$$

Each integral is equal to $F(b) - F(a)$. This simple result (that it does not matter what we call the variable of integration in a definite integral) is used frequently in this chapter.

$$\int_b^a f(x)\,dx = - \int_a^b f(x)\,dx. \qquad \ldots (3)$$

When $a < c < b$, then

$$\int_a^b = \int_a^c f(x)\,dx + \int_c^b f(x)\,dx. \qquad \ldots (4)$$

Results (3) and (4) follow at once from the definition (1).

Example 1

Evaluate $\int_0^3 f(x)\,dx$, where

$$f(x) = \begin{cases} x^2 & \text{for } 0 \leqslant x < 1, \\ 2 - x & \text{for } 1 \leqslant x \leqslant 4. \end{cases}$$

$$\int_0^3 f(x)\,dx = \int_0^1 x^2\,dx + \int_1^4 (2 - x)\,dx$$

$$= \left[\frac{x^3}{3} \right]_0^1 + \left[-\tfrac{1}{2}(2 - x)^2 \right]_1^4 = \tfrac{1}{3} - 2 + \tfrac{1}{2} = -1\tfrac{1}{6}.$$

Example 2

Given that $f(x) = |x(1 - x)|$, show that $\displaystyle\int_0^3 f(x)\,dx = 4\tfrac{5}{6}$.

$$\int_0^3 f(x)\,dx = \int_0^1 x(1 - x)\,dx + \int_1^3 x(x - 1)\,dx$$

$$= \left[\frac{x^2}{2} - \frac{x^3}{3}\right]_0^1 + \left[\frac{x^3}{3} - \frac{x^2}{2}\right]_1^3$$

$$= \tfrac{1}{2} - \tfrac{1}{3} + 9 - \tfrac{9}{2} - \tfrac{1}{3} + \tfrac{1}{2} = 4\tfrac{5}{6}.$$

Result (4) may be generalised to:

When $a_0 < a_1 < a_2 < \ldots < a_r < \ldots < a_n$, then

$$\int_{a_0}^{a_n} f(x)\,dx = \sum_{r=0}^{n-1} \int_{a_r}^{a_{r+1}} f(x)\,dx. \qquad \ldots (5)$$

Example 3

Given that f is a periodic function with period l, prove that

$$\int_0^{nl} f(x)\,dx = n \int_0^l f(x)\,dx,$$

where $n \in \mathbb{Z}^+$.

$$\int_0^{nl} f(x)\,dx = \sum_{r=0}^{n-1} \int_{rl}^{(r+1)l} f(x)\,dx, \qquad \text{(by (5))}.$$

Putting $x = rl + u$, we find

$$\int_{rl}^{(r+1)l} f(x)\,dx = \int_0^l f(rl + u)\,du = \int_0^l f(rl + x)\,dx$$

by (2).

But, since $f(x)$ is periodic with period l, $f(rl + x) = f(x)$ and so

$$\int_0^l f(rl + x)\,dx = \int_0^l f(x)\,dx.$$

The required result follows.

Example 4

Evaluate $\displaystyle\int_0^{12} f(x)\,dx$, where f is a periodic function with period 3 and

$$f(x) = |x(1 - x)| \qquad \text{for } 0 \leqslant x \leqslant 3.$$

By result (5)

$$\int_0^{12} f(x)\,dx = 4\int_0^3 f(x)\,dx$$

and, therefore, using the result of Example 2 above,

$$\int_0^{12} f(x)\,dx = 4(4\tfrac{5}{6}) = 19\tfrac{1}{3}.$$

$$\int_0^a f(x)\,dx = -\int_0^{-a} f(-x)\,dx. \qquad \dots (6)$$

Putting $x = -u$,

$$\int_0^a f(x)\,dx = -\int_0^{-a} f(-u)\,du = -\int_0^{-a} f(-x)\,dx \qquad \text{(by (2))}.$$

$$\int_{-a}^a f(x)\,dx = \int_0^a [f(x) + f(-x)]\,dx. \qquad \dots (7)$$

Proof $\displaystyle\int_{-a}^a f(x)\,dx = \int_{-a}^0 f(x)\,dx + \int_0^a f(x)\,dx.$

But, writing $x = -u$,

$$\int_{-a}^0 f(x)\,dx = -\int_a^0 f(-u)\,du = -\int_a^0 f(-x)\,dx \qquad \text{(by (2))}$$

$$= \int_0^a f(-x)\,dx \qquad \text{(by (3))}.$$

The result follows.

Result (7) immediately leads to two further very important results. If f is an even function, so that $f(-x) = f(x)$, then

$$\int_{-a}^a f(x)\,dx = 2\int_0^a f(x)\,dx. \qquad \dots (8)$$

If f is an odd function so that $f(-x) = -f(x)$, then

$$\int_{-a}^a f(x)\,dx = 0. \qquad \dots (9)$$

Results (8) and (9) are intuitively obvious from graphical considerations, Fig. 7.1.

Example 5

Evaluate I, where $I = \displaystyle\int_{-a}^a |x|\,e^{-b|x|}\,dx$ and a, b are positive constants.

The integrand is an even function. Therefore

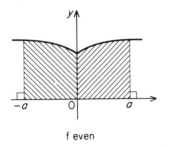

f even

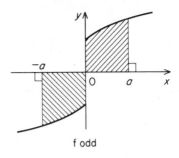

f odd

Fig. 7.1

$$I = 2 \int_0^a x e^{-bx} \, dx$$

$$= 2 \left[-\frac{x}{b} e^{-bx} - \frac{1}{b^2} e^{-bx} \right]_0^a$$

$$= 2(1 - e^{-ab} - abe^{-ab})/b^2$$

Example 6
Show that each of the following integrals is zero:

(a) $\displaystyle\int_{-1}^1 \frac{x^7}{(x^2 + 1)^{10}} \, dx;$ (b) $\displaystyle\int_{-\pi/2}^{\pi/2} \sin^{11} x \, dx;$ (c) $\displaystyle\int_{-\pi}^{\pi} \frac{x^5}{(1 + \sin^2 x)} \, dx.$

Each of the integrands is of the form $f(x)$ where f is odd function. The required results follow from property (9).

$$\int_0^a f(x) \, dx = \int_0^a f(a - x) \, dx. \qquad \dots (10)$$

Putting $x = a - u$ gives

$$\int_0^a f(x) \, dx = -\int_a^0 f(a - u) \, du = \int_0^a f(a - u) \, du \qquad \text{(by (3))},$$

$$= \int_0^a f(a - x) \, dx \qquad \text{(by (2))}.$$

Example 7
Use result (10) to evaluate the definite integrals

(a) $\displaystyle\int_0^{\pi/2} \cos^2 x \, dx,$ (b) $\displaystyle\int_0^{\pi} x \cos^2 x \, dx,$ (c) $\displaystyle\int_0^{\pi/2} \frac{\cos \theta}{\cos \theta + \sin \theta} \, d\theta,$

(d) $\displaystyle\int_0^{\pi} \frac{x}{1 + \sin x} \, dx.$

(a) If $I = \displaystyle\int_0^{\pi/2} \cos^2 x \, dx$, then, by (10),

$$I = \int_0^{\pi/2} \cos^2(\pi/2 - x)\, dx = \int_0^{\pi/2} \sin^2 x\, dx.$$

Then
$$2I = \int_0^{\pi/2} (\cos^2 x + \sin^2 x)\, dx$$

$$= \int_0^{\pi/2} 1\, dx = \pi/2$$

whence
$$I = \pi/4.$$

(b)
$$\int_0^\pi x \cos^2 x\, dx = \int_0^\pi (\pi - x) \cos^2 (\pi - x)\, dx$$

$$= \int_0^\pi (\pi - x) \cos^2 x\, dx$$

$$= \pi \int_0^\pi \cos^2 x\, dx - \int_0^\pi x \cos^2 x\, dx$$

$$\Rightarrow \quad 2 \int_0^\pi x \cos^2 x\, dx = \pi \int_0^\pi \cos^2 x\, dx.$$

But
$$\int_0^\pi \cos^2 x\, dx = \tfrac{1}{2} \int_0^\pi (1 + \cos 2x)\, dx$$

$$= \tfrac{1}{2} \left[x + \tfrac{1}{2} \sin 2x \right]_0^\pi$$

$$= \pi/2$$

$$\therefore \quad \int_0^\pi x \cos^2 x\, dx = \frac{\pi^2}{4}.$$

(c)
$$\int_0^{\pi/2} \frac{\cos \theta}{\cos \theta + \sin \theta}\, d\theta = \int_0^{\pi/2} \frac{\cos (\pi/2 - \theta)}{\cos (\pi/2 - \theta) + \sin (\pi/2 - \theta)}\, d\theta$$

$$= \int_0^{\pi/2} \frac{\sin \theta}{\sin \theta + \cos \theta}\, d\theta.$$

$$\therefore \quad 2 \int_0^{\pi/2} \frac{\cos \theta}{\cos \theta + \sin \theta}\, d\theta = \int_0^{\pi/2} \frac{\cos \theta}{\cos \theta + \sin \theta}\, d\theta + \int_0^{\pi/2} \frac{\sin \theta}{\cos \theta + \sin \theta}\, d\theta$$

$$= \int_0^{\pi/2} 1\, d\theta = \frac{\pi}{2}.$$

$$\therefore \quad \int_0^{\pi/2} \frac{\cos \theta}{\cos \theta + \sin \theta}\, d\theta = \frac{\pi}{4}.$$

(d)
$$I = \int_0^\pi \frac{x}{1 + \sin x}\, dx = \int_0^\pi \frac{(\pi - x)}{1 + \sin (\pi - x)}\, dx$$

$$= \int_0^\pi \frac{\pi - x}{1 + \sin x}\, dx$$

$$\Rightarrow \quad 2I = \pi \int_0^\pi \frac{1}{1 + \sin x}\, dx = \pi J,$$

where $J = \displaystyle\int_0^\pi \frac{1}{1 + \sin x}\, dx.$

Putting $x = \dfrac{\pi}{2} - y,$

$$J = \int_{-\pi/2}^{\pi/2} \frac{1}{1 + \cos y}\, dy \qquad \text{(from (3))}$$

$$= 2 \int_0^{\pi/2} \frac{1}{1 + \cos y}\, dy \qquad \text{(from (8))}$$

$$= 2 \int_0^{\pi/2} \tfrac{1}{2} \sec^2 (y/2)\, dy$$

$$= 2 \left[\tan (y/2) \right]_0^{\pi/2}$$

$$= 2.$$

So $\quad 2I = 2\pi,$

i.e. $\quad I = \pi.$

Example 8
Prove that

(a) $\displaystyle\int_0^a f(x)\, dx = \int_0^{a/2} [f(x) + f(a - x)]\, dx.$

Use this result to show that

(b) $\displaystyle\int_0^\pi x \sin^n x\, dx = \pi \int_0^{\pi/2} \sin^n x\, dx,$

(c) $\displaystyle\int_0^\pi \frac{x \sin x}{\sqrt{(1 + \tan^2 \phi \sin^2 x)}}\, dx = \frac{\pi \phi}{\tan \phi},$ where $0 < \phi < \pi/2.$

(a) $\displaystyle\int_0^a f(x)\, dx = \int_0^{a/2} f(x)\, dx + \int_{a/2}^a f(x)\, dx \qquad \text{(by (4))}.$

Putting $x = a - u$ gives

$$\int_{a/2}^{a} f(x)\,dx = -\int_{a/2}^{0} f(a - u)\,du = \int_{0}^{a/2} f(a - u)\,du \qquad \text{(by (3))}.$$

$$= \int_{0}^{a/2} f(a - x)\,dx \qquad \text{(by (2))}.$$

The required result (a) follows.

(b) By (a), $\displaystyle \int_{0}^{\pi} x \sin^{n} x\,dx = \int_{0}^{\pi/2} [x \sin^{n} x + (\pi - x) \sin^{n} (\pi - x)]\,dx$

$$= \int_{0}^{\pi/2} [x \sin^{n} x + (\pi - x) \sin^{n} x]\,dx$$

$$= \pi \int_{0}^{\pi/2} \sin^{n} x\,dx.$$

(c) By (a),

$$I = \int_{0}^{\pi} \frac{x \sin x}{\sqrt{(1 + \tan^{2} \phi \sin^{2} x)}}\,dx$$

$$= \int_{0}^{\pi/2} \left[\frac{x \sin x}{\sqrt{(1 + \tan^{2} \phi \sin^{2} x)}} + \frac{(\pi - x) \sin (\pi - x)}{\sqrt{[1 + \tan^{2} \phi \sin^{2} (\pi - x)]}} \right] dx$$

$$= \int_{0}^{\pi/2} \frac{x \sin x + (\pi - x) \sin x}{\sqrt{(1 + \tan^{2} \phi \sin^{2} x)}}\,dx$$

$$= \pi \int_{0}^{\pi/2} \frac{\sin x}{\sqrt{(1 + \tan^{2} \phi \sin^{2} x)}}\,dx$$

$$= \pi \int_{0}^{\pi/2} \frac{\sin x}{\sqrt{(\sec^{2} \phi - \tan^{2} \phi \cos^{2} x)}}\,dx$$

$$= -\pi \int_{1}^{0} \frac{1}{\sqrt{(\sec^{2} \phi - u^{2} \tan^{2} \phi)}}\,du \quad \text{(on writing } u = \cos x)$$

$$= \frac{\pi}{\sec \phi} \int_{0}^{1} \frac{1}{\sqrt{(1 - u^{2} \sin^{2} \phi)}}\,du$$

$$= \frac{\pi}{\tan \phi} \left[\sin^{-1} (u \sin \phi) \right]_{0}^{1}$$

$$= \frac{\pi}{\tan \phi} [\sin^{-1} (\sin \phi)] = \frac{\pi\phi}{\tan \phi} \text{ since } 0 < \phi < \pi/2.$$

Exercises 7.1

1 By use of the substitution $y = a + b - x$, or otherwise, prove that

$$\int_a^b f(x)\,dx = \int_a^b f(a + b - x)\,dx.$$

Hence, or otherwise, evaluate

(a) $\displaystyle\int_0^\pi \frac{(\pi - x)}{1 + \sin x}\,dx,$ (b) $\displaystyle\int_{\pi/n}^{(n-1)\pi/n} x\,\sin^3 x\,dx.$

2 Prove that

(a) $\displaystyle\int_a^b f(kx)\,dx = \frac{1}{k}\int_{ka}^{kb} f(x)\,dx,$ where k is a constant;

(b) $\displaystyle\int_0^c x^m(c - x)^n\,dx = \int_0^c x^n\,(c - x)^m\,dx;$

(c) $\displaystyle\int_0^a (ax - x^2)^3\,[x^3 + (x - a)^3]\,dx = 0.$

3 Prove that

$$\int_0^a \frac{x^2}{x^2 + (x - a)^2}\,dx = \int_0^a \frac{(x - a)^2}{x^2 + (x - a)^2}\,dx = \tfrac{1}{2}a.$$

4 Show that, if $f(x) = f(a - x)$, then

$$\int_0^a xf(x)\,dx = \tfrac{1}{2}a\int_0^a f(x)\,dx.$$

Use this result to evaluate

(a) $\displaystyle\int_0^\pi \frac{x}{1 + \cos \phi \sin x}\,dx,$ $(0 < \phi < \pi),$

(b) $\displaystyle\int_0^\pi \frac{x \sin x}{1 + \cos^2 x}\,dx,$

(c) $\displaystyle\int_0^{2\pi} \frac{x \sin x}{1 + \cos^2 x}\,dx,$

5 Prove that

$$\int_0^\pi x\,f(\sin x)\,dx = \frac{\pi}{2}\int_0^\pi f(\sin x)\,dx.$$

Hence evaluate $\displaystyle\int_0^\pi \frac{x \sin^3 x}{1 + \cos^2 x}\,dx.$

6 Prove that $\displaystyle\int_{ra}^{(r+1)a} f(x)\,dx = \int_0^a f(x + ra)\,dx.$

Given that $f(x + a) = c\,f(x)$, where a and c are constants, prove that $f(x + ra) = c^r\,f(x)$ when r is an integer.

Deduce that, when n is a positive integer and c is not equal to unity,

$$(1 - c)\int_0^{na} f(x)\,dx = (1 - c^n)\int_0^a f(x)\,dx.$$

State the corresponding result when $c = 1$. [JMB]

7 By expressing $2\sin x\,(2r - 1)x$ as the difference of two cosines, show that

$$\sum_{r=1}^n \sin\,(2r - 1)x = \frac{\sin^2 nx}{\sin x}.$$

Hence show that $\displaystyle\int_{\pi/4}^{\pi/2} \frac{\sin^2 4x}{\sin x}\,dx = \frac{32\sqrt{2}}{105}.$ [L]

7.2 Reduction formulae

In *Advanced Mathematics 1* we considered integrals such as $\int x\cos x\,dx$, which we evaluated by one application of the method of integration by parts. If, however, we are faced with the problem of evaluating $\int x^6 \cos x\,dx$, we find that

$$\int x^6 \cos x\,dx = \int x^6 \frac{d}{dx}(\sin x)\,dx$$

$$= x^6 \sin x - \int \sin x\,\frac{d}{dx}(x^6)\,dx$$

$$= x^6 \sin x - 6\int x^5 \sin x\,dx.$$

It can be seen that five more applications of the method of integration by parts would be needed before the integral is reduced to $\int \cos x\,dx$.

To avoid the tedious repetition involved in carrying out the same process so many times, we seek to systematise the process. In the example above we would write

$$I_n = \int x^n \cos x\,dx,$$

and find the relation between I_n and I_{n-2}.

$$I_n = x^n \sin x - n\int x^{n-1} \sin x\,dx$$
$$= x^n \sin x - [nx^{n-1}(-\cos x) + n(n - 1)\int x^{n-2} \cos x\,dx].$$

So $\quad I_n = (x^n \sin x + nx^{n-1} \cos x) - n(n - 1)I_{n-2}. \qquad \ldots (1)$

Then we can write

$$I_6 = (x^6 \sin x + 6x^5 \cos x) - 30I_4,$$
$$I_4 = (x^4 \sin x + 4x^3 \cos x) - 12I_2,$$
$$I_2 = (x^2 \sin x + 2x \cos x) - 2I_0,$$

where
$$I_0 = \int \cos x \, dx = \sin x,$$

and hence I_6 can be obtained by successive substitution.
(Note that the arbitrary constant of integration is usually omitted in such discussions and its presence is assumed.)
The above process is referred to as 'integration by use of a reduction formula' (for obvious reasons).

Example 1
Given that $I_n = \int x^n e^{-x} dx$, show that

$$I_n = -x^n e^{-x} + nI_{n-1}.$$

Hence evaluate $\displaystyle\int_0^\infty x^n e^{-x} \, dx$, where $n \in \mathbb{Z}^+$.

$$I_n = \int x^n e^{-x} \, dx$$

$$= -\int x^n \frac{d}{dx}(e^{-x}) \, dx$$

$$= -x^n e^{-x} + \int e^{-x} \frac{d}{dx}(x^n) \, dx$$

$$= -x^n e^{-x} + n \int x^{n-1} e^{-x} dx.$$

But, replacing n by $n - 1$ in the definition of I_n, we have $I_{n-1} = \int x^{n-1} e^{-x} dx$.
Therefore

$$I_n = -x^n e^{-x} + nI_{n-1}. \qquad \ldots (1)$$

Let us now denote by J_n the definite integral

$$\int_0^\infty x^n e^{-x} dx, \quad \text{i.e.} \quad \lim_{X \to \infty} \int_0^X x^n e^{-x} dx.$$

Then, from (1),

$$J_n = \lim_{X \to \infty} \left[-x^n e^{-x} \right]_0^X + nJ_{n-1}$$

$$= \lim_{X \to \infty} (-X^n e^{-X}) + nJ_{n-1}$$

provided that $n > 0$. This last restriction is necessary so that J_{n-1} converges at the lower limit zero (see Section 6.4). But

$$\lim_{X \to \infty} (X^n e^{-X}) = 0 \quad \text{for } n > 0 \text{ (see page 205)}.$$

Therefore $\qquad\qquad\qquad J_n = nJ_{n-1} \quad \text{for } n > 0.$

Similarly $\qquad\qquad\qquad J_{n-1} = (n-1)J_{n-2}, \quad \text{for } n > 1,$

$\Rightarrow \qquad\qquad\qquad J_n = n(n-1)J_{n-2}, \quad \text{for } n > 1.$

Proceeding in this way we find

$$J_n = n(n-1)(n-2) \ldots 2 \cdot 1 J_0.$$

But $\qquad\qquad\qquad J_0 = \int_0^\infty e^{-x}\,dx = \lim_{X \to \infty} \int_0^X e^{-x}\,dx$

$$= \lim_{X \to \infty} \left[-e^{-x} \right]_0^X = \lim_{X \to \infty} (1 - e^{-X}) = 1.$$

Therefore, when $n \in \mathbb{Z}^+$, $J_n = n!$
For example, $J_6 = 6! = 720$.
(In advanced mathematics the formula

$$\int_0^\infty x^n e^{-x}\,dx = n!$$

is used to define $n!$, for $n \geqslant 0$, even when n is not a natural number. Note that, in this case, $0! = \int_0^\infty e^{-x}\,dx = 1$.)

Example 2
Find a reduction formula for J_n, where

$$J_n = \int_0^{\pi/2} x^n \cos x\,dx$$

and evaluate J_6.

Using equation (1) on page 240, we find

$$J_n = \left[x^n \sin x + nx^{n-1} \cos x \right]_0^{\pi/2} - n(n-1)J_{n-2}, \qquad n > 1,$$

i.e. $\qquad J_n = \left(\frac{\pi}{2}\right)^n - n(n-1)J_{n-2}, \qquad n > 1.$

So $\qquad J_6 = \left(\frac{\pi}{2}\right)^6 - 30J_4,$

$$J_4 = \left(\frac{\pi}{2}\right)^4 - 12J_2,$$

$$J_2 = \left(\frac{\pi}{2}\right)^2 - 2J_0 \quad \text{with } J_0 = 1,$$

giving $\quad J_6 = \left(\frac{\pi}{2}\right)^6 - 30\left(\frac{\pi}{2}\right)^4 + 360\left(\frac{\pi}{2}\right)^2 - 720.$

Example 3

Given that $\quad J_n = \displaystyle\int_0^{\pi/4} \tan^n x\, dx$, show that, for $n \geqslant 2$,

$$J_n = \frac{1}{n-1} - J_{n-2}.$$

Find the value of J_7.

In this case we can proceed as follows

$$J_n = \int_0^{\pi/4} \tan^n x\, dx$$

$$= \int_0^{\pi/4} (\tan^{n-2} x \sec^2 x - \tan^{n-2} x)\, dx$$

$$= \frac{1}{n-1}\left[\tan^{n-1} x\right]_0^{\pi/4} - J_{n-2}.$$

So $\quad J_n = \dfrac{1}{n-1} - J_{n-2}$, for $n > 1$, as required.

The condition $n > 1$ is necessary for J_{n-2} to converge at $x = 0$ and for $\tan^{n-1} x = 0$ when $x = 0$.

$$J_7 = \tfrac{1}{6} - J_5,$$
$$J_5 = \tfrac{1}{4} - J_3,$$
$$J_3 = \tfrac{1}{2} - J_1,$$

$$J_1 = \int_0^{\pi/4} \tan x\, dx = \left[\ln \sec x\right]_0^{\pi/4} = \tfrac{1}{2}\ln 2$$

$\Rightarrow \quad J_7 = \tfrac{1}{6} - \tfrac{1}{4} + \tfrac{1}{2} - \tfrac{1}{2}\ln 2 = \tfrac{5}{12} - \tfrac{1}{2}\ln 2.$

Example 4

Derive a reduction formula for J_n, where $J_n = \displaystyle\int_0^{\pi/2} \sin^n \theta\, d\theta$ and $n > 1$.

Hence evaluate

(a) $\displaystyle\int_0^{\pi/2} \sin^6 \theta \, d\theta,$ (b) $\displaystyle\int_0^{\pi/2} \cos^7 \phi \, d\phi.$

$$J_n = -\left[\sin^{n-1}\theta \cos\theta\right]_0^{\pi/2} + \int_0^{\pi/2} \cos\theta \, \frac{d}{d\theta}(\sin^{n-1}\theta) \, d\theta$$

$$= 0 + (n-1)\int_0^{\pi/2} \cos^2\theta \sin^{n-2}\theta \, d\theta$$

$$= (n-1)\int_0^{\pi/2} (1 - \sin^2\theta)\sin^{n-2}\theta \, d\theta$$

$$= (n-1)\int_0^{\pi/2} \sin^{n-2}\theta \, d\theta - (n-1)\int_0^{\pi/2} \sin^n\theta \, d\theta$$

$$= (n-1)J_{n-2} - (n-1)J_n$$

$$\Rightarrow \qquad nJ_n = (n-1)J_{n-2}. \qquad\qquad \dots (1)$$

(Note how the use of the trigonometric identity $\cos^2\theta \equiv 1 - \sin^2\theta$ has enabled us to recover our integral involving only powers of $\sin\theta$.)

The condition $n > 1$ is needed for the convergence of J_{n-2} at the limit $\theta = 0$ and for $\sin^{n-1}\theta \cos\theta$ to vanish when $\theta = 0$. Therefore

$$J_n = \frac{n-1}{n}J_{n-2}, \qquad \text{for } n > 1,$$

$$J_{n-2} = \frac{n-3}{n-2}J_{n-4}, \qquad \text{for } n > 3 \text{ etc.}$$

(a) Therefore
$$\int_0^{\pi/2} \sin^6\theta \, d\theta = J_6 = \frac{5 \cdot 3 \cdot 1}{6 \cdot 4 \cdot 2}J_0.$$

But
$$J_0 = \int_0^{\pi/2} \sin^0\theta \, d\theta = \int_0^{\pi/2} 1 \, d\theta = \frac{\pi}{2}.$$

Therefore
$$J_6 = \frac{5 \cdot 3 \cdot 1}{6 \cdot 4 \cdot 2}\frac{\pi}{2} = \frac{5\pi}{32}.$$

(b) Using result (10), page 235, and putting $\phi = \pi/2 - \theta$,

$$\int_0^{\pi/2} \cos^7\phi \, d\phi = \int_0^{\pi/2} \sin^7\theta \, d\theta = J_7.$$

$$J_7 = \frac{6 \cdot 4 \cdot 2}{7 \cdot 5 \cdot 3}J_1.$$

Also
$$J_1 = \int_0^{\pi/2} \sin\theta \, d\theta = \left[-\cos\theta\right]_0^{\pi/2} = 1.$$

Therefore
$$\int_0^{\pi/2} \cos^7 \phi \, d\phi = \frac{6 \cdot 4 \cdot 2}{7 \cdot 5 \cdot 3} \cdot 1 = \frac{16}{35}.$$

The results of Example 4 are of importance and enable us to write down the values of some definite integrals which occur frequently.

If $I_n = \displaystyle\int_0^{\pi/2} \sin^n \theta \, d\theta, \quad n > 1, \quad$ then $\quad I_n = \dfrac{n-1}{n} I_{n-2}.$

If $J_n = \displaystyle\int_0^{\pi/2} \cos^n \theta \, d\theta, \quad n > 1, \quad$ then $\quad J_n = \dfrac{n-1}{n} J_{n-2}.$

It follows that

$$I_n = J_n = \left[\frac{(n-1)(n-3) \dots 1}{n(n-2) \dots 2} \right] \frac{\pi}{2}, \qquad \text{when } n \text{ is even,}$$

$$I_n = J_n = \frac{(n-1)(n-3) \dots 2}{n(n-2) \dots 3 \cdot 1}, \qquad \text{when } n \text{ is odd.}$$

The formulae are sometimes called *Wallis's formulae*.

Example 5

Given that $I_{m,n} = \displaystyle\int_0^1 x^m(1-x)^n \, dx$, where m and n are positive integers, show that

$$I_{m,n} = \frac{m}{n+1} I_{m-1, n+1}.$$

Hence prove that

$$I_{m,n} = \frac{m! \, n!}{(m+n+1)!}.$$

Here the integral involves two parameters m and n. Proceeding as before, we have

$$I_{m,n} = -\frac{1}{n+1} \int_0^1 x^m \frac{d}{dx}[(1-x)^{n+1}] \, dx$$

$$= \left[-\frac{1}{(n+1)} x^m (1-x)^{n+1} \right]_0^1 + \frac{1}{n+1} \int_0^1 (1-x)^{n+1} \frac{d}{dx}(x^m) \, dx$$

$$= -0 + 0 + \frac{m}{n+1} \int_0^1 x^{m-1}(1+x)^{n+1} \, dx$$

$$\Rightarrow I_{m,n} = \frac{m}{n+1} I_{m-1, n+1}. \qquad \qquad \dots (1)$$

(Note that, since m and n are positive integers, the integrated part vanishes at both ends of the interval of integration.)

Successive use of this result gives

$$I_{m,n} = \left(\frac{m}{n+1}\right)\left(\frac{m-1}{n+2}\right)\left(\frac{m-2}{n+3}\right) \cdots \left(\frac{1}{n+m}\right) I_{0,n+m}.$$

But

$$I_{0,n+m} = \int_0^1 (1-x)^{m+n}\,dx = \left[\frac{-(1-x)^{m+n+1}}{m+n+1}\right]_0^1 = \frac{1}{m+n+1}.$$

Therefore

$$I_{m,n} = \frac{m!}{(m+n+1)!/n!}$$

$$= \frac{m!n!}{(m+n+1)!}$$

Note that if we had integrated by parts as follows:

$$I_{m,n} = \frac{1}{m+1}\int_0^1 (1-x)^n \frac{d}{dx}(x^{m+1})\,dx$$

$$= \left[\frac{1}{m+1}(1-x)^n\, x^{m+1}\right]_0^1 - \frac{1}{m+1}\int_0^1 x^{m+1}\frac{d}{dx}[(1-x)^n]\,dx$$

$$= 0 - 0 + \frac{n}{m+1}\int_0^1 x^{m+1}(1-x)^{n-1}\,dx$$

$$\Rightarrow \qquad I_{m,n} = \frac{n}{m+1}I_{m+1,n-1}, \qquad \qquad \ldots (2)$$

it appears we would have found a different reduction formula. However, writing $m-1$ for m and $n+1$ for n in equation (2) we have

$$I_{m-1,n+1} = \frac{n+1}{m}I_{m,n}$$

$$\Rightarrow \qquad I_{m,n} = \frac{m}{n+1}I_{m-1,n+1}$$

as before.

Example 6
Given that

$$I_n = \int x^n(x^2 + a^2)^{-1/2}\,dx, \qquad \text{where } n > 1,$$

show that, disregarding the constant of integration,

$$nI_n + (n - 1)a^2I_{n-2} = x^{n-1}(x^2 + a^2)^{1/2}. \qquad \dots (1)$$

When a reduction formula for an indefinite integral has to be established it is sometimes convenient to proceed as follows:

$$\frac{d}{dx}[x^{n-1}(x^2 + a^2)^{1/2}] = (n - 1)x^{n-2}(x^2 + a^2)^{1/2} + x^n(x^2 + a^2)^{-1/2}$$

$$= \frac{(n - 1)x^{n-2}(x^2 + a^2) + x^n}{(x^2 + a^2)^{1/2}} \qquad \dots (2)$$

$$\Rightarrow \quad \frac{d}{dx}[x^{n-1}(x^2 + a^2)^{1/2}] = \frac{nx^n}{(x^2 + a^2)^{1/2}} + \frac{(n - 1)a^2x^{n-2}}{(x^2 + a^2)^{1/2}}.$$

Integrating gives

$$x^{n-1}(x^2 + a^2)^{1/2} = n \int x^n(x^2 + a^2)^{-1/2}\,dx$$

$$+ (n - 1)a^2 \int x^{n-2}(x^2 + a^2)^{-1/2}\,dx,$$

which is the required result.

Note that we have differentiated the expression involving x on the right-hand side of the result (1) to be proved and then, in (2), rearranged the result in order to obtain the integrands of the integrals I_n and I_{n-2} of (1).

Exercises 7.2

1 Use Wallis's formulae to evaluate

(a) $\displaystyle\int_0^{\pi/2} \sin^2 x\,dx;$ (b) $\displaystyle\int_0^{\pi/2} \cos^5 x\,dx;$ (c) $\displaystyle\int_0^{\pi/2} \cos^8 x\,dx;$

(d) $\displaystyle\int_0^{\infty} \frac{1}{(1 + x^2)^4}\,dx$ (hint: put $x = \tan \theta$);

(e) $\displaystyle\int_0^{\infty} \frac{1}{[(x - a)^2 + b^2]^2}\,dx$, where $a > 0, b > 0$
   (hint: put $x - a = b \tan \theta$).

2 Given that, for $n \geqslant 0$,

$$I_n = \int_0^1 x^n e^x\,dx,$$

show that, for $n \geqslant 1$,

$$I_n = e - nI_{n-1}.$$

Evaluate I_0 and I_3. [L]

3 Given that

$$I_n = \int_0^\pi x^n \sin{(x/2)}\,dx,$$

prove that, for $n > 1$,

$$I_n + 4n(n - 1)I_{n-2} = 4n(\pi)^{n-1}.$$

Calculate I_3. [JMB]

4 Given that

$$I_n(x, k) = \int_0^x t^n e^{kt}\,dt, \qquad k \neq 0, n \geq 0,$$

show that, for $n \geq 1$,

$$I_n(x, k) = \frac{1}{k}[x^n e^{kx} - nI_{n-1}(x, k)].$$

Find $I_3(-1, 1)$ in terms of e. [JMB]

5 Given that $\quad I_n = \int_0^1 x^n \cosh x\,dx,$

prove that, $\quad$ for $n \geq 2$,

$$I_n = \sinh 1 - n \cosh 1 + n(n - 1)I_{n-2}.$$

Evaluate $\quad$ (a) $\int_0^1 x^4 \cosh x\,dx,$

$\qquad\qquad$ (b) $\int_0^1 x^3 \cosh x\,dx,$

expressing each answer in terms of e. [JMB]

6 Given that, for $n \geq 0$,

$$I_n = \int_0^\pi x^n \sin x\,dx,$$

prove that, for $n \geq 2$,

$$I_n = \pi^n - n(n - 1)I_{n-2}.$$

Hence evaluate

$$\int_0^\pi x^5 \sin x\,dx. \qquad\qquad\text{[L]}$$

7 Let $I_n = \int_0^{\pi/4} \tan^n x\,dx$. Given that n is a positive integer $(n \geq 2)$, prove that

$$I_n + I_{n-2} = \frac{1}{n-1}.$$

Find the value of I_6.

8 Given that $I_n = \displaystyle\int_0^1 (1 - x^3)^n \, dx$, where n is an integer $\geqslant 0$, show that, for $n \geqslant 1$,

$$(3n + 1)I_n = 3nI_{n-1}.$$

9 Given that $I_n = \displaystyle\int_0^2 x^{1/2} (2 - x)^{n+1/2} \, dx$, prove that, for $n > -\frac{1}{2}$,

$$(n + 2)I_n = (2n + 1)I_{n-1}.$$

Hence, or otherwise, evaluate $\displaystyle\int_0^2 x^{3/2} (2 - x)^{5/2} \, dx$.

10 Given that

$$I_n = \int_1^e (\ln x)^n \, dx \qquad n \geqslant 0,$$

show that

$$I_n = e - nI_{n-1}, \qquad n \geqslant 1.$$

Evaluate I_3. [JMB]

11 Given that

$$I_n = \int_0^1 x^p(1 - x)^n \, dx,$$

where p and n are positive, show that

$$(n + p + 1) \, I_n = nI_{n-1}. \qquad \text{[JMB]}$$

12 Given that $I_n = \displaystyle\int_1^2 x(\ln x)^n \, dx$, where n is a non-negative integer, prove that, for $n \geqslant 1$,

$$2I_n + nI_{n-1} = 4(\ln 2)^n.$$

Evaluate

$$\int_1^2 x(\ln x)^3 \, dx. \qquad \text{[L]}$$

13 Given that $I_n = \displaystyle\int_0^{\text{arsinh } 1} \sinh^n x \, dx$, prove that, for $n \geqslant 2$,

$$nI_n = \sqrt{2} - (n - 1)I_{n-2}. \qquad \text{[L]}$$

14 Given that $I_n = \displaystyle\int_{-1}^0 \frac{1}{(x^2 + 2x + 2)^n} \, dx$, prove that

$$2nI_{n+1} = (2n - 1)I_n + 2^{-n}.$$

Find the value of

$$\int_{-1}^0 \frac{1}{(x^2 + 2x + 2)^3} \, dx. \qquad \text{[L]}$$

15 Given that
$$I_n = \int_0^1 x^n(1 - x)^{1/2}\,dx,$$

show that, for $n > 0$, $\quad I_n = \dfrac{2n}{2n + 3}I_{n-1}.$

Hence evaluate $\displaystyle\int_0^1 x^3(1 - x)^{1/2}\,dx.$

16 Given that
$$I_n = \int_0^{\pi/4} \sec^n \theta\,d\theta,$$

show that $\quad (n - 1)I_n = (\sqrt{2})^{n-2} + (n - 2)I_{n-2}.$

Show also that $\quad 8\displaystyle\int_0^{\pi/4} \sec^5 \theta\,d\theta = 7\sqrt{2} + 3\ln(1 + \sqrt{2}),$

and evaluate $\displaystyle\int_0^1 \dfrac{1}{(2 - x^2)^3}\,dx.$ [L]

17 Find $\dfrac{dy}{dx}$ when $y = a^x$, where a is a positive constant.

Given that
$$I_n = \int_0^{\pi} a^x \sin^n x\,dx, \qquad \text{where } n > -1,$$

show that $\quad I_n[(\ln a)^2 + n^2] = n(n - 1)I_{n-2}.$

Hence find $\displaystyle\int_0^{\pi} e^x \sin^3 x\,dx.$

18 Given that
$$I_{m,n} = \int_0^1 (1 - x^m)^n\,dx, \qquad \text{where } m, n > 0,$$

show that $\quad (mn + 1)I_{m,n} = mnI_{m,n-1}.$ [L]

19 Given that
$$I_{m,n} = \int_0^{\pi/2} \sin^m x \cos^n x\,dx,$$

where $m > -1,\ n > 1$, show that
$$(m + n)\,I_{m,n} = (n - 1)I_{m,n-2}.$$

20 Using the result of 19, or otherwise, evaluate:

(a) $\displaystyle\int_0^{\pi/2} \sin^3 x \cos^7 x\,dx;$ (b) $\displaystyle\int_0^{\pi/2} \sin^6 x \cos^4 x\,dx;$

(c) $\displaystyle\int_0^a x^4(a^2 - x^2)^{3/2}\,dx,$ where $a > 0$ (hint: put $x = a\sin\theta$);

(d) $\displaystyle\int_0^b x^5(b - x)^4\,dx$, where $b > 0$ (hint: put $x = b \sin^2 \theta$);

(e) $\displaystyle\int_0^\infty \frac{x^4}{(a^2 + x^2)^4}\,dx$, where $a > 0$ (hint: put $x = a \tan \theta$).

7.3 Applications of integration

In *Advanced Mathematics 1* we used integration to find areas of regions and volumes of solids of revolution, and in *Advanced Mathematics 2* we found centroids of plane regions and solids of revolution. The concept of a definite integral as the limit of a sum will now be used to solve a number of other problems.

The length of a curve

First we give a definition of the length of a plane curve. Suppose (Fig. 7.2) that we wish to find the length of the finite arc AB of the curve shown. We take n intermediate points $P_1, P_2, \ldots, P_n$ on the curve and find the sum of the lengths of the $(n + 1)$ chords $AP_1, P_1P_2, \ldots, P_nB$. Then we *define* the length of the arc to be the limit of this sum as both $n \to \infty$ *and* the length of each chord tends to zero.

Suppose now that R, S are the points (x, y), $(x + \delta x, y + \delta y)$ on the curve $y = f(x)$ (Fig. 7.3). Let δs be the length of the arc RS in Fig. 7.3. Then

$$\frac{\delta s}{\delta x} = \frac{\text{chord RS}}{\delta x} \times \frac{\delta s}{\text{chord RS}}$$

where chord $RS = \sqrt{[(\delta x)^2 + (\delta y)^2]}$.

Assuming that the ratio of the chord RS to δs tends to 1 as $\delta x \to 0$, this gives, in the limit

$$\frac{ds}{dx} = \lim_{\delta x \to 0} \frac{\text{chord RS}}{\delta x} = \lim_{\delta x \to 0} \sqrt{\left[1 + \left(\frac{\delta y}{\delta x}\right)^2\right]}$$

$$= \sqrt{\left[1 + \left(\frac{dy}{dx}\right)^2\right]}.$$

Suppose that s increases with x. Then it follows that the length of the arc AB, where $A \equiv [a, f(a)]$, $B \equiv [b, f(b)]$ is l, where

$$l = \int_{x=a}^{x=b} 1\,ds = \int_a^b \frac{ds}{dx}\,dx = \int_a^b \sqrt{\left[1 + \left(\frac{dy}{dx}\right)^2\right]}\,dx.$$

If the curve is given parametrically by $x = x(t)$, $y = y(t)$, then

$$l = \int_{t_1}^{t_2} \sqrt{\left[\left(\frac{dx}{dt}\right)^2 + \left(\frac{dy}{dt}\right)^2\right]}\,dt,$$

where t_1, t_2 are the parameters corresponding to a, b respectively.

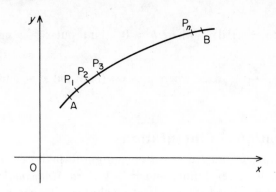

Fig. 7.2

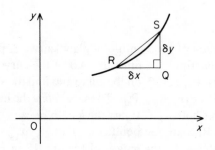

Fig. 7.3

Example 1

Calculate the length of the arc of the curve $y = \ln(\sec x)$ between the points where $x = 0$ and $x = \pi/4$.

Here $\dfrac{\mathrm{d}y}{\mathrm{d}x} = \tan x$.

$$\therefore \qquad \text{Arc length} = \int_0^{\pi/4} \sqrt{(1 + \tan^2 x)}\,\mathrm{d}x = \int_0^{\pi/4} \sec x\,\mathrm{d}x$$

$$= \left[\ln(\sec x + \tan x)\right]_0^{\pi/4} = \ln(1 + \sqrt{2}).$$

Example 2

Calculate the length of the arc of the parabola $x = at^2$, $y = 2at$, where $a > 0$, between the origin and the point where $t = 1$.

$$\text{Arc length} = \int_0^1 \sqrt{[(2at)^2 + (2a)^2]}\,\mathrm{d}t$$

$$= 2a \int_0^1 \sqrt{(t^2 + 1)}\,\mathrm{d}t.$$

Let $t = \sinh u$, so that $\mathrm{d}t/\mathrm{d}u = \cosh u$.

Then arc length $= 2a \displaystyle\int_0^\beta \sqrt{(\sinh^2 u + 1)} \cosh u \, \mathrm{d}u$,

where $\sinh \beta = 1$, $\cosh \beta = \sqrt{2}$, $e^\beta = 1 + \sqrt{2}$, $\beta = \ln (1 + \sqrt{2})$.

Therefore, arc length $= 2a \displaystyle\int_0^\beta \cosh^2 u \, \mathrm{d}u$

$$= a \int_0^\beta (1 + \cosh 2u) \, \mathrm{d}u$$

$$= a \left[u + \sinh u \cosh u \right]_0^\beta$$

$$= a \left[\ln (1 + \sqrt{2}) + \sqrt{2} \right].$$

Example 3   Arc length in polar coordinates
Since $x = r \cos \theta$ and $y = r \sin \theta$,

$$\frac{\mathrm{d}x}{\mathrm{d}\theta} = \frac{\mathrm{d}r}{\mathrm{d}\theta} \cos \theta - r \sin \theta, \qquad \frac{\mathrm{d}y}{\mathrm{d}\theta} = \frac{\mathrm{d}r}{\mathrm{d}\theta} \sin \theta + r \cos \theta$$

$$\Rightarrow \qquad \left(\frac{\mathrm{d}x}{\mathrm{d}\theta}\right)^2 + \left(\frac{\mathrm{d}y}{\mathrm{d}\theta}\right)^2 = r^2 + \left(\frac{\mathrm{d}r}{\mathrm{d}\theta}\right)^2$$

$$\Rightarrow \qquad \frac{\mathrm{d}s}{\mathrm{d}\theta} = \sqrt{\left[r^2 + \left(\frac{\mathrm{d}r}{\mathrm{d}\theta}\right)^2 \right]}.$$

It follows that, if s increases with θ, then the arc of the curve $r = f(\theta)$ from the point where $\theta = \alpha$ and $\theta = \beta$ is

$$\int_\alpha^\beta \sqrt{\left[r^2 + \left(\frac{\mathrm{d}r}{\mathrm{d}\theta}\right)^2 \right]} \, \mathrm{d}\theta.$$

For example, the total length of the cardioid $r = a(1 + \cos \theta)$, where $a > 0$, Fig. 5.16, is

$$2 \int_0^\pi \sqrt{[a^2 (1 + \cos \theta)^2 + a^2 \sin^2 \theta]} \, \mathrm{d}\theta$$

$$= 2a \int_0^\pi \sqrt{(2 + 2 \cos \theta)} \, \mathrm{d}\theta = 4a \int_0^\pi \cos (\theta/2) \, \mathrm{d}\theta$$

$$= 8a \left[\sin (\theta/2) \right]_0^\pi = 8a.$$

Area of a surface of revolution
Referring again to Figs 7.2 and 7.3, when the curve $y = f(x)$ is rotated through 2π about Ox, each of the chords $\mathrm{AP}_1, \mathrm{P}_1\mathrm{P}_2, \ldots$ will generate the curved surface

of the frustum of a right circular cone. The total area of the surface of revolution generated is defined as the limit of the sum of the areas of all the frustums generated by the chords as $n \to \infty$ *and* the length of each chord tends to zero.

The area generated by the chord RS lies between $2\pi y \delta s$ and $2\pi(y + \delta y)\,\delta s$, where δs = chord RS, R is the point (x, y), and S is the point $(x + \delta x, y + \delta y)$. Then the area of the surface of revolution generated by the arc AB is

$$\int_{x=a}^{x=b} 2\pi y \,\mathrm{d}s = 2\pi \int_a^b y \frac{\mathrm{d}s}{\mathrm{d}x}\,\mathrm{d}x = 2\pi \int_a^b y \sqrt{\left[1 + \left(\frac{\mathrm{d}y}{\mathrm{d}x}\right)^2\right]}\,\mathrm{d}x.$$

Example

The area of the surface of revolution generated by the arc of the parabola defined in Example 2 on page 252 is

$$2\pi \int_0^a y \sqrt{\left[1 + \left(\frac{\mathrm{d}y}{\mathrm{d}x}\right)^2\right]}\,\mathrm{d}x.$$

But $x = at^2$, $y = 2at$ $\Rightarrow$ $\dfrac{\mathrm{d}y}{\mathrm{d}x} = \dfrac{1}{t}$.

Therefore, $\qquad\qquad \text{area} = 8\pi a^2 \int_0^1 t\sqrt{(1 + t^2)}\,\mathrm{d}t$

$$= 8\pi a^2 \left[\tfrac{1}{3}(1 + t^2)^{3/2}\right]_0^1$$

$$= 8\pi a^2 \,[2\sqrt{2} - 1]/3.$$

Exercises 7.3(i)

1   Find the length of arc of the curve with parametric equations

$$x = a(1 - 3t^2), \qquad y = a(t - 3t^3),$$

where a is a constant, between the point $(a, 0)$ and the point $(-2a, -2a)$.
[L]

2   Find the arc length of the curve whose parametric equations are

$$x = a\,(t - \sin t),\ y = a\,(1 - \cos t), \qquad 0 \leqslant t \leqslant 2\pi. \quad \text{[JMB]}$$

3   Sketch the curve with parametric equations $x = a\cos^3 t$, $y = a\sin^3 t$ where $a > 0$ and $0 \leqslant t < 2\pi$. Show that the total length of the curve is $6a$.
[L]

4   Show that the length of the arc of the curve

$$9y^2 = 4x^3, \qquad y > 0,$$

between the points for which $x = 3$ and $x = 8$ is $38/3$.
[L]

5 Find the area of the finite region R bounded by the parabola $y^2 = 4x$ and the line $y = 2x - 4$.

Find also the total surface area of the solid formed by rotating that part of R lying in the first quadrant through 2π radians about the x-axis.

[JMB]

6 The arc of the curve $y = \cosh x$ between the points for which $x = 0$ and $x = a$ is rotated through 2π radians about the x-axis to form a surface of area S. The area of the finite region enclosed by this arc, the x-axis, the y-axis and the line $x = a$ is A. Show that

$$\frac{S}{A} = \frac{\pi(2a + \sinh 2a)}{2 \sinh a}.$$

[L]

7 The tangent at the point P on the catenary $y = c \cosh(x/c)$, where $c > 0$, cuts the x-axis in T. The normal at P cuts the x-axis in G. The perpendicular from P to the x-axis cuts the x-axis at N and the perpendicular from N to the line PT cuts the line PT at Q. Show that
(a) the length of NQ is equal to c,
(b) the length of PQ is equal to the length of the arc of the catenary from the point $(0, c)$ to P.

[L]

8 Calculate the length of the arc of the curve with parametric equations

$$\begin{cases} x = a(3 \cos t - \cos 3t) \\ y = a(3 \sin t - \sin 3t) \end{cases}$$

between the points corresponding to $t = 0$ and $t = \pi$.

Find the area of the region enclosed between this arc and the x-axis.

Find also the area of the surface of revolution formed when this arc is rotated through an angle of 2π about the x-axis.

9 A curve is represented parametrically by the equations

$$x = a(\theta - \sin \theta), \quad y = a(1 - \cos \theta).$$

Find the arc-length from the origin O to the point P of parameter θ where $0 \leqslant \theta \leqslant 2\pi$.

Find also the area of the curved surface obtained by rotating the arc OP through four right angles about the x-axis, proving that this area is $32\pi a^2/3$ when P is the point $(a\pi, 2a)$.

10 The parametric equations of a curve are

$$\frac{x}{a} = 3 \cos 2t - 2 \cos 3t, \qquad \frac{y}{a} = 3 \sin 2t - 2 \sin 3t, \qquad (0 \leqslant t < 2\pi),$$

where a is a constant. Show that, at the point P on the curve whose parameter is t, the tangent makes an angle $5t/2$ with the axis of x. Find the length of the arc K traced out as the parameter increases from 0 to $\pi/2$.

Prove that the area of the surface of revolution formed when the arc K is rotated once about the axis of x is $(144\pi a^2 \sqrt{2})/7$. (You need not verify that $y \geqslant 0$ at all points of K.) [JMB]

11 Find the length of the spiral $r = ae^{k\theta}$, where $a > 0$, $k > 0$, from the point where $\theta = 0$ to the point where $\theta = 2\pi$.

12 Find the length of the spiral $r = a\theta$, where $a > 0$, from the the pole to the point where $\theta = 2\pi$.

Mean values and root mean square

In *Advanced Mathematics 1*, Section 10.7, we defined the *mean value* of f(x) with respect to x, for values of x from $x = a$ to $x = b$, as

$$\frac{1}{b - a} \int_a^b f(x)\,dx,$$

i.e. the average height of the curve $y = $ f(x) in the interval $a \leqslant x \leqslant b$. This should be carefully distinguished from the mean value of a probability distribution or a frequency distribution.

Example 1

Find the mean value of $\ln x$ in the interval $1 \leqslant x \leqslant$ e.

$$\text{Mean value} = \frac{1}{e - 1} \int_1^e \ln x\,dx = \frac{1}{e - 1}\left[x \ln x - x\right]_1^e = \frac{1}{e - 1}.$$

Example 2

A particle P moves along Ox so that at time t its displacement x is given by $x = a \sin \omega t$, where a, ω are positive constants. Find, for the interval $t = 0$ to $t = \pi/(2\omega)$, the mean value of the speed of P (a) with respect to t, (b) with respect to x.

The speed v is given by

$$v = \frac{dx}{dt} = \omega a \cos \omega t.$$

(a) Mean value of v with respect to t for $0 \leqslant t \leqslant \pi/(2\omega)$ is

$$\frac{1}{\pi/(2\omega)} \int_0^{\pi/(2\omega)} \omega a \cos \omega t\,dt = \frac{2\omega^2 a}{\pi}\left[\frac{1}{\omega} \sin \omega t\right]_0^{\pi/(2\omega)} = \frac{2\omega a}{\pi}.$$

(b) $\qquad\qquad\qquad v = a \cos \omega t,\; x = a \sin \omega t$

$\Rightarrow \qquad\qquad\qquad v = \omega\sqrt{(a^2 - x^2)}.$

Mean value of v with respect to x for $0 \leqslant t \leqslant \pi/(2\omega)$, i.e. $0 \leqslant x \leqslant a$, is

$$\frac{1}{a} \int_0^a \omega\sqrt{(a^2 - x^2)}\,dx \;=\; \omega a \int_0^{\pi/2} \cos^2\theta\,d\theta, \quad \text{using the substitution } x = a\sin\theta,$$

$$= \pi\omega\,a/4 \qquad\qquad \text{(using Wallis).}$$

(Note that these two mean values are not equal.)

We now define the root mean square (r.m.s.) value of $f(x)$ over an interval to be the square root of the mean value of $[f(x)]^2$ over that interval; i.e. the r.m.s. value of $f(x)$ with respect to x for $a \leqslant x \leqslant b$ is

$$\sqrt{\left(\frac{1}{b-a}\int_a^b [f(x)]^2\ dx\right)}.$$

Example 1
The r.m.s. value of $\cos^3 x$ for $0 \leqslant x \leqslant \pi/2$ is

$$\sqrt{\left(\frac{1}{\pi/2}\int_0^{\pi/2} \cos^6 x\,dx\right)} = \sqrt{\left(\frac{2}{\pi}\cdot\frac{5\cdot3\cdot1}{6\cdot4\cdot2}\frac{\pi}{2}\right)} = \sqrt{\left(\frac{5}{16}\right)} \quad \text{(using Wallis).}$$

Example 2
Given that

$$I = I_1 \sin \omega t + I_2 \sin 2\omega t,$$

where I_1, I_2, ω are constants, find the r.m.s. value of I with respect to t for $0 \leqslant t \leqslant 2\pi/\omega$.

$$\int_0^{2\pi/\omega} I^2\,dt = \int_0^{2\pi/\omega} (I_1{}^2 \sin^2 \omega t + 2I_1 I_2 \sin \omega t \sin 2\omega t + I_2{}^2 \sin^2 2\omega t)\,dt.$$

But
$$\int_0^{2\pi/\omega} \sin^2 \omega t\,dt = \frac{1}{2}\int_0^{2\pi/\omega} (1 - \cos 2\omega t)\,dt$$

$$= \frac{1}{2}\left[t - \frac{1}{2\omega}\sin 2\omega t\right]_0^{2\pi/\omega} = \pi/\omega.$$

Similarly,
$$\int_0^{2\pi/\omega} \sin^2 2\omega t\,dt = \pi/\omega.$$

Also
$$\int_0^{2\pi/\omega} \sin \omega t \sin 2\omega t\,dt$$

$$= \frac{1}{2}\int_0^{2\pi/\omega} (\cos \omega t - \cos 3\omega t)\,dt$$

$$= \frac{1}{2\omega}\left[\sin \omega t - \frac{1}{3}\sin 3\omega t\right]_0^{2\pi/\omega} = 0.$$

Therefore, the r.m.s. value of I is given by

$$\sqrt{\left(\frac{1}{2\pi/\omega} \int_0^{2\pi/\omega} I^2 \, dt\right)} = \sqrt{(\tfrac{1}{2}I_1^2 + \tfrac{1}{2}I_2^2)}.$$

This result can be generalised. The generalisation is left to the reader.

Exercises 7.3(ii)

For the expressions given in **1–4**, find (a) the mean value, (b) the r.m.s. value with respect to x, over the stated interval.

1 $\tan x$ for $0 \leqslant x \leqslant \pi/4$.

2 $x \sin x$ for $0 \leqslant x \leqslant \pi$.

3 $\sqrt{(a^2 - x^2)}$ for $0 \leqslant x \leqslant a$.

4 $4 \cos x - 3 \sin x$ for $0 \leqslant x \leqslant 2\pi$.

5 Given that $y = a(\sin \omega t + |\sin \omega t|)$, find the mean value of y with respect to t for $0 \leqslant t \leqslant 2\pi/\omega$.

6 Given that m and $n \in \mathbb{Z}^+$, show that

(a) $\displaystyle\int_0^{2\pi} \cos mx \cos nx \, dx = 0$, unless $m = n$,

(b) $\displaystyle\int_0^{2\pi} \cos^2 mx \, dx = \pi$.

Hence find the r.m.s. value of $\displaystyle\sum_{r=1}^{k} \cos rx$ with respect to x for $0 \leqslant x \leqslant 2\pi$.

First moments

Suppose that a particle of mass m is situated at the point with coordinates (x, y, z) with respect to a set of cartesian axes Ox, Oy, Oz (in three dimensions). Then the first moments of this particle about the planes $x = 0$, $y = 0$, $z = 0$, are defined to be mx, my, mz respectively. (This definition differs from that of the moment of a force in statics.) This definition can be extended to a system of particles, of which a typical particle is of mass m_i and is situated at (x_i, y_i, z_i), $i = 1, 2 \ldots, n$. In this case the first moments are

$$\sum_{i=1}^{n} m_i x_i, \quad \sum_{i=1}^{n} m_i y_i, \quad \sum_{i=1}^{n} m_i z_i.$$

Denoting $\displaystyle\sum_{i=1}^{n} m_i$ by M,

we define the point $(\bar{x}, \bar{y}, \bar{z})$, where

$$M\bar{x} = \sum_{i=1}^{n} m_i x_i, \quad M\bar{y} = \sum_{i=1}^{n} m_i y_i, \quad M\bar{z} = \sum_{i=1}^{n} m_i z_i,$$

to be the *centre of mass* of the system.

The concept of first moment and the corresponding mean centre can be applied to any entity which is distributed. This distribution may be a distribution in space, e.g. a length, an area or a volume, or over a population such as a statistical frequency distribution. Any sum, discrete or continuous, such as

$$\sum_{i=1}^{n} x_i f_i \quad \text{or} \quad \int_a^b x \, f(x) \, dx$$

is a first moment and can be used to define a mean centre $\bar{x}$ by the equations

$$\bar{x} \sum_{i=1}^{n} f_i = \sum_{i=1}^{n} x_i f_i \quad \text{or} \quad \bar{x} \int_a^b f(x) \, dx = \int_a^b x \, f(x) \, dx.$$

The position of the centre of mass of some simple uniform bodies was found in *Advanced Mathematics 2*, Section 7.7.

The *centroid* of any continuous body is defined to be the centre of mass of the space (arc, area or volume), considered as of uniform density, occupied by that body.

Example 1

Find the coordinates of the centroid of the arc of the catenary $y = c \cosh (x/c)$, where $c > 0$, from the point where $x = 0$ to the point where $x = c$.

Let the centroid be at the point $(\bar{x}, \bar{y})$. Then

$$\bar{x} = \lim_{\delta s \to 0} \left(\frac{\sum x \, \delta s}{\sum \delta s} \right) = \frac{\int_{x=0}^{x=c} x \, ds}{\int_{x=0}^{x=c} ds} = \frac{\int_0^c x \sqrt{[1 + \sinh^2 (x/c)]} \, dx}{\int_0^c \sqrt{[1 + \sinh^2 (x/c)]} \, dx}$$

$$= \frac{\int_0^c x \cosh (x/c) \, dx}{\int_0^c \cosh (x/c) \, dx} = \frac{\left[cx \sinh (x/c) - c^2 \cosh (x/c) \right]_0^c}{\left[c \sinh (x/c) \right]_0^c}$$

$$= c(1 - e^{-1})/\sinh 1.$$

Similarly,

$$\bar{y} = \frac{\int_{x=0}^{x=c} y \, ds}{\int_{x=0}^{x=c} ds} = \frac{c \int_0^c \cosh^2 (x/c) \, dx}{\int_0^c \cosh (x/c) \, dx}$$

$$= \frac{\frac{1}{2} c \left[x + c \cosh (x/c) \sinh (x/c) \right]_0^c}{\left[c \sinh (x/c) \right]_0^c}$$

$$= c \, [1 + \sinh 1 \cosh 1]/(2 \sinh 1).$$

Example 2

Find the coordinates of the centroid of the region bounded by the curve $r = f(\theta)$ and the half-lines $\theta = \alpha$, $\theta = \beta$.

With reference to Fig 5.17, the elemental sector OPR has area $\frac{1}{2}r^2\, \delta\theta$ and its centre of mass is at the point with cartesian coordinates $(\frac{2}{3}r \cos\theta, \frac{2}{3}r \sin\theta)$. Then the coordinates $(\bar{x}, \bar{y})$ of the centroid of the region are given by

$$\bar{x} = \lim_{\delta\theta \to 0} \left(\frac{\Sigma \frac{2}{3}r \cos\theta \times \frac{1}{2}r^2\, \delta\theta}{\Sigma \frac{1}{2}r^2\, \delta\theta} \right) = \frac{\frac{2}{3}\displaystyle\int_{\alpha}^{\beta} r^3 \cos\theta\, d\theta}{\displaystyle\int_{\alpha}^{\beta} r^2\, d\theta},$$

$$\bar{y} = \lim_{\delta\theta \to 0} \left(\frac{\Sigma \frac{2}{3}r \sin\theta \times \frac{1}{2}r^2\, \delta\theta}{\Sigma \frac{1}{2}r^2\, \delta\theta} \right) = \frac{\frac{2}{3}\displaystyle\int_{\alpha}^{\beta} r^3 \sin\theta\, d\theta}{\displaystyle\int_{\alpha}^{\beta} r^2\, d\theta}.$$

The theorems of Pappus

Suppose that the region R bounded by a closed curve C, Fig 7.4, is rotated through an angle α about the line Ox which does not cross C. Then the volume of the solid of revolution generated by that element δS of R which is at a distance y from Ox is the same as that of a cylinder of cross-section δS and length αy, i.e. $\alpha y\, \delta S$. The total volume of the solid of revolution is V, where $V = \lim_{\delta S \to 0} \Sigma\, \alpha y\, \delta S$, the summation being over all elements δS of R.

Therefore $V = \alpha \int y\, dS$, since α is constant. But the y-coordinate $\bar{y}_A$ of the centroid of the region R is given by

$$\bar{y}_A = \frac{\int y\, dS}{\int dS} = \frac{\int y\, dS}{A},$$

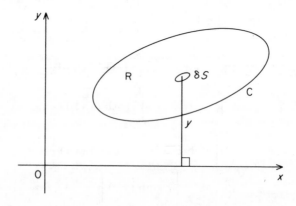

Fig. 7.4

where A is the area of the region R. Therefore

$$V = \alpha \bar{y}_A A,$$

i.e. *the volume of revolution swept out (generated) by the region R enclosed by a plane curve, when that curve rotates about a line l in its plane and not intersecting R, is equal to the area of that region multiplied by the length of the path of the centroid of that region.*

This is the first of the theorems of Pappus or Guldin.

Example 1
A circle of radius a is rotated through 2π about a line in its plane which is at a distance b, where $b > a$, from the centre of the circle. The volume of the anchor ring, or torus, thus generated, is $(2\pi b)(\pi a^2)$, i.e. $2\pi^2 a^2 b$.

Example 2
When the region bounded by the curve $r = f(\theta)$ and the half-lines $\theta = \alpha$, $\theta = \beta$, Fig 5.17, is rotated through 2π about Ox, the volume generated by the elementary sector, whose centroid is at a distance $\frac{2}{3}r \sin \theta$ from Ox, is approximately $2\pi \times \frac{2}{3} r \sin \theta \times \frac{1}{2} r^2 \,\delta\theta$. Hence the total volume of the solid of revolution is

$$\lim_{\delta\theta \to 0} \frac{2\pi}{3} \Sigma r^3 \sin \theta \, \delta\theta = \frac{2\pi}{3} \int_\alpha^\beta r^3 \sin \theta d\theta.$$

We now consider the area of the curved surface generated by the curve C of Fig. 7.4 when rotated through the angle α about Ox. An element of arc, of length δs, generates a 'strip' of width δs and length αy, where y is the distance of the element from Ox. The total area of the curved surface generated is S, where $S = \lim_{\delta s \to 0} \Sigma \alpha y \, \delta s$. But, by definition, $L\bar{y}_c = \lim_{\delta s \to 0} \Sigma y \, \delta s$, where L is the length of the curve C and $\bar{y}_c$ is the distance of the centroid of C from Ox. Therefore

$$S = \alpha \bar{y}_c L,$$

i.e. *the area of the curved surface swept out (generated) by a curve C, when that curve is rotated about a line l in its plane and not intersecting C, is equal to the length of the curve C multiplied by the length of the path of the centroid of that curve.*

This is the second of the theorems of Pappus. Note that the result clearly holds even if C is *not* a closed curve.

Example 1
The area of the surface of the anchor ring generated when a circle of radius a is rotated about a line in its plane, distant b, where $b > a$, from the centre of the circle, is $2\pi b \times 2\pi a = 4\pi^2 ab$.

Example 2

Find the x-coordinate of the centroid of that arc of the curve $y = \cosh x$ which lies between the lines $x = 0$ and $x = \ln 2$. Find also the area of the surface generated when this arc is rotated through 2π about the y-axis.

The first moment about Oy of the element δs of arc shown in Fig 7.5 is $x\,\delta s$. Therefore

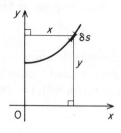

Fig. 7.5

$$\bar{x} = \lim_{\delta s \to 0} \left(\frac{\Sigma x\,\delta s}{\Sigma\,\delta s} \right) = \frac{\displaystyle\int_{x=0}^{x=\ln 2} x\,ds}{\displaystyle\int_{x=0}^{x=\ln 2} ds}$$

$$= \frac{\displaystyle\int_0^{\ln 2} x\sqrt{\left[1 + \left(\frac{dy}{dx}\right)^2\right]}\,dx}{\displaystyle\int_0^{\ln 2} \sqrt{\left[1 + \left(\frac{dy}{dx}\right)^2\right]}\,dx} = \frac{\displaystyle\int_0^{\ln 2} x\sqrt{(1 + \sinh^2 x)}\,dx}{\displaystyle\int_0^{\ln 2} \sqrt{(1 + \sinh^2 x)}\,dx}$$

$$= \frac{\displaystyle\int_0^{\ln 2} x\cosh x\,dx}{\displaystyle\int_0^{\ln 2} \cosh x\,dx} = \frac{\left[x\sinh x - \cosh x\right]_0^{\ln 2}}{\left[\sinh x\right]_0^{\ln 2}}$$

But $\qquad \sinh(\ln 2) = \tfrac{1}{2}(e^{\ln 2} - e^{-\ln 2}) = \tfrac{1}{2}(2 - \tfrac{1}{2}) = \tfrac{3}{4}$.

Therefore $\qquad \bar{x} = (\tfrac{3}{4}\ln 2 - \tfrac{1}{4})/\tfrac{3}{4} = (3\ln 2 - 1)/3$.

The arc length l is $\tfrac{3}{4}$ and so, by Pappus, the area of the surface of revolution is

$$2\pi l\bar{x} = \pi(3\ln 2 - 1)/2.$$

Example 3

Assuming the formula for the surface area of a sphere, deduce by using a theorem of Pappus the distance, from its bounding diameter, of the centroid of a uniform semi-circular arc of radius a.

The centroid of the arc lies, by symmetry, on the perpendicular bisector of the bounding diameter. Let $\bar{y}_A$ be the distance of this centroid from the bounding diameter. The area of the surface of revolution formed by rotation of the arc through 2π about its bounding diameter is the surface area of a sphere of radius a. So, by Pappus, $\qquad (\pi a) \,.\, (2\pi \bar{y}_A) = 4\pi a^2$

$$\Rightarrow \qquad\qquad \bar{y}_A = 2a/\pi.$$

Note that here we have used a Pappus theorem in reverse to derive the position of the centroid of an arc.

Similarly, if we rotate the region bounded by the semi-circular arc and its bounding diameter about this bounding diameter, we obtain a solid of revolution of volume $2\pi \bar{y}_S \times \frac{1}{2}\pi a^2$, where $\bar{y}_S$ is the distance of the centroid of the semi-circular region from the bounding diameter. But this solid of revolution is a sphere of volume $\frac{4}{3}\pi a^3$.

$$\therefore \qquad\qquad \pi^2 \bar{y}_S a^2 = \tfrac{4}{3}\pi a^3$$

$$\Rightarrow \qquad\qquad \bar{y}_S = 4a/(3\pi).$$

Exercises 7.3 (iii)

1 Find the coordinates of the centroid of the region enclosed by the curve $y = a^3/(a^2 + x^2)$ and the lines $y = 0$, $x = a\sqrt{3}$, $x = -a\sqrt{3}$.

2 Find the coordinates of the centroid of the region bounded by the parabola $y^2 = 4ax$ and the line $y = 2x$. Find also the coordinates of the centroid of the volume obtained by rotating this region through 2π about (a) Ox, (b) Oy.

3 Find the position of the centroid of a sector of a circle of radius a which subtends an angle 2α at the centre O of the circle. Deduce that the centroid of the minor segment of the circle cut off by a chord of length $2a \sin \alpha$, $0 < \alpha < \frac{1}{2}\pi$, is at a distance $\frac{2}{3}a \sin^3 \alpha/(\alpha - \sin \alpha \cos \alpha)$ from O.

4 The equation of a curve is

$$y^2 = \frac{x - 1}{x - 2}.$$

Show that no part of the curve lies between $x = 1$ and $x = 2$; and sketch the curve for the remaining values of x.

The arc of the curve between the points where $x = 0$ and $x = 1$ is rotated through $180°$ about the axis of x. Find the volume generated and the distance of the centroid of the volume from the origin.

5 Calculate the area of the region R enclosed between the x-axis and that part of the curve $y^2 = x^2(1 - x)$ which lies in the first quadrant.

Also calculate the volumes swept out when the region R is rotated through four right angles about (a) the x-axis, (ii) the y-axis.

Show further that the coordinates of the centroid of the region R are (4/7, 5/32).

6 Sketch the curve $cy^2 = 4x^2(c - x)$, where $c > 0$, and find the area of the region enclosed by its loop. Find also the x-coordinate of the centroid of this region.

7 The finite region bounded by the arc of a quadrant of a circle of radius a and the tangents at its extremities is rotated about one of these tangents. The volume so generated is filled with material of uniform density. Find the position of the centre of mass of the solid so formed.

8 Show that the area of the surface generated by rotating the arc of the parabola $y^2 = 9x$, $y > 0$, from $x = 0$ to $x = 4$, through 2π radians about the x-axis is 49π. Show also that $\bar{x}$, the distance of the centroid of this surface from the y-axis, may be expressed as

$$\frac{3}{49} \int_0^4 x\sqrt{(4x + 9)} \, dx.$$

Hence, or otherwise, determine the coordinates of the centroid of the surface. [L]

9 Find the area of the surface formed by rotating the arc of the parabola $y^2 = 4ax$ between the origin and the point $(4a, 4a)$ about the x-axis, and prove that the centroid of the surface is on the axis at a distance

$$\frac{a(50\sqrt{5} + 2)}{5(5\sqrt{5} - 1)}$$

from the vertex.

10 Find the y-coordinate of the centroid of the region bounded by the curve $r = a\theta$ $(0 < \theta < \pi/2)$ and the half-line $\theta = \pi/2$.

11 A and B are points on the catenary $y = c \cosh(x/c)$ at which x has the values c and λc respectively with $\lambda > 1$. The region of the plane bounded by the arc AB of the curve, the ordinates at A and B, and the x-axis is rotated through 2π radians about the x-axis. If V is the volume and S is the curved surface area of the solid generated, prove that

$$V = \tfrac{1}{2}cS = \tfrac{1}{2}\pi c^3[\lambda - 1 + \cosh(\lambda + 1)\sinh(\lambda - 1)].$$

Show further that the centroids of the solid and of the curved surface coincide.

12 Sketch the curves $y = e^x$ and $y = 4e^{-x}$. Find the area of the finite region bounded by these curves and the y-axis. Find also the coordinates of the centroid of this region.

13 Prove that the area of the region common to the parabolas $ay = 2x^2$,

$y^2 = 4ax$, where $a > 0$, is $2a^2/3$. Find the coordinates of the centroid of this region and the volume of the solid generated when the region is rotated through 2π about Ox.

14   A circular quadrant of radius $a/2$ and centre B is removed from a square ABCD of side a. The remaining portion is rotated through an angle 360° about the side AD. Prove that the volume generated is

$$\pi a^3\left(\frac{13}{12} - \frac{\pi}{8}\right).$$

15   Sketch the curve $ay^2 = x^2(a - x)$, $a > 0$, and find the area of the region enclosed by its loop. Find also the position of the centroid of this region and deduce the volume of the solid formed when this region is rotated through 2π about a tangent to the curve at the origin.

16   Apply the theorem of Pappus to find the centroid of (a) a uniform wire in the form of one quarter of the circumference of a circle whose centre is at O, which is of radius a and lies in the first quadrant, and (b) a lamina which is one quarter of an elliptic disc, whose centre is at O, whose major and minor semi-axes are a and b respectively, and which lies in the first quadrant..

17   A crescent-shaped region is bounded by the ellipse $(x^2/a^2) + (y^2/b^2) = 1$ and the parabola $y^2 = b(b - x)$, where $a > b$ and x is positive. Calculate the area of this region and also the volume of the solid of revolution formed when the region is rotated through four right angles about the y-axis.
        Deduce the distance of the centroid of the region from the y-axis.

18   Sketch the cycloid $x = a(t - \sin t)$, $y = a(1 - \cos t)$ for values of t from $t = 0$ to $t = 2\pi$. Find the area of the region enclosed by the curve and the x-axis. Find also the position of the centroid of this region. Deduce the volume of the solid obtained by rotating this area through 2π about the x-axis.

19   The region enclosed by the lemniscate

$$r^2 = a^2 \cos 2\theta, \qquad -\pi/4 \leqslant \theta < \pi/4,$$

is rotated through two right angles about the line $\theta = \frac{1}{2}\pi$. Prove that the volume of the solid so formed is $(\pi^2 a^3 \sqrt{2})/8$.

20   Calculate the volume of the solid generated by the complete revolution of the triangle ABC, where A $\equiv$ (3, 4), B $\equiv$ (4, 6), C $\equiv$ (−1, 5) about the line $4x - 3y = 13$.

21   Use the theorems of Pappus to calculate the volume and the surface area of the solid generated when the region for which

$$x^2 + (y - 4a)^2 \leqslant 4a^2,$$

where $a > 0$, is rotated through 2π radians about the x-axis.

Miscellaneous exercises 7

1 (a) Evaluate

(i) $\displaystyle\int_0^1 \arctan x \, dx,$ (ii) $\displaystyle\int_3^{11} \frac{x}{(x+1)(x-2)} \, dx.$

(b) Using the substitution $t = \tan x$, or otherwise, evaluate

$$\int_0^{\pi/4} \frac{1}{4 + 5\cos^2 x} \, dx. \qquad\qquad [L]$$

2 (a) Evaluate $\displaystyle\int_3^4 \frac{x^3 + 4x + 4}{(x-2)^2(x^2+1)} dx.$

(b) Using the substitution $x = \sin u$, or otherwise, evaluate

$$\int_0^{1/2} \frac{\arcsin x}{\sqrt{(1-x^2)}} \, dx. \qquad\qquad [L]$$

3 Evaluate

$$\int_1^2 \frac{1}{\sqrt{(x^2 + 2px + q)}} \, dx,$$

where $p > 0$, $q > 0$, when

(a) $q = p^2$, (b) $q > p^2$ and (c) $q < p^2$.

Show that, as q tends to p^2, the results for (b) and (c) tend to the result for (a). [JMB]

4 A sphere of radius a is divided into two parts by a plane at a distance b from its centre. Prove that the volume and the curved surface area of the smaller part are

$$\frac{\pi}{3}(2a^3 - 3a^2b + b^3) \text{ and } 2\pi a(a - b)$$

respectively.

A cylindrical hole of radius $3r$ is made through a solid sphere of radius $5r$ so that the axis of the cylindrical portion removed is a diameter of the sphere. Find the volume and the *total* surface area of the remaining portion of the sphere. [JMB]

5 (a) Evaluate

$$\int_0^1 \arcsin x \, dx.$$

(b) Express $\tanh x$ in terms of e^x, and hence evaluate

$$\int_0^1 e^x \tanh x \, dx.$$

(c) Given that

$$I_n = \int_0^{\pi/2} \sin^n x\, dx, \qquad n \geqslant 0,$$

show that

$$nI_n = (n - 1) I_{n-2}, \qquad n \geqslant 2.$$

Hence evaluate I_4.   [L]

6 (a) Use the method of integration by parts to evaluate

$$\int_1^e (\ln x)^2 \, dx.$$

(b) Using the substitution $1 + x = \dfrac{1}{y}$, or otherwise, evaluate

$$\int_0^1 \frac{1}{(1 + x)(2 + x - x^2)^{1/2}} \, dx.$$

(c) Evaluate $\displaystyle\int_0^{\pi/4} \sin^5 x \, dx$.   [L]

7                     $$f(x) \equiv \frac{(x - 2)(x + 5)}{(x - 3)(x + 1)}, \qquad x \neq -1, \, x \neq 3.$$

(a) Find the set of values of x for which $f(x) < 0$.
(b) Express $f(x)$ in partial fractions.
(c) Prove that $f'(x) < 0$.
(d) Sketch the curve whose equation is $y = f(x)$.
(e) Evaluate

$$\int_4^5 f(x)\, dx,$$

leaving your answer in terms of natural logarithms.   [L]

8 (a) Show that

$$\int_0^1 \frac{8x + 6}{(x^2 + 1)(x + 2)} dx = \pi + \ln\left(\frac{8}{9}\right).$$

(b) Given that, for $n \geqslant 0$, $I_n = \displaystyle\int_0^{\pi/4} x^n \sin 2x \, dx$, prove that, for $n \geqslant 2$,

$$4I_n = n \left(\frac{\pi}{4}\right)^{n-1} - n(n - 1) I_{n-2}.$$ [L]

9 (a) Evaluate

(i) $\displaystyle\int_0^1 x^3 \arctan x \, dx$,   (ii) $\displaystyle\int_0^{\pi/2} \frac{1}{2 + \cos x} \, dx$.

(b) By means of the substitution $\theta = \dfrac{\pi}{2} - \phi$, or otherwise, show that

$$\int_0^{\pi/2} \cos^4 \theta \sin^2 \theta \, d\theta = \int_0^{\pi/2} \cos^2 \phi \sin^4 \phi \, d\phi,$$

and hence, or otherwise, evaluate

$$\int_0^{\pi/2} \cos^4 \theta \sin^2 \theta \, d\theta. \hspace{3cm} \text{[L]}$$

10 (a) Given that, for $n \geqslant 0$,

$$I_n = \int_0^{\pi/2} x^n \sin x \, dx,$$

show that, for $n \geqslant 2$, $\quad I_n = n \left(\dfrac{\pi}{2} \right)^{n-1} - n(n-1) I_{n-2}$,

and hence evaluate I_4.

(b) Show that the curves with polar equations $r = 2a \cos \theta$, $r = 4\sqrt{2} \, a\theta/\pi$, where $a > 0$, intersect where $r = \sqrt{2} \, a$, and find the area of the smaller region enclosed by these two curves. [L]

11 (a) Evaluate

$$\int_1^2 \cosh^{-1} x \, dx,$$

giving the result in terms of natural logarithms.
(b) Find

$$\int \frac{x^2}{\sqrt{(x^3 + 1)}} \, dx.$$

Given that

$$I_n = \int_0^1 \frac{x^n}{\sqrt{(x^3 + 1)}} \, dx,$$

prove that $\quad (2n - 1)I_n + 2(n - 2)I_{n-3} = 2\sqrt{2} \quad (n > 2)$.

Hence or otherwise evaluate I_8. [JMB]

12 Evaluate the integrals

$$\int_0^{\pi/4} \tan \theta \, d\theta \quad \text{and} \quad \int_0^{\pi/4} \tan^4 \theta \, d\theta.$$

Use these results to show that, for all positive integers n,

$$0 \leqslant \int_0^{\pi/4} \tan^{4n+1} \theta \, d\theta \leqslant \tfrac{1}{2} \ln 2,$$

$$0 \leqslant \int_0^{\pi/4} \tan^{4n} \theta \, d\theta \leqslant \pi/4 - 2/3. \qquad [L]$$

13 (a) Given that $I_n = \displaystyle\int_0^x \frac{t^n}{\sqrt{(1 + t^2)}} \, dt$, show that for $n > 1$,

$$n \, I_n + (n - 1)I_{n-2} = x^{n-1}\sqrt{(1 + x^2)}.$$

(b) By means of the substitution $t = u^2$, find, for $x > 0$,

$$\int_0^x \ln (t + \sqrt{t}) \, dt. \qquad [L]$$

14 Find

(a) $\displaystyle\int \frac{\arctan x}{(1 + x^2)} \, dx$, (b) $\displaystyle\int \frac{1}{\cos^2 x - 4 \sin^2 x} \, dx$. $\qquad [L]$

15 (a) Find (i) $\displaystyle\int x \sec^2 x \, dx$, (ii) $\displaystyle\int \frac{3x - 2}{\sqrt{(1 - 4x^2)}} \, dx$.

(b) Define $\cosh x$ and $\sinh x$ and hence prove that

$$\frac{1}{\cosh 2x + \sinh 2x} = \cosh 2x - \sinh 2x.$$

Hence, or otherwise, show that

$$\int_0^1 \frac{1}{\cosh 2x + \sinh 2x} \, dx = \tfrac{1}{2}(1 - e^{-2}).$$

16 (a) Given that

$$I_n = \int_1^e (\ln x)^n \, dx,$$

where n is a non-negative integer, prove that, for $n \geqslant 1$,

$$I_n + nI_{n-1} = e.$$

Evaluate I_5.
(b) Sketch the curve whose equation in polar coordinates is

$$r = a \cos^2 \theta, \qquad -\pi/2 \leqslant \theta \leqslant \pi/2,$$

where $a > 0$.
   Calculate the area of the region enclosed by the loop of this curve.
$\qquad [L]$

17 (a) By means of the substitution $x = \sin \theta$, or $x = \cos \theta$, or both, evaluate

$$\int_0^1 \frac{1}{x + \sqrt{(1 - x^2)}} \, dx.$$

(b) Given that, for $n \geq 0$,

$$I_n = \int_0^1 x^{n+1/2} (1 - x)^{1/2} \, dx,$$

prove that, for $n \geq 1$,

$$2(n + 2)I_n = (2n + 1) I_{n-1}. \qquad \text{[L]}$$

18 (a) Given that $I_n = \int_0^{\pi/2} x^n \sin x \, dx$, prove that

$$I_n = n \left(\frac{\pi}{2}\right)^{n-1} - n(n - 1) I_{n-2} \text{ for } n \geq 2.$$

Hence find I_4.

(b) Show that $\int_0^1 \frac{(3x + 5)}{x^2 + x + 1} \, dx = \frac{7\pi\sqrt{3}}{18} + \frac{3}{2} \ln 3.$

19 (a) Given that $I_n = \int_0^a \frac{1}{(x^2 + a^2)^n} \, dx$, where $a > 0$, prove that

$$2(n - 1)a^2 I_n = (2n - 3)I_{n-1} + 2^{1-n} a^{3-2n}.$$

Evaluate

$$\int_0^a \frac{1}{(x^2 + a^2)^3} \, dx.$$

(b) Using the substitution $u = \cos^2 x$, or otherwise, evaluate

$$\int_0^{\pi/2} \frac{1}{\tan x + 2 \cot x} \, dx. \qquad \text{[L]}$$

20 A curve is given by the parametric equations

$$x = a \cos^3 t, \ y = a \sin^3 t \quad (-\pi < t \leq \pi).$$

Show that near $t = 0$ the equations reduce approximately to

$$a - x \approx 3at^2/2, \quad y \approx at^3.$$

Sketch the curve.
Show that the total length of the curve is $6a$.
Find the area of the surface generated when the part of the curve which is in the first quadrant is rotated once about the x-axis. [JMB]

21 (a) Show that the length of the curve $y = -\ln(1 - x^2)$ between the origin and the point $(\frac{1}{2}, \ln \frac{4}{3})$ is $\ln 3 - \frac{1}{2}$.

(b) Determine by theorems of Pappus the positions of the centroids of a semi-circular arc and of a semi-circular disc both of radius a. [L]

22 Sketch the curve $ay^2 = x(x - 4a)^2$, $a > 0$.

Show that the area of the finite region enclosed by the loop of the curve is $256a^2/15$. Find the coordinates of the centroid of this region. [L]

23 (a) Find the area of the finite region bounded by arcs of the parabolas $y^2 = x$, $x^2 = y$, and show that the perimeter of this region is

$$\tfrac{1}{2} \ln (2 + \sqrt{5}) + \sqrt{5}.$$

(b) Sketch the curve with polar equation $r = a \cos 3\theta$, where $r > 0$ and $a > 0$, and find the area of the region enclosed by this curve.

24 The parametric coordinates of a point on a curve are given by $x = a(\tan t - t)$, $y = a \ln \sec t$, where $-\pi/2 < t < \pi/2$. Prove that the arc length s of the curve measured from the origin O is $a \sec t - a$.

The arc of length a measured from O is rotated about the x-axis through 2π. Find the surface area of the curved surface so formed.

25 Show that

$$\int_0^\pi \frac{1}{1 + e \cos \theta} \, d\theta = \frac{\pi}{\sqrt{(1 - e^2)}} \qquad (0 < e < 1).$$

Each focal radius vector of an ellipse is produced a constant length c. Show that the area between the curve so formed and the ellipse is $\pi c(2b + c)$, b being the semi-minor axis of the ellipse.

8 Equations and inequalities

8.1 Roots of equations

In *Advanced Mathematics 1* we considered (1) some applications of the remainder theorem (Section 3.8), and (2) symmetric functions of the roots of polynomial equations (Section 2.2). We now discuss some further examples using theorems already considered in this book. First, however, we state (or restate) some important results of algebra.

(1) A polynomial equation $P(x) = 0$ of the nth degree has n roots, real or complex, some or all of which may be repeated. (Throughout this chapter $P(x)$ denotes a polynomial in x.)

(2) If a polynomial equation has real coefficients, then any complex roots must occur in conjugate pairs.

(3) If $P(a) > 0$ and $P(b) < 0$, where $a < b$, then the polynomial equation $P(x) = 0$ has odd number of real roots in the interval $a < x < b$. (Note that this result depends on the fact that the polynomial $P(x)$ is continuous.)

(4) Suppose that the equation $P(x) = 0$, where

$$P(x) \equiv a_n x^n + a_{n-1} x^{n-1} + \ldots + a_1 x + a_0,$$

and $a_n \neq 0$, has roots $\alpha_1, \alpha_2, \ldots, \alpha_n$. Then $(x - \alpha_1), (x - \alpha_2), \ldots, (x - \alpha_n)$ are factors of $P(x)$ and therefore

$$a_n x^n + a_{n-1} x^{n-1} + \ldots a_1 x + a_0 \equiv a_n(x - \alpha_1)(x - \alpha_2) \ldots (x - \alpha_n).$$

Therefore, by equating coefficients of $x^{n-1}, x^{n-2}, \ldots, x^2, x, x^0$ on the two sides of this identity we find

$$\Sigma \, \alpha_1 = -a_{n-1}/a_n,$$
$$\Sigma \, \alpha_1 \alpha_2 = a_{n-2}/a_n,$$
$$\ldots \qquad \ldots$$
$$\alpha_1 \alpha_2 \ldots \alpha_n = (-1)^n a_0/a_n.$$

In the special case when $n = 3$, the equation

$$ax^3 + bx^2 + cx + d = 0,$$

where $a \neq 0$, has roots α, β, γ, where

$$\alpha + \beta + \gamma = -b/a,$$
$$\alpha\beta + \beta\gamma + \gamma\alpha = c/a,$$
$$\alpha\beta\gamma = -d/a.$$

(5) If the equations $f(x) = 0$ and $g(x) = 0$ have a common root, then this root must also satisfy the equation $f(x) + kg(x) = 0$ where k is constant. In particular, this root must satisfy the equation $f(x) - g(x) = 0$.

Example 1
The equations $ax^2 + bx + c = 0$ and $bx^2 + ax + c = 0$, where $a \neq b, c \neq 0$, have a common root. Prove that $a + b + c = 0$.

The common root must satisfy the equation

$$(ax^2 + bx + c) - (bx^2 + ax + c) = 0$$
$$\Rightarrow \qquad (a - b)x^2 + (b - a)x = 0$$
$$\Rightarrow \qquad x^2 - x = 0, \qquad \text{since } a - b \neq 0,$$
$$\Rightarrow \qquad x = 0 \quad \text{or} \quad 1.$$

But, since $c \neq 0$, $x = 0$ cannot satisfy the first of the given equations and, therefore, the common root of the equation must be 1. Substituting in either equation gives $a + b + c = 0$.

Example 2
The polynomial equation $P(x) = 0$ has a repeated root given by $x = a$. Prove that $P'(a) = 0$.

Prove also that, if $P(a) = 0$ and $P'(a) = 0$, then the equation $P(x) = 0$ has a repeated root given by $x = a$.

The cubic equation

$$2x^3 - 15x^2 + 24cx + d = 0,$$

where c and d are real, has a repeated root. Find the set of possible values of c.
Solve the equation completely given that $c = 1$ and $d = 16$.

Since the equation $P(x) = 0$ has a repeated root $x = a$, then

$$P(x) \equiv (x - a)^2 Q(x),$$

where $Q(x)$ is a polynomial. Therefore,

$$P'(x) = 2(x - a)Q(x) + (x - a)^2 Q'(x)$$
$$\Rightarrow \qquad P'(a) = 0.$$

Conversely, suppose that $P(a) = 0 = P'(a)$. Then the Taylor series (Section 6.5) expansion of $P(x)$ in ascending powers of $(x - a)$, given by

$$P(x) = P(a) + (x - a)P'(a) + \tfrac{1}{2}(x - a)^2 P''(a) + \ldots,$$

has its first two terms zero and thus
$P(x) = \tfrac{1}{2}(x - a)^2 P''(a) +$ terms in higher powers of $(x - a)$. Thus, $P(x)$ has a factor $(x - a)^2$, i.e. the equation $P(x) = 0$ has a repeated root $x = a$. (The generalisation of this result is obvious.)

The repeated root of the given cubic equation must satisfy the equation $P'(x) = 0$, i.e. the root must satisfy

$$6x^2 - 30x + 24c = 0,$$

i.e.
$$x^2 - 5x + 4c = 0. \qquad \ldots (1)$$

Further, since c and d are real, this repeated root must be real. (A cubic equation cannot have a repeated complex root with non-zero imaginary parts, otherwise it would have four roots!) Therefore,

$$25 - 16c \geqslant 0 \qquad \Leftrightarrow \qquad c \leqslant \tfrac{25}{16}.$$

The set of possible values of c is $\{c : c \leqslant \tfrac{25}{16}\}$.

When given that $c = 1$ and $d = 16$, at the end of this question, we suspect that the equation may have a repeated root, and from (1) this root must be 1 or 4. Clearly, by the factor theorem 1 is not a root of the equation

$$2x^3 - 15x^2 + 24x + 16 = 0$$

but 4 is a root. Therefore, 4 is a repeated root and, since the product of the roots is -8, the other root is $-\tfrac{1}{2}$. The roots are, therefore, 4, 4 and $-\tfrac{1}{2}$. (Note also as a check that the sum of the roots is $7\tfrac{1}{2}$.)

Example 3
Given that the roots of the equation

$$x^3 + 3bx^2 + 3cx + d = 0 \qquad \ldots (1)$$

are in arithmetic progression, show that the constants b, c and d satisfy the relation $2b^3 = 3bc - d$. State and prove the converse of this result.
Solve the equation

$$x^3 - 12x^2 + 12x + 80 = 0, \qquad \ldots (2)$$

and verify that the roots are in arithmetic progression.

We assume that the roots of equation (1) are $\alpha - \beta$, α and $\alpha + \beta$. Then $3\alpha = \text{sum}$ of roots $= -3b \Rightarrow \alpha = -b$. But α satisfies equation (1) and, so substituting $x = -b$ in (1), we find

$$-b^3 + 3b(-b)^2 + 3c(-b) + d = 0$$

$$\Rightarrow \qquad 2b^3 = 3bc - d. \qquad \ldots (3)$$

The *converse* of this result is as follows: given that $2b^3 = 3bc - d$, then the roots of equation (1) are in arithmetic progression (a.p.).

To prove this converse we substitute $d = 3bc - 2b^3$ in equation (1) obtaining

$$x^3 + 3bx^2 + 3cx + 3bc - 2b^3 = 0$$

$$\Rightarrow \qquad (x + b)[x^2 + 2bx - 2b^2 + 3c] = 0$$

$$\Rightarrow \qquad x = -b \quad \text{or} \quad -b \pm \sqrt{(3b^2 - 3c)}$$

and so the roots of equation (1) are in a.p.

Since we have to verify that the roots of equation (2) are in a.p., we use the first (converse) part of the question (with $b = -4$), which suggests that 4 is one of the roots of equation (2), and the factor theorem confirms this to be so. Long division transforms the equation to

$$(x - 4)(x^2 - 8x - 20) = 0$$
$$\Rightarrow \quad (x - 4)(x + 2)(x - 10) = 0.$$

The roots are -2, 4 and 10 and are clearly in a.p.

Example 4

Show that a necessary and sufficient condition for the roots of the equation

$$x^3 + px^2 + qx + r = 0,$$

where $pqr \neq 0$, to be in geometric progression is that $q^3 - p^3r = 0$.

Necessity Suppose that the roots are in geometric progression (g.p.) (with middle term α and common ratio t). Then we can take the roots as α/t, α and αt. Then the symmetric functions of the roots give

$$\alpha\left(\frac{1}{t} + 1 + t\right) = -p, \qquad \ldots (1)$$

$$\alpha^2\left(\frac{1}{t} + 1 + t\right) = q, \qquad \ldots (2)$$

$$\alpha^3 = -r. \qquad \ldots (3)$$

From equations (1) and (2), $\qquad \alpha = -q/p.$

$$\therefore \qquad -q^3/p^3 = r, \quad \text{i.e.,} \quad q^3 - p^3r = 0.$$

The condition is, therefore, necessary.

Sufficiency Suppose that $q^3 - p^3r = 0$. Then, substituting for r, the cubic equation becomes

$$p^3x^3 + p^4x^2 + p^3qx + q^3 = 0$$
$$\Leftrightarrow \quad (p^3x^3 + q^3) + p^3x(px + q) = 0$$
$$\Leftrightarrow \quad (px + q)[p^2x^2 + (p^3 - pq)x + q^2] = 0.$$

One of the roots of the equation is $-q/p$. Also, the other two roots of the equation must satisfy $p^2x^2 + (p^3 - pq)x + q^2 = 0$ and so their product is q^2/p^2. Therefore they must be of the form $\lambda q/p$ and $q/(p\lambda)$. Hence, the roots are in g.p. and the condition is, therefore, sufficient.

Example 5
The equation

$$x^n + px^2 + qx + r = 0,$$

where $n \geqslant 5$ and $r \neq 0$, has roots $\alpha_1, \alpha_2, \alpha_3, \ldots, \alpha_n$.

Denoting $\sum\limits_{i=1}^{n} \alpha_i^k$ by S_k,

(a) calculate S_2 and deduce that the roots cannot all be real,
(b) prove that

$$S_n + pS_2 + qS_1 + nr = 0$$

and hence find the value of S_n.

Since $r \neq 0$ and the product of the roots is $(-1)^n r$, none of $\alpha_1, \alpha_2, \ldots \alpha_n$ is zero.

(a) $S_2 = \left(\sum\limits_{i=1}^{n} \alpha_i \right)^2 - 2 \sum\limits_{i \neq j} \alpha_i \alpha_j = 0,$

since the coefficients of x^{n-1} and x^{n-2} in the given equation vanish.

Assume that all the roots of the given equation are real. Then, since each α_i is non-zero and real, $S_2 = \sum\limits_{i=1}^{n} \alpha_i^2 > 0$. This contradicts the result $S_2 = 0$ and so our assumption that all the roots are real must be false, i.e. the roots of the given equation cannot all be real.

(Note that this is an example of proof by contradiction.)

(b) Since α_i, $i = 1, 2, \ldots, n$, satisfies the given equation, then

$$\alpha_i^n + p\alpha_i^2 + q\alpha_i + r = 0, \qquad i = 1, 2, \ldots, n.$$

Adding these n equations gives

$$S_n + pS_2 + qS_1 + nr = 0.$$

But $S_2 = 0$, $S_1 = 0$. Therefore

$$S_n = -nr.$$

Example 6
The cubic equation

$$x^3 + px + q = 0, \qquad\qquad \ldots (1)$$

where $q \neq 0$, has roots α, β and γ. Show that an equation whose roots are $\beta^2 \gamma^2$, $\gamma^2 \alpha^2$ and $\alpha^2 \beta^2$ is $f(x) = 0$, where

$$f(x) \equiv x^3 - p^2 x^2 - 2pq^2 x - q^4.$$

Show further that, if p and q are real and $|p + q^2| < 1$, the equation $f(x) = 0$ has at least one root in the interval $0 < x < 1$.

The required equation is

$$(x - \beta^2 \gamma^2)(x - \gamma^2 \alpha^2)(x - \alpha^2 \beta^2) = 0,$$

where $\alpha + \beta + \gamma = 0$, $\alpha\beta + \beta\gamma + \gamma\alpha = p$, $\alpha\beta\gamma = -q$,

i.e. $x^3 - (\beta^2\gamma^2 + \gamma^2\alpha^2 + \alpha^2\beta^2)x^2 + \alpha^2\beta^2\gamma^2(\alpha^2 + \beta^2 + \gamma^2)x - \alpha^4\beta^4\gamma^4 = 0$.

But $\quad \beta^2\gamma^2 + \gamma^2\alpha^2 + \alpha^2\beta^2 \equiv (\alpha\beta + \beta\gamma + \gamma\alpha)^2 - 2\alpha\beta\gamma(\alpha + \beta + \gamma) = p^2$,

$\quad\quad \alpha^2 + \beta^2 + \gamma^2 \equiv (\alpha + \beta + \gamma)^2 - 2(\alpha\beta + \beta\gamma + \gamma\alpha) = -2p$,

and $\quad \alpha^2\beta^2\gamma^2 = q^2$, $\alpha^4\beta^4\gamma^4 = q^4$.

The required equation is, therefore, $f(x) = 0$, where

$$f(x) \equiv x^3 - p^2x^2 - 2pq^2x - q^4.$$

Alternatively, we can proceed as follows. One root of the new equation is $\beta^2\gamma^2$, i.e. $\alpha^2\beta^2\gamma^2/\alpha^2 = q^2/\alpha^2$. The other two roots are q^2/β^2, q^2/γ^2. Therefore if we put $y = q^2/x^2$, i.e. $x^2 = q^2/y$, and eliminate x between this equation and equation (1), we obtain an equation in y whose roots are $\beta^2\gamma^2$, $\gamma^2\alpha^2$, $\alpha^2\beta^2$. This elimination can be performed as follows.

Equation (1) can be written

$$x^3 + px = -q$$
$\Rightarrow \quad\quad\quad\quad\quad (x^3 + px)^2 = q^2$
$\Leftrightarrow \quad\quad\quad\quad\quad x^2(x^2 + p)^2 = q^2.$

Writing $x^2 = q^2/y$ and rearranging, we obtain the required equation $f(y) = 0$. Now consider the signs of $f(0)$ and $f(1)$.

$$f(0) = -q^4 < 0, \text{ since } q \text{ is real.}$$
$$f(1) = 1 - p^2 - 2pq^2 - q^4 = 1 - (p + q^2)^2.$$

But, if $|p + q^2| < 1$, then $(p + q^2)^2 < 1$, since p and q are real, and so $f(1) > 0$.

Since $f(x)$ is a polynomial, it is a continuous function and therefore, since $f(x)$ changes sign in the interval $0 < x < 1$, the equation $f(x) = 0$ must have at least one root in that interval. (There may be one or three roots in this interval.)

Note that the alternative technique for solving the first part of this question can be used in many problems. For example, suppose that we are given the cubic equation

$$x^3 + px^2 + qx + r = 0 \quad\quad\quad \dots (2)$$

with roots α, β, γ.

(a) The equation (in y) with roots $1/\alpha$, $1/\beta$, $1/\gamma$ is obtained by elimination of x between equation (2) and $y = 1/x$.

(b) For the equation with roots $\alpha - k$, $\beta - k$, $\gamma - k$, where k is constant, use $y = x - k$.

(c) For the equation with roots $k\alpha^2$, $k\beta^2$, $k\gamma^2$, use $y = kx^2$.

(d) For the equation with roots $\beta + \gamma$, $\gamma + \alpha$, $\alpha + \beta$, use $y = (\alpha + \beta + \gamma) - x$, i.e. $y = -(p + x)$.

(e) For the equation with roots $(\beta - \gamma)^2$, $(\gamma - \alpha)^2$, $(\alpha - \beta)^2$,

use $\quad\quad y = (\alpha^2 + \beta^2 + \gamma^2) - \alpha^2 - 2\beta\gamma$, i.e. $$y = (\alpha + \beta + \gamma)^2 - 2(\alpha\beta + \beta\gamma + \gamma\alpha) - \alpha^2 - \frac{2\alpha\beta\gamma}{\alpha},$$

i.e. $$y = p^2 - 2q - x^2 + \frac{2r}{x}.$$

Example 7

Given that $f(x) = x^3 - 3qx + r$, where q and r are real, show, by considering the stationary values of $f(x)$, or otherwise, that, if the equation $f(x) = 0$ has three unequal real roots, then $r^2 - 4q^3 < 0$.

State and prove the converse of this result.

In order that the equation $f(x) = 0$ should have three unequal real roots, the curve $y = f(x)$ must have the form illustrated in Fig. 8.1, with its maximum point above Ox and its minimum point below Ox.

(It is immaterial if the curve is moved to the left or the right.)

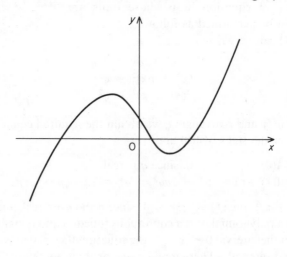

Fig. 8.1

Since $f(x) = x^3 - 3qx + r$, then

$$f'(x) = 3(x^2 - q).$$

The stationary values of $f(x)$ occur when $x = \pm\sqrt{q}$, implying that $q > 0$.
When $x = -\sqrt{q}$, $f(x)$ attains its maximum value $(r + 2q^{3/2})$.
When $x = \sqrt{q}$, $f(x)$ attains its minimum value $(r - 2q^{3/2})$.

For the maximum to be positive and the minimum to be negative we must have

$$r + 2q^{3/2} > 0, \qquad r - 2q^{3/2} < 0$$

giving, by multiplication, $r^2 - 4q^3 < 0$.

This condition is therefore *necessary* for three real unequal roots.

The *converse* of this result is as follows:

'If $r^2 - 4q^3 < 0$ then the equation $f(x) = 0$ has three unequal real roots.'

Since $r^2 > 0$ and $r^2 - 4q^3 < 0$, we must have $q > 0$. Therefore, from our earlier work, the curve $y = f(x)$ has a maximum value, M, and a minimum value, m, and the product Mm of these values is negative. Therefore, $M(>m)$ must be positive and $m(<M)$ must be negative. The curve $y = f(x)$ must, therefore, be as illustrated in Fig. 8.1 and the equation has three real unequal roots.

Note how we have used calculus to solve what appears to be an algebraic problem. The various branches of mathematics should not be isolated but, where convenient, techniques from one branch should be used to solve problems in another.

Exercises 8.1

1 Find the numerical value of k such that the roots of the equation

$$x^3 - 9x^2 + kx - 21 = 0$$

are in arithmetic progression and, when k takes this value, state, in surd form, the two irrational roots of the equation. [JMB]

2 A polynomial f(x), where

$$f(x) \equiv x^3 + px^2 + qx - 24,$$

p and q being constants, leaves remainders 56 and 0 on division by $(x - 5)$ and $(x + 2)$ respectively. Solve the equation

$$f(x) = 0. \qquad [L]$$

3 (a) By considering the stationary values of the expression f(x) where $f(x) = x^3 + px + q, p > 0, q > 0$, or otherwise, show that the equation $f(x) = 0$ has precisely one negative root and no positive root.
(b) Show that the equation

$$x^4 + 4x + 3 = 0$$

has precisely two non-real roots and that these non-real roots have positive real parts. [L]

4 The roots of the equation

$$x^4 + ax^3 + bx^2 + cx + d = 0,$$

where a, b, c, d are constants, are $\alpha, \beta, \gamma, \delta$ and are such that $\alpha + \beta = \gamma + \delta$. Show that

$$a^3 + 8c = 4ab. \qquad [L]$$

Show that the coefficients of the equation

$$16x^4 - 64x^3 + 96x^2 - 64x + 15 = 0$$

satisfy the above condition. Find the values of the roots of the equation. [L]

Given that α, β, γ are the roots of the equation $x^3 + px^2 + qx + r = 0$, where the coefficients p, q, r are real, show that $\alpha^2 + \beta^2 + \gamma^2 = p^2 - 2q$.

Find $(\alpha - \beta)^2 + (\beta - \gamma)^2 + (\gamma - \alpha)^2$ in terms of p and q and show that α, β, γ cannot all be real if $p^2 < 3q$. [L]

6 (a) One root of the equation $x^3 - 3x^2 - 3x + a = 0$ is -2. Find the value of a and prove that the equation has no other real root.

(b) Find the roots of the equation

$$4x^3 + 12x^2 - 63x + 54 = 0,$$

given that it has a repeated root. [L]

7 A quartic equation $f(x) = 0$ has real coefficients and none of its roots is real. Two of the roots are α β, and $\text{Re}(\alpha) = 0$, $|\alpha| = 1$, $\text{Re}(\beta) = 1$, $|\beta| = \sqrt{2}$. Express the other two roots in terms of α and β and find all four roots. Write down $f(x)$ as a polynomial in x.

8 Given that $f(x) \equiv x^3 + px^2 + qx + r$, find the real constants p, q and r when the following conditions are all satisfied:
(a) x is a factor of $f'(x)$,
(b) $(x + 2)$ is a factor of $f(x)$,
(c) when $f(x)$ is divided by $(x - 1)$, the remainder is 6. [L]

9 Given that the equations

$$x^3 - 2x + 4 = 0$$

and

$$x^2 + x + k = 0$$

have a common root, show that

$$k^3 + 4k^2 + 14k + 20 = 0.$$ [JMB]

10 Show that for all real values of k the function f defined by

$$f(x) = x^3 - 6x^2 + 9x - k,$$

where x is real, has a maximum when $x = 1$ and a minimum when $x = 3$.

Illustrate, by means of two separate sketches, the cases when two of the roots of the equation $f(x) = 0$ are real and equal. Find the set of values of k for which the equation has three real distinct roots.

Find the value of k such that the product of two real and distinct roots of the equation $f(x) = 0$ is equal to 1. When k takes this value verify that the three roots of the equation are in arithmetic progression. [JMB]

11 Find the coordinates of the stationary point and of the points of inflexion of the curve

$$y = \frac{x^2}{1 + x^2}.$$

Show that the curve lies entirely in the region $0 \leqslant y < 1$, and that $\dfrac{dy}{dx} > 0$ for $x > 0$.

Sketch the curve.

Find the set of values of k for which the equation

$$\frac{x^2}{1 + x^2} = kx$$

has three real and distinct roots. [JMB]

12 (a) Given that α is a root of the equation

$$x^4 + 6x^3 - 5x^2 + 6x + 1 = 0,$$

prove that $\dfrac{1}{\alpha}$ also is a root of this equation.

Show that the sum of the squares of the reciprocals of the roots of the equation is 46.

By substituting $y = x + \dfrac{1}{x}$, or otherwise, solve the equation.

(b) Solve the equation

$$24x^3 + 28x^2 - 14x - 3 = 0,$$

given that the roots are in geometric progression. [L]

13 The equation $\qquad px^3 + qx^2 + rx + s = 0$

has roots α, $1/\alpha$ and β. Prove that

$$p^2 - s^2 = pr - qs.$$

Solve the equation $\quad 6x^3 + 11x^2 - 24x - 9 = 0.$ [L]

14 Given that α and β are the roots of the equation

$$x^2 - px + q = 0,$$

prove that $\alpha + \beta = p$ and $\alpha\beta = q$.

Prove also that
(a) $\alpha^{2n} + \beta^{2n} = (\alpha^n + \beta^n)^2 - 2q^n$,
(b) $\alpha^4 + \beta^4 = p^4 - 4p^2q + 2q^2$.

Hence, or otherwise, form the quadratic equation whose roots are the fourth powers of those of the equation $x^2 - 3x + 1 = 0$. [L]

15 If the roots of the equation $x^3 - 9x^2 + 3x - 39 = 0$ are α, β, γ, show that an equation whose roots are $\alpha - 3$, $\beta - 3$, and $\gamma - 3$ is

$$x^3 - 24x - 84 = 0.$$

Show also that the equation $x^3 - 24x - 84 = 0$ has only one real root, and show that this root lies between 6 and 7. Sketch the two curves $y = x^3 - 9x^2 + 3x - 39$ and $y = x^3 - 24x - 84$ on the same diagram.

16 (a) If all the roots of the equation

$$x^n + p_1 x^{n-1} + \ldots + p_n = 0$$

are real, prove that $p_1^2 - 2p_2 \geqslant 0$.

(b) An equation in y is formed from the cubic equation

$$a_0 x^3 + a_1 x^2 + a_2 x + a_3 = 0$$

by substituting $y + k$ for x. Find the value of k for which the coefficient of y^2 in the new equation is zero.

   Solve the equation

$$4x^3 - 12x^2 + 13x - 6 = 0. \qquad \text{[JMB]}$$

17 The polynomial $P(x)$ is divided by $x^2 - 4$. Show that the remainder is

$$\tfrac{1}{4}x[P(2) - P(-2)] + \tfrac{1}{2}[P(2) + P(-2)].$$

Given further that
(a) $P(x)$ is of degree 4,
(b) $P(x)$ has coefficient of x^4 equal to unity,
(c) $P(x) = P(-x)$,
(d) $P(2) = -16$,
(e) $P(x)$ has a stationary value at $x = 1$,
find the quotient when $P(x)$ is divided by $x^2 - 4$. State the zeros of $P(x)$
(i) in $\mathbb{R}$,   (ii) in $\mathbb{C}$.

8.2   Further inequalities

In *Advanced Mathematics 1* we considered some simple inequalities. Here we consider some more complicated examples. First, we list the operations which can legitimately be applied to inequalities and indicate where these operations cannot be applied in reverse. (Throughout this section all letters used as symbols denote non-zero real numbers.)

(1) $a > b \iff a + x > b + x$.
In particular $a > b \iff a - b > 0$.
(2) If $a > b$ and $c > d$, then $a + c > b + d$, i.e. we can add corresponding sides of inequalities.
(3) If $x > 0$, then $a > b \iff xa > xb$, i.e. we can multiply both sides of an inequality by the same positive quantity. If $x < 0$, then $a > b \iff xa < xb$, i.e. the direction of the inequality sign must be reversed whenever we multiply both sides of an inequality by the same negative quantity.
   In particular, if $a > 0$ and $b > 0$, then

$$a > b \iff a^2 > b^2.$$

(The left-hand side has been multiplied by a larger positive quantity than the right-hand side.)
But, if $a < 0$ and $b < 0$, then

$$a > b \quad \Leftrightarrow \quad a^2 < b^2$$

(e.g. $\qquad\qquad\qquad -6 > -7 \quad \Leftrightarrow \quad 36 < 49$).

Also, since $|a|$, $|b|$ are positive,

$$|a| > |b| \quad \Leftrightarrow \quad a^2 > b^2 \qquad (\text{since } a^2 = |a|^2).$$

Also, if $a > 0$ and $b > 0$, then

$$a > b \quad \Leftrightarrow \quad \sqrt{a} > \sqrt{b}.$$

(Here, of course, as always, $\sqrt{}$ means the positive square root.)
(4) Care must be taken when a number of quantities are related by inequality and equality symbols. For example, suppose that

$$a > b = c \geqslant d.$$

Then we can deduce that

$$a > c, \qquad a > d, \qquad b \geqslant d.$$

However, if

$$p > q < r,$$

then we cannot deduce any valid inequality between p and r.
(5) We cannot subtract corresponding sides of two inequalities. For, example, if $a > b$ and $c > d$ then no valid deduction relating $(a - c)$ and $(b - d)$ can be derived by subtraction, e.g. $5 > 4$ and $6 > 2$ do not imply $(5 - 6) > (4 - 2)$.

Example 1
Find the set of values of x for which

(a) $\dfrac{2x + 3}{x - 1} < 1$,
$\qquad\qquad$
(b) $\left| \dfrac{2x + 3}{x - 1} \right| < 1$.

(a)
$$\frac{2x + 3}{x - 1} < 1$$

$\Rightarrow \qquad\qquad \dfrac{2x + 3}{x - 1} - 1 < 0 \qquad\qquad\qquad \dots (1)$

$\Rightarrow \qquad\qquad \dfrac{(2x + 3) - (x - 1)}{x - 1} < 0 \qquad\qquad \dots (2)$

$\Rightarrow \qquad\qquad \dfrac{x + 4}{x - 1} < 0 \qquad\qquad\qquad\qquad \dots (3)$

We tabulate for the sign of f(x), where $f(x) = (x + 4)/(x - 1)$, as follows:

	$x < -4$	$-4 < x < 1$	$1 < x$
$x + 4$	$-$	$+$	$+$
$x - 1$	$-$	$-$	$+$
$f(x)$	$+$	$-$	$+$

The solution set is, therefore, $\{x : -4 < x < 1\}$.

Alternatively, we can proceed from (3) by marking the 'critical values' -1 and 4 on the number line

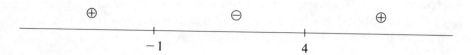

and writing down the result by 'inspection'.

Notes

(1) When solving an inequality such as $f(x) > g(x)$ we transform the problem to $f(x) - g(x) > 0$. It is much easier to solve an inequality when expressed in this form.

(2) In line (2) of our solution we express the left-hand side as one single term, with a common denominator. This reduces, as in line (3), to the quotient of two factors and the signs of the factors and quotient are determined by the use of a table or, with experience, by using the number line. Notice that the intervals in the table are decided by consideration of the zeros of the factors of the numerator and denominator of the left-hand side of (3).

(3) If we multiply (3) by $(x - 1)$, we would have to consider separately the cases $x > 1$ and $x < 1$. Although this will give the correct result it is not to be recommended as a method.

However, an alternative approach is to multiply both sides of the given inequality by $(x - 1)^2$, which is positive if $x \neq 1$. This gives

$$(2x + 3)(x - 1) < (x - 1)^2, \quad x \neq 1.$$
$\Leftrightarrow \quad x^2 + 3x - 4 < 0$
$\Leftrightarrow \quad (x - 1)(x + 4) < 0$
$\Leftrightarrow \quad -4 < x < 1.$

(b) $\left|\dfrac{2x + 3}{x - 1}\right| < 1 \Leftrightarrow |2x + 3| < |x - 1|$

$\Leftrightarrow \quad (2x + 3)^2 < (x - 1)^2$
$\Leftrightarrow \quad (2x + 3)^2 - (x - 1)^2 < 0 \qquad$ (difference of two squares)
$\Leftrightarrow \quad (x + 4)(3x + 2) < 0.$

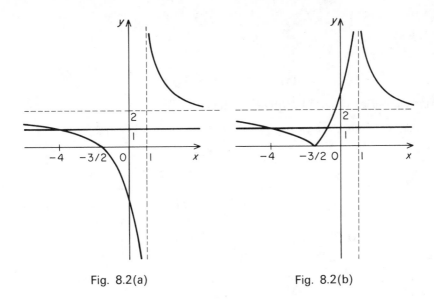

| Fig. 8.2(a) | Fig. 8.2(b) |

The solution set is, therefore, $\{x: -4 < x < -2/3\}$.

The results of this example are illustrated in Fig. 8.2(a), (b) which show the curves $y = (2x + 3)/(x - 1)$ and $y = |(2x + 3)/(x - 1)|$.

Note that the curve in Fig. 8.2(b) is obtained by 'rectification' of the curve in Fig 8.2(a), i.e. reflecting in the x-axis those parts of the curve below the x-axis.

Note also that reasonable sketches of the two given expressions provide yet another method of solution.

Example 2

Find the solution set of the inequality $f(x) > 8$, where

$$f(x) \equiv x + |x| - 2x|x|.$$

When $x < 0$, $\quad |x| = -x$, $\quad f(x) = 2x^2$.
When $x > 0$, $\quad |x| = x$, $\quad f(x) = 2x - 2x^2$.

Therefore,
for $x < 0$,

$$f(x) > 8 \iff 2x^2 > 8 \iff x^2 > 4 \implies x < -2,$$

for $x > 0$,

$$f(x) > 8 \iff 2x - 2x^2 > 8 \iff 2x^2 - 2x + 8 < 0$$
$$\iff x^2 - x + 4 < 0 \iff (x - \tfrac{1}{2})^2 + \tfrac{15}{4} < 0,$$

which is not true for any value of x.

The required solution set is, therefore, $\{x: x < -2\}$.

Example 3

Sketch the graph of $y = f(x)$, where

$$f(x) \equiv |2x - 3| - |x + 2|.$$

Find the set of values of x for which $f(x) > 2$.

The critical values are clearly given by $x = -2$ and $x = 3/2$.

	$x \leqslant -2$	$-2 < x \leqslant 3/2$	$x > 3/2$
$\|2x - 3\|$	$-(2x - 3)$	$-(2x - 3)$	$2x - 3$
$\|x + 2\|$	$-(x + 2)$	$x + 2$	$x + 2$
$f(x)$	$5 - x$	$1 - 3x$	$x - 5$

The graph in Fig. 8.3 consists of two half-lines and a straight line segment as shown.

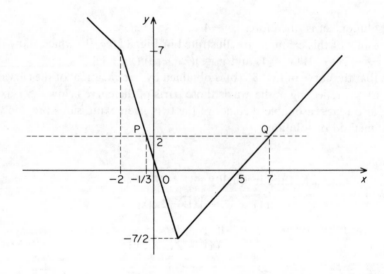

Fig. 8.3

The line $y = 2$ cuts the graph at P and Q where $x = -\frac{1}{3}$ and where $x = 7$ respectively. The required solution set is therefore

$$\{x : x < -\tfrac{1}{3}\} \cup \{x : x > 7\}.$$

Example 4

Find the solution set of the inequality $f(x) \leqslant 1$, where

$$f(x) \equiv |\sin x + \cos x|.$$

$$f(x) \leqslant 1 \quad \Leftrightarrow \quad (\sin x + \cos x)^2 \leqslant 1$$
$$\Leftrightarrow \quad \sin^2 x + \cos^2 x + 2 \sin x \cos x \leqslant 1$$
$$\Leftrightarrow \quad \sin 2x \leqslant 0$$
$$\Leftrightarrow \quad k\pi - \tfrac{1}{2}\pi \leqslant x \leqslant k\pi, \qquad \text{for } k \in \mathbb{Z}.$$

Alternatively, $\qquad\qquad f(x) \equiv \sqrt{2}|\sin (\pi/4 + x)|$

and therefore $\qquad\qquad |\sin (\pi/4 + x)| \leqslant 1/\sqrt{2}$

giving the same result as before.

Example 5
Find the set of values of x for which $f(x) > 1$, where

$$f(x) \equiv 4 \cosh^2 x - 7 \sinh x.$$

$$f(x) > 1 \quad \Leftrightarrow \quad 4 \cosh^2 x - 7 \sinh x - 1 > 0$$
$$\Leftrightarrow \quad 4(1 + \sinh^2 x) - 7 \sinh x - 1 > 0$$
$$\Leftrightarrow \quad 4 \sinh^2 x - 7 \sinh x + 3 > 0$$
$$\Leftrightarrow \quad (\sinh x - 1)(4 \sinh x - 3) > 0$$
$$\Leftrightarrow \quad \sinh x > 1 \quad \text{or} \quad \sinh x < 3/4$$
$$\Leftrightarrow \quad x > \ln (1 + \sqrt{2}) \quad \text{or} \quad x < \ln 2.$$

The solution set is $\{x : x < \ln 2\} \cup \{x : x > \ln (1 + \sqrt{2})\}$.

We now consider some further inequalities which hold *for all* (real) values of the variables involved. A fundamental inequality is derived from

$$(a - b)^2 \geqslant 0; \qquad a, b \in \mathbb{R}.$$

This implies that

$$a^2 - 2ab + b^2 \geqslant 0$$
i.e. $\qquad\qquad a^2 + b^2 \geqslant 2ab.$ $\qquad\qquad\qquad$ ... (1)
Also, $\qquad\qquad (a + b)^2 = a^2 + b^2 + 2ab$
$\Rightarrow \qquad\qquad a^2 + 2ab + b^2 \geqslant 2ab + 2ab = 4ab$
$\Rightarrow \qquad\qquad \tfrac{1}{2}(a + b) \geqslant \sqrt{(ab)}, \qquad \text{when } a > 0, b > 0.$ $\qquad$ ... (2)

Equality only occurs when $a = b$.

This result (2) may be stated in words in the form: '*The arithmetic mean of two positive numbers is greater than or equal to their geometric mean.*'

Example 1
(This is a generalisation of the preceding result.)

Defining the arithmetic mean, A, and the geometric mean, G, of the n positive numbers $a_1, a_2, \ldots, a_n$ by

$$A = \frac{1}{n}(a_1 + a_2 + \ldots + a_n),$$

$$G = \sqrt[n]{(a_1 a_2 \ldots a_n)},$$

then $A \geqslant G$ with equality only holding when $a_1 = a_2 = \ldots = a_n$.

We establish this result by using the result of Example 1, page 296, that, for $x > 0$, $\ln x \leqslant x - 1$. Substituting $x = (a_1/A), (a_2/A), \ldots, (a_n/A)$ in turn and adding we find

$$\sum_{i=1}^{n} \ln (a_i/A) \leqslant \sum_{i=1}^{n} (a_i/A) - n$$

$\Leftrightarrow$ $\quad \displaystyle\sum_{i=1}^{n} \ln (a_i/A) \leqslant n - n = 0, \qquad$ since $A = \displaystyle\sum_{i=1}^{n} a_i/n,$

$\Leftrightarrow$ $\quad \ln \left(\dfrac{a_1 a_2 \ldots a_n}{A^n} \right) \leqslant 0$

$\Leftrightarrow$ $\quad \dfrac{a_1 a_2 \ldots a_n}{A^n} \leqslant 1$

$\Leftrightarrow$ $\quad A^n \geqslant a_1 a_2 \ldots a_n$

$\Leftrightarrow$ $\quad A \geqslant \sqrt[n]{(a_1 a_2 \ldots a_n)} = G$

as required. Note that, when $a_1 = a_2 = \ldots = a_n$, then $A = a_1$, $G = a_1$ and $A = G$.

We use the result of this example in Examples 2 and 3 following.

Example 2

Given that a, b and c are positive, prove that

(a) $(a + b + c)(a^2 + b^2 + c^2) \geqslant 9abc$,

(b) $ab^2 + ac^2 + bc^2 + ba^2 + ca^2 + cb^2 \geqslant 6abc$.

(a) By the result of Example 1,

$$\frac{a + b + c}{3} \geqslant (abc)^{1/3}, \quad \frac{a^2 + b^2 + c^2}{3} \geqslant (a^2 b^2 c^2)^{1/3}$$

$\Rightarrow$ $\qquad (a + b + c)(a^2 + b^2 + c^2) \geqslant 9abc.$

(b) Similarly

$$\frac{ab^2 + ac^2 + bc^2 + ba^2 + ca^2 + cb^2}{6}$$

$$\geqslant [(ab^2)(ac^2)(bc^2)(ba^2)(ca^2)(cb^2)]^{1/6}$$
$$= (a^6 b^6 c^6)^{1/6} = abc,$$

giving the required result.

Example 3

Given that x, y and z are positive and $x + y + z = 1$, show that

$$\frac{1}{x^2} + \frac{1}{y^2} + \frac{1}{z^2} \geqslant 27.$$

By the result of Example 1,

$$\frac{1}{3}\left(\frac{1}{x^2} + \frac{1}{y^2} + \frac{1}{z^2}\right) \geqslant \frac{1}{(x^2y^2z^2)^{1/3}} = \frac{1}{(xyz)^{2/3}}.$$

Also

$$(xyz)^{1/3} \leqslant \tfrac{1}{3}(x + y + z) = \tfrac{1}{3},$$

The required result follows at once.

Exercises 8.2

1 Show that, for real x, the expression

$$\frac{x}{x^2 - 1}$$

can take all real values. [L]

2 Given that $$y = \frac{9x}{(x - 1)(4 - x)},$$

find the set of values of y which correspond to real values of x. Sketch the graph of y against x, showing its asymptotes and giving the coordinates of its turning points. [JMB]

3 Find the set of values of x for which

$$\frac{(x + 2)}{(x + 1)(x - 2)} > 0.$$ [L]

4 Find the set of values of x for which

$$|2x - 5| < |x|.$$ [L]

5 Show that, for all positive values of a and b,

$$a^3 + 2b^3 \geqslant 3ab^2.$$ [L]

6 Find the set of values of x for which

$$|x^2 - 3| > |x^2 - 2|.$$ [L]

7 Find graphically, or otherwise, the set of values of x for which

$$\sqrt{(16 - x^2)} > |x|.$$ [L]

8 Find the sets of values of x for which

(a) $0 < \dfrac{(x-1)(x+3)}{(3x-1)} < 1$,

(b) $|x^2 + 1| < |x^2 - 4|$. [L]

9 Sketch the curve whose equation is

$$y = \frac{(x^2 - 3x + 3)}{(x - 2)(x - 3)},$$

giving the coordinates of any points where the curve cuts the coordinate axes, and clearly marking all the asymptotes.

From your graph, or otherwise, find the set of values of x for which

$$\frac{x}{x - 3} > \frac{1}{x - 2}.$$

Give the first three terms in the expansion of y as a series in ascending powers of x, stating the set of values of x for which this series expansion is valid. [L]

10 Find the sets of values of x for which

(a) $2 \sinh x - \cosh x > 1$,

(b) $\cos^2 x < 3 \sin^2 x$, $0 \leqslant x \leqslant \pi$,

(c) $\sinh x > \text{sech } x$.

11 Find the set of values of x for which

(a) $\tanh x + 4 \text{ sech } x \geqslant 4$, (b) $7 \text{ cosech } x + 2 \coth x < 6$.

12 Prove that

(a) $2(a^2 + b^2) \geqslant (a + b)^2$, (b) $a^2 + b^2 + c^2 \geqslant ab + bc + ca$,

(c) $(a^4 + b^4)(a^2 + b^2) \geqslant (a^3 + b^3)^2$.

13 Given that a, b and c are positive prove that

(a) $(a + b)(b + c)(c + a) \geqslant 8\,abc$, (b) $a^3b + ab^3 \leqslant a^4 + b^4$.

14 Given that $a > 0$, $b > 0$, prove that

$$\frac{1}{a^2} + \frac{1}{b^2} \geqslant \frac{8}{(a + b)^2}.$$

15 Given that a and b are positive, show that

$$(1 - a)(1 - b) > 1 - a - b.$$

Hence, show that, if c is also positive and at least one of a, b and c is less than unity, then

$$(1 - a)(1 - b)(1 - c) > 1 - a - b - c.$$

8.3 Inequalities and related loci

Consideration of the implications of inequalities in two variables is often helped by consideration of related loci in the x-y plane. In general, a geometrical approach is to be recommended rather than an attempt to manipulate inequalities. The techniques used are illustrated by the following examples.

Example 1
The variables x, y satisfy the simultaneous inequalities

$$x - y \leqslant 3,$$
$$y^2 \leqslant 4x.$$

Find the greatest value of x and the least value of y.

First consider the loci $x - y - 3 = 0$ and $y^2 - 4x = 0$ as shown in Figs 8.4(a) and (b) respectively. Figure 8.4(a) shows that the straight line $x - y - 3 = 0$ divides the x-y plane into two regions, in one of which all points (x, y) are such that $x - y - 3 < 0$, and in the other, all points are such that $x - y - 3 > 0$. (We determine whether $x + y - 3$ is positive or negative in a region by finding the value of this expression at some suitably chosen point in the region, say $(0, 0)$ or $(5, 0)$.)

Similarly, as shown in Fig. 8.4(b), the parabola $y^2 - 4x = 0$ divides the x-y plane into two regions, in one of which $y^2 - 4x > 0$, and in the other $y^2 - 4x < 0$. In Figs 8.4(a) and (b) the respective regions for which $x - y - 3 < 0$ and $y^2 - 4x < 0$ are shown shaded.

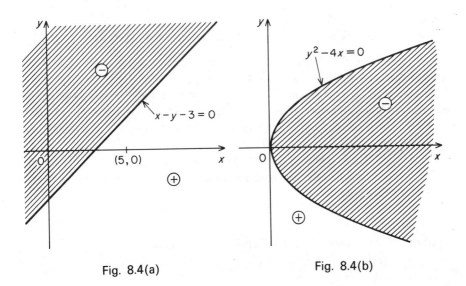

Fig. 8.4(a)   Fig. 8.4(b)

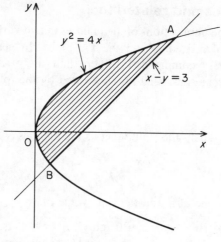

Fig. 8.4(c)

Figure 8.4(c) shows both loci on one diagram, and the region shown shaded clearly contains all points (x, y) for which $x + y < 3$ and $y^2 < 4x$.

By including points on the boundary of this region we have the complete set of points (x, y) for which $x - y \leqslant 3$ and $y^2 \leqslant 4x$. (When a boundary is to be included in a region, it is usually drawn heavier than the rest of the curve.)

Obviously the greatest value of x and the least value of y occur respectively at the points of intersection A and B of the curves. The points A, B are easily found to be $(9, 6)$, $(1, -2)$ respectively, and so the greatest value of x is 9 and the least value of y is -2.

Example 2
Given that $x, y \in \mathbb{R}$ and $x^2 + y^2 \leqslant 2$, show that $|x + y| \leqslant 2$.

The circle C, $x^2 + y^2 = 2$, divides the x-y plane into two regions, with $x^2 + y^2 < 2$ inside C and $x^2 + y^2 > 2$ outside C.

The value of $|x + y|$ is constant on each of the lines $x + y = \lambda$, where λ is a parameter. (The value of this constant is $|\lambda|$.) The circle and some of the family of lines are shown sketched in Fig. 8.5 in which the $\oplus$ and $\ominus$ indicate the signs of $x^2 + y^2 - 2$.

The perpendicular distance from O to the lines is $|\lambda|/\sqrt{2}$. The greatest value, 2, of $|x + y|$ occurs when the line touches the circle, either at the point $(1, 1)$ or at the point $(-1, -1)$.

Example 3
Find the set of values of x for which f$(x) > 0$, where

$$\text{f}(x) \equiv x^3 - 4x^2 - x + 4.$$

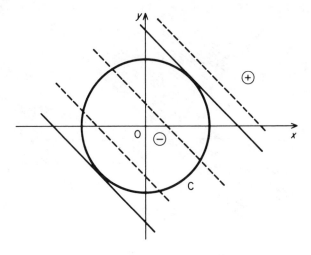

Fig. 8.5

Shade in a diagram of the x-y plane the regions for which $g(x, y) > 0$, where

$$g(x, y) \equiv x^3 - 4x^2y - xy^2 + 4y^3.$$

$$f(x) \equiv (x - 4)(x^2 - 1) \equiv (x + 1)(x - 1)(x - 4).$$
$$\therefore \qquad f(x) > 0 \qquad \text{for} \qquad -1 < x < 1 \qquad \text{and} \qquad x > 4.$$

Using the factors already obtained for $f(x)$, we find

$$g(x, y) \equiv (x + y)(x - y)(x - 4y).$$

The locus $g(x, y) = 0$, therefore, consists of three straight lines through the origin, i.e.

$$x + y = 0, \qquad x - y = 0, \qquad x - 4y = 0.$$

Using the technique already described, we find

$$
\begin{array}{lll}
x + y > 0 & \text{above the line} & x + y = 0, \\
x - y < 0 & \text{above the line} & x - y = 0, \\
x - 4y < 0 & \text{above the line} & x - 4y = 0.
\end{array}
$$

Clearly $g(x, y) > 0$ on the positive x-axis and, as the point (x, y) moves round a circle with centre O, $g(x, y)$ will change sign each time a line is crossed. The lines and appropriate regions (shaded) are shown in Fig. 8.6.

Exercises 8.3

1 Shade on a sketch the region(s) of the x–y plane in which the three inequalities

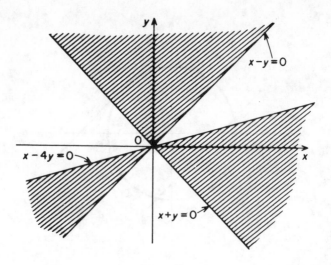

Fig. 8.6

$$x + y < 1,$$
$$x - 2y < 1,$$
$$y - 2x < 1$$

are simultaneously satisfied. [L]

2 Sketch on the same diagram the line $x + y = 1$ and the parabola $y^2 = 4x$. Shade on your sketch the regions of the x-y plane in which

$$(x + y - 1)(y^2 - 4x) > 0.$$ [L]

3 Show, by shading in a diagram, the region or regions of the x-y plane in which

$$x(2x^2 - xy - y^2) < 0.$$ [L]

4 Sketch on the same diagram the curves whose equations are

$$x^2 + y^2 = 16 \text{ and } x^2 = 6y.$$

(You need not find the coordinates of the points of intersection.)
Shade in your diagram the regions of the x-y plane in which

$$(x^2 + y^2 - 16)(x^2 - 6y) < 0.$$ [L]

5 Sketch in the same diagram the curves with equations
(a) $x^2 + y^2 = 36$, (b) $y^2 = 5x$, (c) $y^2 = 6(x + 6)$,

and shade the region of the x-y plane determined by the *simultaneous* inequalities

$$x^2 + y^2 \leqslant 36,$$
$$y^2 \geqslant 5x,$$
$$y^2 \leqslant 6(x + 6). \qquad \text{[L]}$$

6 In one diagram, sketch the three lines

$$x + y - 3 = 0, \quad y - 2x - 4 = 0, \quad y - \frac{x}{2} + 2 = 0.$$

Indicate, by shading on your diagram, the region in which the following three inequalities are all satisfied, marking this region with a large X:

$$x + y - 3 > 0, \quad y - 2x - 4 > 0, \quad y - \frac{x}{2} + 2 > 0. \qquad \text{[L]}$$

7 Sketch in one diagram
(a) the curve with equation

$$y = \frac{(x - 1)}{x(x - 2)}$$

showing its asymptotes and intersections with the axes,
(b) the line with equation $y = x - 1$,
(c) the circle with equation $(x - 1)^2 + y^2 = 1$.
Show that the area of the region in which the inequalities

$$y > \frac{(x - 1)}{x(x - 2)}, \quad y < x - 1, \quad (x - 1)^2 + y^2 < 1$$

are simultaneously satisfied is less than $3\pi/8$. [L]
8 The domain D is the semi-circular region of the plane defined by $x^2 + y^2 \leqslant 1$, $x \geqslant 0$. Find the largest and smallest values attained in D by each of the following expressions: (a) $x + y$; (b) $(x + y)^2$.

8.4 The calculus applied to inequalities

Inequalities involving non-algebraic functions of a single variable may often be conveniently established using methods involving the calculus.

The use of differentiation
Suppose that, for $a \leqslant x \leqslant b$, the greatest and least values of f(x) are M and m respectively. (These values can usually be found by calculating the maxima and minima of f(x) in the interval and also the values f(a), f(b) of f(x) at the end points of the interval.) Then, for $a \leqslant x \leqslant b$,

$$m \leqslant f(x) \leqslant M.$$

This technique is illustrated in Examples 1 and 2 below.

Example 1
Show that, for $x > 0$,

$$\ln x \leqslant x - 1.$$

Consider f(x), where f(x) $\equiv \ln x - x + 1$.
Then f$'(x) = \dfrac{1}{x} - 1$, showing that f(x) is stationary when $x = 1$. Also, as
x increases through unity, f$'(x)$ changes from positive through zero to negative
and therefore f(x) takes its maximum value 0 when $x = 1$.
   Hence f(x) $\leqslant 0$ for all $x > 0$,

i.e. $\ln x \leqslant x - 1$ for all $x > 0$.

The result is illustrated in Figs 8.7(a) and (b), showing respectively the tangent
$y = x - 1$ to the curve $y = \ln x$ and the curve $y = \ln x + 1 - x$.

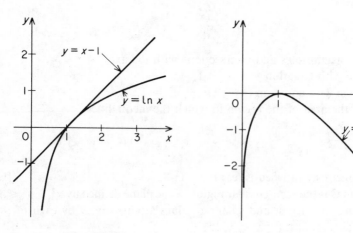

Fig. 8.7(a)   Fig. 8.7(b)

Example 2
The period T of a compound pendulum is given by

$$T = 2\pi \sqrt{\left[\frac{6a^2 + 4x^2}{(a + 4x)g} \right]},$$

where a and g are positive constants and $0 \leqslant x \leqslant 2a$. Show that

$$2\pi \sqrt{(2a/g)} \leqslant T \leqslant 2\pi \sqrt{(6a/g)}.$$

Here we consider the behaviour, for $0 \leqslant x \leqslant 2a$, of f($x$), where

$$f(x) \equiv \frac{6a^2 + 4x^2}{a + 4x}.$$

$$f'(x) = \frac{8x(a + 4x) - 4(6a^2 + 4x^2)}{(a + 4x)^2} = \frac{16x^2 + 8ax - 24a^2}{(a + 4x)^2}$$

$$= \frac{8(2x^2 + ax - 3a^2)}{(a + 4x)^2} = \frac{8(2x + 3a)(x - a)}{(a + 4x)^2}.$$

Since $0 \leqslant x \leqslant 2a$, the relevant stationary value of f(x) occurs when $x = a$ and, since f'(x) changes sign from negative to positive as x increases through the value a, it follows that f(x) has a minimum value $2a$ when $x = a$.

The graph of f(x), for $0 \leqslant x \leqslant 2a$, is sketched in Fig. 8.8.
At A, f(x) = $6a$. At B, f(x) = $22a/9$.

Therefore $\qquad\qquad 2a \leqslant f(x) \leqslant 6a,$

i.e. $\qquad\qquad 2\pi\sqrt{(2a/g)} \leqslant T \leqslant 2\pi\sqrt{(6a/g)}.$

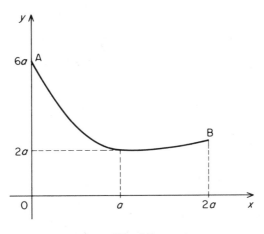

Fig. 8.8

Some inequalities can be established using the following result.
Suppose that f'(x) > 0 for all values of x in the interval $a < x < b$. Then, if $a < x_1 < b$,

$$f(a) < f(x_1) < f(b).$$

The corresponding result when f'(x) $\geqslant$ 0 is

$$f(a) \leqslant f(x_1) \leqslant f(b).$$

Example 1
Prove that, for $x \in \mathbb{R}^+$,

$$x > \tan^{-1} x > x - \tfrac{1}{3}x^3.$$

Consider f(x), where f(x) $\equiv x - \tan^{-1} x$.

Then
$$f'(x) = 1 - \frac{1}{1 + x^2} = \frac{x^2}{1 + x^2} > 0 \qquad \text{for } x > 0.$$

Therefore, for $x > 0$, $f(x)$ increases as x increases.
But $f(0) = 0$.

$\therefore$ $\qquad\qquad\qquad f(x) > 0 \qquad\qquad$ for $x > 0$

$\Rightarrow$ $\qquad\qquad\qquad x > \tan^{-1} x \qquad$ for $x > 0$.

Next consider $g(x)$, where $g(x) \equiv \tan^{-1} x - x + \frac{1}{3}x^3$.

$$g'(x) = \frac{1}{1 + x^2} - 1 + x^2 = \frac{x^4}{1 + x^2} > 0 \qquad \text{for } x > 0.$$

Therefore $g(x)$ increases as x increases for $x > 0$, and, since $g(0) = 0$, $g(x) > 0$ for $x > 0$,

$\Rightarrow$ $\qquad\qquad\qquad \tan^{-1} x > x - \frac{1}{3}x^3 \qquad$ for $x > 0$.

Example 2
Prove that, for $x \in \mathbb{R}$,

$$1 - \frac{x^2}{2} \leqslant \cos x \leqslant 1 - \frac{x^2}{2} + \frac{x^4}{24}.$$

First consider $f(x)$, where

$$f(x) = 1 - \frac{x^2}{2} - \cos x.$$

Then $\qquad\qquad\qquad f'(x) = \sin x - x,$
and $\qquad\qquad\qquad f''(x) = \cos x - 1.$

Therefore $f''(x) \leqslant 0$ for $x \geqslant 0$ and so $f'(x)$ decreases or is zero for $x \geqslant 0$.

But $f'(0) = 0$, and so $f'(x) \leqslant 0$ for $x \geqslant 0$,
i.e. $f(x)$ decreases as x increases for $x \geqslant 0$.
But $f(0) = 0$, and so $f(x) \leqslant 0$ for $x \geqslant 0$

$\Rightarrow$ $\qquad\qquad\qquad 1 - \frac{x^2}{2} \leqslant \cos x \qquad$ for $x \geqslant 0$.

Similarly, considering $g(x)$, where

$$g(x) \equiv \cos x - 1 + \frac{x^2}{2} - \frac{x^4}{24},$$

we find $g^{(4)}(x) = \cos x - 1 \leqslant 0$ for $x \geqslant 0$, and $g'''(0) = 0$
$\Rightarrow$ $g'''(x) \leqslant 0$ for $x \geqslant 0$, and so on, leading to $g(x) \leqslant 0$ for $x \geqslant 0$.
Since f and g are even functions, the inequalities must be true for $x < 0$ also.

Inequalities involving integrals

Suppose that $f(x) \geqslant 0$ for $a \leqslant x \leqslant b$ and $a < b$. Then $\int_a^b f(x)\, dx \geqslant 0$. This result follows at once by considering the definite integral as the limit of a sum or as the area under a curve. (Note that, if strict inequality holds, i.e. $f(x) > 0$, as opposed to $f(x) \geqslant 0$, then $\int_a^b f(x)\, dx > 0$.)

Some simple applications of this result are given below.

(1)
$$\int_a^b [|f(x)| - f(x)]\, dx \geqslant 0$$

$\Leftrightarrow$
$$\int_a^b |f(x)|\, dx \geqslant \int_a^b f(x)\, dx.$$

(2) If $g(x) \geqslant h(x)$ for $a \leqslant x \leqslant b$, then

$$\int_a^b [g(x) - h(x)]\, dx \geqslant 0$$

$\Leftrightarrow$
$$\int_a^b g(x)\, dx \geqslant \int_a^b h(x)\, dx.$$

(3) Suppose that, for $a \leqslant x \leqslant b$,

$$m \leqslant f(x) \leqslant M,$$

where m and M are constants, which could be, but need not necessarily be, the least and greatest values respectively of $f(x)$ for $a \leqslant x \leqslant b$.

Then, from application (2) above $\int_a^b m\, dx \leqslant \int_a^b f(x)\, dx \leqslant \int_a^b M\, dx$

$\Rightarrow$
$$m(b - a) \leqslant \int_a^b f(x)\, dx \leqslant M(b - a).$$

Note that this result is of value when evaluating integrals numerically.

Example 1
Show, without evaluating the integral, that

$$1 < \int_0^1 (1 + x^3)^{1/4}\, dx < 1 \cdot 19.$$

For $0 < x < 1$, $1 + x^3$ increases as x increases from 1 to 2.

$\therefore$ $\qquad\qquad 1 < 1 + x^3 < 2 \qquad$ for $0 < x < 1$.

Taking fourth roots,

$$1 < (1 + x^3)^{1/4} < 2^{1/4}.$$

Integrating with respect to x from 0 to 1 gives

$$\int_0^1 1 \, dx < \int_0^1 (1 + x^3)^{1/4} \, dx < \int_0^1 2^{1/4} \, dx$$

$$\Rightarrow \qquad 1 < \int_0^1 (1 + x^3)^{1/4} \, dx < 2^{1/4}.$$

But $2^{1/4} \approx 1 \cdot 1892$ and so the required result follows.

Example 2
Prove that, for $x > 0$,
(a) $\cosh x > \sinh x > x$,   (b) $1 + \sinh x > \cosh x > 1 + x^2/2$,
(c) $\sinh x - x > \cosh x - 1 - x^2/2 > x^4/24$.

(a) The results follow at once from the definitions of the hyperbolic functions thus:

$$\cosh t - \sinh t = e^{-t} > 0,$$

$$\sinh t = t + \frac{t^3}{3!} + \ldots > t \qquad \text{for } t > 0,$$

$$\Rightarrow \qquad \cosh t > \sinh t > t \qquad \text{for } t > 0. \qquad \ldots (1)$$

Note that we have used t in place of x, but the results are equivalent to those required.

(b) We integrate with respect to t the inequalities (1) from 0 to x, giving, for $x > 0$,

$$\int_0^x \cosh t \, dt > \int_0^x \sinh t \, dt > \int_0^x t \, dt$$

$$\Rightarrow \qquad \sinh x > \cosh x - 1 > x^2/2$$
$$\Leftrightarrow \qquad 1 + \sinh x > \cosh x > 1 + x^2/2. \qquad \ldots (2)$$

(c) Writing (b) in terms of t and integrating as before, we have, for $x > 0$,

$$\int_0^x (1 + \sinh t) \, dt > \int_0^x \cosh t \, dt > \int_0^x (1 + t^2/2) \, dt$$

$$\Rightarrow \qquad x + \cosh x - 1 > \sinh x > x + x^3/6.$$

Integrating once more over the same interval we find, for $x > 0$,

$$\frac{x^2}{2} + \sinh x - x > \cosh x - 1 > x^2/2 + x^4/24$$

$$\Rightarrow \qquad \sinh x - x > \cosh x - 1 - x^2/2 > x^4/24$$

as required.

These results can also be established by the differentiation technique described on page 298.

Exercises 8.4

1 Prove that, for $x \in \mathbb{R}$,
   (a) $x(1 - x) \leqslant \frac{1}{4}$,   (b) $xe^{1-x} \leqslant 1$,   (c) $e^x > 1 + x$.

2 Prove that, for $x \in \mathbb{R}^+$,

   (a) $\dfrac{2 + 3 \sinh x}{2 + 3 \cosh x} > \frac{2}{5}$,   (b) $\ln x \geqslant (x - 1)/x$.

3 Prove that, for $x \in \mathbb{R}^+$,

   (a) $x > \dfrac{3 \sin x}{(2 + \cos x)}$,   (b) $x > \ln (1 + x) > x - \frac{1}{2}x^2$.

4 Prove that, for $x > 4$, $x^2 < 2^x$.

5 Prove that
$$1 > \int_0^1 \frac{1}{\sqrt{(1 + 3x^5)}} \, dx > \frac{1}{2}.$$

6 Given that $n > 1$, prove that
$$\int_0^{\pi/2} \sin^{n+1} x \, dx < \int_0^{\pi/2} \sin^n x \, dx.$$

   By putting $n = 9$ and 10 in this inequality and using Wallis's formula (page 245) prove that $3 < \pi < 3 \cdot 3$.

7 Prove that $\displaystyle\int_0^1 \frac{\sqrt{(1 - x^2)}}{1 + x^2} \, dx = \frac{1}{2}\pi(\sqrt{2} - 1)$.

   Hence prove that
$$\tfrac{1}{2}\pi(\sqrt{2} - 1) < \int_0^1 \frac{\sqrt{(1 - x^2)}}{1 + x^4} \, dx < \tfrac{1}{4}\pi. \qquad\qquad \text{[JMB]}$$

8 Prove that, if $0 \leqslant x < 1$,
$$\left(1 + \frac{x^2}{2}\right)^2 \leqslant \frac{1}{1 - x^2}.$$

   Deduce that
$$x + \frac{x^3}{6} \leqslant \sin^{-1} x$$

   if $0 \leqslant x < 1$.

9 The continuous functions f, g and h are such that
   (a) $f(x) \geqslant 0$,   (b) $g(x) \geqslant h(x)$,
   when $a \leqslant x \leqslant b$. Prove that
$$\int_a^b f(x)g(x) \, dx \geqslant \int_a^b f(x)h(x) \, dx.$$

Use this result to prove that

$$0{\cdot}78 \leqslant \int_0^1 \frac{x}{\sqrt{(1 - x^3)}} \, dx \leqslant 1{\cdot}00.$$

10 Prove that $\displaystyle\int_0^1 \frac{x^4(1 - x)^4}{1 + x^2} \, dx = \frac{22}{7} - \pi.$

Show that $\displaystyle\frac{1}{2} \int_0^1 x^4(1 - x)^4 \, dx < \int_0^1 \frac{x^4(1 - x)^4}{1 + x^2} \, dx < \int_0^1 x^4(1 - x)^4 \, dx,$

and deduce that $\displaystyle\frac{22}{7} - \frac{1}{1260} > \pi > \frac{22}{7} - \frac{1}{630}.$

11 By considering the derivatives of $(x - \sin x)$ and $[\sin x - x + x^3/(3!)]$, show that, for all $x > 0$,

$$x > \sin x > x - x^3/6.$$

Show also, by integration, that, for $x > 0$,

$$1 - \frac{x^2}{2} < \cos x < 1 - \frac{x^2}{2} + \frac{x^4}{24}.$$

12 Show that, for $0 < x < \pi/2$, $\displaystyle\frac{2x}{\pi} < \sin x < x.$

Deduce that, for $0 < x < \pi/2$,

$$e^{-x} < e^{-\sin x} < e^{-2x/\pi}$$

Hence show that

$$1 - e^{-\pi/2} < \int_0^{\pi/2} e^{-\sin x} \, dx < \frac{\pi}{2}\left(\frac{e - 1}{e}\right).$$

13 Show that, for $x > 0$, the expression $\left(1 + \dfrac{1}{x}\right)^{1+x}$ decreases as x increases.

Miscellaneous exercises 8

1 Express E(x), where

$$E(x) \equiv 4x^3 - 9x^2 - 16x + 36$$

as a product of three linear factors. State the zeros of E(x) in (a) $\mathbb{Z}$, (b) $\mathbb{Q}$. Use the factors of E(x) to express F(x), where

$$F(x) \equiv 4x^6 - 9x^4 - 16x^2 + 36$$

as a product of six linear factors. State the zeros of F(x) in (c) $\mathbb{Q}$, (d) $\mathbb{R}$, (e) $\mathbb{C}$.

2 The roots of the equation

$$\sum_{r=0}^{n} a_r x^{n-r} = 0$$

where $a_0 = 1$, are the first n positive integers.

Find and simplify expressions for a_1 and a_n in terms of n.

Show that $a_2 = \dfrac{1}{24}n(n + 1)(n - 1)(3n + 2)$. [JMB]

3 (a) Show that $(a + b)$ is a root of the equation

$$x^3 - 3abx - (a^3 + b^3) = 0.$$

Express the equation $x^3 - 6x - 6 = 0$ in the above form, giving your values for a^3 and b^3. Hence find a real root of the equation $x^3 - 6x + 6 = 0$, expressing your answer in the form $\sqrt[3]{m} + \sqrt[3]{n}$, where m and n are positive integers.

(b) Given that the equation

$$x^3 + px^2 + qx + r = 0$$

has three roots α, β, γ where $\alpha + \beta = \gamma$, show that

$$p^3 + 8r = 4pq.$$ [L]

4 Sketch the curve $y = x^3 + x^2 - 1$.

Show graphically that the equation $x^3 + x^2 = kx + 1$ has three real roots if $k > 1$ and only one real root if $k < 1$.

5 The equation $x^2 + px + q = 0$, where p, q are given non-zero constants, has roots α and β. Prove that, if $u_n = A\alpha^n + B\beta^n$, where A and B are arbitrary constants, then, for $n \geqslant 0$,

$$u_{n+2} + pu_{n+1} + qu_n = 0.$$

The sequence whose first four terms are 1, 11, 49, 179 is known to have a general term of the form u_n, where $u_n = A\alpha^n + B\beta^n$, where u_1 is the first term. Find α, β, A and B, and hence an expression for u_n. [L]

6 p is the statement:

$(x - \alpha)^2$ is a factor of the polynomial f(x).

q is the statement:

$(x - \alpha)$ is a factor of the polynomial f$'(x)$.

(a) Prove that $p \Rightarrow q$.

(b) Show by a counter-example that $q \not\Rightarrow p$.

Given that

$$f(x) = 4x^4 + 4x^3 - 11x^2 - 6x + 9,$$

write down $f'(x)$. Given also that $f(x)$ has a repeated linear factor, factorise $f'(x)$ and $f(x)$. [L]

7 (a) Use the remainder theorem to show that if two polynomials in x of degree n have equal values when x takes each of $n + 1$ different values, then they are identical.

Given that a, b, c are all different, prove that

$$\frac{(x + a + b)(x + a + c)}{(b - a)(c - a)} + \frac{(x + b + c)(x + b + a)}{(c - b)(a - b)}$$

$$+ \frac{(x + c + a)(x + c + b)}{(a - c)(b - c)} \equiv 1.$$

(b) Given that p, q, r are real, prove that
(i) $(p + q + r)^2 \geqslant 3(qr + rp + pq)$,
(ii) $p^4 - p^3q - pq^3 + q^4 \geqslant 0$. [L]

8 Show that the polynomial equation $f(x) = 0$ has a repeated root $x = \beta$ if, and only if, $f(\beta) = 0$ and $f'(\beta) = 0$. Hence, or otherwise, show that, if the equation $x^3 + ax^2 + bx + c = 0$ has a repeated root and $a^2 \neq 3b$, then the value of this root may be expressed in the form

$$\frac{9c - ab}{2(a^2 - 3b)}.$$

Find the roots of the equation

$$x^3 + 2(1 - \sqrt{3})x^2 + (3 - 4\sqrt{3})x + 6 = 0,$$

given that it has a repeated root. [L]

9 A polynomial $P(z)$, of the fourth degree and with real coefficients, is zero when $z = 2 + 3i$. Find a quadratic factor $Q(z)$ of $P(z)$.

Given the additional information that
(a) the coefficient of z^4 in $P(z)$ is 1,
(b) the sum of the six products of pairs of roots of the equation $P(z) = 0$ is 7,
(c) when $P(z)$ is divided by $z + 1$, the remainder is -36,
find numbers f and g such that

$$P(z) = (z^2 + fz + g)Q(z).$$

Hence, or otherwise, obtain all the roots of the equation $P(z) = 0$. [JMB]

10 (a) Given that the roots of the equation

$$x^3 + bx^2 + cx + d = 0,$$

where b, c and d are non-zero constants, are in geometric progression, find a relation between b, c and d.

Solve the equation

$$x^3 - 2x^2 - 4x + 8 = 0,$$

and verify that your solutions are in geometric progression.

(b) Solve the equation

$$4x^3 - 24x^2 + 23x + 18 = 0,$$

given that the roots are in arithmetic progression. [L]

11 (a) Find the set of values of x for which

$$x(x - 4)(x^2 - 4) < 0.$$

(b) Given that p, q, r and s are positive and unequal, show that
(i) $(pq + rs)^2 \leqslant (p^2 + r^2)(q^2 + s^2)$,
(ii) $p^3q + pq^3 < p^4 + q^4$.
   Show also that

$$p^4 + q^4 + r^4 + s^4 > 4pqrs.$$ [L]

12 Find, graphically or otherwise, the set of values of x for which

$$|2x - 1| + |4 - x| > 5.$$ [L]

13 Given that the equation p(x) = 0, where

$$p(x) \equiv x^4 - 4x^3 + 7x^2 - 12x + 12,$$

has a repeated integer root, find all the factors of p(x).

Hence express $\dfrac{4x - 1}{x^4 - 4x^3 + 7x^2 - 12x + 12}$

in partial fractions, and obtain

$$\int_0^1 \frac{4x - 1}{x^4 - 4x^3 + 7x^2 - 12x + 12}\,dx.$$ [L]

14 In the same diagram sketch the graphs of

$$3y + 4x = 12, \qquad \frac{x^2}{25} + \frac{y^2}{16} = 1.$$

Indicate, by shading, the region for which both the inequalities

$$3y + 4x \geqslant 12, \qquad \frac{x^2}{25} + \frac{y^2}{16} \leqslant 1$$

are satisfied.

Write down the coordinates of the point in this region for which $(x - 3)^2 + y^2$ is greatest. [L]

15 Given that the stationary values of the polynomial h(x), where

$$h(x) \equiv 3x^4 - 20x^3 + 12x^2 + 96x - 100,$$

all occur at integer values of x, sketch the curve $y = $ h(x), indicating the coordinates of each turning point, and locate, for each root of h(x) = 0, an interval between two consecutive integers within which the root must lie.
    Find the sum of the squares of the roots of this equation.   [L]

16 Sketch the curve $y = x^3 e^{-x}$, indicating the points of inflection. Find the equations of the tangents to the curve that pass through the origin.
    Show that, for $a > 6$,

$$5 < \int_0^a x^3 e^{-x} \, dx < 6.$$ [L]

17 (a) Find the set of values of the real number x for which the following inequalities hold:

(i) $\dfrac{1}{x + 6} < \dfrac{2}{2 - 3x}$,   (ii) $|5 - 3x| \leqslant |x + 1|$.

(b) Show, by shading on a sketch of the x-y plane, the region for which $x^2 + y^2 \leqslant 1$, $y \geqslant x$ and $y \leqslant x + 1$.
    Hence find
(i) the greatest value of y,   (ii) the least value of $x + y$
for which these inequalities hold.

18 By considering the graphs of $x^2 + y^2 - x = 0$ and $x^2 + y^2 - 1 = 0$, or otherwise, prove that if $x^2 + y^2 < x$, then $x^2 + y^2 < 1$.

19 (a) By considering separately the three intervals

$$x < 0, \, 0 < x < 1, \, 1 < x,$$

sketch the graph of $y = |2x| - |1 - x|$.
(b) By considering separately the signs of $x - 1$ and $x - y$ in the four regions into which the plane is divided by the straight lines $x - 1 = 0$ and $x - y = 0$, sketch the graph of

$$2y = |x - 1| + |x - y|.$$ [JMB]

20 (a) Prove that the inequalities

$$3x + 2y > 6, \qquad x - 2y > 2, \qquad -2x + y > 2$$

cannot simultaneously be satisfied.
(b) Find the set of values of a for which the inequalities

$$0 < x + 2y < a, \qquad xy > 3$$

can be simultaneously satisfied.

21 For the set of curves

$$(x^2 - y^2)(x^2 + y^2 - 1) = k$$

in the plane of the coordinate axes Oxy, show in a diagram the regions in which k is positive.

22 Sketch in the same diagram the two curves
(a) $4x^2 - 3y^2 = 4$,   (b) $xy = 4$.

Shade the areas of the plane in which the inequality

$$(4x^2 - 3y^2 - 4)(xy - 4) \geqslant 0 \qquad\qquad \ldots (1)$$

holds. Indicate the points of these areas for which

$$y^2 + 2x = 8. \qquad\qquad \ldots (2)$$

Find the greatest value of x for points satisfying (1) and (2) simultaneously. Show that there is no least value of x for such points.

23 Prove that

$$a^4 + b^4 + c^4 \geqslant b^2c^2 + c^2a^2 + a^2b^2 \geqslant abc\,(a + b + c). \qquad \text{[L]}$$

24 (a) Given that $0 \leqslant \theta \leqslant 2\pi$, find the set of values of θ for which

$$\cos 2\theta < \cos 4\theta.$$

(b) Find the set of values of x for which

$$0 < \frac{x(2 + x)}{2x + 1} < 1. \qquad \text{[L]}$$

25 Use the fact that

$$\int_a^b [f(x) + \lambda g(x)]^2 \, dx \geqslant 0, \qquad a < b,$$

for $\lambda \in \mathbb{R}$, to show that

$$\int_a^b [f(x)]^2 \, dx \, . \int_a^b [g(x)]^2 \, dx \geqslant \left[\int_a^b f(x)\, g(x) \, dx \right]^2.$$

Deduce that

$$\int_0^{2\pi} \sqrt{(1 - k^2 \sin^2 \theta)} \, d\theta \leqslant 2\pi \sqrt{(1 - \tfrac{1}{2}k^2)},$$

where $0 \leqslant k \leqslant 1$.

9 Numerical solution of differential equations

9.1 Isoclines

It was seen in Chapter 4 that the general solution of a first-order differential equation contains one arbitrary constant.

Thus, the general solution of the differential equation

$$2x\frac{dy}{dx} = y$$

is given by

$$y^2 = cx,$$

where c is an arbitrary constant.

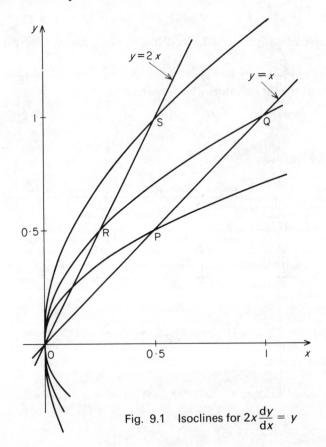

Fig. 9.1   Isoclines for $2x\dfrac{dy}{dx} = y$

This is the equation of a family of parabolas which are the integral curves of the differential equation. Each value of c gives a particular parabola corresponding to a particular solution of the differential equation.

On each parabola there is a point at which the gradient takes a given value m. A curve which cuts each integral curve at the point at which the gradient takes the value m is said to be the isocline for gradient m.

When $\dfrac{dy}{dx} = m$, the differential equation $2x\dfrac{dy}{dx} = y$ becomes $y = 2mx$. In this case, the isocline for gradient m is the straight line $y = 2mx$. Figure 9.1 shows the isoclines $y = x$ (for gradient $\frac{1}{2}$) and $y = 2x$ (for gradient 1) with the three integral curves $y^2 = \frac{1}{2}x$, $y^2 = x$ and $y^2 = 2x$. At the points P and Q the gradient of the integral curves is $\frac{1}{2}$, while at the points R and S the gradient of the integral curves is 1.

By the use of isoclines it is possible to sketch an integral curve before the general solution of the differential equation has been found.

Example

Sketch the integral curve through the point $(0, 0\cdot5)$ for the differential equation

$$\frac{dy}{dx} = x^2 + y^2.$$

The isocline for gradient m will be the circle $x^2 + y^2 = m$. Figure 9.2 shows the circles given by $m = 0\cdot25, 0\cdot5, 0\cdot75$ and 1. On each circle, short line segments are

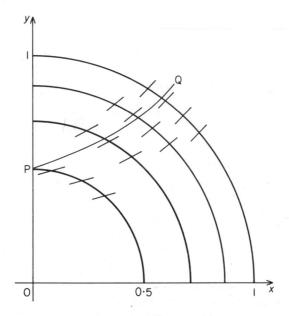

Fig. 9.2   An integral curve for $\dfrac{dy}{dx} = x^2 + y^2$

drawn with gradient equal to the value of m corresponding to that circle. At the point of intersection of an integral curve with a circle, the tangent to the integral curve will be parallel to the line segments on that circle.

At the point P(0, 0·5) the gradient of the integral curve is 0·25. With the line segments as guides, the integral curve PQ can be sketched and shown in Fig. 9.2.

Exercises 9.1

1 Show that the integral curves of the differential equation

$$y\frac{dy}{dx} + x = 0$$

are circles, and that the isoclines are straight lines.

2 Use the isoclines of the differential equation $dy/dx = x - y$, to sketch the integral curve through the point (0, 1).

3 Show that the isoclines of the differential equation

$$\frac{dy}{dx} + 2xy = 0$$

are rectangular hyperbolae. Sketch the integral curve which passes through the point (0, 2).

4 Sketch the integral curve of the differential equation

$$\frac{dy}{dx} = \frac{1}{\sqrt{(x^2 + y^2)}}$$

which passes through the point (0, 1).

5 Show that the isoclines of the differential equation $3x\frac{dy}{dx} = y$ are straight lines. Sketch the integral curve which passes through the point (1, 1). Check your estimate of the curve by finding its equation by integration.

9.2 The Euler method

Let the curve through the point $P_0(x_0, y_0)$ in Fig. 9.3 be an integral curve of the differential equation

$$\frac{dy}{dx} = f(x, y).$$

The Euler method involves the calculation of approximations to the values of y on this integral curve at the points at which x equals $x_0 + h$, $x_0 + 2h$, ..., $x_0 + rh$, ..., where the constant h is known as the *step-length*. It is usual to denote $x_0 + h$ by x_1, $x_0 + 2h$ by x_2 and $x_0 + rh$ by x_r. At the point P_0 the gradient of the integral curve is $f(x_0, y_0)$. The equation of the tangent at P_0 to this curve is

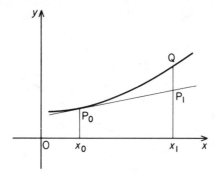

Fig. 9.3

$$y - y_0 = f(x_0, y_0)(x - x_0).$$

Let this tangent meet the line $x = x_1$ at the point $P_1(x_1, y_1)$.

Then
$$y_1 = y_0 + hf(x_0, y_0).$$

This is the value of y at P_1. Let the line $x = x_1$ meet the integral curve at Q. Then y_1 is an approximation to the value of y at Q. Whether it is a good approximation or not depends on the size of h and on whether the gradient of the integral curve is changing rapidly.

The equation of the tangent at P_1 to the integral curve through P_1 is

$$y - y_1 = f(x_1, y_1)(x - x_1).$$

This tangent meets the line $x = x_2$ at the point $P_2(x_2, y_2)$, where

$$y_2 = y_1 + f(x_1, y_1)(x_2 - x_1)$$

$\Rightarrow$
$$y_2 = y_1 + hf(x_1, y_1).$$

Similarly, the tangent at P_2 to the integral curve through P_2 meets the line $x = x_2 + h = x_3$ at the point $P_3(x_3, y_3)$, where

$$y_3 = y_2 + hf(x_2, y_2).$$

By repeated use of the iteration formula

$$y_{n+1} = y_n + hf(x_n, y_n),$$

a sequence of values of y is calculated corresponding to the values $x_1, x_2, \ldots, x_n,$ $\ldots$ of x.

In Fig. 9.4 the tangent to the integral curve through P_0 is represented by P_0P_1, the tangent to the integral curve through P_1 by P_1P_2, and so on.

The Euler method is an example of a step-by-step method. Unless the step-length h is kept small, the discrepancies between the calculated values of y and the corresponding values of y on the integral curve through P_0 may increase rapidly as x increases.

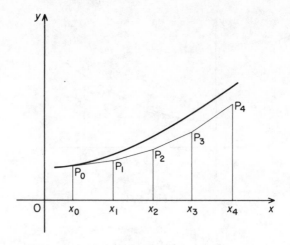

Fig. 9.4

It is convenient to use the notation $x = a(h)b$ to denote the values of x from a to b by steps of length h. Thus the values $x = 0, 0.25, 0.5, 0.75, 1$ are denoted by $x = 0(0.25)1$.

Example 1

Given that $\dfrac{dy}{dx} = x - y$ and that $y = 1$ when $x = 0$, estimate the values of y for $x = 0(0.01)0.04$.

The iteration formula is

$$y_{n+1} = y_n + hf(x_n, y_n),$$

where in this case $f(x, y) = x - y$.

Step 1 $x_0 = 0$, $y_0 = 1$, $h = 0.01$.

$$f(x_0, y_0) = x_0 - y_0 = -1.$$
$$y_1 = y_0 + hf(x_0, y_0)$$
$$\Rightarrow \qquad y_1 = 1 + (0.01)(-1) = 0.99.$$

Step 2 $x_1 = 0.01$, $y_1 = 0.99$, $h = 0.01$.

$$f(x_1, y_1) = x_1 - y_1 = -0.98.$$
$$y_2 = y_1 + hf(x_1, y_1)$$
$$\Rightarrow \qquad y_2 = 0.99 + (0.01)(-0.98) = 0.9802.$$

Step 3 $x_2 = 0.02$, $y_2 = 0.9802$, $h = 0.01$.

$$f(x_2, y_2) = x_2 - y_2 = -0.9602.$$
$$y_3 = y_2 + hf(x_2, y_2)$$
$$\Rightarrow \qquad y_3 = 0.9802 + (0.01)(-0.9602)$$
$$= 0.9706 \text{ to four decimal places.}$$

Step 4 $x_3 = 0.03$, $y_3 = 0.9706$, $h = 0.01$.

$$f(x_3, y_3) = x_3 - y_3 = -0.9406.$$
$$y_4 = y_3 + hf(x_3, y_3)$$
$\Rightarrow$
$$y_4 = 0.9706 + (0.01)(-0.9406)$$
$$= 0.9612 \text{ to four decimal places.}$$

Since the general solution of the differential equation $\dfrac{dy}{dx} = x - y$ can be found explicitly, it is possible to compare the estimated values found above with the true values.

The particular solution such that $y = 1$ when $x = 0$ is given by

$$y = x - 1 + 2e^{-x}.$$

The following table shows the true values (to four decimal places) and the estimated values, together with a column giving the error defined by

$$\text{error} = \text{true value} - \text{estimated value.}$$

x	True value	Estimated value	Error
0·01	0·9901	0·99	0·0001
0·02	0·9804	0·9802	0·0002
0·03	0·9709	0·9706	0·0003
0·04	0·9616	0·9612	0·0004

Since in many cases it is not possible to find the general solution of the differential equation, a method is needed for estimating the error in approximations to the values of y.

Assuming that the Maclaurin series for y exists, it is given by

$$y = y_0 + xy_0' + \frac{x^2}{2!}y_0'' + \ldots + \frac{x^n}{n!}y_0^{(n)} + \ldots,$$

where y_0 denotes the value of y when $x = 0$,

y_0' denotes the value of $\dfrac{dy}{dx}$ when $x = 0$,

and y_0'' denotes the value of $\dfrac{d^2y}{dx^2}$ when $x = 0$.

When $x = h$ this gives

$$y_1 = y_0 + hy_0' + \tfrac{1}{2}h^2y_0'' + \ldots,$$

where y_1 denotes the value of y when $x = h$.

In the formula

$$y_1 = y_0 + hf(x_0, y_0),$$

$f(x_0, y_0)$ is the value of dy/dx when $x = 0$, i.e. $f(x_0, y_0) = y_0'$. This gives

$$y_1 = y_0 + hy_0'.$$

Comparison of this equation with the Maclaurin series

$$y_1 = y_0 + hy_0' + \tfrac{1}{2}h^2y_0'' + \dots$$

shows that the first term neglected is $\tfrac{1}{2}h^2y''_0$, where y_0'' is the value of d^2y/dx^2 at $x = 0$.

In Example 1 above, $\quad \dfrac{dy}{dx} = x - y$

$$\frac{d^2y}{dx^2} = 1 - \frac{dy}{dx} = 1 - x + y.$$

When $x = 0$, $y = 1$ and hence $\dfrac{d^2y}{dx^2} = 2$, i.e. $y_0'' = 2$.

Since the step-length h is $0{\cdot}01$, this gives

$$\tfrac{1}{2}h^2y_0'' = \tfrac{1}{2}(0{\cdot}01)^2 \times 2 = 0{\cdot}0001.$$

This indicates that the estimated value of y_1 is too small by approximately $0{\cdot}0001$. Assuming that the value of d^2y/dx^2 is changing slowly, the error at each step is approximately $0{\cdot}0001$. As the table shows, these errors are cumulative.

The modified Euler method

A very serious disadvantage of the Euler method is that the step-length h must be kept small to prevent the error becoming large.

One method of overcoming this difficulty is based on the trapezium rule

$$\int_{x_0}^{x_1} y \, dx \approx \tfrac{1}{2}h(y_0 + y_1),$$

where $x_1 = x_0 + h$.

Let y be replaced by y', i.e. by $\dfrac{dy}{dx}$. This gives

$$\int_{x_0}^{x_1} y' \, dx \approx \tfrac{1}{2}h(y_0' + y_1')$$

$$\Rightarrow \qquad y_1 - y_0 \approx \tfrac{1}{2}h(y_0' + y_1')$$

$$\Rightarrow \qquad y_1 \approx y_0 + \tfrac{1}{2}h(y_0' + y_1').$$

If the approximation sign $\approx$ is replaced by an equals sign, giving

$$y_1 = y_0 + \tfrac{1}{2}h(y_0' + y_1'),$$

this formula can be used to find an estimated value for y_1.

Similarly, the formula

$$y_2 = y_1 + \tfrac{1}{2}h(y_1' + y_2')$$

can be used to calculate a value for y_2.

In this way, the iterative formula

$$y_{n+1} = y_n + \tfrac{1}{2}h(y_n' + y_{n+1}')$$

can be used to find a sequence of values of y.

In order to calculate y_{n+1} by this formula, it is necessary to substitute a value for y_{n+1}' in the right-hand side. The way in which this difficulty is overcome is explained in the following example.

Example 2

Given that $\dfrac{dy}{dx} = x - y$ and that $y = 1$ when $x = 0$, estimate the value of y when $x = 0.04$, using a step-length 0.04.

The formula to be used is

$$y_1 = y_0 + \tfrac{1}{2}h(y_0' + y_1'),$$

with $x_0 = 0$, $y_0 = 1$ and $h = 0.04$.

From $\dfrac{dy}{dx} = x - y, \quad y_0' = x_0 - y_0 = -1.$

Since the value of y_1' is so far unknown, assume that

$$y_1' = y_0' = -1.$$

Let y_1^* denote the first approximation to the value of y_1. Then

$$y_1^* = 1 + \tfrac{1}{2}(0.04)(-1 - 1) = 0.96.$$

If $y_1 = 0.96$, $y_1' = x_1 - y_1$

$$\Rightarrow \qquad y_1' = 0.04 - 0.96 = -0.92.$$

Let y_1^{**} denote the second approximation to the value of y_1. Then, with the value -0.92 for y_1',

$$y_1^{**} = 1 + \tfrac{1}{2}(0.04)(-1 - 0.92) = 0.9616.$$

If $y_1 = 0.9616$, $y_1' = x_1 - y_1$

$$\Rightarrow \qquad y_1' = 0.04 - 0.9616 = -0.9216.$$

Let y_1^{***} denote the third approximation to the value of y_1. Then

$$y_1^{***} = 1 + \tfrac{1}{2}(0.04)(-1 - 0.9216)$$
$$= 0.9616 \text{ to four decimal places.}$$

As the second and third approximations agree to four decimal places, the value of y when $x = 0.04$ is taken to be 0.9616.

An estimate of the error involved in taking the value of y_1 to be given by $y_0 + \frac{1}{2}h(y_0' + y_1')$ can be obtained by considering the Maclaurin series

$$y = y_0 + xy_0' + \frac{x^2}{2!}y_0'' + \frac{x^3}{3!}y_0''' + \ldots.$$

Differentiation of this series gives

$$y' = y_0' + xy_0'' + \frac{x^2}{2!}y_0''' + \ldots.$$

When $x = h$, this becomes

$$y_1' = y_0' + hy_0'' + \frac{h^2}{2!}y_0''' + \ldots.$$

It follows that

$$y_0 + \frac{1}{2}h(y_0' + y_1') = y_0 + \frac{1}{2}h(2y_0' + hy_0'' + \frac{1}{2}h^2 y_0''' + \ldots)$$
$$= y_0 + hy_0' + \frac{1}{2}h^2 y_0'' + \frac{1}{4}h^3 y_0''' + \ldots.$$

But the substitution of h for x in the Maclaurin series gives

$$y_1 = y_0 + hy_0' + \frac{1}{2}h^2 y_0'' + \frac{1}{6}h^3 y_0''' + \ldots.$$

These two series differ in the coefficients of the terms involving h^3.

Hence the error involved in taking the value of y_1 to be $y_0 + \frac{1}{2}h(y_0' + y_1')$ is approximately equal to

$$\frac{1}{6}h^3 y_0''' - \frac{1}{4}h^3 y_0''', \qquad \text{i.e.} \qquad -\frac{1}{12}h^3 y_0'''.$$

In Example 2 above, $y' = x - y$
$$y'' = 1 - y' = 1 - x + y$$
$$y''' = -y'' = -1 + x - y.$$

When $x = 0$, $y = 1$ and hence $y_0''' = -2$.

The step-length h was 0.04, giving

$$-\frac{1}{12}h^3 y_0''' = -\frac{1}{12}(0.04)^3(-2) \approx 10^{-5}.$$

It is therefore not surprising that the value 0.9616 obtained for y when $x = 0.04$ was correct to four decimal places.

Example 3

Given that $\dfrac{dy}{dx} = x^2 + y^2$ and that $y = 0.5$ when $x = 0$, use the modified Euler method to estimate to four decimal places the value of y when $x = 0.05$ and when $x = 0.1$.

Step 1 $x_0 = 0$, $y_0 = 0.5$, $y'_0 = 0.25$, $h = 0.05$.

$$y_1^* = y_0 + \tfrac{1}{2}h(y'_0 + y'_1).$$

If initially y'_1 is taken equal to y'_0,

$$y_1^* = 0.5 + \tfrac{1}{2}(0.05)(0.25 + 0.25)$$
$\Rightarrow$ $\qquad\qquad y_1^* = 0.5125.$
Then $\qquad\qquad y'_1 = x_1^2 + y_1^2 = (0.05)^2 + (0.5125)^2$
$\Rightarrow$ $\qquad\qquad y'_1 = 0.2652.$
$$y_1^{**} = 0.5 + \tfrac{1}{2}(0.05)(0.25 + 0.2652)$$
$\Rightarrow$ $\qquad\qquad y_1^{**} = 0.5129.$
$$y'_1 = (0.05)^2 + (0.5129)^2 = \underline{0.2656.}$$
$$y_1^{***} = 0.5 + \tfrac{1}{2}(0.05)(0.25 + 0.2656)$$
$\Rightarrow$ $\qquad\qquad y_1^{***} = \underline{0.5129.}$

Step 2 $x_1 = 0.05$, $y_1 = 0.5129$, $y'_1 = 0.2656$, $h = 0.05$.

$$y_2^* = y_1 + \tfrac{1}{2}h(y'_1 + y'_2).$$

If initially y'_2 is taken equal to y'_1,

$$y_2^* = 0.5129 + \tfrac{1}{2}(0.05)(0.2656 + 0.2656)$$
$\Rightarrow$ $\qquad\qquad y_2^* = 0.5262.$
$$y'_2 = x_2^2 + y_2^2 = 0.01 + 0.2769 = 0.2869.$$
$$y_2^{**} = 0.5129 + \tfrac{1}{2}(0.05)(0.2656 + 0.2869)$$
$\Rightarrow$ $\qquad\qquad y_2^{**} = 0.5267.$
$$y'_2 = 0.01 + 0.2774 = \underline{0.2874.}$$
$$y_2^{***} = 0.5129 + \tfrac{1}{2}(0.05)(0.2656 + 0.2874)$$
$\Rightarrow$ $\qquad\qquad y_2^{***} = \underline{0.5267.}$

From the differential equation,

$$y' = x^2 + y^2$$
$$y'' = 2x + 2yy'$$
$$y''' = 2 + 2y'^2 + 2yy''.$$

A short calculation shows that when $x = 0$, $y''' \approx 2.5$.

This gives $\qquad\qquad -\tfrac{1}{12}h^3 y''' \approx -3 \times 10^{-5},$

so that the error at each step is of this magnitude.

The error in the value found for y_2 will be approximately -6×10^{-5}. This means that the fourth decimal place in the value 0.5267 is unreliable.

Example 4

Given that $x\dfrac{dy}{dx} = y + \sqrt{(x^2 + y^2)}$ and that $y = -3$ when $x = 4$, use the modified Euler method to estimate the value of y when $x = 4.4$.

When $h = 0.4$, the formula $y_1 = y_0 + \frac{1}{2}h(y_0' + y_1')$

becomes $\qquad\qquad y_1 = y_0 + (0.2)(y_0' + y_1')$.

When $x_0 = 4$ and $y_0 = -3$, the differential equation gives

$$4y_0' = -3 + \sqrt{(16 + 9)}$$
$$\Rightarrow \qquad y_0' = 0.5.$$

For the first approximation, assume that $y_1' = y_0'$.

$$y_1^* = -3 + (0.2)(0.5 + 0.5)$$
$$\Rightarrow \qquad y_1^* = -2.8.$$

When $x = 4.4$ and $y = -2.8$, the differential equation gives

$$4.4y_1' = -2.8 + \sqrt{[(4.4)^2 + (2.8)^2]}$$
$$\Rightarrow \qquad y_1' = 0.5489 \text{ to four decimal places.}$$
$$y_1^{**} = -3 + (0.2)(0.5 + 0.5489)$$
$$\Rightarrow \qquad y_1^{**} = -2.7902 \text{ to four decimal places.}$$

When $x = 4.4$ and $y = -2.7902$, the differential equation gives

$$4.4y_1' = -2.7902 + \sqrt{[(4.4)^2 + (2.7902)^2]}$$
$$\Rightarrow \qquad y_1' = 0.5500 \text{ to four decimal places.}$$
$$y_1^{***} = -3 + (0.2)(0.5 + 0.5500)$$
$$\Rightarrow \qquad y_1^{***} = -2.7900.$$

This value will not be changed by another iteration.

(The particular solution of the given differential equation is $x^2 = 8y + 8\sqrt{(x^2 + y^2)}$. It will be found that this solution is satisfied exactly by the values $x = 4.4$, $y = -2.79$.)

Exercises 9.2

1 Given that $\dfrac{dy}{dx} = y$ and that $y = 1$ when $x = 0$, use the iteration formula

$y_{n+1} = y_n + hy_n$ to estimate to four decimal places the values of y for $x = 0(0.02)0.1$.

Show that the particular solution of this differential equation with the given condition is $y = e^x$, and hence find the errors in your estimates. Compare these errors with those deduced from the expression $\frac{1}{2}h^2 y_0''$.

2 Given that $\dfrac{dy}{dx} = x^2 + y^2$ and that $y = 1$ when $x = 0$, use the Euler

method to estimate the values of y for $x = 0(0.01)0.04$. Give your answers to four decimal places, and show that the error per step is approximately 10^{-4}.

3 Use the Euler method to estimate the values of y for $x = 0(0.02)0.1$, given

that $5\dfrac{dy}{dx} = x + y$ and that $y = 1$ when $x = 0$. Give your answers to three decimal places.

4   Use the modified Euler method to estimate to three decimal places the value of y when $x = 0.2$, given that $\dfrac{dy}{dx} = x + y$ and that $y = 1$ when $x = 0$.

5   Use the modified Euler method with $h = 0.2$ to estimate to four decimal places the value of y when $x = 0.2$, given that $\dfrac{dy}{dx} = 1 + xy$ and that $y = 1$ when $x = 0$.

6   Given that $4\dfrac{dy}{dx} + 2y + 1 = 0$ and that $y = 0.5$ when $x = 0$, use the Euler method with step-length 0.02 to estimate the value of y when $x = 0.1$ (in five steps).

Use the modified Euler method with step-length 0.1 to estimate the same value of y. Give both answers to four decimal places.

9.3   Taylor series method

Under certain conditions, f(x) can be expanded as a series of powers of $(x - a)$, known as the *Taylor* series for f(x), as in Section 6.5.

If $x - a = h$, this series is given by

$$f(a + h) = f(a) + hf'(a) + \frac{h^2}{2!}f''(a) + \frac{h^3}{3!}f'''(a) + \ldots + \frac{h^n}{n!}f^{(n)}(a) + \ldots.$$

Let $y = $ f(x) be a particular solution of a first-order differential equation, the value of f(a) being given. Then the value of f$'(a)$ can be found by substituting $x = a$ and $y = $ f(a) in the differential equation.

The values of f$''(a)$, f$'''(a)$, $\ldots$ can be found in succession by differentiating the differential equation with respect to x and substituting known values.

In this way, the coefficients of h, h^2, h^3, $\ldots$ in the Taylor series above can be evaluated. The first few terms in the series can then be used to give an approximation to the value of f($a + h$).

The size of the step-length h will depend on the accuracy required and on the number of terms used in the series.

The next stage is the calculation of an approximation to the value of f($a + 2h$). This is based on the series

$$f(a + 2h) = f(a + h) + hf'(a + h) + \frac{h^2}{2!}f''(a + h) + \ldots.$$

To construct this series, first the value of f$'(a + h)$ is found by substituting

$x = a + h$ and $y = f(a + h)$ in the differential equation. The other coefficients are then found as before, by successive differentiation.

In this way, the solution is advanced to give the values of $f(a + 3h)$, $f(a + 4h)$, ..., $f(a + nh)$.

Example

Given that $\dfrac{dy}{dx} = x - y$ and that $y = 1$ when $x = 0$, estimate the value of y when $x = 0.2$, 0.4 and 0.6.

Let $y = f(x)$. Then $f(0) = 1$ and $h = 0.2$.

Step 1 To find $f(0.2)$:

$$
\begin{aligned}
y' &= x - y &&\Rightarrow & f'(0) &= -1. \\
y'' &= 1 - y' &&\Rightarrow & f''(0) &= 2. \\
y''' &= -y'' &&\Rightarrow & f'''(0) &= -2. \\
y^{(4)} &= -y''' &&\Rightarrow & f^{(4)}(0) &= 2.
\end{aligned}
$$

With $a = 0$ and $h = 0.2$, the Taylor series gives

$$f(0.2) = 1 + (0.2)(-1) + \frac{1}{2!}(0.2)^2(2) + \frac{1}{3!}(0.2)^3(-2) + \frac{1}{4!}(0.2)^4(2) \ldots$$

$$= \underline{0.8375} \text{ four decimal places.}$$

Step 2 To find $f(0.4)$, given that $f(0.2) = 0.8375$:

$$
\begin{aligned}
y' &= x - y &&\Rightarrow & f'(0.2) &= 0.2 - 0.8375 = -0.6375. \\
y'' &= 1 - y' &&\Rightarrow & f''(0.2) &= 1.6375. \\
y''' &= -y'' &&\Rightarrow & f'''(0.2) &= -1.6375. \\
y^{(4)} &= -y''' &&\Rightarrow & f^{(4)}(0.2) &= 1.6375.
\end{aligned}
$$

With $a = 0$, $h = 0.2$ and $a + 2h = 0.4$, the Taylor series gives

$$f(0.4) = 0.8375 + (0.2)(-0.6375)$$

$$+ (1.6375)\left[\frac{1}{2!}(0.2)^2 - \frac{1}{3!}(0.2)^3 + \frac{1}{4!}(0.2)^4 - \ldots \right]$$

$$= \underline{0.7407} \text{ to four decimal places.}$$

Step 3 To find $f(0.6)$, given that $f(0.4) = 0.7407$:

$$
\begin{aligned}
y' &= x - y &&\Rightarrow & f'(0.4) &= 0.3407. \\
y'' &= 1 - y' &&\Rightarrow & f''(0.4) &= 1.3407. \\
y''' &= -y'' &&\Rightarrow & f'''(0.4) &= -1.3407. \\
y^{(4)} &= -y''' &&\Rightarrow & f^{(4)}(0.4) &= 1.3407.
\end{aligned}
$$

With $a + 3h = 0.6$, the Taylor series gives

$$f(0.6) = 0.7407 + (0.2)(-0.3407) +$$

$$+ (1.3407)\left(\frac{1}{2!}(0.2)^2 - \frac{1}{3!}(0.2)^3 + \frac{1}{4!}(0.2)^4 - \ldots\right)$$

$= \underline{0.6977}$ to four decimal places.

The values found for f(0·4), f'(0·4), f"(0·4), ... can be checked by substituting $h = -0.2$ in the Taylor series

$$f(0.4 + h) = f(0.4) + hf'(0.4) + \frac{h^2}{2!}f''(0.4) + \frac{h^3}{3!}f'''(0.4) + \frac{h^4}{4!}f^{(4)}(0.4) + \ldots.$$

It will be found that to four decimal places the sum of this series is 0·8375, which is the value of f(0·4 − 0·2).

When the first n terms of a Taylor series are used to give an approximation to the sum of the series, the first neglected term is $h^n f^{(n)}(a)/n!$. It can be shown that the magnitude of the error due to the terms neglected is not greater than $|h|^n M/n!$, where M is the maximum value of $|f^{(n)}(a)|$ for values of a in the interval concerned.

In the example above, the first neglected term is $h^5 f^{(5)}(a)/5!$. Since $f^{(5)}(0) = -2$ and $f^{(5)}(0.4) = -1.3407$, the error at each step from this source is less than $(0.2)^5 \times 2/5!$, i.e. less than 10^{-5}.

Figure 9.5 shows three integral curves for the differential equation $dy/dx = x - y$. The isoclines are straight lines given by $x - y = m$. As x increases, the integral curves draw closer together, indicating that any small inaccuracy incurred at any step will not be subsequently magnified.

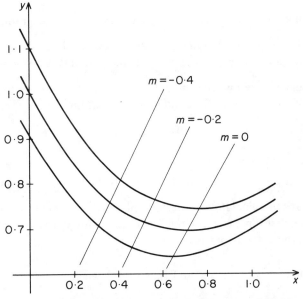

Fig. 9.5   Integral curves for $\dfrac{dy}{dx} = x - y$

Exercises 9.3

1 Given that $\dfrac{dy}{dx} = x^2 - xy$, and that $y = 1$ when $x = 0$, show that the first four terms in the expansion of y as a series of powers of x are

$$1 - \tfrac{1}{2}x^2 + \tfrac{1}{3}x^3 + \tfrac{1}{8}x^4.$$

Estimate to three decimal places the value of y when $x = 0.1$ and when $x = 0.2$.

2 Given that $\dfrac{dy}{dx} = 1 + xy^2$, and that $y = 1$ when $x = 0$, show that the first six terms of the Maclaurin series for y are

$$1 + x + \tfrac{1}{2}x^2 + \tfrac{2}{3}x^3 + \tfrac{1}{2}x^4 + 7x^5/15.$$

Estimate to three decimal places the value of y when $x = 0.2$.

3 Given that $y = 0$ when $x = 0$, show that the series solution of the differential equation $\dfrac{dy}{dx} = 1 - 2xy$ is

$$y = x - \frac{2}{3}x^3 + \frac{4}{15}x^5 - \frac{8}{105}x^7 + \frac{16}{945}x^9 - \ \ldots .$$

Show that to three decimal places the value of y when $x = 0.5$ is 0.424. By finding the Taylor series for y in powers of $(x - 0.5)$, show that to three decimal places the value of y when $x = 1$ is 0.539.

4 Given that $\dfrac{dy}{dx} = 1 - 2xy$, show that the isocline for gradient m is the curve $2xy = 1 - m$. Use the isoclines to sketch the integral curve which passes through the origin. Estimate the maximum value of y on this curve.

5 Given that $y = 0.5$ when $x = 0$, find the first five terms in the Maclaurin series for the solution of the differential equation

$$\frac{dy}{dx} = 2 - 2xy.$$

6 Given that $2\dfrac{dy}{dx} = 1 - xy$, and that $y = 0$ when $x = 0$, use the Maclaurin series for y to calculate to 3 decimal places the value of y when $x = 1$.

Use the first three terms in the Taylor series for y in powers of $(x - 1)$ to estimate the value of y when $x = 2$.

7 Given that $\dfrac{dy}{dx} = 1 - xy$, and that $y = 1$ when $x = 0$, use the first six terms in the Maclaurin series for y to calculate to three decimal places the value of y when $x = 1$.

Obtain the first four terms in the Taylor series for y in powers of $(x - 1)$, and hence estimate to two decimal places the value of y when $x = 1 \cdot 5$.

8   Given that

$$(1 - x^2) \frac{dy}{dx} = xy + 1$$

and that $y = 0$ when $x = 0$, express y as a series of ascending powers of x up to and including the term in x^5.

9   Given that

$$2 \frac{dy}{dx} + y^2 = 4x$$

and that $y = 1$ when $x = 0$, express y as a series of ascending powers of x up to and including the term in x^3.

10   Given that

$$\frac{dy}{dx} = x + \cos y$$

and that $y = 0$ when $x = 2$, express y as a series of ascending powers of $(x - 2)$ up to and including the term in $(x - 2)^4$.

9.4   Second-order equations

The general solution of a second-order differential equation of the form

$$\frac{d^2 y}{dx^2} + p \frac{dy}{dx} + qy = r,$$

where p, q and r are functions of x or constants, will contain two arbitrary constants.

Two conditions are needed in order to evaluate these constants. When the value of y is known for each of two values of x, a step-by-step method of solution can be used.

When the values of y and dy/dx are known for $x = 0$, the Maclaurin series for y can be found. This series can then be used to estimate the value of y for a suitable value h of x. The numerical solution can then be continued either by a step-by-step method or by the Taylor series method.

Example 1
Find the first four non-zero terms in the Maclaurin series for y given that

$$\frac{d^2 y}{dx^2} - 2x \frac{dy}{dx} + 2y = 0,$$

and that $y = 1$, $dy/dx = 0$ when $x = 0$.

Let $y = f(x)$.

The Maclaurin series for f(x) is given by

$$f(x) = f(0) + xf'(0) + \frac{x^2}{2!}f''(0) + \frac{x^3}{3!}f'''(0) + \ldots + \frac{x^n}{n!}f^{(n)}(0) + \ldots.$$

Since $y = 1$ when $x = 0$, f(0) = 1.
Since $dy/dx = 0$ when $x = 0$, f'(0) = 0.
The differential equation gives

$$\frac{d^2y}{dx^2} = 2x\frac{dy}{dx} - 2y. \qquad \ldots (1)$$

Hence, when $x = 0$, f''(0) = -2.
By differentiating equation (1),

$$\frac{d^3y}{dx^3} = 2x\frac{d^2y}{dx^2} + 2\frac{dy}{dx} - 2\frac{dy}{dx}$$

$$\Rightarrow \qquad \frac{d^3y}{dx^3} = 2x\frac{d^2y}{dx^2}. \qquad \ldots (2)$$

This gives f'''(0) = 0.
From equation (2)

$$\frac{d^4y}{dx^4} = 2x\frac{d^3y}{dx^3} + 2\frac{d^2y}{dx^2}. \qquad \ldots (3)$$

This gives f$^{(4)}$(0) = -4.
From equation (3)

$$\frac{d^5y}{dx^5} = 2x\frac{d^4y}{dx^4} + 4\frac{d^3y}{dx^3}. \qquad \ldots (4)$$

This gives f$^{(5)}$(0) = 0.
From equation (4)

$$\frac{d^6y}{dx^6} = 2x\frac{d^5y}{dx^5} + 6\frac{d^4y}{dx^4}.$$

Hence f$^{(6)}$(0) = -24.
With these values of f(0), f'(0), f''(0), $\ldots$ the Maclaurin series for y is

$$y = 1 + x^2(-2)/2! + x^4(-4)/4! + x^6(-24)/6! + \ldots$$
$$= 1 - x^2 - x^4/6 - x^6/30 + \ldots.$$

The values of the coefficients in this series can be found more quickly by the use of Leibnitz's theorem.

When the given differential equation is differentiated n times, it becomes

$$\frac{d^{n+2}y}{dx^{n+2}} - 2x\frac{d^{n+1}y}{dx^{n+1}} - 2n\frac{d^ny}{dx^n} + 2\frac{d^ny}{dx^n} = 0.$$

When $x = 0$ this gives

$$f^{(n+2)}(0) = (2n - 2)f^{(n)}(0).$$

Since $f'(0) = 0$, this shows that for all odd values of n

$$f^{(n)}(0) = 0.$$

When $n = 2$, $f^{(4)}(0) = 2f''(0) = -4$.
When $n = 4$, $f^{(6)}(0) = 6f^{(6)}(0) = -24$.
For all even values of n, $f^{(n)}(0)$ will be negative.
Consider the ratio of the term in x^{2n+2} to the term in x^{2n} in the series for y.

$$\left| \frac{\text{Term in } x^{2n+2}}{\text{Term in } x^{2n}} \right| = \left| \frac{f^{(2n+2)}(0)}{f^{(2n)}(0)} \right| \times \frac{x^2}{(2n+1)(2n+2)}$$

$$= (4n - 2)x^2/[(2n + 1)(2n + 2)].$$

For all values of x, this expression tends to zero as n tends to infinity. It follows by the ratio test that the Maclaurin series converges for all values of x.

The coefficient of x^8 in the series is easily shown to be $-1/168$, and when $x = 0.5$,

$$-x^8/168 \approx -2 \times 10^{-5}.$$

The coefficients decrease rapidly in magnitude, and hence the Maclaurin series up to and including the term in x^6 will give the value of y correct to three decimal places when $x = 0.5$.

Exercises 9.4(i)

1   Find the first four non-zero terms in the Maclaurin series for y, given that

$$\frac{d^2 y}{dx^2} + x^2 y = 0$$

and that $y = 1$, $\dfrac{dy}{dx} = 1$ when $x = 0$.

2   Find the first three non-zero terms in the Maclaurin series for y, given that

$$\frac{d^2 y}{dx^2} = xy$$

and that $y = 0$, $\dfrac{dy}{dx} = 1$ when $x = 0$.

3   Find the first three non-zero terms in the Maclaurin series for y, given that

$$\frac{d^2 y}{dx^2} + x\frac{dy}{dx} + y = 0$$

(a) when $y = 1$, $\dfrac{dy}{dx} = 0$ when $x = 0$;

(b) when $y = 0$, $\dfrac{dy}{dx} = 1$ when $x = 0$.

4 Given that

$$\frac{d^2y}{dx^2} = x\frac{dy}{dx} - y^2$$

and that $y = 1$, $\dfrac{dy}{dx} = 2$ when $x = 0$, show that

$$y = 1 + 2x - x^2/3 - x^3/3 - x^4/3 + \dots .$$

5 Given that

$$(1 - x^2)y'' - xy' + 4y = 0$$

and that $y = 0$, $y' = 2$ when $x = 0$, express y as a series of ascending powers of x up to and including the term in x^5.

6 Given that

$$y'' = x^2 - y^2 - y'^2$$

and that $y = 1$, $y' = -1$ when $x = 1$, express y as a series of ascending powers of $(x - 1)$ up to and including the term in $(x - 1)^4$.

7 Obtain the first four terms in the Maclaurin series for y, given that

$$xy'' + y' + y = 0$$

and that $y = 1$, $y' = -1$ when $x = 0$.

8 Given that $y'' = 2x + 2yy'$ and that $y = 1$, $y' = 0$ when $x = 1$, obtain the expansion of y in a series of powers of $(x - 1)$ up to and including the term in $(x - 1)^5$.

Step-by-step method

Let $y = f(x)$, and let y_0 denote f(0), y_1 denote f(h) and y_{-1} denote f($-h$), where h is the step-length. Using the Maclaurin series for f(x), an approximation can be obtained which expresses y_1 in terms of y_0 and y_{-1}.

Provided that the Maclaurin series for f(x) exists, the expansion of f(h) in a series of powers of h is given by

$$f(h) = f(0) + hf'(0) + \frac{h^2}{2!}f''(0) + \frac{h^3}{3!}f'''(0) + \frac{h^4}{4!}f^{(4)}(0) + \dots .$$

When h is replaced by $-h$, this becomes

$$f(-h) = f(0) - hf'(0) + \frac{h^2}{2!}f''(0) - \frac{h^3}{3!}f'''(0) + \frac{h^4}{4!}f^{(4)}(0) - \ldots$$

By addition,

$$f(h) + f(-h) = 2f(0) + h^2f''(0) + \tfrac{1}{12}h^4f^{(4)}(0) + \ldots$$

When h is sufficiently small for terms in h^4 and higher powers of h to be neglected,

$$h^2f''(0) \approx f(h) - 2f(0) + f(-h),$$
$$h^2y_0'' \approx y_1 - 2y_0 + y_{-1},$$

where y_0'' denotes the value of $\dfrac{d^2y}{dx^2}$ when $x = 0$.

If the value of x is increased from 0 to h, this approximation becomes

$$h^2y_1'' \approx y_2 - 2y_1 + y_0,$$

where $y_2 = f(2h)$ and y_1'' denotes the value of $\dfrac{d^2y}{dx^2}$ when $x = h$.

When this approximation is put in the form

$$y_2 \approx 2y_1 - y_0 + h^2y_1'',$$

it can be used to find an approximation to the value of y_2.

In general, the approximation

$$y_{n+1} \approx 2y_n - y_{n-1} + h^2y_n'',$$

where y_r denotes $f(rh)$, can be used to find a value for y_{n+1}. This provides the basis for a step-by-step method for the numerical solution of second-order differential equations.

Example 1

Given that $\dfrac{d^2y}{dx^2} = y - x^2$ and that $y = 1$ when $x = 0$, $y = 1{\cdot}005$ when $x = 0{\cdot}1$, estimate the value of y when $x = 0{\cdot}2$ and when $x = 0{\cdot}3$.

Step 1

$$x_0 = 0, \quad y_0 = 1, \quad x_1 = 0{\cdot}1, \quad y_1 = 1{\cdot}005, \quad h = 0{\cdot}1.$$

From the differential equation,

$$y_1'' = y_1 - x_1^2 = 1{\cdot}005 - 0{\cdot}01 = 0{\cdot}995$$
$\Rightarrow \qquad h^2y_1'' = 0{\cdot}009\,95$ to five decimal places.

From the approximation

$$y_2 \approx 2y_1 - y_0 + h^2 y''_1,$$

the estimated value of y_2 is given by

$$y_2 = 2\cdot010 - 1 + 0\cdot009\,95$$
$$= \underline{1\cdot019\,95} \text{ to five decimal places.}$$

Step 2

$$x_1 = 0\cdot1, \quad y_1 = 1\cdot005, \quad x_2 = 0\cdot2, \quad y_2 = 1\cdot019\,95, \quad h = 0\cdot1.$$

From the differential equation,

$$y''_2 = y_2 - x_2^2 = 1\cdot019\,95 - 0\cdot04 = 0\cdot979\,95$$
$$\Rightarrow \qquad h^2 y''_2 = 0\cdot009\,80 \text{ to five decimal places.}$$

From the approximation

$$y_3 \approx 2y_2 - y_1 + h^2 y''_2,$$

the estimated value of y_3 is given by

$$y_3 = 2\cdot039\,90 - 1\cdot005 + 0\cdot009\,80$$
$$= \underline{1\cdot044\,70} \text{ to five decimal places.}$$

Example 2

Given that $\dfrac{d^2 y}{dx^2} + y = 0$ and that $y = 0$ when $x = 0$, $y = 1$ when $x = 1$, estimate the values of y when $x = 0\cdot2, 0\cdot4, 0\cdot6$ and $0\cdot8$.

The differential equation can be expressed as $h^2 y'' + h^2 y = 0$.
When $x = h = 0\cdot2$, this gives

$$(y_2 - 2y_1 + y_0) + h^2 y_1 = 0.$$

Hence $y_2 = (2 - h^2)y_1 = 1\cdot96 y_1$, since $y_0 = 0$.
When $x = 2h = 0\cdot4$,

$$(y_3 - 2y_2 + y_1) + h^2 y_2 = 0.$$

Hence $y_3 = 1\cdot96 y_2 - y_1 = 2\cdot84 y_1$.
When $x = 3h = 0\cdot6$,

$$(y_4 - 2y_3 + y_2) + h^2 y_3 = 0.$$

Hence $y_4 = 1\cdot96 y_3 - y_2 = 3\cdot61 y_1$.
When $x = 4h = 0\cdot8$,

$$(1 - 2y_4 + y_3) + h^2 y_4 = 0.$$

Hence $1 - 1\cdot96 y_4 + y_3 = 0$.
Since $y_4 = 3\cdot61 y_1$ and $y_3 = 2\cdot84 y_1$, this gives

$$(1\cdot96)(3\cdot61)y_1 - 2\cdot84 y_1 = 1$$
$$\Rightarrow \qquad\qquad\qquad\qquad y_1 = 0\cdot236.$$

Hence $y_2 = 0.463$, $y_3 = 0.671$, $y_4 = 0.852$.
These values agree closely with those found from the exact solution
$y = (\sin x)/(\sin 1)$.

Consider now the error in the approximation

$$f(h) \approx 2f(0) - f(-h) + h^2f''(0).$$

From the equation

$$f(h) - 2f(0) + f(-h) = h^2f''(0) + \tfrac{1}{12}h^4f^{(4)}(0) + \ldots,$$

it can be seen that the error is approximately $\tfrac{1}{12}h^4f^{(4)}(0)$.
Let f be replaced by f'' in the approximation itself. This gives

$$f''(h) \approx 2f''(0) - f''(-h) + h^2f^{(4)}(0)$$
$$\Rightarrow \qquad \tfrac{1}{12}h^4f^{(4)}(0) \approx \tfrac{1}{12}h^2[f''(h) - 2f''(0) + f''(-h)].$$

This expression is used to provide a correction term in the following way.
Let y_0 and y_1 be known. Then the initial approximation to y_2 is given by

$$y_2 = 2y_1 - y_0 + h^2y_1''.$$

This value of y_2 is then used to calculate y_2''.
Then the correction term $\tfrac{1}{12}h^2(y_2'' - 2y_1'' + y_0'')$ can be evaluated and added to
the initial approximation to y_2.

Example 3
Estimate to five decimal places the value of y when $x = 0.2$, given that
$\dfrac{d^2y}{dx^2} + 2x^2y = 0$ and that $y = 1$ when $x = 0$ and $y = 1.1$ when $x = 0.1$.

$$x_0 = 0, \, y_0 = 1, \, y_1 = 1.1, \, h = 0.1.$$
$$y_2 = 2y_1 - y_0 + h^2y_1''$$
$$\Rightarrow \qquad y_2 = 2.2 - 1 - 2(0.01)(0.1)^2(1.1) = 1.19978$$

$$\text{Correction term} = \tfrac{1}{12}h^2(y_2'' - 2y_1'' + y_0'')$$
$$= \tfrac{1}{12}(0.01)(-2x_2^2y_2 + 4x_1^2y_1)$$
$$= -0.00004 \text{ to five decimal places.}$$

Hence a better approximation is given by

$$y_2 = 1.19978 - 0.00004$$
$$= 1.19974.$$

Exercises 9.4(ii)

1 Given that $\dfrac{d^2y}{dx^2} = xy$ and that $y = 0$ when $x = 0$, and $y = 0.2$ when

$x = 0.2$, estimate to four decimal places the value of y
(a) when $x = 0.4$,   (b) when $x = 0.6$.

2 Given that $\dfrac{d^2y}{dx^2} = 2x + y$ and that $y = 1$ when $x = 0$, and $y = 0.9$ when $x = 0.1$, estimate the value of y to 3 decimal places
(a) when $x = 0.2$,   (b) when $x = 0.3$.

3 Given that $\dfrac{d^2y}{dx^2} + y = 0$, $y = 0$ when $x = 0$, and $y = 0.09983$ when $x = 0.1$, estimate to four decimal places the value of y when $x = 0.2$.

4 Continue Example 4 above to estimate the value of y when $x = 0.3$, employing the correction term.

5 Given that $\dfrac{d^2y}{dx^2} + 5xy = 0$, $y = 1$ when $x = 0$, and $y = 1.1$ when $x = 0.1$, estimate to four decimal places the value of y when $x = 0.2$.

6 Given that $\dfrac{d^2y}{dx^2} = xy$, $y = 0$ when $x = 0$, and $y = 1$ when $x = 1$, estimate to three decimal places the value of y for $x = 0(0.2)1$.

7 Given that $\dfrac{d^2y}{dx^2} + y = 0$, $y = 1$ when $x = 0$, and $y = 0$ when $x = 1$, estimate y to three decimal places for $x = 0.25, 0.5$ and 0.75.

8 Given that $\dfrac{d^2y}{dx^2} = y$, $y = 1$ when $x = 0$, and $y = 1.54308$ when $x = 1$, estimate to 3 decimal places the value of y for $x = 0.25, 0.5$ and 0.75. Compare your values with the corresponding values of $\cosh x$.

9.5 Order of convergence

Let the members $a_1, a_2, a_3, \ldots$ of a sequence of real terms satisfy the condition

$$|a_{n+1} - l| \leqslant k|a_n - l|, \qquad n \in \mathbb{Z}^+,$$

where l is a constant and k is a positive constant less than 1.

Then
$$|a_2 - l| \leqslant k|a_1 - l|,$$
$$|a_3 - l| \leqslant k|a_2 - l| \leqslant k^2|a_1 - l|,$$

and, by induction,

$$|a_n - l| \leqslant k^{n-1}|a_1 - l|.$$

Since $0 < k < 1$, k^{n-1} tends to zero as n tends to infinity, and so $|a_n - l|$ will also tend to zero.
Hence a_n tends to the limit l.
This shows that the condition

$$|a_{n+1} - l| \leqslant k|a_n - l|, \qquad n \in \mathbb{Z}^+,$$

where $0 < k < 1$, is a sufficient condition for convergence of the sequence $\{a_n\}$.

When this condition is satisfied, the sequence $\{a_n\}$ is said to possess first-order convergence (or linear convergence).

Not all convergent sequences satisfy this condition, i.e. the condition is not necessary for convergence. For example, the sequence in which the nth term a_n equals $(2n + 1)/n$ converges to the limit 2. However,

$$a_n - 2 = 1/n, \qquad a_{n+1} - 2 = 1/(n + 1),$$

and it is impossible to find a fixed value of k between 0 and 1 such that, for all positive values of n,

$$\frac{1}{n + 1} \leqslant \frac{k}{n}.$$

Example 1

Show that the sequence $\{a_n\}$ in which $a_n = \dfrac{2^{n+1} + 1}{2^n}$ possesses first-order convergence.

Since $a_n = 2 + 1/2^n$, a_n tends to the limit 2 as n tends to infinity.

$$a_n - 2 = 1/2^n, \qquad a_{n+1} - 2 = 1/2^{n+1}$$
$$\Rightarrow \qquad\qquad |a_{n+1} - 2| = \tfrac{1}{2}|a_n - 2|.$$

Thus the sequence satisfies the condition

$$|a_{n+1} - l| \leqslant k|a_n - l|, \qquad n \in \mathbb{Z}^+,$$

with $l = 2$ and $k = \tfrac{1}{2}$, proving that the sequence possesses first-order convergence.

Example 2

Show that the sequence $\{a_n\}$ defined by $a_{n+1} = (6a_n - 5)/a_n$, $a_1 = 2$, possesses first-order convergence.

The first step is to find the possible values for the limit of the sequence.

Assuming that a_n tends to l as n tends to infinity, a_{n+1} will also tend to l, and the relation $a_{n+1} = (6a_n - 5)/a_n$ becomes

$$l = (6l - 5)/l$$
$$\Rightarrow \qquad\qquad l^2 - 6l + 5 = 0$$
$$\Rightarrow \qquad\qquad l = 5 \text{ or } 1.$$

These are the possible values for the limit.

(a) Consider whether the sequence is increasing or not.

$$a_{n+1} - a_n = (6a_n - 5)/a_n - a_n$$
$$\Rightarrow \qquad\qquad a_{n+1} - a_n = -\frac{(a_n - 5)(a_n - 1)}{a_n}.$$

Since this expression is positive when $1 < a_n < 5$, a_{n+1} is greater than a_n when $1 < a_n < 5$.

(b) Consider the sign of $a_{n+1} - 5$.

$$a_{n+1} - 5 = \frac{6a_n - 5}{a_n} - 5 = \frac{a_n - 5}{a_n}.$$

This shows that, when $1 < a_n < 5$, $(a_{n+1} - 5)$ is negative, i.e. $a_{n+1} < 5$.

The results of (a) and (b) combined show that, with $a_1 = 2$,

$$2 < a_n < a_{n+1} < 5, n \geqslant 2.$$

Since a_n is never less than 2, the equation

$$a_{n+1} - 5 = (a_n - 5)/a_n$$

gives

$$|a_{n+1} - 5| \leqslant \tfrac{1}{2}|a_n - 5|, n \geqslant 1.$$

Thus the sequence satisfies the condition for first-order convergence. Also

$$|a_n - 5| \leqslant \frac{1}{2^{n-1}} |a_1 - 5|,$$

showing that, as n tends to infinity, a_n tends to the limit 5.

Example 3

The sequence $\{a_n\}$ is such that

$$(a_{n+2} - a_{n+1}) = k(a_{n+1} - a_n), \qquad n \in \mathbb{Z}^+,$$

where $|k| < 1$. Show that the sequence possesses first-order convergence. The sum of the $n - 1$ equations

$$a_3 - a_2 = k(a_2 - a_1),$$
$$a_4 - a_3 = k^2(a_2 - a_1),$$
$$\cdots \qquad \cdots$$
$$a_{n+1} - a_n = k^{n-1}(a_2 - a_1)$$

gives

$$a_{n+1} - a_2 = (k + k^2 + \ldots + k^{n-1})(a_2 - a_1)$$

$$= \left(\frac{k - k^n}{1 - k}\right)(a_2 - a_1).$$

As n tends to infinity, k^n tends to zero since $|k| < 1$.

Hence a_{n+1} tends to a limit l, where

$$l - a_2 = \frac{k}{1 - k}(a_2 - a_1).$$

It follows that

$$a_{n+1} - l = -\frac{k^n}{1 - k}(a_2 - a_1)$$

$$\Rightarrow \qquad a_n - l = -\frac{k^{n-1}}{1 - k}(a_2 - a_1).$$

Hence

$$|a_{n+1} - l| = |k| \times |a_n - l|, \qquad n \in \mathbb{Z}^+.$$

Since $|k|$ is a positive constant less than 1, the condition for first-order convergence is satisfied.

This result has an application to the modified Euler method, which was used in Example 2, Section 9.2, to solve the differential equation $\dfrac{dy}{dx} = x - y$, where $y = 1$ when $x = 0$.

The formula

$$y_1 = y_0 + \tfrac{1}{2}h(y_0' + y_1')$$

was employed to produce a sequence of approximations $y_1^*, y_1^{**}, y_1^{***}, \ldots$ to the value of y_1.

These approximations satisfied the equations

$$y_1^{**} = y_0 + \tfrac{1}{2}h(y_0' + x_1 - y_1^*),$$
$$y_1^{***} = y_0 + \tfrac{1}{2}h(y_0' + x_1 - y_1^{**}),$$

giving

$$y_1^{***} - y_1^{**} = -\tfrac{1}{2}h(y_1^{**} - y_1^*).$$

Since $|-\tfrac{1}{2}h| < 1$, the sequence of approximations behaves like the sequence in Example 3 above. Hence the sequence $y_1^*, y_1^{**}, y_1^{***}, \ldots$ possesses first-order convergence.

Second-order convergence
Let the members of the real sequence $\{a_n\}$ satisfy the condition

$$|a_{n+1} - l| \leqslant k|a_n - l|^2, \qquad n \in \mathbb{Z}^+,$$

where l and k are constants and $k > 0$.

Then

$$|a_2 - l| \leqslant k|a_1 - l|^2,$$
$$|a_3 - l| \leqslant k|a_2 - l|^2 \leqslant k^3|a_1 - l|^4,$$
$$|a_4 - l| \leqslant k|a_3 - l|^2 \leqslant k^7|a_1 - l|^8.$$

By induction,

$$|a_n - l| \leqslant k^{m-1}|a_1 - l|^m,$$

where $m = 2^{n-1}$.

Hence, provided that $k|a_1 - l| < 1$, a_n converges to the limit l. When these conditions are satisfied, the sequence is said to have second-order convergence or quadratic convergence. To compare first-order and second-order convergence, consider the following two sequences:

(a) $1, 1/2, 1/4, 1/8, 1/16, \ldots, 1/2^{n-1}, \ldots$, defined by $a_{n+1} = \frac{1}{2}a_n$, $a_1 = 1$. This sequence has first-order convergence.

(b) $1, 1/2, 1/8, 1/128, 1/32\,768, \ldots, 1/2^{m-1}, \ldots$, where $m = 2^{n-1}$. This sequence is defined by $a_{n+1} = \frac{1}{2}a_n^2$, $a_1 = 1$, and has second-order convergence.

Example 4

Show that the sequence defined by $a_{n+1} = (a_n^2 + 5)/(2a_n)$, $a_1 = 2$, converges to $\sqrt{5}$ and has second-order convergence.

Since $a_1 > 0$, every member of the sequence will be positive.

$$a_{n+1} - \sqrt{5} = (a_n^2 - 2\sqrt{5}a_n + 5)/(2a_n) = (a_n - \sqrt{5})^2/(2a_n).$$

This shows that a_{n+1} is greater than $\sqrt{5}$ for $n \geqslant 1$.

It follows that

$$|a_{n+1} - \sqrt{5}| < |a_n - \sqrt{5}|^2/4, \qquad n \geqslant 1.$$

The conditions for second-order convergence are satisfied, with the limit l equal to $\sqrt{5}$.

It will be found that, to six decimal places,

$$a_2 = 2{\cdot}25, \quad a_3 = 2{\cdot}236\,111, \quad a_4 = 2{\cdot}236\,068,$$

the value of a_4 giving $\sqrt{5}$ to this degree of accuracy.

Example 5

The Newton–Raphson method is used to give a sequence of approximations converging to a root α of the equation $f(x) = 0$. Show that the sequence has second-order convergence.

Let x_1 and x_2 be the first and second approximations to α. Then

$$x_2 = x_1 - f(x_1)/f'(x_1)$$
$$\Rightarrow \qquad f(x_1) = (x_1 - x_2)f'(x_1).$$

The expansion of $f(x)$ in powers of $(x - x_1)$ is given by

$$f(x) = f(x_1) + (x - x_1)f'(x_1) + \tfrac{1}{2}(x - x_1)^2 f''(x_1) + \ldots.$$

When $f(x_1)$ is replaced by the expression above, this becomes

$$f(x) = (x - x_2)f'(x_1) + \tfrac{1}{2}(x - x_1)^2 f''(x_1) + \ldots$$

Since $f(\alpha) = 0$, the substitution $x = \alpha$ gives

$$0 = (\alpha - x_2)f'(x_1) + \tfrac{1}{2}(\alpha - x_1)^2 f''(x_1) + \ldots$$

If the terms involving $(\alpha - x_1)^3$ and higher powers can be neglected, this gives

$$(x_2 - \alpha)f'(x_1) = \tfrac{1}{2}(x_1 - \alpha)^2 f''(x_1)$$

$\Rightarrow$
$$|x_2 - \alpha| \leqslant k|x_1 - \alpha|^2,$$

where k is the maximum value of $|\tfrac{1}{2}f''(x)/f'(x)|$ in the neighbourhood of the root.
The same method shows that

$$|x_{n+1} - \alpha| \leqslant k|x_n - \alpha|^2.$$

Therefore, providing that the initial approximation x_1 is sufficiently close to the root, the sequence will have second-order convergence.

Example 6
Under certain conditions, a sequence converging to a root α of the equation $x = F(x)$ can be found by means of the interative formula $x_{n+1} = F(x_n)$. Show that such a sequence will have first-order convergence.

Explain how three consecutive approximations x_1, x_2 and x_3 can be used to increase the rate of convergence, and apply your method to calculate to five decimal places the real root of the equation $x^3 - x^2 - 1 = 0$.

It was shown in *Advanced Mathematics 2*, Section 1.6, that, when the initial approximation x_1 is sufficiently close to the root α of the equation $x = F(x)$,

$$(x_{n+1} - \alpha) \approx (x_n - \alpha)F'(\alpha).$$

Provided that $|F'(x)|$ is less than 1, the sequence $\{x_n\}$ will converge to the limit α.
Let k be a constant such that $|F'(\alpha)| < k < 1$. Then

$$|x_{n+1} - \alpha| < k|x_n - \alpha|,$$

showing that the sequence has first-order convergence.
Let x_1, x_2 and x_3 be consecutive members of the sequence.
Then

$$(x_2 - \alpha) \approx (x_1 - \alpha)F'(\alpha),$$
$$(x_3 - \alpha) \approx (x_2 - \alpha)F'(\alpha).$$

It follows that

$$(x_3 - \alpha)(x_1 - \alpha) \approx (x_2 - \alpha)^2$$

$\Rightarrow$
$$\alpha \approx \frac{x_1 x_3 - x_2^2}{x_3 - 2x_2 + x_1}.$$

For purposes of calculation, this is much better expressed in the equivalent form

$$\alpha \approx x_3 - \frac{(x_3 - x_2)^2}{x_3 - 2x_2 + x_1}.$$

The equation $x^3 - x^2 - 1 = 0$, which has a root near $1\cdot5$, can be put in the form $x = (1 + x^2)^{1/3}$.

This is of the type $x = F(x)$, with $F(x) = (1 + x^2)^{1/3}$. Differentiation gives

$$F'(x) = (2x/3)(1 + x^2)^{-2/3}$$
$$\Rightarrow \qquad F'(1\cdot5) \approx 0\cdot46.$$

Since $|F'(1\cdot5)| < 1$, the sequence defined by

$$x_{n+1} = (1 + x_n^2)^{1/3}, \qquad x_1 = 1\cdot5,$$

will converge.

It will be found that, to six decimal places,

$$x_2 = 1\cdot481\,248, \qquad x_3 = 1\cdot472\,706.$$

Substitution in the expression for α above gives

$$\alpha \approx 1\cdot465\,560,$$

to six decimal places. With this value, the iteration can be continued to give the result $1\cdot465\,57$ to five decimal places.

Exercises 9.5

1 Show that the sequence given by $a_n = (2^{n+2} + 5)/2^n$ has first-order convergence.

2 The first term in a geometric series is a, and the common ratio r is such that $|r| < 1$. Show that the sequence $S_1, S_2, S_3, \ldots$ of partial sums of the series has first-order convergence.

3 Find which of the following sequences have first-order convergence.
   (a) $1, 1/8, 1/27, \ldots, 1/n^3, \ldots$,
   (b) $1, 1/3, 1/9, \ldots, 1/3^{n-1}, \ldots$,
   (c) $1, 8/3, 3, \ldots, n^3/3^{n-1}, \ldots$.

4 A sequence is defined by $x_{n+1} = (x_n + 8)/(x_n + 3)$, $x_1 = 4$. Show that
   (a) $x_n > 2 \Rightarrow x_{n+1} < 2$,
   (b) x_n tends to 2 as n tends to infinity,
   (c) the sequence has first-order convergence.

5 Write down the first four terms in the sequence given by $a_{n+1} = (2a_n + 1)/(a_n + 2)$ (a) when $a_1 = 2$, (b) when $a_1 = 1/2$.
   Show that in each case the sequence has first-order convergence.

6 Show that the sequence defined by

$$20a_{n+1} = 30a_n - a_n^3, \quad a_1 = 3$$

converges to $\sqrt{10}$ and that it possesses second-order convergence. Show also that a_4 gives $\sqrt{10}$ correct to six decimal places.

7 Show that the sequence defined by

$$3x_{n+1} = 4x_n - x_n^4/12, \quad x_1 = 2$$

converges to the cube root of 12, and that x_5 gives the value of this limit correct to six decimal places.

8 By applying the Newton–Raphson method to the equation

$$f(x) = 1/x - a = 0,$$

obtain the iteration formula

$$x_{n+1} = 2x_n - ax_n{}^2.$$

Show that when $0 < a < 1$ and $x_1 = 1$ the sequence has second-order convergence to the limit $1/a$.

Show also that when $a = 0 \cdot 7$ the limit is given correct to five decimal places by x_5.

9 The equation $\cos x - \ln x = 0$ has a real root close to $1 \cdot 3$. Show that with initial approximation $1 \cdot 3$ one application of the Newton–Raphson gives the root correct to five decimal places.

10 Use the iterative formula $x_{n+1} = (2 + x_n)^{1/4}$, with $x_1 = 1 \cdot 3$, to calculate to five decimal places the positive root of the equation $x^4 - x - 2 = 0$ by the method of Example 6 above.

Miscellaneous exercises 9

1 The gradient of a curve which passes through the point $(0, 1)$ is given by

$$\frac{dy}{dx} = \frac{1}{x + y + 4}.$$

Estimate the value of y at $x = 1$ by using iteratively the approximation

$$y(h) \approx y(0) + \tfrac{1}{2}h[y'(0) + y'(h)],$$

taking $h = 1$ and giving your result to three decimal places. [L]

2 Explain graphically why the approximation $y_{n+1} \approx y_n + hy'_n$ is in general less accurate than the approximation

$$y_{n+1} \approx y_n + \tfrac{1}{2}h(y'_n + y'_{n+1}).$$

Given that

$$\frac{dy}{dx} = x^2 + y^2,$$

and that $y = 1$ when $x = 0$, employ the second relation above to estimate to five decimal places the value of y when $x = 0\cdot1$. [L]

3 Show that Euler's method for solving numerically the differential equation $\frac{dy}{dx} = x + y$ leads to the relation

$$y_{n+1} = (1 + h)y_n + hx_n.$$

Evaluate y_2 taking $x_0 = 1$, $y_0 = 0\cdot5$ and $h = 0\cdot1$. [L]

4 Calculate to four decimal places the value of ln $2\cdot05$ by applying the approximation

$$y(a + h) \approx y(a) + hy'(a)$$

to the differential equation $dy/dx = 1/x$, where $y = 0\cdot6931$ when $x = 2$.
    Explain with the help of a diagram why the calculated value is greater than the true value.

5 Given that

$$5\frac{dy}{dx} = \ln(x - y + 2)$$

and that $y = 1$ when $x = 0$, use the approximation

$$y_{n+1} \approx y_n + \tfrac{1}{2}h(y'_n + y'_{n+1})$$

to estimate to three decimal places the value of y at $x = 0\cdot5$ and at $x = 1$.

6 Show that the approximation $y(x + h) \approx y(x) + hy'(x)$ is exact when y is a linear function of x. Show also that the approximation

$$y(x + h) \approx y(x) + \tfrac{1}{2}h[y'(x) + y'(x + h)]$$

is exact when y is a quadratic function of x.
    Given that $dy/dx = 1 + xy$ and that $y(0) = 1$, estimate to three decimal places the value of $y(0\cdot2)$
    (a) using the first approximation twice with $h = 0\cdot1$,
    (b) using the second approximation iteratively with $h = 0\cdot2$. [L]

7 Use the approximation

$$f(x + h) \approx f(x) + \tfrac{1}{2}h[f'(x) + f'(x + h)]$$

with $h = 0\cdot5$ to estimate to four decimal places the value of y when $x = 0\cdot5$, given that $dy/dx = 1/(xy + 2)$ and that $y = 0$ at $x = 0$.

8 Use the Euler method to estimate the value of y for $x = 0(0{\cdot}02)0{\cdot}1$, given that $dy/dx = 1 + x - y$ and that $y = 0{\cdot}5$ when $x = 0$. Give your values to four decimal places.

Compare your answers with values calculated from the solution $y = x + \frac{1}{2}e^{-x}$.

9 Use the modified Euler method with $h = 0{\cdot}2$ to estimate to four decimal places the value of y when $x = 0{\cdot}2$ and when $x = 0{\cdot}4$, given that $dy/dx = \ln (y - x + 1)$ and that $y = 1$ when $x = 0$.

10 Given that $dy/dx = 1 + x + y$ and that $y = 0$ when $x = 0$, use the Maclaurin series for y to show that when $x = 0{\cdot}2$ the value of y to four decimal places is $0{\cdot}2428$.

By finding the first four terms in the expansion of y in powers of $(x - 0{\cdot}2)$, show that the value of y to four decimal places when $x = 0{\cdot}4$ is $0{\cdot}5836$.

11 Given that

$$\frac{d^2y}{dx^2} = (x - 1)(x - 2)y$$

and that $y = 1$ when $x = 0$ and $y = 1{\cdot}05$ when $x = 0{\cdot}2$, estimate to two decimal places the value of y when $x = 0{\cdot}4$ and when $x = 0{\cdot}6$.

12 Use the approximation $y_{n+1} \approx 2y_n - y_{n-1} + h^2y_n''$ to estimate to three decimal places the value of y when $x = 1$ and when $x = 1{\cdot}5$, given that

$$\frac{d^2y}{dx^2} = xe^{-y},$$

and that $y(0) = 0$, $y(0{\cdot}5) = 0{\cdot}2$.

13 The function y satisfies the differential equation

$$\frac{d^2y}{dx^2} = (1 + y)x^3,$$

and is such that $y = 0$, $dy/dx = 1$ at $x = 0$.

Obtain the first three non-zero terms in the expansion of y in a series of ascending powers of x.

By substituting these three terms as an approximation for y in the differential equation and then integrating, show that the next non-zero term in the series is $x^{10}/1800$.

14 The function y satisfies the differential equation

$$\frac{d^2y}{dx^2} - x\frac{dy}{dx} + xy = 0.$$

Find the first three non-zero terms in the Maclaurin series for y
(a) given that $y = 0$, $dy/dx = 1$ at $x = 0$,
(b) given that $y = 1$, $dy/dx = 0$ at $x = 0$.

15 Obtain the Maclaurin series for y up to and including the term in x^4, given that

$$(1 - x)\frac{d^2y}{dx^2} - 2\frac{dy}{dx} + y = 0,$$

(a) when $y = 1$, $dy/dx = 0$ at $x = 0$,
(b) when $y = 1$, $dy/dx = 1$ at $x = 0$.

16 Given that $\dfrac{d^2y}{dx^2} = 1 + y^2$ and that $y(0) = 0$ and $y'(0) = 1$, obtain the Maclaurin series for y as far as the term in x^6. Show, using these terms, that $y(0\cdot1)$ is approximately equal to $0\cdot105\,01$.

Use the relation

$$f''(a) \approx \frac{1}{h^2}[f(a + h) - 2f(a) + f(a - h)]^{\cdot}$$

with a step-length h equal to $0\cdot1$ and working to 5 decimal places, to estimate the value of $y(0\cdot2)$. [L]

17 Show that in the Maclaurin expansion of $\cosh x \cos x$ the coefficients of x, x^2, x^3 are zero, and that the coefficient of x^4 is $-1/6$. Obtain the expansion as far as the term in x^8.

Show that $\quad 1\cdot933\,33 < \displaystyle\int_{-1}^{1} \cosh x \cos x \, dx < 1\cdot933\,43$. [L]

18 A solution of the equation $dy/dx = f(x)$ is required such that $y = 1$ when $x = 0$. Draw up a flow diagram showing how to calculate values of y at $x = 0\cdot1, 0\cdot2, \ldots, 1\cdot0$ using the approximation

$$y_{n+2} - y_n \approx \tfrac{1}{3}h[f(x_{n+2}) + 4f(x_{n+1}) + f(x_n)],$$

where $h = 0\cdot05$ and y_n denotes the value of y at $x = x_n = nh$.

Using the above method, calculate to three decimal places the values of y at $x = 0\cdot1$ and $x = 0\cdot2$ when $f(x) = \sec(\pi x)$. [L]

19 Show that the equation $x^4 = 6x + 5$ has only two real roots and that the sum of the other two roots is real. Find the sum of the squares of all the roots of the equation.

By writing the equation
(a) in the form $x = (6x + 5)^{1/4}$,
(b) in the form $x = (x^4 - 5)/6$,
evaluate the two real roots to five decimal places by iteration.

20 Find graphically the number of roots of the equation

$$4x^2 - 1 = \tan x$$

in the interval $-1.5 < x < 1.5$.

Taking 0.65 as the initial approximation to one root, obtain a better approximation

(a) by applying the Newton–Raphson method once to the equation

$$4x^2 - 1 - \tan x = 0,$$

(b) by writing the equation in the form $x = \frac{1}{2}\sqrt{(1 + \tan x)}$, and using an iterative method.

Give your answers to four decimal places. [L]

21 Find graphically the number of real roots of the equation

$$e^x - 20 \ln x = 0.$$

By applying Newton's method to this equation, with 1 as the first approximation, find a second and a third approximation to the smaller root, giving your answers to four decimal places.

Verify graphically that if the first approximation is taken greater than 2.21, the sequence of approximations will converge to the larger root. Find this root to three decimal places. [L]

22 Explain the meaning of the statement that the Newton–Raphson method has quadratic convergence.

By sketching the graphs of $y = e^x - 3$ and $y = \ln(x + 3)$, show that the equation

$$e^x - \ln(x + 3) - 3 = 0$$

has two real roots. Use the Newton–Raphson method with 1.5 as the first approximation to evaluate to five decimal places the positive root of this equation. [L]

23 Show graphically that the equation $2 \ln x + 1 - x = 0$ has two real roots. Taking 3 as a first approximation and working to three decimal places, find by applying Newton's method to this equation a second and a third approximation to the larger root. Show that the method will fail to give the larger root if the first approximation is taken to be less than 2. [L]

24 State without proof the condition for convergence of the method of successive substitution, in which a root of the equation $x = f(x)$ is found by iteration using the relation $x_{n+1} = f(x_n)$. Use this method to calculate to four decimal places the positive root of the equation $x^4 = 2x + 10$ by writing this equation in the form $x = (2x + 10)^{1/4}$.

Show that it is not possible to calculate this root by writing the equation in the form $x = \frac{1}{2}(x^4 - 10)$. [L]

25 Explain what is meant by linear convergence of a sequence. Show that the sequence defined by

$$x_{n+1} = \frac{2x_n + 1}{2x_n + 2}, \qquad x_1 = 1$$

has linear convergence.

   Find how many members of the sequence differ from the limit by more than 10^{-6}.

26 Explain the meaning of quadratic convergence of a sequence. Show that the sequence defined by

$$x_{n+1} = \frac{x_n^2 + 2}{2x_n}, \qquad x_1 = 1$$

has quadratic convergence.

   Find how many members of the sequence differ from the limit by more than 10^{-6}.

27 Show graphically that the equation $15x = 10 \sinh x + 1$ has a positive root, and evaluate it to six significant figures.   [AEB]

28 Show graphically that the equation $(\sqrt{x}) \tanh x + 2x = 5 \cdot 8$ has a real root and evaluate it to six significant figures.   [AEB]

29 A sequence is defined by the equations

$$x_1 = a, \qquad x_{n+1} = \tfrac{1}{2}(x_n + b/x_n),$$

where a and b are positive numbers. If $x_n = b^{1/2}(1 + \delta_n)$, where $|\delta_n| < 1$, show that

$$\delta_{n+1} = \tfrac{1}{2}\delta_n^2(1 - \delta_n + \delta_n^2 - \delta_n^3 + \ldots),$$

and state the limit of the sequence.   [AEB]

30 Solve the equation $(2x^2 + 2)\dfrac{dy}{dx} = y$, given that $y = 1$ when $x = 0$.

   Calculate to four decimal places the value of y when $x = 0 \cdot 4$
   (a) from the exact solution,
   (b) by a step-by-step method based on the approximation

$$y(a + h) \approx y(a) + hy'(a),$$

using a step-length $0 \cdot 1$.

10 Coordinate geometry

10.1 The rectangular hyperbola $xy = c^2$, $x = ct$, $y = c/t$

In *Advanced Mathematics 1* we considered certain curves such as the parabola $y^2 = 4ax$. Since the point with coordinates $x = at^2$, $y = 2at$ lies on the curve $y^2 = 4ax$ for all values of t, we were able to express the equation of the curve in the parametric form $x = at^2$, $y = 2at$, where t is a parameter which can take all real values.

The coordinates $x = ct$, $y = c/t$ satisfy the equation $xy = c^2$ for all values of t except $t = 0$, i.e. the point with coordinates $x = ct$, $y = c/t$ lies on the curve $xy = c^2$. Thus the equation of the curve $xy = c^2$ can be expressed in parametric form by $x = ct$, $y = c/t$.

From the cartesian equation $xy = c^2$ or from the parametric equations $x = ct$, $y = c/t$ we can deduce certain properties of the curve.

(1) If the point (a, b) is on the curve, then $ab = c^2$ and so the point $(-a, -b)$ is also on the curve since $(-a)(-b) = ab = c^2$. Thus the curve has point symmetry about the origin. It will also be shown later in this chapter that the curve is symmetrical about the lines $y = x$ and $y = -x$.
(2) Since $y = c^2/x$ we see that y tends to 0 as x tends to infinity. Thus the x-axis is an asymptote of the curve. Similarly x tends to 0 as y tends to infinity and so the y-axis is also an asymptote.

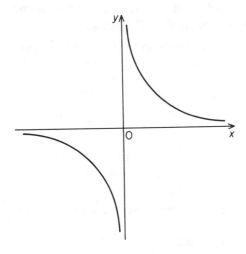

Fig. 10.1

(3) Differentiating $y = c^2/x$ gives $\dfrac{dy}{dx} = -c^2/x^2$. Therefore $\dfrac{dy}{dx}$ is negative for all

real values of x. Thus the gradient of the curve is always negative. Also $\dfrac{dy}{dx}$ is

never zero and so the curve has no turning points.

As x tends to infinity, $\dfrac{dy}{dx}$ tends to 0, and as x tends to 0, $\dfrac{dy}{dx}$ tends to infinity.

Using these properties of the curve and the ideas outlined in *Advanced Mathematics 1*, Section 9.4, a sketch of the curve is as shown in Fig. 10.1.

Since $\dfrac{d^2y}{dx^2} = \dfrac{2c^2}{x^3}$, we see that $\dfrac{d^2y}{dx^2}$ also is never equal to zero and so the curve has no point of inflexion.

This curve, whose asymptotes are at right angles to each other, is known as a *rectangular hyperbola*. Later in this chapter we shall consider hyperbolas with asymptotes which are not at right angles.

Exercises 10.1

1 Using the same axes, sketch the curves with equations
 (a) $xy = 4$, (b) $xy = 9$, (c) $xy = 24$, (d) $xy = 60$.

2 Sketch the curves with parametric equations
 (a) $x = t, y = 1/t$, (b) $x = 4t, y = 4/t$,
 (c) $x = 5t, y = 5/t$.

10.2 Tangent and normal to the rectangular hyperbola $xy = c^2$ at the point P $(cp, c/p)$

In Fig. 10.2, P $(cp, c/p)$ is a point on the rectangular hyperbola $xy = c^2$, where $c > 0$, PT is the tangent at P and PN is the normal at P.

| $xy = c^2$ | Or | $x = ct, \qquad y = c/t.$ |

$\therefore \quad y = \dfrac{c^2}{x} = c^2 x^{-1}.$ $\qquad \therefore \quad \dfrac{dx}{dt} = c, \quad \dfrac{dy}{dt} = -\dfrac{c}{t^2}.$

$\therefore \quad \dfrac{dy}{dx} = -\dfrac{c^2}{x^2}.$ $\qquad \therefore \quad \dfrac{dy}{dx} = \dfrac{dy}{dt} \Big/ \dfrac{dx}{dt} = -\dfrac{1}{t^2}.$

At P, $x = cp$ and the parameter t has the value p, and so $\dfrac{dy}{dx} = -\dfrac{1}{p^2}$, i.e. the

gradient of the tangent at P is $-\dfrac{1}{p^2}$.

Therefore the equation of the tangent at P is

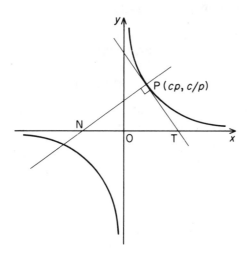

Fig. 10.2

$$y - \frac{c}{p} = -\frac{1}{p^2}(x - cp),$$

i.e. $x + p^2y = 2cp.$

The gradient of the tangent at P is $-\frac{1}{p^2}$ and so the gradient of the normal at P is p^2, since for perpendicular lines $m_1 m_2 = -1$.

Therefore the equation of the normal at P is

$$y - \frac{c}{p} = p^2(x - cp),$$

i.e. $p^3x - py = c(p^4 - 1).$

Example 1

The tangent to the rectangular hyperbola $xy = c^2$ at the point P $(cp, c/p)$ cuts the axes at the points A and B. Show that

(a) PA = PB, (b) the area of the triangle OAB is independent of p.

Find, as p varies, the least possible length of AB.

Figure 10.3 shows the rectangular hyperbola $xy = c^2$ with its tangent at the point P crossing the axes at A and B.

(a) The equation of the tangent at P is $x + p^2y = 2cp.$

At A, $y = 0$ and $x = 2cp$, i.e. the coordinates of A are $(2cp, 0)$.

At B, $x = 0$ and $y = 2c/p$, i.e. the coordinates of B are $(0, 2c/p)$.

The mid-point of AB has coordinates

$$x = \tfrac{1}{2}(2cp + 0) = cp \qquad \text{and} \qquad y = \tfrac{1}{2}(0 + 2c/p) = c/p,$$

i.e. the mid-point of AB is at the point P $(cp, c/p)$.

Thus $$PA = PB$$

(b) The area of the triangle AOB $= \tfrac{1}{2}(OA)(OB) = \tfrac{1}{2}(2\,cp)\left(\dfrac{2c}{p}\right) = 2c^2.$

This is independent of p, the parameter for the point P, and so for all positions of P the area of the triangle OAB will be $2c^2$.

$$AB^2 = OA^2 + OB^2 = (2cp)^2 + \left(\frac{2c}{p}\right)^2$$

$$= 4c^2\left(p^2 + \frac{1}{p^2}\right)$$

$$= 4c^2\left[\left(p - \frac{1}{p}\right)^2 + 2\right]$$

and so the least possible length of AB is $2c\sqrt{2}$ and this occurs when $p = 1$ or when $p = -1$.

This result could, of course, be obtained by finding, by calculus, the minimum values of f(p), where f$(p) \equiv p^2 + \dfrac{1}{p^2}$.

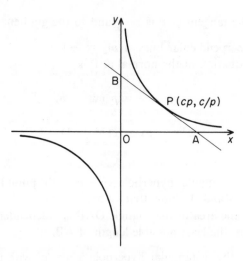

Fig. 10.3

Example 2

The normal at P $(cp, c/p)$ to the rectangular hyperbola $xy = c^2$ meets the hyperbola again at Q. Find the coordinates of Q. Find also the equation of the locus of M, the mid-point of PQ.

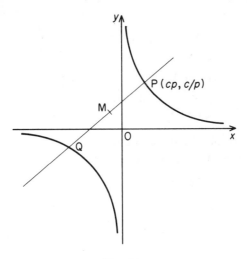

Fig. 10.4

The equation of the normal at P in Fig. 10.4 is $p^3x - py = c(p^4 - 1)$.

To find the coordinates of Q, the point where the normal at P meets the hyperbola again, we solve the simultaneous equations

$$p^3x - py = c(p^4 - 1)$$

and $x = ct$, $y = c/t$, the parametric equations of the hyperbola.
Putting $x = ct$ and $y = c/t$ in $p^3x - py = c(p^4 - 1)$

gives
$$p^3ct - \frac{pc}{t} = c(p^4 - 1),$$

i.e.
$$p^3t^2 + (1 - p^4)t - p = 0.$$

We know that $t = p$ is one solution.

Therefore
$$(t - p)(p^3t + 1) = 0,$$

giving $p^3t + 1 = 0$, i.e. $t = -\dfrac{1}{p^3}$ is the other solution.

At P, $t = p$ and at Q, $t = -1/p^3$.

Thus the coordinates of Q are $x = -c/p^3$, $y = -cp^3$.

The coordinates (X, Y) of M, the mid-point of PQ, are

$$X = \frac{1}{2}\left[cp - \frac{c}{p^3}\right] = \frac{c(p^4 - 1)}{2p^3},$$

$$Y = \frac{1}{2}\left[\frac{c}{p} - cp^3\right] = \frac{-c(p^4 - 1)}{2p}.$$

At M, $c(p^4 - 1) = 2p^3X = -2pY,$

giving

$$p^2 = -Y/X \quad \text{and} \quad Y^2 = \frac{c^2(p^4 - 1)^2}{4p^2}.$$

Eliminating p gives $\quad 4\left(-\dfrac{Y}{X}\right)Y^2 = c^2\left[\dfrac{Y^2}{X^2} - 1\right]^2,$

giving $\qquad\qquad 4X^3Y^3 + c^2(Y^2 - X^2)^2 = 0.$

This equation is satisfied by the coordinates of M for all values of p, i.e. the cartesian equation of the locus of M, as p varies, is

$$4x^3y^3 + c^2(y^2 - x^2)^2 = 0.$$

Example 3

Show that the line $y = -\lambda^2 x + \mu$ is a tangent to the rectangular hyperbola $xy = c^2$ if $\mu = \pm 2\lambda c$.

The line $y = -\lambda^2 x + \mu$ meets the rectangular hyperbola $xy = c^2$,

where $\qquad\qquad\qquad x(-\lambda^2 x + \mu) = c^2,$
i.e. where $\qquad\qquad \lambda^2 x^2 - \mu x + c^2 = 0.$

If the line is tangent, there will be only one point of contact of the line and the hyperbola, i.e. this quadratic equation will have equal roots.

Hence the required condition is $\mu^2 - 4\lambda^2 c^2 = 0,$

i.e. $\qquad\qquad\qquad\qquad \mu = \pm 2\lambda c.$

An alternative solution:

The given line has a negative gradient $-\lambda^2$. Suppose that it is the tangent at P $(cp, c/p)$. Then $-\dfrac{1}{p^2} = -\lambda^2$ and so $p = \pm 1/\lambda$.

Taking $p = 1/\lambda$, we require, if the line is to pass through P, $c\lambda = -\dfrac{\lambda^2 c}{\lambda} + \mu,$

i.e. $\mu = 2c\lambda$, and if $p = -1/\lambda$, we have $\mu = -2c\lambda$.

Exercises 10.2

1 Find the equations of the tangents and normals to the rectangular hyperbolas
(a) $xy = 4$ at the points $(1, 4)$ and $(2, 2)$,
(b) $xy = 9$ at the points $(3, 3)$ and $(-3, -3)$,
(c) $xy = 36$ at the points $(-4, -9)$ and $(-2, -18)$.

2 Find the equations of the tangents and normals to the rectangular hyperbolas
(a) $x = 2t, y = 2/t$ at the points where $t = 1$ and $t = 4$,
(b) $x = 3t, y = 3/t$ at the points where $t = 2$ and $t = -1$,
(c) $x = 4t, y = 4/t$ at the points where $t = -2$ and $t = -5$.

3 Find the coordinates of the points of intersection of the tangents and normals in **1** with the coordinate axes.

4 Find the coordinates of the points of intersection of the tangents and normals in **2** with the coordinate axes.

5 The tangent at the point (2, 8) to the rectangular hyperbola $xy = 16$ cuts the coordinate axes at A and B. Calculate
(a) the length AB,   (b) the area of the triangle OAB.

6 The tangent to the rectangular hyperbola $xy = c^2$ at the point P $(cp, c/p)$ cuts the coordinate axes at A and B. Find the coordinates of G, the centroid of the triangle OAB. Find also the cartesian equation of the locus of G as p varies.

7 The tangent at P $(3p, 3/p)$ to the rectangular hyperbola $xy = 9$ cuts the x-axis at A and the normal at P cuts the y-axis at D. Find the cartesian equation of the locus, as p varies, of
(a) the mid-point of AP,
(b) the mid-point of PD,
(c) the mid-point of AD.

8 The tangent at P $(p, 1/p)$ to the rectangular hyperbola $xy = 1$ meets the coordinate axes at A and B. The normal at P meets the coordinate axes at C and D. Find the area of
(a) the triangle OAB,   (b) the triangle OCD.

9 The normal at P $(2p, 2/p)$ to the rectangular hyperbola $xy = 4$ meets the hyperbola again at Q. Find the coordinates of Q. The line PO meets the hyperbola again at P'. Write down the coordinates of P'. Show that the angle PP'Q $= 90°$.
   (Note that the straight line POP', a line passing through the origin and joining two points on the hyperbola, is called a *diameter* of the hyperbola.)

10 Given that the line $y = -4x + \lambda$ is a tangent to the rectangular hyperbola $xy = 9$, find the possible values of λ.

11 Find the equations of the tangents from the point (4, 8) to the rectangular hyperbola $xy = 36$.

12 Find the equations of the tangents from the point $(-5, -3)$ to the rectangular hyperbola $xy = 64$.

10.3 The hyperbola $\dfrac{x^2}{a^2} - \dfrac{y^2}{b^2} = 1$

If all the points in the x-y plane are rotated about the origin through an angle of $-45°$ (i.e. a clockwise rotation of 45°), the curve $xy = c^2$ shown in Fig. 10.5(a) is transformed into the curve shown in Fig. 10.5(b).

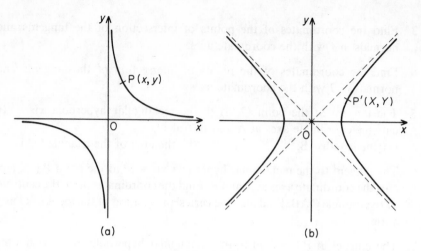

Fig. 10.5

The point $P(x, y)$ becomes the point $P'(X, Y)$. The length of OP = the length of OP' and the angle $POP' = 45°$.

The unit vectors $\begin{pmatrix} 1 \\ 0 \end{pmatrix}$ and $\begin{pmatrix} 0 \\ 1 \end{pmatrix}$ become $\begin{pmatrix} \cos 45° \\ -\sin 45 \end{pmatrix}$ and $\begin{pmatrix} \sin 45° \\ \cos 45° \end{pmatrix}$ respectively so that the transformation is given by

$$\begin{pmatrix} X \\ Y \end{pmatrix} = \begin{pmatrix} \cos 45° & \sin 45° \\ -\sin 45° & \cos 45° \end{pmatrix} \begin{pmatrix} x \\ y \end{pmatrix}.$$

Then $X = \dfrac{x}{\sqrt{2}} + \dfrac{y}{\sqrt{2}}$, giving $X^2 = \dfrac{x^2}{2} + \dfrac{y^2}{2} + xy$

and $Y = -\dfrac{x}{\sqrt{2}} + \dfrac{y}{\sqrt{2}}$, giving $Y^2 = \dfrac{x^2}{2} + \dfrac{y^2}{2} - xy$

$\therefore$ $\qquad\qquad X^2 - Y^2 = 2xy = 2c^2.$

Thus the equation of the rectangular hyperbola shown in Fig. 10.5(b) is

$$x^2 - y^2 = 2c^2.$$

This rectangular hyperbola is the rectangular hyperbola $xy = c^2$, rotated clockwise through an angle of $45°$ about the origin O. The asymptotes of the hyperbola $xy = c^2$ are the coordinate axes; the asymptotes of the hyperbola $x^2 - y^2 = 2c^2$ are the lines $y = x$ and $y = -x$.

The equation $x^2 - y^2 = 2c^2$ can be written as

$$\frac{x^2}{2c^2} - \frac{y}{2c^2} = 1.$$

This is a special case, with $b^2 = a^2 = 2c^2$, of the more general equation

$$\frac{x^2}{a^2} - \frac{y^2}{b^2} = 1.$$

In *Advanced Mathematics 1*, page 185, we considered the set of points, each of which is the same distance from a fixed point as it is from a fixed straight line. Now we consider the set of points, the distance of each of which from a fixed point is e times its distance from a fixed straight line, e being a constant such that $e > 1$.

In Fig. 10.6 the fixed point is S and l is the fixed straight line. P (x, y) is any point in the set, i.e. PS $= e$PL $(e > 1)$. The points A and A′ are such that AS $= e$AN and A′S $= e$A′N, i.e. A and A′ are elements of the set of points.

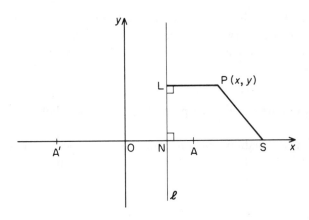

Fig. 10.6

Let AA′ $= 2a$. Then SA $+$ SA′ $= e($AN $+$ A′N$) = 2ae$,

i.e. $\qquad\qquad$ 2SA $+$ AA′ $= 2ae$

so that $\qquad\qquad$ AS $= a(e - 1)$

and $\qquad\qquad$ AN $= \dfrac{AS}{e} = a\left(1 - \dfrac{1}{e}\right)$.

Taking O, the origin, to be the mid-point of AA′ and the axes Ox, Oy to be as shown in Fig. 10.6, then

$$OS = OA + AS = a + a(e - 1) = ae$$
and $\qquad$ $ON = OA - AN = a - a(1 - 1/e) = a/e$
Then $\qquad$ $PL^2 = (x - a/e)^2$
and $\qquad$ $PS^2 = (x - ae)^2 + y^2.$
But $\qquad\qquad$ $PS = e$PL,
∴ $\qquad$ $(x - ae)^2 + y^2 = e^2(x - a/e)^2,$
giving $\quad$ $x^2(e^2 - 1) - y^2 = a^2(e^2 - 1)$

or

$$\frac{x^2}{a^2} - \frac{y^2}{a^2(e^2 - 1)} = 1$$

Writing $b^2 = a^2(e^2 - 1)$ gives

$$\frac{x^2}{a^2} - \frac{y^2}{b^2} = 1.$$

Thus the set of points, the distance of each of which from the fixed point S $(ae, 0)$ is e times its distance from the fixed straight line $x = a/e$, where $e > 1$ and $b^2 = a^2(e^2 - 1)$, can be defined as $\left\{(x, y) : \frac{x^2}{a^2} - \frac{y^2}{b^2} = 1\right\}$. The cartesian equation of the curve formed by this set of points is $\frac{x^2}{a^2} - \frac{y^2}{b^2} = 1$. The fixed point S $(ae, 0)$ is called the *focus* and the fixed straight line $x = a/e$ is called the *directrix*. The curve is a hyperbola whose axis is along the x-axis. The constant e is known as the *eccentricity* of the hyperbola and is given by $b^2 = a^2(e^2 - 1)$.

We now examine some of the properties of the hyperbola $\frac{x^2}{a^2} - \frac{y^2}{b^2} = 1$.

(1) Since the equation contains only an even power of x, the curve is symmetrical about the y-axis. Similarly the equation contains only an even power of y and so the curve is symmetrical about the x-axis. Thus the equation gives a curve which is symmetrical about both coordinate axes.

(2) When $y = 0$, $x = \pm a$, i.e. the curve cuts the x-axis at the points $(a, 0)$ and $(-a, 0)$.

When $x = 0$, $y^2 = -b^2$, i.e. when $x = 0$ there are no real values of y and so the curve does not cross or touch the y-axis.

(3) The equation of the curve can be expressed in the form

$$\frac{y^2}{x^2} = \frac{b^2}{a^2} - \frac{b^2}{x^2}.$$

When x is very large,

$$\frac{y^2}{x^2} \approx \frac{b^2}{a^2}, \qquad \text{i.e.} \qquad y^2 \approx \frac{b^2}{a^2}x^2.$$

Thus the asymptotes of the curve are $y = +\frac{b}{a}x$ and $y = -\frac{b}{a}x$. Figure 10.7 shows the hyperbola $\frac{x^2}{a^2} - \frac{y^2}{b^2} = 1$ with

(a) the points of intersection of the hyperbola and the x-axis, A $(a, 0)$ and A$'(-a, 0)$,

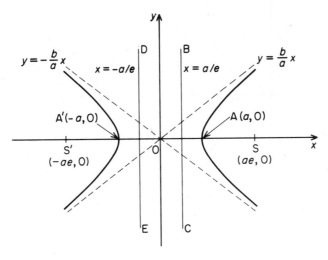

Fig. 10.7

(b) the focus of the hyperbola, S $(ae, 0)$,

(c) the directrix of the hyperbola, BC with equation $x = a/e$,

(d) the asymptotes (dotted lines) with equations $y = \dfrac{b}{a}x$ and $y = -\dfrac{b}{a}x$.

From the symmetry of the curve we see that the point S′ $(-ae, 0)$ is also a focus of the hyperbola and that the line $x = -a/e$ is also a directrix of the hyperbola.

In (3) above we saw that the hyperbola $\dfrac{x^2}{a^2} - \dfrac{y^2}{b^2} = 1$ had asymptotes with equations $y = \dfrac{b}{a}x$ and $y = -\dfrac{b}{a}x$. When these asymptotes are at right angles to each other, we have

$$\left(\frac{b}{a}\right)\left(-\frac{b}{a}\right) = -1, \qquad \text{i.e.} \qquad b^2 = a^2.$$

In this special case the curve is called a *rectangular hyperbola* (see Sections 10.1 and 10.2).

Figure 10.8 shows two hyperbolas, one of which is a rectangular hyperbola and the other which is a hyperbola whose asymptotes are not at right angles.

Figure 10.8(a) shows the hyperbola $\dfrac{x^2}{4} - \dfrac{y^2}{4} = 1$ and since $b^2 = a^2$, this hyperbola is rectangular.

Figure 10.8(b) shows the hyperbola $\dfrac{x^2}{2} - \dfrac{y^2}{1} = 1$ in which $a^2 = 2$ and $b^2 = 1$. The equations of the asymptotes of this hyperbola are $y = +\dfrac{1}{\sqrt{2}}x$ and

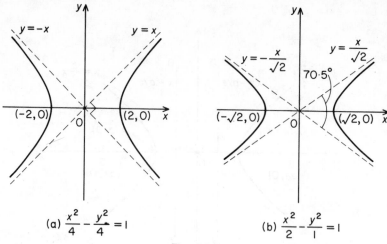

(a) $\dfrac{x^2}{4} - \dfrac{y^2}{4} = 1$ (b) $\dfrac{x^2}{2} - \dfrac{y^2}{1} = 1$

Fig. 10.8

$y = -\dfrac{1}{\sqrt{2}}x$. The angle between the asymptotes is $2 \tan^{-1} (1/\sqrt{2})$ or approximately $70 \cdot 5°$.

The equation $\dfrac{x^2}{a^2} - \dfrac{y^2}{b^2} = 1$ can be expressed in parametric form.

Since $\sec^2 \theta - \tan^2 \theta = 1$, the point with coordinates $(a \sec \theta, b \tan \theta)$ will lie on the hyperbola with equation $\dfrac{x^2}{a^2} - \dfrac{y^2}{b^2} = 1$ for all values of θ. Thus the equation of this hyperbola can be expressed parametrically by

$$x = a \sec \theta, \qquad y = b \tan \theta,$$

where θ is a parameter.

The right-hand branch of the hyperbola is given by $-\pi/2 < \theta < \pi/2$ and the left-hand branch is given by $\pi/2 < \theta < 3\pi/2$.

Also $\cosh^2 u - \sinh^2 u = 1$ and so the point with coordinates $(a \cosh u, b \sinh u)$ lies on the hyperbola $\dfrac{x^2}{a^2} - \dfrac{y^2}{b^2} = 1$ for all values of u. Thus an alternative parametric form for this hyperbola is

$$x = a \cosh u, \qquad y = b \sinh u,$$

where u is a parameter which can take all real values.

A drawback with this parametric form is that real values of u give only one branch of the hyperbola, the branch for which x is positive.

Example 1
Find the foci, the eccentricity, the directrices and the asymptotes of the hyperbola $16x^2 - 9y^2 = 144$.

Writing the equation of the hyperbola in the form $\dfrac{x^2}{a^2} - \dfrac{y^2}{b^2} = 1$ gives

$\dfrac{x^2}{9} - \dfrac{y^2}{16} = 1$ so that $a^2 = 9$ and $b^2 = 16$.

Using $b^2 = a^2(e^2 - 1)$, we have $16 = 9(e^2 - 1)$, giving $e = 5/3$. Then the foci, $(ae, 0)$ and $(-ae, 0)$, are the points $(5, 0)$ and $(-5, 0)$.

The equations of the directrices, $x = a/e$ and $x = -a/e$, are $x = 9/5$ and $x = -9/5$.

The equations of the asymptotes, $y = +\dfrac{b}{a}x$ and $y = -\dfrac{b}{a}x$, are $y = \dfrac{4}{3}x$ and $y = -\dfrac{4}{3}x$.

Example 2

Given that P is any point on the hyperbola $\dfrac{x^2}{a^2} - \dfrac{y^2}{b^2} = 1$, with foci S $(ae, 0)$ and S' $(-ae, 0)$, show that $S'P - SP = 2a$.

Figure 10.9 shows the hyperbola with its foci, S and S', and with the perpendicular from P to the corresponding directrices meeting them at Q and R respectively.

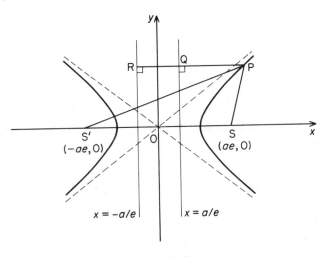

Fig. 10.9

From the focus–directrix property

$$SP = ePQ \quad \text{and} \quad S'P = ePR.$$
$$\therefore \quad S'P - SP = ePR - ePQ = eQR = e(2a/e) = 2a.$$

Note in this example how by using the focus–directrix property we have avoided the direct computation of the lengths SP and S'P.

Exercises 10.3

1 Sketch the curves with equations

   (a) $x^2 - y^2 = 1$,   (b) $x^2 - y^2 = 4$,   (c) $3x^2 - 3y^2 = 4$.

2 Sketch the curves with equations

   (a) $\dfrac{x^2}{4} - \dfrac{y^2}{9} = 1$,   (b) $\dfrac{x^2}{3} - \dfrac{y^2}{2} = 1$,   (c) $5x^2 - 4y^2 = 1$.

3 Find (a) the eccentricity,

      (b) the coordinates of the foci,

      (c) the equations of the directrices,

      (d) the equations of the asymptotes

  of each hyperbola in **1**.

4 Find (a) the eccentricity,

      (b) the coordinates of the foci,

      (c) the equations of the directrices,

      (d) the equations of the asymptotes

  of each hyperbola in **2**.

5 Find the equation of the hyperbola which has

   (a) its foci at $(\pm 5, 0)$ and eccentricity 5/3,

   (b) its foci at $(\pm 2\sqrt{2}, 0)$ and eccentricity $\sqrt{2}$,

   (c) eccentricity $\dfrac{1}{2}\sqrt{29}$ and directrices with equations $x = \pm\dfrac{4}{\sqrt{29}}$,

   (d) eccentricity 13/12 and directrices with equations $x = \pm\dfrac{144}{13}$.

6 Find the equation of the hyperbola with eccentricity 2 which crosses the x-axis at the points $(4, 0)$ and $(-4, 0)$ and is symmetrical about the x-axis.

7 Find the equation of the hyperbola with foci $(\pm 10, 0)$ and asymptotes with equations $y = \pm\frac{3}{4}x$.

10.4   Tangent and normal to the hyperbola $\dfrac{x^2}{a^2} - \dfrac{y^2}{b^2} = 1$

In Fig. 10.10, $P_1 (x_1, y_1)$ is a point on the hyperbola $\dfrac{x^2}{a^2} - \dfrac{y^2}{b^2} = 1$, $P_1 T$ is the tangent at P_1 and $P_1 N$ is the normal at P_1.
Differentiating,

$$\frac{2x}{a^2} - \frac{2y}{b^2}\frac{dy}{dx} = 0, \quad \text{i.e.} \quad \frac{dy}{dx} = \frac{b^2 x}{a^2 y}$$

so that, at $P_1 (x_1, y_1)$,

$$\frac{dy}{dx} = \frac{b^2 x_1}{a^2 y_1}.$$

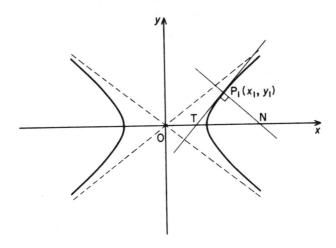

Fig. 10.10

The equation of the tangent at P_1 is

$$y - y_1 = \frac{b^2 x_1}{a^2 y_1}(x - x_1).$$

Multiplying by y_1/b^2 and rearranging gives

$$\frac{xx_1}{a^2} - \frac{yy_1}{b^2} = \frac{x_1^2}{a^2} - \frac{y_1^2}{b^2}.$$

$P_1(x_1, y_1)$ is a point on the hyperbola and so $\dfrac{x_1^2}{a^2} - \dfrac{y_1^2}{b^2} = 1$.

Thus the equation of the tangent at the point P_1 (x_1, y_1) is

$$\frac{xx_1}{a^2} - \frac{yy_1}{b^2} = 1.$$

Putting $x_1 = a \sec \alpha$, $y_1 = b \tan \alpha$, gives the equation of the tangent at P_1 ($a \sec \alpha$, $b \tan \alpha$) as

$$\frac{xa \sec \alpha}{a^2} - \frac{yb \tan \alpha}{b^2} = 1$$

or $$bx \sec \alpha - ay \tan \alpha = ab.$$

This equation can also be obtained as shown in Example 1 which follows.

Example 1
Find the equation of the tangent to the hyperbola given parametrically by

$$x = a \sec \theta, \qquad y = b \tan \theta$$

at the point where $\theta = \alpha$.

$$x = a \sec \theta. \qquad \therefore \qquad \frac{dx}{d\theta} = a \sec \theta \tan \theta.$$

$$y = b \tan \theta. \qquad \therefore \qquad \frac{dy}{d\theta} = b \sec^2 \theta.$$

$$\therefore \qquad \frac{dy}{dx} = \frac{dy/d\theta}{dx/d\theta} = \frac{b \sec^2 \theta}{a \sec \theta \tan \theta} = \frac{b}{a} \operatorname{cosec} \theta.$$

At the point P_1, where the parameter θ takes the value α, $\dfrac{dy}{dx} = \dfrac{b}{a} \operatorname{cosec} \alpha$, the gradient of tangent at P_1. The equation of the tangent to the hyperbola at P_1 ($a \sec \alpha$, $b \tan \alpha$) is

$$y - b \tan \alpha = \frac{b}{a} \operatorname{cosec} \alpha(x - a \sec \alpha),$$

i.e. $\qquad\qquad bx \sec \alpha - ay \tan \alpha = ab.$

Example 2

Given that $y = mx + c$ is a tangent to the hyperbola $\dfrac{x^2}{a^2} - \dfrac{y^2}{b^2} = 1$, show that $c^2 = a^2 m^2 - b^2$.

Suppose that $y = mx + c$ is the equation of the tangent at the point (x_1, y_1). The equation of the tangent at this point is

$$\left(\frac{x_1}{a^2}\right)x - \left(\frac{y_1}{b^2}\right)y = 1 \qquad \ldots (1)$$

The equation $y = mx + c$ can be written as

$$-mx + y = c. \qquad \ldots (2)$$

Equations (1) and (2) represent the same line, the tangent at (x_1, y_1), and so the coefficients of the two equations must be in the same ratio.

Therefore $\qquad \dfrac{x_1/a^2}{-m} = \dfrac{-y_1/b^2}{1} = \dfrac{1}{c},$

giving $\qquad\qquad x_1 = \dfrac{-ma^2}{c} \qquad$ and $\qquad y_1 = \dfrac{-b^2}{c}.$

The point (x_1, y_1) lies on the hyperbola $\dfrac{x^2}{a^2} - \dfrac{y^2}{b^2} = 1$ and so

$$\frac{m^2 a^4}{a^2 c^2} - \frac{b^4}{b^2 c^2} = 1,$$

giving
$$c^2 = a^2 m^2 - b^2.$$

Alternatively, and less neatly, putting $y = mx + c$ into $b^2 x^2 - a^2 y^2 = a^2 b^2$, the equation of the hyperbola, we get

$$b^2 x^2 - a^2 (mx + c)^2 = a^2 b^2,$$

i.e. $$(b^2 - a^2 m^2) x^2 - 2a^2 mcx - a^2 (b^2 + c^2) = 0.$$

This quadratic equation gives the x-coordinates of the points of intersection of the hyperbola and the line $y = mx + c$. Since this line is a tangent to the hyperbola, it meets the hyperbola in just one point and so this quadratic equation has equal roots.

$$\therefore \qquad (-2a^2 mc)^2 + 4(b^2 - a^2 m^2) a^2 (b^2 + c^2) = 0,$$

giving, on simplification, $c^2 = a^2 m^2 - b^2$.

Example 3

The tangent to the hyperbola $\dfrac{x^2}{a^2} - \dfrac{y^2}{b^2} = 1$ at the point P ($a \sec \theta$, $b \tan \theta$) crosses the coordinate axes at the points A and B. Find the equation of the locus, as θ varies, of G, the centroid of the triangle OAB.

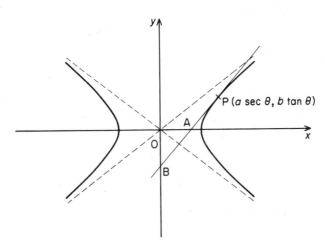

Fig. 10.11

Figure 10.11 shows the hyperbola and the tangent PAB. The equation of the tangent at P ($a \sec \theta$, $b \tan \theta$) is

$$bx \sec \theta - ay \tan \theta = ab.$$

At A, $y = 0$, giving $x = a \cos \theta$ and at B, $x = 0$, giving $y = -b \cot \theta$.

The centroid G has coordinates (X, Y), where

$$X = \tfrac{1}{3}(a \cos \theta) \quad \text{and} \quad Y = \tfrac{1}{3}(-b \cot \theta).$$

$\therefore \qquad \cos \theta = \dfrac{3X}{a} \quad \text{and} \quad Y^2 = \tfrac{1}{9}(b^2 \cot^2 \theta)$

$\therefore \qquad Y^2 = \dfrac{b^2}{9} \cdot \dfrac{9X^2}{a^2 - 9X^2}$

a, $\sqrt{(a^2 - 9X^2)}$, θ, $3X$

i.e. $\qquad a^2 Y^2 - 9X^2 Y^2 = b^2 X^2.$

This does not contain θ and so, for all values of θ, the coordinates of G (X, Y) satisfy the equation

$$a^2 y^2 - 9x^2 y^2 = b^2 x^2$$

and so this is the equation of the locus of G.

An alternative approach is to take the point P to have coordinates (x_1, y_1).

The equation of the tangent at P is $\dfrac{xx_1}{a^2} - \dfrac{yy_1}{b^2} = 1.$

At A, $y = 0$, giving $x = a^2/x_1$, and at B, $x = 0$ giving $y = -b^2/y_1$.

The centroid G has coordinates (X, Y) where

$$X = a^2/(3x_1) \quad \text{and} \quad Y = -b^2/(3y_1).$$

Since

$$\dfrac{x_1^2}{a^2} - \dfrac{y_1^2}{b^2} = 1 \quad \text{we have} \quad \dfrac{a^2}{9X^2} - \dfrac{b^2}{9Y^2} = 1,$$

giving the equation of the locus of G as

$$\dfrac{a^2}{9x^2} - \dfrac{b^2}{9y^2} = 1 \quad \text{or} \quad a^2 y^2 - 9x^2 y^2 = b^2 x^2.$$

Example 4

Find the equation of the normal to the hyperbola $\dfrac{x^2}{a^2} - \dfrac{y^2}{b^2} = 1$ at the point $P_1 (x_1, y_1)$.

The equation of the tangent at the point $P_1 (x_1, y_1)$ is

$$\dfrac{xx_1}{a^2} - \dfrac{yy_1}{b^2} = 1.$$

$\therefore$ the equation of the normal at P_1 is of the form

$$\dfrac{xy_1}{b^2} + \dfrac{yx_1}{a^2} = c.$$

P_1 lies on this line and so $c = \dfrac{x_1 y_1}{b^2} + \dfrac{y_1 x_1}{a^2} = x_1 y_1 \left(\dfrac{1}{a^2} + \dfrac{1}{b^2} \right)$, i.e. the equation of the normal at $P_1(x_1, y_1)$ is

$$\frac{y_1 x}{b^2} + \frac{x_1 y}{a^2} = x_1 y_1 \left(\frac{1}{a^2} + \frac{1}{b^2} \right).$$

Exercises 10.4

1 Find the equations of the tangents and normals to the hyperbolas
   (a) $2x^2 - y^2 = 7$ at the point $(4, 5)$,
   (b) $x^2 - y^2 = 5$ at the point $(-3, -2)$,
   (c) $4x^2 - 3y^2 = 73$ at the point $(-5, 3)$,
   (d) $3x^2 - 2y^2 = 76$ at the point $(6, -4)$.

2 Find the coordinates of the points of intersection of the tangents in **1** with the coordinate axes.

3 Find the coordinates of the points of intersection of the normals in **1** with the coordinate axes.

4 Find the values of c for which the line $y = 2x + c$ is a tangent to the hyperbola $2x^2 - y^2 = 9$. Hence write down the equations of the tangents to the hyperbola which are parallel to the line $y = 2x$.

5 Find the equations of the tangents to the hyperbola $x^2 - 2y^2 = a^2$ which are parallel to the line $y = x$.

6 Find the equations of the tangents from the point $(0, -3)$ to the hyperbola $x^2 - y^2 = 3$.

7 Find the equations of the tangents from the point $(0, -1)$ to the hyperbola $x^2 - y^2 = 2$.

8 Find the gradients of the tangents from the point $(0, -1)$ to the hyperbola $3x^2 - y^2 = 3$. Deduce the tangent of the acute angle between these two tangents.

9 The tangent at the point P $(a \sec \theta, b \tan \theta)$ to the hyperbola $b^2 x^2 - a^2 y^2 = a^2 b^2$ cuts the asymptotes of the hyperbola at the points E and F. Find the coordinates of E and F and show that $PE = PF$.

10 One of the asymptotes of the hyperbola $b^2 x^2 - a^2 y^2 = a^2 b^2$ cuts the directrix $x = \dfrac{a}{e}$ at the point L. Given that S is the focus with coordinates $(ae, 0)$, show that the angle $OLS = 90°$.

11 The tangent at the point P $(a \sec \theta, b \tan \theta)$ to the hyperbola $b^2 x^2 - a^2 y^2 = a^2 b^2$ cuts the coordinate axes at the points A and B. Find the equation of the locus, as θ varies, of M, the mid-point of AB.

12 The normal at the point P $(a \sec \theta, b \tan \theta)$ to the hyperbola $b^2x^2 - a^2y^2 = a^2b^2$ cuts the coordinate axes at the points C and D. Find the equation of the locus, as θ varies, of
(a) the mid-point of CD,
(b) G, the centroid of the triangle OCD.

13 Show that the point Q $(a \csc \theta, b \cot \theta)$ lies on the hyperbola $b^2x^2 - a^2y^2 = a^2b^2$ for all values of θ. The normal at Q meets the normal at P $(a \sec \theta, b \tan \theta)$ at R. Show that the locus of R, as θ varies, is a straight line parallel to the x-axis.

14 The normal at the point P on the hyperbola $x^2/a^2 - y^2/b^2 = 1$ cuts the x-axis at C and N is the foot of the perpendicular from P to the x-axis. The line NP produced meets an asymptote at Q. Show that the angle OQC $= 90°$.

15 The line $y = mx + c$ cuts the hyperbola $b^2x^2 - a^2y^2 = a^2b^2$ at points P and Q and cuts the asymptote $y = \dfrac{b}{a}x$ at E and the asymptote $y = -\dfrac{b}{a}x$ at F. Find the coordinates of the mid-points of PQ and EF. Hence show that EP = FQ.

10.5 The ellipse $\dfrac{x^2}{a^2} + \dfrac{y^2}{b^2} = 1$

We next consider the set of points, the distance of each of which from a fixed point is e times its distance from a fixed straight line, where $e < 1$.

In Fig. 10.12, the fixed point is S, l is the fixed straight line and P (x, y) is any point in the set, i.e. PS $= e$PL $(e < 1)$. The points A and A′ are such that AS $= e$AN and A′S $= e$A′N, i.e. A and A′ are both elements of the set of points. Let AA′ $= 2a$. Then

$$AS + A'S = 2a$$
and
$$AN + A'N = 2a + 2\,AN,$$
$$AS + A'S = e(AN + A'N).$$
∴
$$2a = 2ae + 2e\,AN.$$

∴
$$AN = a(1 - e)/e \quad \text{and} \quad AS = a(1 - e).$$

Taking O, the origin, to be the mid-point of AA′ and the axes Ox, Oy to be as shown in Fig. 10.12, then

$$ON = OA + AN = a + \frac{a}{e}(1 - e) = \frac{a}{e}$$
and
$$OS = OA - AS = a - a(1 - e) = ae.$$

Then

$$PL^2 = \left(\frac{a}{e} - x\right)^2 \quad \text{and} \quad PS^2 = (ae - x)^2 + y^2.$$

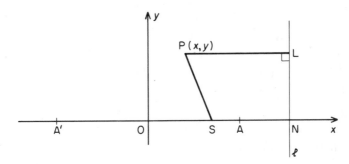

Fig. 10.12

But PS $=$ ePL and so

$$(ae - x)^2 + y^2 = e^2\left(\frac{a}{e} - x\right)^2,$$

i.e.

$$x^2(1 - e^2) + y^2 = a^2(1 - e^2),$$

or

$$\frac{x^2}{a^2} + \frac{y^2}{a^2(1 - e^2)} = 1,$$

Writing $b^2 = a^2(1 - e^2)$ gives

$$\frac{x^2}{a^2} + \frac{y^2}{b^2} = 1.$$

Thus the set of points, the distance of each of which from a fixed point S $(ae, 0)$ is e times its distance from the fixed straight line $x = a/e$, where $e < 1$, $b^2 = a^2(1 - e^2)$ and $0 < b < a$, can be defined as $\left\{(x, y):\dfrac{x^2}{a^2} + \dfrac{y^2}{b^2} = 1\right\}$.

The cartesian equation of the curve formed by this set of points is $\dfrac{x^2}{a^2} + \dfrac{y^2}{b^2} = 1$.

The fixed point S $(ae, 0)$ is called the *focus* and the fixed straight line $x = a/e$ is called the *directrix*. The curve is an ellipse and the constant e is known as the *eccentricity* of the ellipse and is given by $b^2 = a^2(1 - e^2)$.

We now examine some of the properties of the ellipse $\dfrac{x^2}{a^2} + \dfrac{y^2}{b^2} = 1$.

(1) Since the equation contains only an even power of x, the curve is symmetrical about the y-axis. Similarly, the equation contains only an even power of y and so the curve is symmetrical about the x-axis. Thus the equation gives a curve which is symmetrical about both coordinate axes.

(2) When $y = 0$, $x = \pm a$, i.e. the curve cuts the x-axis at the points $(a, 0)$ and $(-a, 0)$.

When $x = 0$, $y = \pm b$, i.e. the curve cuts the y-axis at the points $(0, b)$ and $(0, -b)$.

(3) From the equation $\dfrac{x^2}{a^2} + \dfrac{y^2}{b^2} = 1$ we obtain

$$b^2 x^2 = a^2(b^2 - y^2) \qquad \text{and} \qquad bx = \pm a\sqrt{(b^2 - y^2)}.$$

∴ there are no real values of x if $y^2 > b^2$, i.e. if $|y| > b$.
∴ there are no real values of x if $y < -b$ or if $y > b$.
Similarly there are no real values of y if $x < -a$ or if $x > a$. Thus the curve lies wholly within the region defined by

$$-a \leqslant x \leqslant a \qquad \text{and} \qquad -b \leqslant y \leqslant b.$$

Thus the curve has no asymptotes.

From these properties the curve can be drawn and is shown in Fig. 10.13.

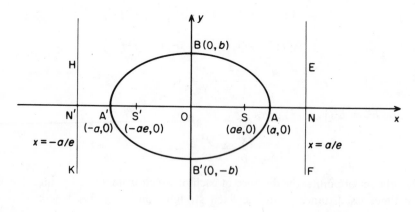

Fig. 10.13

This shows the ellipse $\dfrac{x^2}{a^2} + \dfrac{y^2}{b^2} = 1$ with

(1) the points of intersection of the ellipse and x-axis, A $(a, 0)$ and A' $(-a, 0)$,
(2) the points of intersection of the ellipse and the y-axis, B $(0, b)$ and B' $(0, -b)$,
(3) the focus of the ellipse, S $(ae, 0)$,
(4) the directrix of the ellipse, EF with equation $x = a/e$.
(The lines AA' and BB' are the *major and minor axes* respectively of the ellipse.)
From the symmetry of the curve we see that the point S' $(-ae, 0)$ is also a focus of the ellipse and that the line $x = -a/e$ is also a directrix of the ellipse.

In the special case of $b = a$, the equation $\dfrac{x^2}{a^2} + \dfrac{y^2}{b^2} = 1$ becomes $x^2 + y^2 = a^2$

which is the equation of the circle of radius a and centre the origin.

The equation $\dfrac{x^2}{a^2} + \dfrac{y^2}{b^2} = 1$ can be expressed in parametric form. Since

$\cos^2 \theta + \sin^2 \theta = 1$, the point with coordinates ($a \cos \theta$, $b \sin \theta$) will lie on the ellipse for all values of θ. Thus the equation of this ellipse can be expressed parametrically by

$$x = a \cos \theta, \qquad y = b \sin \theta,$$

where θ is a parameter, $0 \leqslant \theta < 2\pi$ giving all points on the ellipse.

We now see that for the three curves, the parabola, the hyperbola and the ellipse

$$\frac{\text{the distance of any point of the curve from a fixed point (focus)}}{\text{the distance of the point from a fixed straight line (directrix)}} = e,$$

where e is a constant and,

in the case of the parabola $\qquad e = 1,$
in the case of the hyperbola $\qquad e > 1,$
in the case of the ellipse $\qquad e < 1.$

This property is known as the *focus–directrix* property of the curves.

Example 1
Find (a) the eccentricity,
  (b) the coordinates of the foci,
  (c) the equations of the corresponding directrices
of the ellipse $\dfrac{x^2}{16} + \dfrac{y^2}{9} = 1.$

(a) Comparing the equation with the general equation $\dfrac{x^2}{a^2} + \dfrac{y^2}{b^2} = 1$, we have
$a^2 = 16$ and $b^2 = 9$, i.e. $a = 4$ and $b = 3$.
Using $b^2 = a^2(1 - e^2)$ gives $9 = 16(1 - e^2)$.
Therefore $e = \frac{1}{4}\sqrt{7}$.
(b) The coordinates of the foci are ($\pm ae$, 0), i.e. ($\sqrt{7}$, 0) and ($-\sqrt{7}$, 0).
(c) The equations of the directrices are $x = \pm a/e$, i.e. $x = 16/\sqrt{7}$ and $x = -16/\sqrt{7}$.

Example 2
Given that P is any point on the ellipse $\dfrac{x^2}{a^2} + \dfrac{y^2}{b^2} = 1$, which has foci S and S′,
show that SP + PS′ is constant for all positions of P.

The ellipse is shown in Fig. 10.14 in which EF and HK are the directrices and the line HPE is perpendicular to EF and HK.
From the focus–directrix property we have

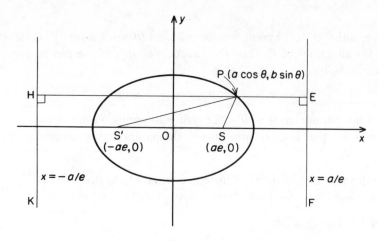

Fig. 10.14

$$SP = ePE = e\left(\frac{a}{e} - a\cos\theta\right) = a - ae\cos\theta,$$

$$PS' = ePH = e\left(\frac{a}{e} + a\cos\theta\right) = a + ae\cos\theta.$$

$$\therefore \qquad SP + PS' = 2a.$$

This result is independent of θ and so is true for all positions of P on the ellipse.

Exercises 10.5

1 Sketch the curves with equations

   (a) $\dfrac{x^2}{9} + \dfrac{y^2}{4} = 1$, (b) $\dfrac{x^2}{4} + \dfrac{y^2}{1} = 1$,

   (c) $25x^2 + 36y^2 = 900$, (d) $9x^2 + 4y^2 = 36$.

2 Sketch the ellipses with
   (a) foci $(\pm 2, 0)$ and corresponding directrices $x = \pm 8$,
   (b) foci $(\pm 3, 0)$ and corresponding directrices $x = \pm 27$.

3 Find
   (a) the eccentricities,
   (b) the coordinates of the foci,
   (c) the equations of the corresponding directrices
   of the ellipses in **1**.

4 Find the equations of the ellipses of the form $x^2/a^2 + y^2/b^2 = 1$ which pass through the points

(a) $(6, 0)$, $(-6, 0)$, $(0, 4)$, $(0, -4)$;

(b) $(4, 0)$, $(-4, 0)$, $(0, 2)$, $(0, -2)$;

(c) $(4, 0)$, $(-4, 0)$, $(0, 5)$, $(0, -5)$.

5 Find

(a) the eccentricities,

(b) the coordinates of the foci,

(c) the equations of the corresponding directrices of the ellipses in **4**.

10.6 Tangent and normal to the ellipse $\dfrac{x^2}{a^2} + \dfrac{y^2}{b^2} = 1$

In Fig. 10.15, $P_1 (x_1, y_1)$ is a point on the ellipse $\dfrac{x^2}{a^2} + \dfrac{y^2}{b^2} = 1$, $P_1 T$ is the tangent at P_1 and $P_1 N$ is the normal at P_1.

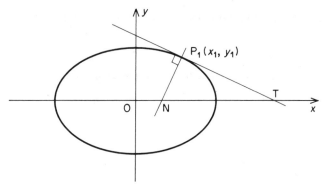

Fig. 10.15

Differentiating,

$$\frac{2x}{a^2} + \frac{2y}{b^2}\frac{dy}{dx} = 0, \quad \text{i.e.} \quad \frac{dy}{dx} = -\frac{b^2 x}{a^2 y}.$$

At $P_1 (x_1, y_1)$,

$$\frac{dy}{dx} = -\frac{b^2 x_1}{a^2 y_1}.$$

The equation of the tangent at $P_1 (x_1, y_1)$ is

$$y - y_1 = -\left(\frac{b^2 x_1}{a^2 y_1}\right)(x - x_1)$$

Multiplying by y_1/b^2 and rearranging gives

$$\frac{xx_1}{a^2} + \frac{yy_1}{b^2} = \frac{x_1^2}{a^2} + \frac{y_1^2}{b^2} = 1$$

since (x_1, y_1) is on the ellipse.

Thus the equation of the tangent to the ellipse $\dfrac{x^2}{a^2} + \dfrac{y^2}{b^2} = 1$ at the point $P_1\,(x_1, y_1)$ is

$$\frac{xx_1}{a^2} + \frac{yy_1}{b^2} = 1.$$

Putting $x_1 = a \cos \alpha$, $y_1 = b \sin \alpha$ gives the equation of the tangent at $P_1\,(a \cos \alpha, b \sin \alpha)$ as

$$\frac{xa \cos \alpha}{a^2} + \frac{yb \sin \alpha}{b^2} = 1$$

or $\qquad\qquad\qquad bx \cos \alpha + ay \sin \alpha = ab.$

This equation can also be obtained as shown in Example 1 which follows.

Example 1
Find the equation of the tangent to the ellipse given parametrically by

$$x = a \cos \theta, \qquad y = b \sin \theta$$

at the point where $\theta = \alpha$.

$$x = a \cos \theta \qquad \therefore \qquad \frac{dx}{d\theta} = -a \sin \theta.$$

$$y = b \sin \theta \qquad \therefore \qquad \frac{dy}{d\theta} = b \cos \theta.$$

$$\therefore \qquad \frac{dy}{dx} = \frac{dy/d\theta}{dx/d\theta} = -\frac{b \cos \theta}{a \sin \theta}.$$

At the point P_1, where the parameter θ takes the value α,

$$\frac{dy}{dx} = -\frac{b \cos \alpha}{a \sin \alpha} = \text{gradient of the tangent at } P_1.$$

$\therefore$ the equation of the tangent at $P_1\,(a \cos \alpha, b \sin \alpha)$ is

$$y - b \sin \alpha = -\frac{b \cos \alpha}{a \sin \alpha}(x - a \cos \alpha),$$

i.e. $\qquad\qquad bx \cos \alpha + ay \sin \alpha = ab$

or $\qquad\qquad \dfrac{x}{a} \cos \alpha + \dfrac{y}{b} \sin \alpha = 1.$

Example 2

Find the equation of the normal to the ellipse $\dfrac{x^2}{a^2} + \dfrac{y^2}{b^2} = 1$ at the point $P_1\,(x_1, y_1)$.

The equation of the tangent at $P_1\,(x_1, y_1)$ is

$$\frac{xx_1}{a^2} + \frac{yy_1}{b^2} = 1.$$

Therefore the equation of the normal at P_1 is of the form

$$\frac{xy_1}{b^2} - \frac{yx_1}{a^2} = c,$$

and, since P_1 lies on this line,

$$c = \frac{x_1 y_1}{b^2} - \frac{y_1 x_1}{a^2},$$

i.e. the equation of the normal at $P_1\,(x_1, y_1)$ is

$$\frac{xy_1}{b^2} - \frac{yx_1}{a^2} = x_1 y_1\left(\frac{1}{b^2} - \frac{1}{a^2}\right).$$

Example 3

Find the equation of the normal to the ellipse given parametrically by

$$x = a\cos\theta, \qquad y = b\sin\theta$$

at the point where $\theta = \alpha$.

The equation of the tangent at the point $(a\cos\alpha, b\sin\alpha)$ is

$$bx\cos\alpha + ay\sin\alpha = ab.$$

$\therefore$ the equation of the normal is

$$ax\sin\alpha - by\cos\alpha = c = a^2\sin\alpha\cos\alpha - b^2\sin\alpha\cos\alpha$$

i.e. $\qquad ax\sin\alpha - by\cos\alpha = (a^2 - b^2)\sin\alpha\cos\alpha$

or $\qquad ax\sec\alpha - by\csc\alpha = a^2 - b^2.$

Example 4

The tangent at the point P $(a\cos\theta, b\sin\theta)$ to the ellipse $\dfrac{x^2}{a^2} + \dfrac{y^2}{b^2} = 1$ meets the coordinate axes at A and B. Find, as θ varies, the least possible area of the triangle OAB.

Figure 10.16 shows the ellipse and the tangent APB.

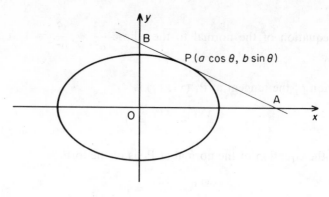

Fig. 10.16

The equation of the tangent at P $(a \cos \theta, b \sin \theta)$ is

$$\frac{x}{a} \cos \theta + \frac{y}{b} \sin \theta = 1.$$

At A, $y = 0$ giving $x = a/\cos \theta$ and at B, $x = 0$ giving $y = b/\sin \theta$.

∴ the area of △OAB $= \dfrac{1}{2}\left(\dfrac{a}{\cos \theta}\right)\left(\dfrac{b}{\sin \theta}\right)$

$$= \frac{ab}{\sin 2\theta}.$$

The area will have its least value when $\sin 2\theta$ has its greatest value. The greatest value of $\sin 2\theta$ is 1. Thus the least possible area of the triangle OAB is ab.

Example 5
The normal at the point P $(a \cos \theta, b \sin \theta)$ to the ellipse $b^2 x^2 + a^2 y^2 = a^2 b^2$ meets the coordinate axes at points C and D. Show that the locus, as θ varies, of M, the mid-point of CD, is another ellipse.

Figure 10.17 shows the ellipse and the normal PCD.

The equation of the normal at P $(a \cos \theta, b \sin \theta)$ is

$$ax \sec \theta - by \operatorname{cosec} \theta = a^2 - b^2.$$

At C, $y = 0$, giving $x = \dfrac{(a^2 - b^2) \cos \theta}{a}$ and

at D, $x = 0$, giving $y = -\dfrac{(a^2 - b^2) \sin \theta}{b}$.

M, the mid-point of CD, has coordinates (X, Y), where

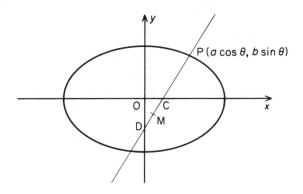

Fig. 10.17

$$X = \frac{(a^2 - b^2)\cos\theta}{2a} \quad \text{and} \quad Y = -\frac{(a^2 - b^2)\sin\theta}{2b}.$$

$$\therefore \qquad \cos\theta = \frac{2aX}{a^2 - b^2} \quad \text{and} \quad \sin\theta = \frac{-2bY}{a^2 - b^2}.$$

Eliminating θ by using $\sin^2\theta + \cos^2\theta = 1$, we get

$$\frac{4a^2 X^2}{(a^2 - b^2)^2} + \frac{4b^2 Y^2}{(a^2 - b^2)^2} = 1.$$

This equation does not involve θ, and so, for all values of θ, the coordinates of M (X, Y) satisfy the equation

$$\frac{4a^2 x^2}{(a^2 - b^2)^2} + \frac{4b^2 y^2}{(a^2 - b^2)^2} = 1$$

or

$$\frac{x^2}{(a^2 - b^2)^2/4a^2} + \frac{y^2}{(a^2 - b^2)^2/4b^2} = 1.$$

This is the equation of the locus of M.

Its form shows that it is the equation of an ellipse, with semi-axes of length $(a^2 - b^2)/2a$ and $(a^2 - b^2)/2b$.

Example 6
Show that the area of the region enclosed by the ellipse $x^2/a^2 + y^2/b^2 = 1$ is πab.

From the symmetry of the ellipse, the required area is four times the area of the finite region in the first quadrant enclosed by the ellipse and the coordinate axes.

$$\text{Total area} = 4\int_0^a y \, dx.$$

Let $x = a\cos\theta$ and $y = b\sin\theta$.

Then $\dfrac{dx}{d\theta} = -a \sin \theta$.

Also $\theta = \pi/2$ when $x = 0$, and $\theta = 0$ when $x = a$.

$$\text{Required area} = 4 \int_{\pi/2}^{0} (b \sin \theta)(-a \sin \theta)\, d\theta$$

$$= -4ab \int_{\pi/2}^{0} \sin^2 \theta \, d\theta$$

$$= 4ab \int_{0}^{\pi/2} \sin^2 \theta \, d\theta$$

$$= 2ab \int_{0}^{\pi/2} (1 - \cos 2\theta) \, d\theta$$

$$= 2ab \left[\theta - \frac{1}{2} \sin 2\theta \right]_{0}^{\pi/2}$$

$$= \pi ab.$$

When $b = a$, the equation $\dfrac{x^2}{a^2} + \dfrac{y^2}{b^2} = 1$ becomes $x^2 + y^2 = a^2$, i.e. the equation of a circle of radius a. The area πab then becomes πa^2.

Exercises 10.6

1 Find the equations of the tangents and normals to the ellipses

(a) $\dfrac{x^2}{100} + \dfrac{y^2}{25} = 1$ at the point $(8, 3)$,

(b) $x^2 + 2y^2 = 17$ at the point $(3, 2)$,

(c) $\dfrac{x^2}{39^2} + \dfrac{y^2}{13^2} = 1$ at the point $(-15, 12)$,

(d) $3x^2 + 5y^2 = 8$ at the point $(-1, -1)$.

2 The line $y = mx + c$ is a tangent to the ellipse $\dfrac{x^2}{a^2} + \dfrac{y^2}{b^2} = 1$. Show that $c = \pm \sqrt{(a^2 m^2 + b^2)}$. (See Example 2 of Section 10.4)

3 Find the equations of the tangents to the ellipse $\dfrac{x^2}{16} + \dfrac{y^2}{9} = 1$ which are parallel to the line $y = x$.

4 Find the equations of the tangents to the ellipse $5x^2 + 16y^2 = 80$ which are parallel to the line $y = \frac{1}{2}x$.

5 Find the equations of the tangents to the ellipse $x^2 + 2y^2 = 9$ from the point $(5, -1)$.

6 Find the gradients of the tangents from the point $(5, 2)$ to the ellipse $x^2 + 2y^2 = 22$. Hence find the tangent of the acute angle between these two tangents.

7 The tangent to the ellipse $\dfrac{x^2}{a^2} + \dfrac{y^2}{b^2} = 1$ at the point P cuts the coordinate axes at A and B. Find the equation of the locus of
(a) the mid-point of AB,
(b) the centroid of the triangle OAB.

8 The normal at P to the ellipse $b^2 x^2 + a^2 y^2 = a^2 b^2$ intersects the coordinate axes at C and D. Find the equation of the locus of the centroid of the triangle OCD.

9 The tangent and normal at the point P on the ellipse $\dfrac{x^2}{a^2} + \dfrac{y^2}{b^2} = 1$ meet the x-axis at T and G respectively. N is the foot of the perpendicular from P to the x-axis. Show that $|OT . GN| = b^2$.

10 The perpendicular from the focus S $(ae, 0)$ to the tangent at P to the ellipse $b^2 x^2 + a^2 y^2 = a^2 b^2$ meets the line OP produced at K. Show that, for all positions of P, the point K lies on the directrix whose equation is $x = a/e$.

11 The line $x = a$ meets the tangent at P to the ellipse $b^2 x^2 + a^2 y^2 = a^2 b^2$ at U and the line $x = -a$ meets the tangent at P at V. Given that S is the focus $(ae, 0)$, show that $\angle USV = 90°$.

12 The tangents to the ellipse $\dfrac{x^2}{a^2} + \dfrac{y^2}{b^2} = 1$ at P $(a \cos \theta, b \sin \theta)$ and Q $(-a \sin \theta, b \cos \theta)$ meet at R. Find the coordinates of R. Find also the equation of the locus, as θ varies, of R.

13 The points S and S' are the foci of the ellipse $b^2 x^2 + a^2 y^2 = a^2 b^2$ and P is any point on the ellipse.
Show that SP and S'P are equally inclined to the tangent at P.

14 The line $y = mx + c$ cuts the ellipse $b^2 x^2 + a^2 y^2 = a^2 b^2$ at the points P and Q. Obtain a quadratic equation in x whose roots are the x-coordinates of P and Q. Hence, without solving this equation, obtain the coordinates of M, the mid-point of PQ.
Show that the locus of M, as c varies, is a straight line with gradient $-b^2/(a^2 m)$.

15 Write down the equation of a line with gradient m which passes through the point P (X, Y). Given that this line is a tangent to the ellipse $\dfrac{x^2}{a^2} + \dfrac{y^2}{b^2} = 1$, show that the gradients m_1, m_2 of the two tangents from P to the ellipse are the roots of the equation

$$m^2(a^2 - X^2) + 2mXY + (b^2 - Y^2) = 0.$$

Deduce that the locus of the point from which two perpendicular tangents can be drawn to the ellipse is the circle

$$x^2 + y^2 = a^2 + b^2.$$

10.7 Change of axes

It should be emphasised at once that, in this section, we are *not* considering transformations of the plane (e.g. reflections, translations, shears, rotations etc., in which the axes of reference remain fixed but the points are transformed). Here we are concerned with the effect on a given point, or a set of points, of referring the point, or points, to a different set of axes.

In Fig. 10.18 the point P has coordinates (x, y) referred to coordinate axes Ox, Oy and coordinates (X, Y) referred to coordinate axes AX, AY. The point A has coordinates (a, b) referred to axes Ox, Oy. The axes AX, AY are parallel to the axes Ox, Oy respectively.

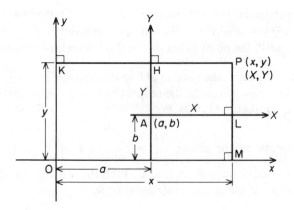

Fig. 10.18

and
$$x = PH + HK = X + a$$
$$y = PL + LM = Y + b.$$

Example 1
Find the equation of the curve $y = x^2 - 2x + 3$, referred to axes Ox, Oy, with reference to the axes AX, AY, where A has coordinates $(1, 2)$ referred to Ox, Oy and where AX and AY are parallel to Ox and Oy respectively.

$$x = X + 1 \quad \text{and} \quad y = Y + 2$$

Substituting in $y = x^2 - 2x + 3$, the equation of the curve referred to axes Ox, Oy, we get

$$Y + 2 = (X + 1)^2 - 2(X + 1) + 3,$$

i.e. $$Y = X^2,$$

which is the equation of the curve referred to the axes AX, AY. In this example the axes have undergone a translation $(1, 2)$. In the general case considered above the axes underwent a translation (a, b).

We now consider the change of axes produced by rotating the axes in their own plane through a positive (i.e. anticlockwise) angle while keeping the origin fixed.

In the Fig. 10.19 the point P has coordinates (x, y) referred to coordinate axes Ox, Oy and coordinates (X, Y) referred to coordinate axes OX, OY. The angle $xOX = \alpha$.

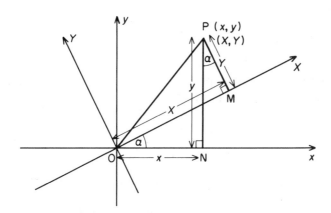

Fig. 10.19

$$\overrightarrow{OP} = \overrightarrow{ON} + \overrightarrow{NP} = \overrightarrow{OM} + \overrightarrow{MP}.$$

Considering components parallel to Ox we get

$$x = X \cos \alpha - Y \sin \alpha.$$

Considering components parallel to Oy we get

$$y = X \sin \alpha + Y \cos \alpha.$$

Example 2
A curve has equation $9x^2 + 24xy + 16y^2 - 40x + 30y = 0$ referred to axes Ox, Oy. These axes are rotated *clockwise* about O through an acute angle α, where $\tan \alpha = \frac{3}{4}$, to the positions OX, OY. Find the equation of the curve referred to the axes OX, OY.

The axes have been rotated through an angle $-\alpha$.

$$\therefore \qquad x = X \cos(-\alpha) - Y \sin(-\alpha)$$
$$= X \cos \alpha + Y \sin \alpha$$
$$= X(4/5) + Y(3/5) = \frac{1}{5}(4X + 3Y),$$

and
$$y = X \sin(-\alpha) + Y \cos(-\alpha)$$
$$= -X \sin \alpha + Y \cos \alpha$$
$$= -X(3/5) + Y(4/5) = \frac{1}{5}(-3X + 4Y)$$

Substituting in the equation $9x^2 + 24xy + 16y^2 - 40x + 30y = 0$ gives

$$\frac{9}{25}(4X + 3Y)^2 + \frac{24}{25}(4X + 3Y)(-3X + 4Y) + \frac{16}{25}(-3X + 4Y)^2$$

$$-\frac{40}{5}(4X + 3Y) + \frac{30}{5}(-3X + 4Y) = 0$$

which simplifies to $\qquad Y^2 = 2X$.

This example shows the simplification which may be effected in the equation to a given curve by choosing axes suitably related to the curve.

Example 3

A straight line has equation $x + y = 4$ referred to axes Ox, Oy. The axes are given a translation $\begin{pmatrix} 1 \\ 2 \end{pmatrix}$ so that the origin moves to the point A $(1, 2)$. The axes are then rotated *anticlockwise* about A through an acute angle α, where $\tan \alpha = 5/12$. Find the equation of the straight line referred to the new axes.

Let the axes after the first transformation, the translation, be AX, AY. Putting $x = X + 1$, $y = Y + 2$ in the equation $x + y = 4$ gives $X + 1 + Y + 2 = 4$, i.e. $X + Y = 1$.

Let the axes after the second transformation, the rotation, be Ax', Ay'. Putting

$$X = x' \cos \alpha - y' \sin \alpha, \qquad Y = x' \sin \alpha + y' \cos \alpha,$$

into $X + Y = 1$ gives

$$x' \cos \alpha - y' \sin \alpha + x' \sin \alpha + y' \cos \alpha = 1.$$

Substituting for $\sin \alpha$ and $\cos \alpha$ gives

$$\frac{17}{13}x' + \frac{7}{13}y' = 1, \qquad \text{i.e.} \qquad 17x' + 7y' = 13,$$

which is the equation of the straight line referred to the new axes Ax', Ay'.

Exercises 10.7

1 The coordinate axes are given a translation so that the origin moves to the point A (a, b). Find the equations, referred to the new axes, of
   (a) $y = x$, when $a = 2, b = 0$,
   (b) $y = x$, when $a = 0, b = 3$,
   (c) $y = x$, when $a = 2, b = 3$,
   (d) $y^2 = 4x$, when $a = 3, b = 0$,
   (e) $y^2 = 4x$, when $a = 0, b = 1$,
   (f) $y^2 = 4x$, when $a = 3, b = 1$,
   (g) $x^2 + y^2 = 4$, when $a = 5, b = 4$,
   (h) $x^2 + y^2 = 9$, when $a = -5, b = -4$,
   (i) $x^2 + y^2 - 2x - 3 = 0$, when $a = 4, b = -2$,
   (j) $x^2 + y^2 - 4x + 6y + 4 = 0$, when $a = -5, b = 3$,
   (k) $x^2 + y^2 - 6x - 8y - 25 = 0$, when $a = -3, b = -4$.

2 The coordinate axes are rotated anticlockwise about the origin through an acute angle α. Find the equations, referred to the new axes, of
   (a) $y = x$, when $\alpha = \tan^{-1} \frac{3}{4}$,
   (b) $y = 2x + 3$, when $\alpha = \tan^{-1} \frac{3}{4}$,
   (c) $y = x^2$, when $\alpha = \tan^{-1} \frac{3}{4}$,
   (d) $y = x^2 + 1$, when $\alpha = -\tan^{-1} \frac{3}{4}$,
   (e) $y^2 = 2x$, when $\alpha = -\tan^{-1} \frac{3}{4}$,
   (f) $x^2 - y^2 = 4$, when $\alpha = 45°$,
   (g) $x^2 - y^2 = 4$, when $\alpha = -45°$,
   (h) $x^2 + y^2 = 1$, when $\alpha = \tan^{-1} \frac{5}{12}$.

3 The coordinate axes are translated so that the origin moves to the point A (a, b) and are then rotated anticlockwise about A through an acute angle α. Find the equation, referred to the new axes, of
   (a) $y = 2x$, when $a = 4, b = -3$ and $\alpha = \tan^{-1} \frac{3}{4}$,
   (b) $y^2 = x$, when $a = -2, b = 0$ and $\alpha = -\tan^{-1} \frac{3}{4}$,
   (c) $x^2 - y^2 = 1$, when $a = 2, b = -1$ and $\alpha = \tan^{-1} \frac{5}{12}$,
   (d) $x^2 + y^2 = 4$, when $a = -3, b = -5$ and $\alpha = -\tan^{-1} \frac{5}{12}$.

10.8 The equation $ax^2 + 2hxy + by^2 = 0$

It will be noted that change of axes, as in the preceding section, does not alter the degree of the curve. All the curves previously studied (circle, parabola, hyperbola and ellipse) are all second-degree curves and are particular cases of curves given by the *general equation* of the second degree, namely

$$ax^2 + 2hxy + by^2 + 2gx + 2fy + c = 0,$$

where a, b, c, f, g and h are constants.

The particular curve given when $g = f = c = 0$ in this equation has not yet been considered. We now consider this case when the general equation becomes

$$ax^2 + 2hxy + by^2 = 0.$$

Provided that $b \neq 0$, this equation may be written as

$$b\left[y^2 + \frac{2h}{b}xy + \frac{a}{b}x^2\right] = 0$$

i.e.

$$b[(y - m_1 x)(y - m_2 x)] = 0,$$

where

$$m_1 + m_2 = -\frac{2h}{b}, \qquad m_1 m_2 = \frac{a}{b},$$

and, provided that

$$\left(\frac{2h}{b}\right)^2 \geqslant 4\left(\frac{a}{b}\right), \qquad \text{i.e.} \qquad h^2 \geqslant ab,$$

the values of m_1 and m_2 are real.

Thus the equation $ax^2 + 2hxy + by^2 = 0$ represents two straight lines through the origin provided that $h^2 \geqslant ab$ and $b \neq 0$. Similarly, provided that $a \neq 0$ and $h^2 \geqslant ab$, the equation still represents two straight lines through the origin.

If $a = 0$ and $b = 0$, the equation is $2hxy = 0$ and represents the coordinate axes.

In the general case the two straight lines are clearly at right angles if $m_1 m_2 = -1$, i.e. if $a + b = 0$.

Example 1

Show that the equation $6x^2 + xy - 2y^2 = 0$ represents a pair of straight lines through the origin and find the tangent of the acute angle between them.

The equation $6x^2 + xy - 2y^2 = 0$ clearly represents two straight lines, since the equation may be written as

$$(3x + 2y)(2x - y) = 0.$$

Thus the equation represents the lines $y = -\frac{3}{2}x$ and $y = 2x$. Let θ be the angle between the lines. Then

$$\tan \theta = \frac{-\frac{3}{2} - 2}{1 + (-\frac{3}{2})(2)} = \frac{7}{4}.$$

Exercises 10.8

1   State which of the following equations represent a pair of real straight lines through the origin:
  (a) $3x^2 + 4xy + 2y^2 = 0$,   (b) $2x^2 + 3xy - 2y^2 = 0$,
  (c) $6x^2 + 11xy + 4y^2 = 0$,   (d) $5x^2 - 6xy + 2y^2 = 0$,
  (e) $4x^2 - 12xy + 9y^2 = 0$.

2 Find the tangents of the acute angles between the lines represented by the equations

(a) $2x^2 + 9xy - y^2 = 0$,   (b) $6x^2 - 11xy + 4y^2 = 0$,

(c) $3x^2 - 6xy + 2y^2 = 0$,   (d) $5x^2 + 4xy - 3y^2 = 0$,

(e) $7x^2 - 12xy + 5y^2 = 0$.

3 A curve is given by the equation

$$8x^2 + 4xy + 5y^2 = 1.$$

Show that when the coordinate axes are rotated anticlockwise through the angle $\tan^{-1}(1/2)$ this equation becomes $9X^2 + 4Y^2 = 1$.

4 Transform the equation $7x^2 - 8xy + y^2 = 0$ by turning the coordinate axes anticlockwise through the angle $\tan^{-1} 2$.

5 Show that the equation $5x^2 - 24xy - 5y^2 = 0$ represents a pair of straight lines.

Find the form taken by this equation when the coordinate axes

(a) are rotated anticlockwise through the angle $\tan^{-1}(3/2)$,

(b) are rotated clockwise through the angle $\tan^{-1}(2/3)$.

6 When the coordinate axes are rotated anticlockwise through the angle α, the equation

$$ax^2 + 2hxy + by^2 = 0$$

takes the form $AX^2 + BY^2 = 0$. Prove that

(a) $\tan 2\alpha = \dfrac{2h}{a - b}$,   (b) $A + B = a + b$,   (c) $AB = ab - h^2$.

10.9 The equation $ax^2 + 2hxy + by^2 = 1$

There are three cases to consider, depending on the values of the constants a, b and h.

Case 1 $h^2 > ab$

It was shown in Section 10.7 that the transformation given by

$$x = X \cos \alpha - Y \sin \alpha, \qquad y = X \sin \alpha + Y \cos \alpha$$

has the effect of rotating the coordinate axes anticlockwise through an angle α.

Consider the effect of this change of axes on the equation

$$\frac{x^2}{a^2} - \frac{y^2}{b^2} = 1. \qquad\qquad \dots (1)$$

Equation (1) becomes

$$b^2(X \cos \alpha - Y \sin \alpha)^2 - a^2(X \sin \alpha + Y \cos \alpha)^2 = a^2 b^2$$

$\Rightarrow \qquad\qquad AX^2 + 2HXY + BY^2 = 1, \qquad\qquad \dots (2)$

where
$$A = (b^2 \cos^2 \alpha - a^2 \sin^2 \alpha)/(a^2 b^2),$$
$$B = (b^2 \sin^2 \alpha - a^2 \cos^2 \alpha)/(a^2 b^2),$$
$$H = -(a^2 + b^2)(\sin \alpha \cos \alpha)/(a^2 b^2).$$

The same substitution will convert the equation

$$\frac{x^2}{a^2} - \frac{y^2}{b^2} = 0 \qquad \ldots (3)$$

to the form
$$AX^2 + 2HXY + Y^2 = 0. \qquad \ldots (4)$$

Equation (4) represents in terms of X and Y the pair of lines represented by equation (3) in terms of x and y. It follows that H^2 must be greater than AB. This can be verified by showing that $H^2 - AB = 1/(a^2 b^2)$.

Equation (2) represents the same hyperbola as equation (1). Provided that $H^2 > AB$, any equation of the form

$$AX^2 + 2HXY + BY^2 = 1$$

can be converted to the form of equation (1) by a rotation of the coordinate axes.

It follows that the equation

$$AX^2 + 2HXY + BY^2 = 1,$$

in which $H^2 > AB$, represents a hyperbola with asymptotes given by the equation

$$AX^2 + 2HXY + BY^2 = 0.$$

For example, the equation

$$x^2 - 4xy + 3y^2 = 1$$

represents a hyperbola with asymptotes given by

$$x^2 - 4xy + 3y^2 = 0,$$
i.e. $$(x - 3y)(x - y) = 0.$$

Case 2 $h^2 < ab$

The same rotation of the axes as in case 1 will convert the equation

$$\frac{x^2}{a^2} + \frac{y^2}{b^2} = 1,$$

which represents an ellipse, to the form

$$AX^2 + 2HXY + BY^2 = 1,$$

where
$$A = (b^2 \cos^2 \alpha + a^2 \sin^2 \alpha)/(a^2 b^2),$$
$$B = (b^2 \sin^2 \alpha + a^2 \cos^2 \alpha)/(a^2 b^2),$$
$$H = (a^2 - b^2)(\sin \alpha \cos \alpha)/(a^2 b^2).$$

Then $H^2 - AB$ simplifies to $-1/(a^2 b^2)$, showing that $H^2 < AB$. Provided that $H^2 < AB$, any equation of the form

$$AX^2 + 2HXY + BY^2 = 1$$

represents an ellipse, assuming that the equation can be satisfied by real values of X and Y. This rules out an equation such as $-4X^2 - 9Y^2 = 1$.

Case 3 $h^2 = ab$
When $h^2 = ab$, the equation

$$ax^2 + 2hxy + by^2 = 1$$

can be satisfied by real values of x and y only if a and b are positive.
   When this is so, the equation will represent a pair of parallel lines given by

$$px + qy = \pm 1,$$

where $p^2 = a$, $q^2 = b$ and $pq = h$.

Example 1
Show that the equation

$$7x^2 - 48xy - 7y^2 = 3$$

represents a rectangular hyperbola.

Comparison with the equation $ax^2 + 2hxy + by^2 = 1$ gives

$$a = 7/3, \qquad b = -7/3, \qquad h = -8.$$

Since $h^2 > ab$, the equation represents a hyperbola.
   The asymptotes are given by the equation

$$7x^2 - 48xy - 7y^2 = 0$$
$$\Rightarrow \qquad (7x + y)(x - 7y) = 0.$$

Since the asymptotes $7x + y = 0$ and $x - 7y = 0$ are at right angles, the curve is a rectangular hyperbola.
   The general equation of the second degree is

$$ax^2 + 2hxy + by^2 + 2gx + 2fy + c = 0, \qquad \ldots (5)$$

where a, b, c, f, g and h are constants.
   Let the substitution $x = X + p$, $y = Y + q$ be made, moving the origin to the point (p, q).
   In the resulting equation, the coefficients of X and Y will be zero if

$$ap + hq + g = 0 \qquad \text{and} \qquad bq + hp + f = 0.$$

Provided that h^2 is not equal to ab, these two equations can be solved for p and q. With these values for p and q, equation (5) will be converted to the form

$$aX^2 + 2hXY + bY^2 = k,$$

where k is a constant.

The case when $k = 0$ has been considered in Section 10.8. When k is not zero, it follows from case 1 above that when $h^2 > ab$ then equation (5) represents a hyperbola with asymptotes parallel to the pair of lines given by

$$ax^2 + 2hxy + by^2 = 0.$$

Example 2
Show that the equation

$$2x^2 + 5xy - 3y^2 + 3x + 16y - 8 = 0$$

represents a hyperbola, and find the equations of its asymptotes.

Comparison of the terms of second degree with the expression $ax^2 + 2hxy + by^2$ gives

$$a = 2, \qquad b = -3, \qquad h = 5/2.$$

Since $h^2 - ab = 25/4 + 6 > 0$, the equation represents a hyperbola.

When the substitution $x = X + p$, $y = Y + q$ is made, the given equation becomes

$$2(X + p)^2 + 5(X + p)(Y + q) - 3(Y + q)^2 + 3(X + p)$$
$$+ 16(Y + q) - 8 = 0$$

$\Rightarrow$ $\qquad 2X^2 + 5XY - 3Y^2 + (4p + 5q + 3)X + (5p - 6q + 16)Y$
$$+ (2p^2 + 5pq - 3q^2 + 3p + 16q - 8) = 0.$$

The coefficients of X and Y will be zero if

$$4p + 5q + 3 = 0$$
and $\qquad\qquad 5p - 6q + 16 = 0.$

These equations are satisfied when $p = -2$ and $q = 1$. The equation of the curve then becomes

$$2X^2 + 5XY - 3Y^2 = 3$$
$\Rightarrow$ $\qquad\qquad (2X - Y)(X + 3Y) = 3.$

The equations of the asymptotes will be

$$2X - Y = 0, \qquad X + 3Y = 0.$$

Referred to the original axes, these equations become

$$2(x + 2) - (y - 1) = 0, \qquad x + 2 + 3(y - 1) = 0$$
i.e. $\qquad\qquad 2x - y + 5 = 0, \qquad x + 3y - 1 = 0.$

An alternative approach uses the fact that the equation of the asymptotes differs from the equation of the hyperbola only in the constant term.

Since the factors of the terms of second degree $2x^2 + 5xy - 3y^2$ are $2x - y$ and $x + 3y$, the equations of the asymptotes will be of the form

$$2x - y + m = 0, \qquad x + 3y + n = 0,$$

where m and n are constants.

Compare the equation of the hyperbola

$$2x^2 + 5xy - 3y^2 + 3x + 16y - 8 = 0$$

with the equation

$$(2x - y + m)(x + 3y + n) = 0.$$

The coefficients of x will be the same in the two equations if

$$m + 2n = 3.$$

The coefficients of y will be the same if

$$3m - n = 16.$$

These equations are satisfied by the values $m = 5, n = -1$. Hence the equations of the asymptotes are

$$2x - y + 5 = 0, \qquad x + 3y - 1 = 0.$$

Example 3

Find the equations of the asymptotes of the hyperbola

$$x^2 - xy + x + 2y - 4 = 0.$$

Consider the terms of second degree.

Since $x^2 - xy = x(x - y)$, one asymptote is parallel to the line $x = 0$, and the other is parallel to the line $y = x$.

The given equation can be put in the form

$$y(x - 2) = x^2 + x - 4$$

$$\Rightarrow \qquad y = \frac{x^2 + x - 4}{x - 2}.$$

As x tends to the value 2, y tends to infinity. This shows that the line $x = 2$ is an asymptote.

Also

$$y = \frac{(x^2 - 2x) + (3x - 4)}{x - 2}$$

$$\Rightarrow \qquad y = x + 3 + \frac{2}{x - 2}$$

$$\Rightarrow \qquad y - x - 3 = \frac{2}{x - 2}.$$

As x tends to infinity, the right-hand side of this equation tends to zero, showing that the curve gets closer and closer to the line $y - x - 3 = 0$.

This line is the other asymptote.

The result can be confirmed by comparing the equation of the hyperbola with the equation

$$(x - 2)(y - x - 3) = 0,$$

i.e.
$$x^2 - xy + x + 2y - 6 = 0.$$

This equation differs from the equation of the hyperbola only in the constant term.

Exercises 10.9

1 Determine the nature of the locus given by each of the following equations. Give the equations of the asymptotes of each hyperbola.

(a) $6x^2 - xy - 12y^2 = 6$.   (b) $x^2 + xy + y^2 = 3$.

(c) $6x^2 - 5xy - 6y^2 = 2$.   (d) $3x^2 + xy + y^2 = -1$.

(e) $x^2 + 6xy + 9y^2 = 1$.   (f) $x^2 + 2xy + 4y^2 = 5$.

(g) $9x^2 - 25y^2 = 4$.   (h) $2x^2 + 2xy + 3y^2 = 2$.

(i) $2x^2 + 3xy - 2y^2 = 1$.   (j) $4x^2 - 12xy + 9y^2 = 4$.

2 Show that each of the following equations represents a hyperbola. In each case obtain the equations of the asymptotes.

(a) $(x + 4)(y - 3) = 5$.

(b) $x^2 - xy - 2x - y + 1 = 0$.

(c) $xy + y^2 - x = 0$.

(d) $2x^2 + 3xy - 2y^2 + 2x - y = 4$.

(e) $8x^2 + 10xy - 3y^2 - 2x + 4y = 0$.

10.10 Tangents at the origin

Consider the equation

$$ax^2 + 2hxy + by^2 + 2gx + 2fy = 0, \qquad \dots (1)$$

in which the constants f and g are not zero.

This equation contains no constant term, and represents a curve which passes through the origin O.

Near the origin, x^2 is small compared with x, y^2 is small compared with y, and xy is small compared with both x and y.

This implies that an approximation to the curve near O is given by the first-degree terms equated to zero, i.e.

$$gx + fy = 0.$$

This is the equation of the tangent to the curve at O.

This can be shown by differentiating equation (1) with respect to x.
When $x = 0$ and $y = 0$, the equation

$$2ax + 2h\left(y + x\frac{dy}{dx}\right) + 2by\frac{dy}{dx} + 2g + 2f\frac{dy}{dx} = 0$$

reduces to

$$\frac{dy}{dx} = \frac{-g}{f}.$$

Hence the equation of the tangent to the curve at the origin is

$$gx + fy = 0.$$

Consider next the curve given by the equation

$$ax^2 + 2hxy + by^2 + 2gx = 0. \qquad \ldots (2)$$

Near the origin O, the terms x^2 and xy are small compared with x. It follows that the equation

$$by^2 + 2gx = 0$$

gives an approximation near O to the curve given by equation (2).

This approximation is the equation of a parabola which touches the y-axis at its vertex.

Exercises 10.10
State the type of curve represented by each of the following equations, and give the equation of the tangent at the origin.

1 $5x^2 + 4xy + y^2 + x - 2y = 0$.

2 $x^2 + y^2 - 5x + 2y = 0$.

3 $x^2 - y^2 - 2x = 0$.

4 $y = 2x/(x + 1)$.

5 $y = x^2/(x - 2)$.

6 $(x + y)^2 = (2x + 1)(y - 2) + 2$.

7 $(x + y)^2 = 4(x - y)$.

8 $x^2 = y(x - 4y + 1)$.

Miscellaneous exercises 10
1 Show that the line joining the points $(ct_1, c/t_1)$ and $(ct_2, c/t_2)$ on the hyperbola $xy = c^2$ has equation $x + t_1 t_2 y = c(t_1 + t_2)$.

Deduce the equation of the tangent at the point $(ct, c/t)$. The tangents at two points, P and Q, on the hyperbola meet at a point R. Show that the line joining R to the origin bisects PQ. [MEI]

2 Prove that the equation of the chord joining the points $P\left(cp, \dfrac{c}{p}\right)$ and $Q\left(cq, \dfrac{c}{q}\right)$ on the rectangular hyperbola $xy = c^2$ is $x + pqy = c(p + q)$.

If this chord passes through the point $(cp + cq - c, c)$, show that the tangents at P and Q meet on the line $y = x$. [L]

3 Prove that the gradient of the chord joining the point $P(cp, c/p)$ and the point $Q(cq, c/q)$ on the rectangular hyperbola $xy = c^2$ is $-1/pq$.

The points P, Q and R lie on a rectangular hyperbola, the angle QPR being a right angle. Prove that the angle between QR and the tangent at P is also a right angle. [L]

4 Obtain the equation of the chord joining the points $(t_1, 1/t_1)$, $(t_2, 1/t_2)$ on the hyperbola $xy = 1$.

Find the equation satisfied by the coordinates of the mid-point of this chord

(a) when the gradient of the chord is equal to 2,

(b) when one end of the chord is held at the point $(1, 1)$.

Illustrate each case by a diagram. [L]

5 Prove that the equation of the tangent at the point $P(ct, c/t)$ on the rectangular hyperbola $xy = c^2$ is $x + t^2y = 2ct$.

The tangent at P meets the x-axis at Q and the y-axis at R. The line through Q parallel to the y-axis meets the hyperbola at S; and the line through R parallel to the x-axis meets the hyperbola at T. Prove that, as P varies, the locus of the mid-point of ST is the rectangular hyperbola $16xy = 25c^2$. [JMB]

6 Find the equations of the tangent and normal to the rectangular hyperbola $xy = c^2$ at the point $P(cp, c/p)$.

Assuming that $p > 1$, find in terms of p and c the perpendicular distances of the tangent and normal at P from the origin O. Prove that, if the tangent and normal at P and the lines drawn from O perpendicular to this tangent and normal form a square, then $p^2 = \sqrt{2} + 1$. [C]

7 Show that the gradient of the chord joining the points $Q(cq, c/q)$, $R(cr, c/r)$ on the hyperbola $xy = c^2$ is $-1/qr$.

If $P(ct, c/t)$ is a fixed point on the curve, and QR subtends a right angle at P, show that

(a) Q and R are on different branches of the hyperbola,

(b) the direction of QR is fixed.

(c) the mid-point of QR lies on a fixed straight line through the origin. [L]

8 The chord through two variable points P and Q on the rectangular hyperbola $xy = c^2$ cuts the x-axis at R. If S is the mid-point of PQ and O is the origin, prove that the triangle OSR is isosceles. Show that, if

OP, OQ and OS make angles θ_1, θ_2 and θ_3 respectively with OR, then $\tan^2 \theta_3 = \tan \theta_1 \tan \theta_2$. [AEB]

9 Show that the equation of the normal at $P(ct, c/t)$ to the rectangular hyperbola $xy = c^2$ is $t^3 x - ty = c(t^4 - 1)$.

The tangent at P meets the x-axis at A and the y-axis at B and the normal at P meets the y-axis at C. If M is the mid-point of AC find the equation of the locus of M. Show that the area of the triangle ABC is four times the area of the triangle AMP. [L]

10 The tangent to the hyperbola $xy = c^2$ at the point $P\left(ct, \dfrac{c}{t}\right)$ intersects the x-axis at T and the y-axis at T'. The normal to the hyperbola at P intersects the x-axis at N and the y-axis at N'. The areas of the triangles PNT and PN'T' are Δ and Δ' respectively. Prove that

$$\frac{1}{\Delta} + \frac{1}{\Delta'} = \frac{2}{c^2}.$$ [L]

11 Two points $P(4p, 4/p)$ and $Q(4q, 4/q)$ lie on one branch of the rectangular hyperbola $xy = 16$. If the line LPQM meets the axes at L and M, show that LP = QM.

The tangent at a point T on the other branch of the rectangular hyperbola meets the axes at R and S. Prove that TR = TS.

The tangents to the hyperbola at P and Q meet at U. Show that if PQ and RS are parallel, the points T and U are collinear with the origin. [L]

12 The line $y = mx$ meets the hyperbola $xy = c^2$, where $c > 0$, at the points R and S. Prove that the tangents to the hyperbola at R and S are parallel. Find the distance between the parallel tangents and show that, as m varies, the maximum distance between them is $2c\sqrt{2}$.

The tangents and normals at R and S together form a rectangle. Find the area of the rectangle and show that, when $m = 3$, the area is $6\cdot4 c^2$. [L]

13 The gradient m of the chord PQ of the hyperbola $xy = c^2$ is constant and positive. Show that there are two fixed points through which the circle on PQ as diameter passes for all positions of PQ.

Show also that if the chord RS is perpendicular to PQ, the circle on RS as diameter cuts orthogonally the circle on PQ as diameter. [L]

14 Show that the circle on the line joining the points (x_1, y_1), (x_2, y_2) as diameter has the equation

$$(x - x_1)(x - x_2) + (y - y_1)(y - y_2) = 0.$$

Prove that the normal to the rectangular hyperbola $xy = c^2$ at the point $A(ct, c/t)$ meets the hyperbola again at $B(-c/t^3, -ct^3)$.

Find the point C at which the circle on AB as diameter meets the rectangular hyperbola again and prove that the origin is the mid-point of AC. [L]

15 The foot of the perpendicular from a point P to the straight line $x + y = \sqrt{2}$ is the point R, and Q is the point with coordinates $(\sqrt{2}, \sqrt{2})$. If P varies in such a way that $PQ^2 = 2PR^2$, show that its locus is the rectangular hyperbola $xy = 1$.

Find the equation of the tangent to this hyperbola at the point $(t, 1/t)$.

The tangent cuts the x-axis at A and the y-axis at B, and C is the point on AB such that $AC:CB = a:b$. Show that the locus of C as t varies is the rectangular hyperbola

$$xy = \frac{4ab}{(a + b)^2}.$$ [JMB]

16 A point P moves so that its distances from $A(a, 0)$, $A'(-a, 0)$, $B(b, 0)$, $B'(-b, 0)$ are related by the equation $AP \cdot PA' = BP \cdot PB'$. Show that the locus of P is a hyperbola and find the equations of its asymptotes. [L]

17 Write down the coordinates of the centre and the equations of the asymptotes of the rectangular hyperbola

$$(x - h)(y - k) = c^2.$$

Sketch the hyperbolae $2x(y - 2) = 3$ and $2y(x - 1) = 3$ and find the coordinates of the points P and Q in which they intersect.

Show that the tangents to the hyperbolae at P and Q form a parallelogram. [L]

18 Find the equation of the normal at the point P $(ct, c/t)$ on the rectangular hyperbola $xy = c^2$.

Show that, if the normals at the points P_1, P_2, P_3, P_4 with parameters t_1, t_2, t_3, t_4 respectively are concurrent, then $t_1 t_2 t_3 t_4 = -1$.

Show also that, in this case, a line joining any two of the four points is perpendicular to the line joining the other two and deduce in terms of t_1, t_2, t_3 the coordinates of the orthocentre of triangle $P_1 P_2 P_3$. [L]

19 The line $x = 1$ is a directrix of the hyperbola $x^2/a^2 - y^2/b^2 = 1$, and the point $(4, 0)$ is the corresponding focus. Find the values of a and b. [L]

20 Write down the equations of the two asymptotes of the hyperbola $x^2/9 - y^2/16 = 1$.

The tangent to the hyperbola at the point $P(3 \sec \theta, 4 \tan \theta)$ meets the asymptotes at X and Y. Show that
(a) P is the mid-point of XY,
(b) if O is the origin, the area of the $\triangle XOY$ is independent of θ. [L]

21 A hyperbola of the form $x^2/\alpha^2 - y^2/\beta^2 = 1$ has asymptotes $y^2 = m^2 x^2$ and passes through the point $(a, 0)$. Find the equation of the hyperbola in terms of x, y, a and m.

A point P on this hyperbola is equidistant from one of its asymptotes and the x-axis. Prove that, for all values of m, P lies on the curve.

$$(x^2 - y^2)^2 = 4x^2(x^2 - a^2).$$ [L]

22 State the equations of the asymptotes of the hyperbola

$$\frac{x^2}{a^2} - \frac{y^2}{b^2} = 1.$$

The point P on the curve lies in the first quadrant. The line through P parallel to Oy meets an asymptote of the curve at the point Q which also lies in the first quadrant. The normal at P meets the x-axis at G. Prove that QG is perpendicular to the asymptote. [L]

23 Show that the equations of the tangents with gradient m to the hyperbola $x^2 - 4y^2 = 4$ are $y = mx \pm \sqrt{(4m^2 - 1)}$, where $|m| > \frac{1}{2}$.
   Deduce that the points of intersection of perpendicular tangents to the hyperbola lie on the circle

$$x^2 + y^2 = 3.$$ [L]

24 The tangents to the hyperbola $b^2x^2 - a^2y^2 = a^2b^2$ at points A and B on the curve meet at T. If M is the mid-point of AB, prove that TM passes through the centre of the hyperbola. Prove that the product of the gradients of AB and TM is constant. [L]

25 Find the equation of the normal at the point $P[a(t + 1/t), a(t - 1/t)]$ on the hyperbola $x^2 - y^2 = 4a^2$. N is the foot of the perpendicular from P to the x-axis and the normal at P meets the x-axis at G. If NP produced meets the line $y = x$ at Q_1 and the line $y = -x$ at Q_2, prove that GQ_1 and GQ_2 are parallel to the two asymptotes. Show that, if O is the origin, then GQ_2OQ_1 is a square, and prove that, as P varies, the locus of the mid-point of GQ_1 is a straight line.

26 Find the equation of the normal to the hyperbola $x^2/a^2 - y^2/b^2 = 1$ at the point $P(a \sec \theta, b \tan \theta)$.
   If there is a value of θ such that the normal at P passes through the point $(2a, 0)$, show that the eccentricity of the hyperbola cannot be greater than $\sqrt{2}$. Show that in this case, the parameter ϕ of the point on the hyperbola where the normal passes through the point $(0, -2b)$, is such that $-1 \leqslant \tan \phi < 0$.
   If the normal at any point P meets the y-axis at L, find the locus, as θ varies, of the mid-point of PL. [AEB]

27 Find the equation of the hyperbola with eccentricity $\sqrt{2}$ which has one focus at the point $(10, 0)$ and the line with equation $x = 5$ as the corresponding directrix.
   Determine also, for this hyperbola,
   (a) the length of the semi-major axis,

(b) the angle between the asymptotes,

(c) the locus of the points of intersection of tangents drawn at the ends of a chord passing through the given focus as the chord varies. [L]

28 Obtain the equation of the ellipse having major axis of length 10 units and foci at the points (3, 0) and (− 3, 0). [L]

29 Show that the equation

$$\frac{x^2}{29 - c} + \frac{y^2}{4 - c} = 1$$

represents

(a) an ellipse if c is any constant less than 4,

(b) a hyperbola if c is any constant between 4 and 29.

Show that the foci of each ellipse in (a) and each hyperbola in (b) are independent of the value of c.

If $c = 13$ find the coordinates of the vertices A and B of the hyperbola.

If P and Q are points on this hyperbola such that PQ is a double ordinate, prove that the locus of the intersection of AP and BQ is an ellipse. [AEB]

30 Show that the equation of the chord of the ellipse

$$\frac{x^2}{a^2} + \frac{y^2}{b^2} = 1$$

which joins the points whose coordinates are $(a \cos \theta, b \sin \theta)$ and $(a \cos \phi, b \sin \phi)$ is

$$bx \cos \frac{\theta + \phi}{2} + ay \sin \frac{\theta + \phi}{2} = ab \cos \frac{\theta - \phi}{2}.$$

If the centre of the above ellipse is O and chords of the ellipse are drawn all passing through the point $P\left(\frac{a}{2}, 0\right)$, show that their middle points lie on an ellipse whose centre bisects OP. [AEB]

31 Prove that the equation of the chord of the ellipse

$$\frac{x^2}{a^2} + \frac{y^2}{b^2} = 1$$

joining the points $(a \cos \theta, b \sin \theta)$ and $(a \cos \phi, b \sin \phi)$ is

$$\frac{x}{a} \cos \tfrac{1}{2}(\theta + \phi) + \frac{y}{b} \sin \tfrac{1}{2}(\theta + \phi) = \cos \tfrac{1}{2}(\theta - \phi).$$

Prove that, if this chord touches the ellipse,

$$\frac{x^2}{a^2} + \frac{y^2}{b^2} = \frac{1}{2},$$

θ and ϕ differ by an odd multiple of $\frac{\pi}{2}$. [AEB]

32 Show that the equation of the tangent to the ellipse $x^2/a^2 + y^2/b^2 = 1$ at the point $(a \cos \theta, b \sin \theta)$ is

$$\frac{x \cos \theta}{a} + \frac{y \sin \theta}{b} = 1.$$

P is any point on the ellipse and the tangent at P meets the coordinate axes at Q, R. If P is the mid-point of QR, show that P lies on a diagonal of the rectangle which circumscribes the ellipse and has its sides parallel to the axes of coordinates.
  Find the equation of the locus of the mid-point of QR. [L]

33 The tangent at $P(a \cos \theta, b \sin \theta)$ to the ellipse $b^2 x^2 + a^2 y^2 = a^2 b^2$ cuts the y-axis at Q. The normal at P is parallel to the line joining Q to one focus S'. If S is the other focus, show that PS is parallel to the y-axis. [L]

34 Prove that the equation of the tangent to the ellipse

$$\frac{x^2}{a^2} + \frac{y^2}{b^2} = 1$$

at the point $P_1(x_1, y_1)$ is

$$\frac{xx_1}{a^2} + \frac{yy_1}{b^2} = 1.$$

The tangent at P_1 meets the tangent at $P_2(x_2, y_2)$ at T. Show that the line

$$\frac{xx_1}{a^2} + \frac{yy_1}{b^2} = \frac{xx_2}{a^2} + \frac{yy_2}{b^2}$$

passes through T and through the mid-point of $P_1 P_2$.
  Prove that if $P_1 T P_2$ is a right angle then

$$\frac{x_1 x_2}{a^4} + \frac{y_1 y_2}{b^4} = 0.$$ [L]

35 Show that the equation of the tangent to the ellipse $x^2/a^2 + y^2/b^2 = 1$ at the point $P(a \cos \theta, b \sin \theta)$ is $bx \cos \theta + ay \sin \theta = ab$.
  If the line joining P to $Q(a \cos \phi, b \sin \phi)$ passes through a focus of this ellipse, show that

$$\sin (\theta - \phi) = \pm e(\sin \theta - \sin \phi),$$

where e is the eccentricity.

Deduce that tangents at the ends of a focal chord meet on the corresponding directrix. [L]

36 The tangent and the normal at a point $P(3\sqrt{2}\cos\theta, 3\sin\theta)$ on the ellipse $x^2/18 + y^2/9 = 1$ meet the y-axis at T and N respectively. If O is the origin, prove that OT . ON is independent of the position of P. Find the co-ordinates of X, the centre of the circle through P, T and N. Find also the equation of the locus of the point Q on PX such that X is the mid-point of PQ. [AEB]

37 Find the equation of the tangent at a general point θ on the ellipse $x = 2\cos\theta, y = 3\sin\theta$.

The tangents at the points θ and $\theta + \dfrac{\pi}{4}$ intersect at P. Show that the locus of P is an ellipse and find its equation.

If the foci of the given ellipse are S and R find the maximum area of triangle PSR. [AEB]

38 Show that if the line $y = mx + c$ is a tangent to the ellipse

$$\frac{x^2}{a^2} + \frac{y^2}{b^2} = 1,$$

then $c^2 = a^2 m^2 + b^2$.

Find the gradients of the tangents to this ellipse through the point $(\sqrt{(a^2 + b^2)}, 0)$. Hence or otherwise determine the coordinates of the vertices of a square whose sides touch the ellipse. [C]

39 Show that, if $y = mx + c$ is a tangent to the ellipse $b^2 x^2 + a^2 y^2 = a^2 b^2$, then $c^2 = a^2 m^2 + b^2$.

Find the equations of the tangents to the ellipse $9x^2 + 16y^2 = 144$ from the point $P(4, 9)$. These tangents touch the ellipse at the points A and B, and C is the mid-point of AB. Find the coordinates of A and B, and show that the straight line PC goes through the origin. [AEB]

40 Show that for all values of m the straight lines with equations $y = mx \pm \sqrt{(b^2 + a^2 m^2)}$ are tangents to the ellipse

$$\frac{x^2}{a^2} + \frac{y^2}{b^2} = 1.$$

Hence show that, if the tangents from an external point P to the ellipse meet at right angles, the locus of P is a circle and find its equation. [L]

41 Find the equation of the tangent and of the normal to the ellipse $b^2 x^2 + a^2 y^2 = a^2 b^2$ at the point $P(a\cos\theta, b\sin\theta)$ when $a > b$.

The normal at P meets the x-axis at G. The tangent at P meets the y-axis at T. The foot of the perpendicular from T to OP is L, O being the origin.

Prove that OL . OP $= b^2$.

Find the cartesian equation of the locus of the mid-point of TG.   [AEB]

42 The ellipse

$$\frac{x^2}{16} + \frac{y^2}{4} = 1$$

cuts the y-axis at A and C and the negative x-axis at B. Show that the circle passing through A, B and C has the equation $x^2 + y^2 + 3x - 4 = 0$.

Find the equation of the tangent at the point $\left(\frac{12}{7}, \frac{4\sqrt{10}}{7}\right)$ on the ellipse and show that this tangent also touches the circle.   [L]

43 P is any point on the ellipse $9x^2 + 25y^2 = 225$, and A and B are the points $(-4, 0)$ and $(4, 0)$ respectively. Prove that PA $+$ PB $= 10$.

Prove also that the normal at P bisects the angle APB.   [L]

44 Show that the equations of the tangent and normal to the ellipse $x = a \cos \theta$, $y = b \sin \theta$ at the point $\theta = t$ are

$$xb \cos t + ya \sin t = ab$$

and $$xa \sin t - ya \cos t = (a^2 - b^2) \sin t \cos t$$

respectively.

The normal at P intersects the x-axis at N, and the perpendicular drawn from the origin O to the tangent at P intersects the tangent at M. Prove that OM . PN $= b^2$.   [L]

45 Prove that the normal at the point P$(a \cos \theta, b \sin \theta)$ on the ellipse

$$\frac{x^2}{a^2} + \frac{y^2}{b^2} = 1 \quad \text{has the equation} \quad \frac{ax}{\cos \theta} - \frac{by}{\sin \theta} = a^2 - b^2.$$

The normal at P cuts the axis of x at G, and PG is produced to Q so that GQ $= 2$PG. Express the coordinates of Q in terms of a, b and θ and deduce that, as θ varies, Q lies on a fixed ellipse. Give the (x, y) equation of this ellipse.   [C]

46 Prove that the normal at P$(a \cos \theta, b \sin \theta)$ on the ellipse

$$\frac{x^2}{a^2} + \frac{y^2}{b^2} = 1$$

has the equation

$$\frac{ax}{\cos \theta} - \frac{by}{\sin \theta} = a^2 - b^2.$$

This normal meets the line OQ, where O is the origin and Q is the point

$(a \cos \theta, -b \sin \theta)$, at K. Find the coordinates of K, and prove that as θ varies K moves on a fixed ellipse. State the (x, y) equation of this ellipse.   [C]

47 Show that the equation of the normal at P $(a \cos \theta, b \sin \theta)$ to the ellipse $b^2x^2 + a^2y^2 = a^2b^2$ is

$$ax \sin \theta - by \cos \theta = (a^2 - b^2) \sin \theta \cos \theta.$$

The normal at P meets the x-axis at A and the y-axis at B. Show that the area of triangle OAB, where O is the origin, cannot exceed $(a^2 - b^2)^2/(4ab)$. Find the equation of the locus of the centroid of the triangle OAB.   [L]

48 Show that the equation of the normal to the ellipse $x^2/a^2 + y^2b^2 = 1$ at the point P $(a \cos \theta, b \sin \theta)$ is

$$ax \sin \theta - by \cos \theta = (a^2 - b^2) \sin \theta \cos \theta.$$

If the normal to the ellipse $x^2/25 + y^2/9 = 1$ at a point Q meets the coordinate axes at A and B respectively, show that, as Q varies, the locus of the mid-point of AB is another ellipse, and give the coordinates of the foci of this second ellipse.   [L]

49 Obtain the equation of the chord PQ of the ellipse $x^2/a^2 + y^2/b^2 = 1$, given that the coordinates of its mid-point M are (h, k).
Find the equation satisfied by the coordinates of M
(a) if PQ passes through the point (a, b),
(b) if the perpendicular bisector of PQ passes through the point (a, b).   [L]

50 Show that the coordinates of any point P on a given straight line through the point A (h, k) can be expressed in the form $(h + r \cos \theta, k + r \sin \theta)$, indicating the quantities r and θ on a figure.
Using the above coordinates for P, find, as a quadratic equation in r, the condition that P lies on the ellipse

$$\frac{x^2}{a^2} + \frac{y^2}{b^2} = 1.$$

Deduce, or prove otherwise, that if a line through A with gradient $\tan \theta$ meets the ellipse at two points Q and R such that A is the mid-point of QR, then

$$\tan \theta = -\frac{hb^2}{ka^2},$$

and obtain the equation of the chord of the ellipse which is bisected at the point (h, k).   [C]

51 Show that the two straight lines represented by the equation $ax^2 + 2hxy + by^2 = 0$ contain an angle β such that

$$\tan \beta = \frac{2\sqrt{(h^2 - ab)}}{(a + b)}.$$

(a) Show that, if both lines lie in the first and third quadrants only, and if the angle made by one line with the x-axis is equal to the angle made by the other line with the y-axis, then $a = b$.

(b) Find the condition for the two lines to be at right angles.   [L]

52 Show that the line-pair represented by the equation

$$(x^2 + y^2) \cos 2\alpha + 2xy = 0$$

is always real, and find the angle between the lines.   [L]

53 Show that the equation

$$kx^2 + 2\lambda xy - ky^2 = 0$$

represents two perpendicular straight lines.

Show that the pair of straight lines joining the origin O to the intersections A and B of the line $lx + my = 1$ with the conic $a^2 x^2 + b^2 y^2 = 1$ has the equation

$$(a^2 - l^2)x^2 - 2lmxy + (b^2 - m^2)y^2 = 0.$$

Deduce that if AOB is a right angle then the line AB touches the circle $(a^2 + b^2)(x^2 + y^2) = 1$.   [L]

54 Write down the equation of the tangent at the origin to the curve with equation

$$x^2 + 2xy + 3y^2 + 2x - y = 0.$$

By considering only the second-degree terms in the equation, establish whether the curve has real asymptotes, and hence determine whether it is an ellipse or a hyperbola.   [L]

11 Algebraic structure

11.1 Groups

In *Advanced Mathematics 1* Section 1.7, the following definitions were given.

Binary operation: A binary operation defined on a set S is a rule which assigns to each ordered pair (a, b), where $a, b \in S$, a unique element c.

Closure: A set S is said to be closed under an operation $\circ$ if, for every ordered pair (a, b), where $a, b \in S$, the element $a \circ b \in S$. The operation $\circ$ is then said to be a closed binary operation on S. Throughout this chapter, an operation will be assumed to be a binary operation unless otherwise stated.

Commutativity: An operation $\circ$ defined on a set S is said to be commutative if, for every ordered pair (a, b), where $a, b \in S$, $a \circ b = b \circ a$.

Associativity: A closed operation $\circ$ defined on a set S is said to be associative if, for every $a, b, c \in S$, $(a \circ b) \circ c = a \circ (b \circ c)$.

Identity element: For an operation $\circ$ defined on a set S an identity element e is said to exist if, for every element $a \in S$, $a \circ e = e \circ a = a$, where $e \in S$.

Inverse element: For an operation $\circ$ defined on a set S which has an identity element e, an element $a \in S$ has an inverse element denoted by a^{-1}, if $a \circ a^{-1} = a^{-1} \circ a = e$ and $a^{-1} \in S$.

The set S is said to form a *group* under the binary operation $\circ$ if

(1) the set S is closed under the operation $\circ$,
(2) the operation $\circ$ defined on the set S is associative,
(3) there exists an identity element e for the operation $\circ$ in the set S,
(4) there is a unique inverse element for every element of the set S under the operation $\circ$.

If, in addition to these four properties, the operation $\circ$ defined on the set S is also commutative, then the group is said to be a commutative (or Abelian) group.

The group formed by a set S with a binary operation $\circ$ is denoted by $(S, \circ)$.

Let $S = \{x : x \in \mathbb{R}, x \neq -1\}$ and let the operation $\circ$ be defined on S by $a \circ b = a + b + ab$.

(1) The sum of two real numbers is a real number and the product of two real numbers is also a real number. Thus for all $a, b \in S$, $a + b + ab \in S$, i.e. the set S is closed under the operation $\circ$.

(2)
$$(a \circ b) \circ c = (a + b + ab) \circ c$$
$$= a + b + ab + c + (a + b + ab)c$$
$$= a + b + c + ab + bc + ca + abc$$

and
$$a \circ (b \circ c) = a + (b + c + bc) + a(b + c + bc)$$
$$= a + b + c + ab + bc + ca + abc,$$

i.e. for all $a, b, c \in S, (a \circ b) \circ c = a \circ (b \circ c)$ and so the operation $\circ$ is associative.

(3) If an identity element e exists, then

$$a \circ e = e \circ a = a,$$

i.e. $a + e + ae = e + a + ea = a$,
which is true when $e = 0$ and so 0 is the identity element.

(4) a^{-1} is the inverse of a if

$$a \circ a^{-1} = a^{-1} \circ a = 0 \quad \text{and} \quad a^{-1} \in S.$$

This gives

$$a^{-1}(1 + a) = -a \quad \text{and} \quad a^{-1} = \frac{-a}{1 + a}.$$

Thus a has a unique inverse $\dfrac{-a}{1 + a}$ provided that $a \neq -1$. $\dfrac{-a}{1 + a}$ is real and cannot equal -1.

Therefore, for each element $a \in S$, a unique inverse $\dfrac{-a}{1 + a}$ exists, which is itself a member of S.

Thus the set S forms a group under the operation $\circ$.

Furthermore, $a \circ b = a + b + ab = b + a + ba = b \circ a$ and so the operation $\circ$ is also commutative. Thus the S under the operation $\circ$ forms a commutative (or Abelian) group.

The set of integers $\mathbb{Z}$, where $\mathbb{Z} = \{0, \pm 1, \pm 2, \ldots\}$, forms a group under addition since

(1) the sum of any two integers is an integer, i.e. $\mathbb{Z}$ is closed under $+$,
(2) the operation of addition on $\mathbb{Z}$ is known to be associative,
(3) there is in $\mathbb{Z}$ an identity element 0,
(4) the inverse of each element n of $\mathbb{Z}$ is $(-n)$ since $n + (-n) = 0$.

Also, addition is commutative and so $(\mathbb{Z}, +)$ is a commutative group.

Since $\mathbb{Z}$ is an infinite set, the group $(\mathbb{Z}, +)$ is said to have *infinite order*.

Consider the sets S_0, S_1, S_2 and S_3, where

$S_0 = \{0, \pm 4, \pm 8, \pm 12, \ldots\}$, or $\{n: n = 4k, k \in \mathbb{Z}\}$
$S_1 = \{\ldots, -11, -7, -3, 1, 5, 9, 13, \ldots\}$, or $\{n: n = 4k + 1, k \in \mathbb{Z}\}$
$S_2 = \{\ldots, -10, -6, -2, 2, 6, 10, 14, \ldots\}$, or $\{n: n = 4k + 2, k \in \mathbb{Z}\}$
$S_3 = \{\ldots, -9, -5, -1, 3, 7, 11, 15, \ldots\}$, or $\{n: n = 4k + 3, k \in \mathbb{Z}\}$

S_0 is the set of integers which, when divided by 4, give a remainder or residue of 0.

S_1 is the set of integers which, when divided by 4, give a residue of 1.

S_2 is the set of integers which, when divided by 4, give a residue of 2.

S_3 is the set of integers which, when divided by 4, give a residue of 3.

These four subsets of $\mathbb{Z}$, the set of integers, are known as the *residue classes, modulo* 4 (often abbreviated to *mod* 4).

We shall denote S_0, S_1, S_2, S_3, the residue classes, mod 4, by **0, 1, 2, 3** and the residue classes, mod n, by **0, 1, 2, 3, ..., $n - 1$**.

If any member of one element of the set S, where $S = \{0, 1, 2, ..., n - 1\}$, is added to any member of another element (or the same element) of S, the sum will belong to one element of S. This latter element is defined as the sum, mod n, of the two original elements of S, e.g. the sum, mod 4, of **2** and **3** of the set $\{0, 1, 2, 3\}$ is **1** since the sum of any member of the set $\{..., -10, -6, -2, 2, 6, 10 ...\}$ and any member of the set $\{..., -9, -5, -1, 3, 7, 11, ...\}$ belongs to the set $\{..., -11, -7, -3, 1, 5, 9 ...\}$. Thus we have a definition of $+$, mod n, on the set $\{0, 1, 2, ..., n - 1\}$.

In a similar way the operation of multiplication, modulo n ($\times$, mod n) can be defined on the set $\{0, 1, 2, ..., n - 1\}$.

Example

Investigate whether or not $(S, \circ)$ forms a group when

(a) $S = \{0, 1, 2, 3\}$ and $\circ$ is $+$, mod 4,

(b) $S = \{1, 2, 3\}$ and $\circ$ is $\times$, mod 4.

If a group is formed, state the identity element and the inverse element of each element of S.

If a group is not formed, state which of the group-defining properties are not satisfied.

(a) The operation (or combination) table is

+	0	1	2	3
0	0	1	2	3
1	1	2	3	0
2	2	3	0	1
3	3	0	1	2

Each sum shown in the table belongs to S and so clearly the set S is closed under the operation.

Consider any element, $4k_1 + 1$, of the residue class **1**. Consider also any element, $4k_2 + 2$, of the residue class **2** and any element, $4k_3 + 3$, of the residue class **3**.

Then $\qquad [(4k_1 + 1) + (4k_2 + 2)] + (4k_3 + 3)$
and $\qquad (4k_1 + 1) + [(4k_2 + 2) + (4k_3 + 3)]$
both equal $\qquad (4k_1 + 1) + (4k_2 + 2) + (4k_3 + 3),$

i.e. $4(k_1 + k_2 + k_3) + 1 + 2 + 3$ or $4(k_1 + k_2 + k_3 + 1) + 2$ which is of the form $4k + 2$ and so

$$(1 + 2) + 3 = 1 + (2 + 3) = 2.$$

Similarly, we can treat any three elements of the set S and thus show that the operation of addition, mod 4, is associative on the set S, where $S = \{0, 1, 2, 3\}$.

There is an identity element, **0**. Each element has an inverse as follows:

$$0^{-1} = 0, \quad 1^{-1} = 3, \quad 2^{-1} = 2, \quad 3^{-1} = 1.$$

Thus, in this case, $(S, \circ)$ forms a group. Since the set S has 4 members, the order of the group is 4.

(b) The table for the operation of multiplication, mod 4, on S is

$\times$	1	2	3
1	**1**	**2**	**3**
2	**2**	**0**	**2**
3	**3**	**2**	**1**

The set S is not closed under the operation since $2 \times 2 = 0$ and **0** is not an element of S. There is an identity element, **1**, but **2** does not have an inverse element because $2 \times 1 \neq 1$, $2 \times 2 \neq 1$ and $2 \times 3 \neq 1$.

Note that properties of a set S under an operation $\circ$ can be investigated by the use of an operation table such as those shown in the example above.
(1) Closure exists if each element in the operation table belongs to the set S.
(2) The operation $\circ$ is commutative if the operation table is symmetrical about the diagonal which goes from the top left corner to the bottom right corner, since this shows that $a \circ b = b \circ a$.
(3) If an identity element exists, the identity-element row will be identical to the S row of the table and the identity-element column will be identical to the S column of the table as shown in the following table:

$\circ$	1	2	3	$\leftarrow S$ row
1	**1**	**2**	**3**	$\leftarrow$ identity-element row
2	**2**	**0**	**2**	
3	**3**	**2**	**1**	

$\quad\quad\quad\;\; \uparrow \quad\; \uparrow$
$\quad\quad S$ column $\quad$ identity-element column

(4) Each element of the set S will have a **unique** inverse if the identity element appears once and only once in each row and column of the combination table.

If $(S, \circ)$ forms a group of order n, then each of the n elements of S will appear once and only once in each of the n rows and the n columns of the combination table.

It should be noted that the property of associativity is not easily established from consideration of an operation table.

Exercises 11.1

In each of the following cases investigate whether or not $(S, \circ)$ is a group. If it is, state the identity element and the inverse elements of all the elements of S. If it is not a group, state which of the group-defining properties $(S, \circ)$ lacks. In each case state whether or not the operation $\circ$ is commutative. (In **5**, **6** and **7** associativity may be assumed.)

1 $S = \{0, 1, 2\}$ and $\circ$ is $+$, mod 3.

2 $S = \{0, 1, 2\}$ and $\circ$ is $\times$, mod 3.

3 $S = \{1, 2\}$ and $\circ$ is $\times$, mod 3.

4 $S = \{1, 3, 5, 7\}$ and $\circ$ is $\times$, mod 8.

5 $S = \{a, b, c, d\}$ and $\circ$ is defined by

$\circ$	a	b	c	d
a	a	b	c	d
b	b	a	d	c
c	c	d	a	b
d	d	c	b	a

6 $S = \{p, q, r\}$ and $\circ$ is defined by

$\circ$	p	q	r
p	p	q	r
q	q	r	p
r	r	p	q

7 $S = \{1, 2, 3, 4\}$ and $\circ$ is defined by

$\circ$	1	2	3	4
1	1	2	3	4
2	2	1	4	3
3	3	4	1	2
4	4	3	2	1

11.2 Symmetry groups

A transformation of a plane or space which preserves distance, i.e. where the distance between any two points in the plane or space is unchanged by the transformation, is an *isometry*. An isometry of a plane or space which leaves a plane figure, or a three-dimensional figure, unchanged is a *symmetry* of the figure. For instance, the reflection of a rhombus in a diagonal is a symmetry of the rhombus. It can be shown that the symmetries of each figure form a group under the operation of the composition of transformations.

As an example, consider the symmetries of the rhombus ABCD. These are
I – the identity transformation which maps every point to itself,
H – the transformation which maps every point to its reflection in the diagonal BD,
V – the transformation which maps every point to its reflection in the diagonal AC,
R – the rotation of every point through π radians about the centre of ABCD.

The compilation of the operation table for this set of symmetries under the operation of the composition of transformations is illustrated in the following examples.

H followed by *V* is shown in Fig. 11.1. Thus, *H* followed by *V* is equivalent to the transformation *R*.

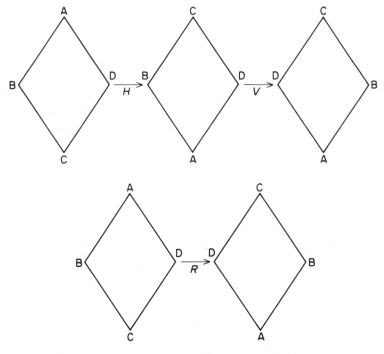

Fig. 11.1

Similarly, V followed by R is shown in Fig. 11.2. Thus V followed by R is equivalent to H. The complete group table is

	I	H	V	R
I	I	H	V	R
H	H	I	R	V
V	V	$\textcircled{R}$	I	H
R	R	V	$\textcircled{H}$	I

The transformations I, H, V, R are listed in the row above the table and also in the column on the left of the table.

The transformation R equivalent to VH (H followed by V) is entered in the row level with V and in the column vertically below H, and is shown ringed.

The transformation H equivalent to RV (V followed by R) is entered in the row level with R and in the column vertically below V. This is also shown ringed.

Note that we use no explicit operation symbol for composition of transformations. We relay on 'ordered juxtaposition', following the notation of composition of functions. Thus H followed by V (as illustrated in Fig. 11.1) is denoted by VH and V followed by R (as illustrated in Fig. 11.2) is denoted by RV.

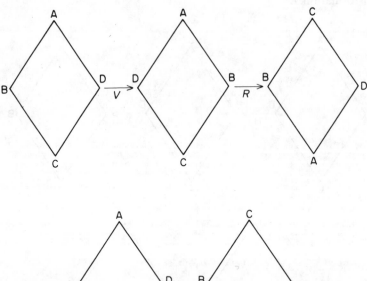

Fig. 11.2

It can be seen that the set $\{I, H, V, R\}$ is closed under the operation of composition of transformations. Since *all* compositions of operations are associative the sequence of operations A followed by B followed by C, i.e. CBA, is clearly independent of whether it is regarded as the sequence C(BA) or the sequence (CB)A. There is an identity element, I, and each element is its own inverse. From the fact that the table is symmetrical about the dotted line, we can see that the operation is commutative. Thus the symmetries of the rhombus form a commutative group.

Exercises 11.2

1 Complete the group tables for the symmetries of the following figures. In each case give the inverse of each symmetry and state whether or not the group is an Abelian (or commutative) group.
   (a) The parallelogram.   (b) The rectangle.
   (c) The isosceles triangle.   (d) The letter H.
   (e) The letter X.

2 Complete the group tables for
   (a) the rotational symmetries of the equilateral triangle,
   (b) the rotational symmetries of the regular pentagon,
   (c) the rotational symmetries of the regular hexagon.

11.3 Subgroups and cyclic groups

The symmetries of the equilateral triangle ABC (see Fig. 11.3) are
I – the identity transformation which maps every point to itself,
R_1 – the anticlockwise rotation through $2\pi/3$ radians about O,
R_2 – the anticlockwise rotation through $4\pi/3$ radians about O,

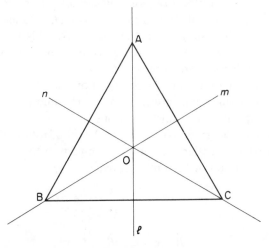

Fig. 11.3

M_1 – the transformation which maps every point to its reflection in l,
M_2 – the transformation which maps every point to is reflection in m,
M_3 – the transformation which maps every point to its reflection in n,

The group table is:

(The axes of symmetry l, m, n are fixed and do not move with the triangle.)

	I	R_1	R_2	M_1	M_2	M_3
I	I	R_1	R_2	M_1	M_2	M_3
R_1	R_1	R_2	I	M_3	M_1	M_2
R_2	R_2	I	R_1	M_2	M_3	M_1
M_1	M_1	M_2	M_3	I	R_1	R_2
M_2	M_2	M_3	M_1	R_2	I	R_1
M_3	M_3	M_1	M_2	R_1	R_2	I

This set of symmetries forms a group with R_1 being the inverse of R_2 and with M_1, M_2 and M_3 being their own inverses.

It will be seen from the table that the operation is not commutative on this set of symmetries. For example, the transformation R_1 followed by the transformation M_1 gives the transformation M_2, whereas the transformation M_1 followed by the transformation R_1 gives the transformation M_3. Thus this group is not an Abelian group. From the table it will also be seen that the set $\{I, R_1, R_2\}$ forms a group under the composition of transformations. This group is known as a *subgroup* of the original group $\{I, R_1, R_2, M_1, M_2, M_3\}$ under the operation of the composition of transformations.

In general, if $(S, \circ)$ is a group and (1) $S_1 \subset S$ and (2) $(S_1, \circ)$ is a group, then $(S_1, \circ)$ is a subgroup of the group $(S, \circ)$.

Two other subgroups of the group of S, where $S = \{I, R_1, R_2, M_1, M_2, M_3\}$, under $\circ$, the composition of transformations, are $(I, \circ)$ and $(S, \circ)$. However these are known as trivial subgroups. A subgroup $(S_1, \circ)$ is a trivial subgroup of $(S, \circ)$ when $S_1 = S$ or when $S_1 = I$, where I is the identity element of the group $(S, \circ)$. All other subgroups are known as proper subgroups. The set $\{I, R_1, R_2\}$ forms a proper subgroup under composition of transformations. Other examples of proper subgroups are $\{I, M_1\}$, $\{I, M_2\}$ and $\{I, M_3\}$, all under the operation of composition of transformations.

Next we consider the group consisting of the set of rotational symmetries of the equilateral triangle under the composition of transformations. The group table, using the notation of the previous page, is

	I	R_1	R_2
I	I	R_1	R_2
R_1	R_1	R_2	I
R_2	R_2	I	R_1

If we denote R_1 by r, then R_2 is rr and this is often denoted by r^2. (Note that the index 2 indicates that the rotation r has been performed twice just as the index 2 in $\dfrac{d^2 y}{dx^2}$ indicates that the differentiation is performed twice.) The group table then becomes

	I	r	r^2
I	I	r	r^2
r	r	r^2	I
r^2	r^2	I	r

and $I = r^3$. Thus all the elements of the group are 'powers' of r. This group is known as a cyclic group of order 3 and r is called the generator of the group. A cyclic group of order n is one in which the elements are the identity element e and $x, x^2, x^3, \ldots, x^{n-1}$, with $x^n = e$. A generator of the group is x.

From the group table of the symmetries of the equilateral triangle we see that

$$R_1 R_1 = R_1^2 = R_2, \qquad R_1 R_1 R_1 = R_1^3 = R_1^2 R_1 = R_2 R_1 = I$$

and also that

$$R_1^4 = R_1, \qquad R_1^5 = R_2, \qquad R_1^6 = I \quad \text{etc.}$$

Thus the 'powers' of the element R_1 produce only three values, I, R_1 and R_2. The *order* of the element R_1 is said to be 3. Similarly, the order of the element R_2 is also 3.

Exercises 11.3

1 List the proper subgroups of the symmetries of
  (a) the rhombus,   (b) the rectangle,
  (c) the letter H,   (d) the letter X,
  (e) the equilateral triangle,   (f) the square.

2 State which of the subgroups of **1** are cyclic groups and give their generators.

3 State the order of the elements of the cyclic subgroups of **1**.

11.4 Permutation groups

There are six permutations of the letters a, b, c, i.e. the three letters can be arranged in six different orders. Thus there are six permutations of the elements of the set S, where $S = \{a, b, c\}$. Each permutation is a mapping of S onto itself. These permutations can be expressed as follows:

$$\begin{pmatrix} a & b & c \\ a & b & c \end{pmatrix}, \quad \begin{pmatrix} a & b & c \\ c & a & b \end{pmatrix}, \quad \begin{pmatrix} a & b & c \\ b & c & a \end{pmatrix}, \text{ etc.}$$

In the first of these each element of S is mapped to itself. In the second a is mapped to c, b is mapped to a and c is mapped to b. In the third a is mapped to b, b is mapped to c and c is mapped to a. The six permutations, $P_1, P_2, P_3, P_4, P_5, P_6$, are

$$P_1\begin{pmatrix} a & b & c \\ a & b & c \end{pmatrix}, \quad P_2\begin{pmatrix} a & b & c \\ c & a & b \end{pmatrix}, \quad P_3\begin{pmatrix} a & b & c \\ b & c & a \end{pmatrix},$$

$$P_4\begin{pmatrix} a & b & c \\ c & b & a \end{pmatrix}, \quad P_5\begin{pmatrix} a & b & c \\ a & c & b \end{pmatrix}, \quad P_6\begin{pmatrix} a & b & c \\ b & a & c \end{pmatrix}.$$

The composition (or 'product') of the two permutations P_3 and P_4, denoted by $P_4 P_3$, is the permutation given by replacing the letters a, b, c by b, c, a respectively (as given by P_3) and then replacing the letters b, c, a by b, a, c respectively (as given by P_4). Thus

$$a \xrightarrow{P_3} b \xrightarrow{P_4} b$$

$$b \xrightarrow{P_3} c \xrightarrow{P_4} a$$

$$c \xrightarrow{P_3} a \xrightarrow{P_4} c$$

Thus a, b, c has been changed to b, a, c i.e. $P_4 P_3 = P_6$. Other 'products' of permutations are similarly obtained. The complete table for the group $(S, *)$, where $S = \{P_1, P_2. P_3, P_4, P_5, P_6\}$, and $*$ is the operation of forming a 'product' of permutations as just described, is

$*$	P_1	P_2	P_3	P_4	P_5	P_6
P_1	P_1	P_2	P_3	P_4	P_5	P_6
P_2	P_2	P_3	P_1	P_6	P_4	P_5
P_3	P_3	P_1	P_2	P_5	P_6	P_4
P_4	P_4	P_5	P_6	P_1	P_2	P_3
P_5	P_5	P_6	P_4	P_3	P_1	P_2
P_6	P_6	P_4	P_5	P_2	P_3	P_1

Note that the above table justifies the term 'permutation group' since the table indicates closure, the existence of the identity element P_1 and the existence of an inverse for each element. Associativity exists since the operation is a combination of transformations.

Exercise 11.4

1 Compile the group table for the permutations of the numbers 1 and 2.
2 Show that the permutation group of two letters is commutative but that the permutation group of three letters is not.

11.5 Isomorphism

Group table for the symmetries of the equilateral triangle

	I	R_1	R_2	M_1	M_2	M_3
I	I	R_1	R_2	M_1	M_2	M_3
R_1	R_1	R_2	I	M_3	M_1	M_2
R_2	R_2	I	R_1	M_2	M_3	M_1
M_1	M_1	M_2	M_3	I	R_1	R_2
M_2	M_2	M_3	M_1	R_2	I	R_1
M_3	M_3	M_1	M_2	R_1	R_2	I

Group table for the permutations of the letters a, b, c

	P_1	P_2	P_3	P_4	P_5	P_6
P_1	P_1	P_2	P_3	P_4	P_5	P_6
P_2	P_2	P_3	P_1	P_6	P_4	P_5
P_3	P_3	P_1	P_2	P_5	P_6	P_4
P_4	P_4	P_5	P_6	P_1	P_2	P_3
P_5	P_5	P_6	P_4	P_3	P_1	P_2
P_6	P_6	P_4	P_5	P_2	P_3	P_1

Comparing the above two group tables we see that not only do they have the same number of elements but they have the same pattern. If we

replace $\qquad\qquad I,\quad R_1,\quad R_2,\quad M_1,\quad M_2,\quad M_3$

by $\qquad\qquad\quad P_1,\quad P_2,\quad P_3,\quad P_4,\quad P_5,\quad P_6$

respectively, the group table for the symmetries of the equilateral triangle becomes identical to the group table for the permutations of the letters a, b, c. The two groups are said to be *isomorphic*. In general, two groups $(S, \circ)$ and $(T, *)$ are said to be isomorphic when there is a one–one mapping $f: S \to T$ such that, for all $s_1, s_2 \in S$, we have $f(s_1 \circ s_2) = f(s_1) * f(s_2)$.

In the case just considered

$$f(I) = P_1, \quad f(R_1) = P_2, \quad f(R_2) = P_3, \quad f(M_1) = P_4,$$
$$f(M_2) = P_5, \quad f(M_3) = P_6.$$

An example from the above table gives $M_3 R_2 = M_2$, i.e. R_2 followed by M_3 gives M_2.

$$f(M_3) = P_6 \qquad \text{and} \qquad f(R_2) = P_3,$$
$$f(M_3)f(R_2) = P_6 P_3 = P_5 \qquad \text{and} \qquad f(M_3 R_2) = f(M_2) = P_5,$$

i.e. $\qquad\qquad\qquad f(M_3 R_2) = f(M_3)f(R_2).$

Example
Is the group of symmetries of the rhombus isomorphic to the group with the following group table?

$\circ$	a	b	c	d
a	a	b	c	d
b	b	a	d	c
c	c	d	a	b
d	d	c	b	a

In Section 11.2 we obtained the group table of the symmetries of the rhombus as:

	I	H	V	R
I	I	H	V	R
H	H	I	R	V
V	V	R	I	H
R	R	V	H	I

Putting $a = I, b = H, c = V, d = R$, the given table becomes identical to the group table for the symmetries of the rhombus. Thus the group whose group table was given is isomorphic to the group of symmetries of the rhombus.

Exercises 11.5

1 Show that the groups $(S, \circ)$ and $(T, *)$ are isomorphic when

$$S = \{1, 2, 3, 4\} \quad \text{and} \quad \circ = \times, \text{mod } 5,$$
$$T = \{0, 1, 2, 3\} \quad \text{and} \quad * = +, \text{mod } 4.$$

2 Find which of the following groups are isomorphic:
(a) the rotational symmetries of the square,
(b) the symmetries of the rhombus,
(c) the symmetries of the rectangle,
(d) the symmetries of the letter H,
(e) the symmetries of the letter X,
(f) $\{1, 3, 5, 7\}$ under $\times$, mod 8,
(g) $\{2, 4, 6, 8\}$ under $\times$, mod 10.

11.6 Further examples of groups

Groups appear in many branches of mathematics. We have considered some groups in modular arithmetic, symmetry groups and permutation groups. Here are some further examples of groups.

Example 1
Show that the set S, where $S = \{1, i, -1, -i\}$ forms a group under multiplication of complex numbers.

The combination table for S under multiplication is

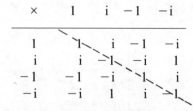

This table shows that $(S, \times)$ has closure and that the identity element is 1. The operation of multiplication is associative. The inverse elements of i, -1 and $-i$

are $-i$, -1 and i respectively. Multiplication is commutative and this can be seen from the symmetry of the above table about the dotted line. Thus $(S, \times)$ is a commutative group. The group is also cyclic with i as generator. The group has order 4 and the orders of the elements 1, i, -1, $-i$ are 1, 4, 2, 4 respectively.

Example 2
Show that the set S, where $S = \{A, B, C, D\}$ and

$$A = \begin{pmatrix} 1 & 0 \\ 0 & 1 \end{pmatrix}, \quad B = \begin{pmatrix} -1 & 0 \\ 0 & 1 \end{pmatrix}, \quad C = \begin{pmatrix} 1 & 0 \\ 0 & -1 \end{pmatrix}, \quad D = \begin{pmatrix} -1 & 0 \\ 0 & -1 \end{pmatrix},$$

forms a group under matrix multiplication.

The multiplication table for the set of matrices is

$\times$	A	B	C	D
A	A	B	C	D
B	B	A	D	C
C	C	D	A	B
D	D	C	B	A

The table shows that the set S is closed under multiplication and that **A** is the identity element. It also shows that each element is its own inverse. Matrix multiplication is associative. Thus the set S forms a group under matrix multiplication. Although, in general, matrix multiplication is not commutative, the fact that the table above is symmetrical about the dotted line shows that matrix multiplication is commutative in this particular set S. Thus the set S forms a commutative group under matrix multiplication. The group has order 4 and the orders of the elements **A**, **B**, **C**, **D** are 1, 2, 2, 2 respectively.

Example 3
The functions f_1, f_2, f_3, f_4, each with domain $\mathbb{R}$ excluding 0, are defined by

$$f_1 : x \mapsto x, \quad f_2 : x \mapsto -x, \quad f_3 : x \mapsto \frac{1}{x}, \quad f_4 : x \mapsto -\frac{1}{x}.$$

Show that the set $\{f_1, f_2, f_3, f_4\}$ forms a group under the composition of functions.

The combination table is

	f_1	f_2	f_3	f_4
f_1	f_1	f_2	f_3	f_4
f_2	f_2	f_1	f_4	f_3
f_3	f_3	f_4	f_1	f_2
f_4	f_4	f_3	f_2	f_1

The table shows that the set is closed under the composition of functions and that f_1 is the identity element. The composition of functions is clearly an associative operation. Each element is its own inverse. Thus the set forms a group under the composition of functions. The symmetry of the table about the dotted line shows that the operation is commutative under the given operation so that the set $\{f_1, f_2, f_3, f_4\}$ forms a commutative group under the composition of functions. The group is of order 4 and the element f_1 has order 1 and the elements f_2, f_3, f_4 each have order 2.

Notice that the group in this example and the group in Example 2 are isomorphic, f_1, f_2, f_3, f_4 corresponding to **A, B, C, D**.

Nearly all the groups, $(S, \circ)$, we have considered so far have been finite, i.e. S has contained a finite number of elements. There are many examples of infinite groups, i.e. when an operation is defined on an infinite set. For instance $\mathbb{Z}$, the set of integers, forms a group under addition. The sum of any two integers is an integer and so the set $\mathbb{Z}$ is closed under addition. The identity element is 0 and the inverse of any element n is $(-n)$. Addition is both associative and commutative. Thus $\mathbb{Z}$ forms a commutative group under addition. The set $\mathbb{Q}$, the set of rational numbers, $\mathbb{R}$, the set of real numbers, and $\mathbb{C}$, the set of complex numbers, all form groups under addition. The sets $\mathbb{Q}$, $\mathbb{R}$ and $\mathbb{C}$, all with 0 excluded, also form groups under multiplication.

Sets of vectors can also form groups. For example, the infinite set of vectors in a plane forms a group under vector addition.

Example 4
Show that the set of complex numbers with modulus 1 forms a group under multiplication.

Any element of this set of complex numbers can be expressed in the form $\cos \theta + i \sin \theta$. Consider any two elements $\cos \theta_1 + i \sin \theta_1$ and $\cos \theta_2 + i \sin \theta_2$. Then

$$(\cos \theta_1 + i \sin \theta_1)(\cos \theta_2 + i \sin \theta_2) = \cos(\theta_1 + \theta_2) + i \sin(\theta_1 + \theta_2)$$

which is also a member of the given set of complex numbers with modulus 1, i.e. the given set is closed under multiplication. Multiplication is both associative and commutative. The identity element is $\cos 0 + i \sin 0$, i.e. 1.

If $\cos \phi + i \sin \phi$ is the inverse of $\cos \theta + i \sin \theta$, then

$$(\cos \theta + i \sin \theta)(\cos \phi + i \sin \phi) = \cos 0 + i \sin 0$$

so that $\theta + \phi = 0$, i.e. $\phi = -\theta$.

Therefore the inverse of $\cos \theta + i \sin \theta$ is $\cos(-\theta) + i \sin(-\theta)$. Thus each element of the given set of complex numbers with modulus 1 has a unique inverse which is an element of the set. The set of complex numbers with modulus 1 therefore forms a commutative group under multiplication.

Example 5

Show that the infinite set of matrices of the form $\begin{pmatrix} x & 0 \\ 0 & 1 \end{pmatrix}$, where $x \in \mathbb{Q}$ excluding 0, forms a group under matrix multiplication.

Consider any two elements of the given set of matrices, with x taking the values x_1 and x_2.

$$\begin{pmatrix} x_1 & 0 \\ 0 & 1 \end{pmatrix}\begin{pmatrix} x_2 & 0 \\ 0 & 1 \end{pmatrix} = \begin{pmatrix} x_1 x_2 & 0 \\ 0 & 1 \end{pmatrix}$$

which is also of the form $\begin{pmatrix} x & 0 \\ 0 & 1 \end{pmatrix}$ and so the set is closed under multiplication.

Matrix multiplication is associative and the identity element is $\begin{pmatrix} 1 & 0 \\ 0 & 1 \end{pmatrix}$. If the inverse of any element $\begin{pmatrix} x & 0 \\ 0 & 1 \end{pmatrix}$ is $\begin{pmatrix} y & 0 \\ 0 & 1 \end{pmatrix}$, then

$$\begin{pmatrix} x & 0 \\ 0 & 1 \end{pmatrix}\begin{pmatrix} y & 0 \\ 0 & 1 \end{pmatrix} = \begin{pmatrix} y & 0 \\ 0 & 1 \end{pmatrix}\begin{pmatrix} x & 0 \\ 0 & 1 \end{pmatrix} = \begin{pmatrix} 1 & 0 \\ 0 & 1 \end{pmatrix}$$

i.e. $xy = 1$ and $y = 1/x$, since x is not zero.

$1/x$ belongs to the set $\mathbb{Q}$ excluding 0 and so each element has a unique inverse in the set.

Although matrix multiplication is not generally commutative, the operation is commutative on this particular set since

$$\begin{pmatrix} x & 0 \\ 0 & 1 \end{pmatrix}\begin{pmatrix} y & 0 \\ 0 & 1 \end{pmatrix} = \begin{pmatrix} y & 0 \\ 0 & 1 \end{pmatrix}\begin{pmatrix} x & 0 \\ 0 & 1 \end{pmatrix} = \begin{pmatrix} xy & 0 \\ 0 & 1 \end{pmatrix}.$$

Thus the given set, $\mathbb{Q}$ excluding 0, forms a commutative group under matrix multiplication.

Exercises 11.6

1 Show that the three cube roots of 1 form a group under multiplication. Find whether:
   (a) the set of the nth roots of 1, where $n \in \mathbb{Z}^+$, under multiplication, forms a group,
   (b) the set of the nth roots of a, where $n \in \mathbb{Z}^+$ and $a \in \mathbb{Z}^+$ excluding 1, under multiplication, forms a group.

2 Investigate whether the following form groups:
   (a) the set of complex numbers $x + iy$, where $x, y \in \mathbb{R}$, under multiplication,
   (b) the set of complex numbers $x + iy$, where $x, y \in \mathbb{R}$ excluding 0, under multiplication.

3 Show that the following sets form groups under matrix multiplication:

(a) $\begin{pmatrix} 1 & 0 \\ 0 & 1 \end{pmatrix}$, $\begin{pmatrix} 0 & 1 \\ -1 & 0 \end{pmatrix}$, $\begin{pmatrix} -1 & 0 \\ 0 & -1 \end{pmatrix}$, $\begin{pmatrix} 0 & -1 \\ 1 & 0 \end{pmatrix}$,

(b) $\begin{pmatrix} 1 & 0 \\ 0 & 1 \end{pmatrix}$, $\begin{pmatrix} -1 & 0 \\ 0 & 1 \end{pmatrix}$, $\begin{pmatrix} 1 & 0 \\ 0 & -1 \end{pmatrix}$, $\begin{pmatrix} -1 & 0 \\ 0 & -1 \end{pmatrix}$,

(c) $\begin{pmatrix} 1 & 0 \\ 0 & 1 \end{pmatrix}$, $\begin{pmatrix} -1 & 1 \\ -1 & 0 \end{pmatrix}$, $\begin{pmatrix} 0 & -1 \\ 1 & -1 \end{pmatrix}$,

$\begin{pmatrix} -1 & 0 \\ 0 & -1 \end{pmatrix}$, $\begin{pmatrix} 1 & -1 \\ 1 & 0 \end{pmatrix}$, $\begin{pmatrix} 0 & 1 \\ -1 & 1 \end{pmatrix}$,

(d) $\begin{pmatrix} 1 & 0 \\ 0 & 1 \end{pmatrix}$, $\begin{pmatrix} 0 & 1 \\ -1 & 0 \end{pmatrix}$, $\begin{pmatrix} -1 & 0 \\ 0 & -1 \end{pmatrix}$, $\begin{pmatrix} i & 0 \\ 0 & -i \end{pmatrix}$,

$\begin{pmatrix} -i & 0 \\ 0 & i \end{pmatrix}$, $\begin{pmatrix} 0 & i \\ i & 0 \end{pmatrix}$, $\begin{pmatrix} 0 & -1 \\ 1 & 0 \end{pmatrix}$, $\begin{pmatrix} 0 & -i \\ -i & 0 \end{pmatrix}$.

4 Investigate whether matrices of the form $\begin{pmatrix} z & w \\ w* & z* \end{pmatrix}$, where $z, w \in \mathbb{C}$, form a group under matrix addition.

5 The functions f_1, f_2, f_3, each having domain $\mathbb{R}$ excluding 0 and 1, are defined by

$$f_1 : x \mapsto x, \qquad f_2 : x \mapsto \frac{1}{1-x}, \qquad f_3 : x \mapsto \frac{x-1}{x}.$$

Show that the set $\{f_1, f_2, f_3\}$ forms a group under the composition of functions.

6 The functions $f_1, f_2, f_3, f_4, f_5, f_6$, each having domain $\mathbb{R}$ excluding 0 and 1, are defined by

$$f_1(x) = x, \qquad f_2(x) = \frac{1}{1-x}, \qquad f_3(x) = \frac{x-1}{x},$$

$$f_4(x) = \frac{1}{x}, \qquad f_5(x) = \frac{x}{x-1}, \qquad f_6(x) = 1 - x.$$

Show that the set $\{f_1, f_2, f_3, f_4, f_5, f_6\}$ forms a group under the composition of functions. Show also that the group is isomorphic to the group of the permutations of the letters a, b, c.

11.7 Rings and fields

In the preceding sections of this chapter we have considered a number of cases of a set and the effect of one particular operation on the elements of that set. All readers will be familiar from their knowledge of arithmetic with the situation of one set under two operations, e.g. the set of real numbers under addition and multiplication.

Consider the finite set S, where $S = \{0, 1, 2, 3, 4, 5\}$ under the two operations of $+$, mod 6, and $\times$, mod 6. The combination tables for these two operations on the set S are:

+	0	1	2	3	4	5
0	0	1	2	3	4	5
1	1	2	3	4	5	0
2	2	3	4	5	0	1
3	3	4	5	0	1	2
4	4	5	0	1	2	3
5	5	0	1	2	3	4

×	0	1	2	3	4	5
0	0	0	0	0	0	0
1	0	1	2	3	4	5
2	0	2	4	0	2	4
3	0	3	0	3	0	3
4	0	4	2	0	4	2
5	0	5	4	3	2	1

$(S, +, \text{mod } 6)$ is a commutative group.

$(S, \times, \text{mod } 6)$ is not a group but it has two group properties; the set S is closed under $\times$, mod 6, and the operation $\times$, mod 6, is associative on the set S.

Furthermore, for any three elements, a, b, c, of the set S, it can be shown that
$$a \times (b + c) = a \times b + a \times c$$
and
$$(b + c) \times a = b \times a + c \times a,$$

e.g. $4 \times (3 + 5) = 4 \times 2 = 2$ and $(4 \times 3) + (4 \times 5) = 0 + 2 = 2$;
also $(3 + 5) \times 4 = 2 \times 4 = 2$ and $(3 \times 4) + (5 \times 4) = 0 + 2 = 2$.

When the equations
$$a \times (b + c) = a \times b + a \times c$$
and
$$(b + c) \times a = b \times a + c \times a$$

are satisfied, the operation $\times$ is said to be distributive over the operation $+$.

Note that, in this case, the operation $+$ is not distributive over the operation $\times$ since

$$a + (b \times c) \neq (a + b) \times (a + c).$$
For example $\qquad\qquad 4 + (3 \times 5) = 4 + 3 = 1$
but $\qquad\qquad (4 + 3) \times (4 + 5) = 1 \times 3 = 3.$

The set S under the operations $+$, mod 6, and $\times$, mod 6, has the following properties:

(1) $(S, +, \text{mod } 6)$ is a commutative group.
(2) $(S, \times, \text{mod } 6)$ has closure.
(3) $(S, \times, \text{mod } 6)$ has associativity.
(4) The operation $\times$, mod 6, is distributive over the operation $+$, mod 6.

The set S is then said to form a *ring* under the operations $+$, mod 6, and $\times$, mod 6. This ring may be denoted by $(S, +, \text{mod } 6, \times, \text{mod } 6)$. By examining the above table for the set S under the operation $\times$, mod 6, we see that, even if we exclude 0 from S and consider S_1, where $S_1 = \{1, 2, 3, 4, 5\}$, S_1 is not a group under $\times$, mod 6.

However, let us now examine the set T, where $T = \{0, 1, 2, 3, 4\}$, under the operation $+$, mod 5, and $\times$, mod 5. The combination tables are:

+	0	1	2	3	4
0	0	1	2	3	4
1	1	2	3	4	0
2	2	3	4	0	1
3	3	4	0	1	2
4	4	0	1	2	3

×	0	1	2	3	4
0	0	0	0	0	0
1	0	1	2	3	4
2	0	2	4	1	3
3	0	3	1	4	2
4	0	4	3	2	1

$(T, +, \text{mod } 5, \times, \text{mod } 5)$ has the properties of a ring.

In this case the set T excluding 0 forms a commutative group under the operation $\times$, mod 5. The set T with the two operations $+$, mod 5, and $\times$, mod 5, is said to form a *field*. $(T, +, \text{mod } 5, \times, \text{mod } 5)$ is a field.

Formal definitions of a ring and a field may be expressed as follows:

Ring   A set S with two operations $\circ$ and $*$ defined on it forms a ring if
(1) $(S, \circ)$ is a commutative group,
(2) for all $a, b \in S, a * b \in S$,
(3) for all $a, b, c \in S, (a * b) * c = a * (b * c)$,
(4) for all $a, b, c \in S, a * (b \circ c) = (a * b) \circ (a * c)$
and $\qquad\qquad (b \circ c) * a = (b * a) \circ (c * a)$.

Field   A set S with two operations $\circ$ and $*$ defined on it forms a field if
(1) $(S, \circ)$ is a commutative group,
(2) the set consisting of S less the identity element of the group $(S, \circ)$ forms a commutative group under the operation $*$,
(3) for all $a, b, c \in S, a * (b \circ c) = (a * b) \circ (a * c)$.

Exercises 11.7
Which of the following form (a) a ring, (b) a field?

1   $S = \{0, 1, 2\}$ with $+$, mod 3 and $\times$, mod 3.

2   $S = \{0, 1, 2, 3\}$ with $+$, mod 4, and $\times$, mod 4.

3   $S = \{0, 1, 2, 3, 4, 5, 6\}$ with $+$, mod 7, and $\times$, mod 7.

4   $S = \{0, 1, 2, 3, 4, 5, 6, 7\}$ with $+$, mod 8, and $\times$, mod 8.

5   $S = \{0, 1, 2, 3, 4, 5, 6, 7, 8\}$ with $+$, mod 9, and $\times$, mod 9.

6   $\mathbb{Z}$ with addition and multiplication.

7   $\mathbb{R}$ with addition and multiplication.

8   $\mathbb{C}$ with addition and multiplication.

9 The set of matrices of the form $\begin{pmatrix} a & b \\ 0 & c \end{pmatrix}$, where $a, b, c \in \mathbb{R}$, with the operations of matrix addition and matrix multiplication.

Miscellaneous exercises 11

1 Show that the symmetries of a rectangle, of sides a, b ($a \neq b$), form a group V (the symmetry group of the rectangle).

Identify all the proper subgroups of V. [C]

2 (a) An operation * is defined on the set $\mathbb{R}$ of real numbers by

$$a * b = a + b - 1.$$

Show that the system $(\mathbb{R}, *)$ is a group, giving its identity element and the inverse of a general element g.

(b) Draw four quadrilaterals with symmetry groups of order 1, 2, 4, 8 respectively. Label your diagrams clearly and state, in each case, the transformations under which the quadrilateral is invariant. [L]

3 Show that the rational numbers do not form a group under the operation * defined by

$$a * b = a + ab.$$ [L]

4 $\mathbb{Q}$ is the set of rational numbers and two operations * and $\circ$ are defined on $\mathbb{Q}$ by

$$x * y = x + y + 1$$

and

$$x \circ y = 1$$

for all $x, y \in \mathbb{Q}$.

(a) Show that the operations * and $\circ$ are both associative.

(b) Determine whether the systems $(\mathbb{Q}, *)$ and $(\mathbb{Q}, \circ)$ possess identity elements.

(c) Show that just one of the two systems is a group. [L]

5 A set of S elements $\{a, b, c, d\}$ is closed under an associative binary operation *: that is, if $x, y, z \in S$, then $x * y \in S$ and $(x * y) * z = x * (y * z)$. One of the elements of S is the identity element under the operation *. Given that

$$a * a = a, \quad b * b = d, \quad c * c = d, \quad d * d = a,$$

determine whether $(S, *)$ is a group. [C]

6 Let the set $G = \{0, 1, 2\}$. Show that G under the binary operation 'addition modulo 3' forms a group P, but that under the binary operation 'multiplication modulo 3' G does not form a group.

Let H be the rotation group of an equilateral triangle. Is H isomorphic to P? Give reasons for your answer.

Examine whether or not the set G forms a group under the binary operation *, where

$$A * B = |A^2 - B^2|, \text{(mod 3)}. \qquad \text{[L]}$$

7 (a) The operation $\circ$ is defined on $\mathbb{R}^+$, the set of positive real numbers, by

$$a \circ b = \frac{a^2 + b^2}{a + b}, \ (a, b \in \mathbb{R}^+).$$

State, with reasons, which of the defining properties of a commutative group are possessed by $(\mathbb{R}^+, \circ)$ and which are not.

(b) **0, 1, 2, 3** are the residue classes modulo 4 (i.e. **2** is the set of integers which leave a remainder 2 when divided by 4). T is the set $\{1, 2, 3\}$ and the operation * denotes multiplication modulo 4. State which group defining properties $(T, *)$ possesses and which it does not. If $x \in 2$ and $y \in 3$, find the remainders when $(x + y)$ and xy are divided by 4. [L]

8 Show that the set A, where $A = \{1, 2, 3, 4, 5, 6\}$, forms a group under the operation of multiplication modulo 7. State the inverse of each element.

An element a of the set A is said to have order n if n is the smallest positive integer such that

$$a^n \equiv 1 \, (\text{mod } 7).$$

Find the order of each of the six elements of A and write 5 as a product of two elements of order 2 and 3. [L]

9 Given that S is the set of all matrices of the form $\begin{pmatrix} a & b \\ 2b & a \end{pmatrix}$, where a and b are real, prove that S is closed under addition and under multiplication.

If a and b can take all real values, show that (a) S forms a group under addition, (b) S does not form a group under multiplication.

(The associative properties for addition and for multiplication may be assumed.)

If a and b are rational and a is not zero, show that (c) S does not form a group under addition, (d) no member of S has zero determinant, (e) S forms a group under multiplication. [JMB]

10 Let G be the set of all matrices of the form $\begin{pmatrix} 1 & x \\ 0 & 1 \end{pmatrix}$ with x a real number.

(a) Show that G does *not* form a group under matrix addition.
(b) Show that G does form a group under matrix multiplication.
(You may assume associativity of matrix addition and multiplication.)
[JMB]

11 T is the set of triangular matrices of the form $\begin{pmatrix} 1 & a \\ 0 & 1 \end{pmatrix}$ where $a \in \mathbb{Z}$ (the set of integers). Prove that T is a group with respect to matrix multiplication and that this group is isomorphic with the additive group of integers. (The associativity of matrix multiplication may be assumed.)

Is the same true for the set of triangular matrices of the form $\begin{pmatrix} a & 1 \\ 0 & a \end{pmatrix}$ with $a \in \mathbb{Z}$? Justify your answer. [L]

12 (a) Prove that the set of non-zero real numbers forms a group under the operation $\circ$ defined by $x \circ y = 2xy$.

State the identity element for this system, and give the inverse of x.

(b) A set of four 2×2 matrices forms a group under matrix multiplication. Two members of the set are $\begin{pmatrix} -1 & 0 \\ 0 & 1 \end{pmatrix}$ and $\begin{pmatrix} 1 & 0 \\ 0 & -1 \end{pmatrix}$. Find the other members and write out the group table. [L]

13 The matrix $\mathbf{A}$ is defined by $\mathbf{A} = \begin{pmatrix} 0 & 1 \\ -1 & 0 \end{pmatrix}$.

By verifying that all the group axioms are satisfied, show that the set $\{\mathbf{A}, \mathbf{A}^2, \mathbf{A}^3, \mathbf{A}^4\}$ forms a commutative group under matrix multiplication. (The associativity of matrix multiplication may be assumed.)

Show also that the set of complex numbers $\{1, -1, i, -i\}$ forms a group under multiplication. Show further that the two groups are isomorphic. [L]

14 Show that H, the set of all non-singular 2×2 real matrices, forms a group under matrix multiplication. (It may be assumed that matrix multiplication is associative and that det $\mathbf{A} \times$ det $\mathbf{B} =$ det $\mathbf{AB}$.)

Given that S is the set of all non-singular 2×2 matrices which commute with the matrix $\begin{pmatrix} 1 & 0 \\ 1 & 1 \end{pmatrix}$, show that each member of S must be of the form $\begin{pmatrix} a & 0 \\ b & a \end{pmatrix}$, where $a \neq 0$.

Show that, under the operation of matrix multiplication, S is a commutative group. [L]

15 Prove that the integers $\{2, 4, 6, 8\}$ form a group under the operation of multiplication modulo 10.

Prove that the numbers $\{1, -1, i, -i\}$ form a group under the operation of complex multiplication.

Prove that these groups are isomorphic. [C]

16 Given that $z = r(\cos \frac{1}{3}\pi + i \sin \frac{1}{3}\pi)$, where $r > 0$, and that $S = \{1, z, z^2, z^3, z^4, z^5\}$, show that $r = 1$ is a necessary and sufficient condition for S to form a group under multiplication. (The associative property for multiplication of complex numbers may be assumed.)

If $r = 1$, list the proper sub-groups of S.

Show that this group is isomorphic to the group $G = \{1, 2, 3, 4, 5, 6\}$ of residue classes modulo 7 under multiplication.　　　　　[JMB]

17　Define a *group* and explain what is meant by saying that two groups are *isomorphic*.

Compile the combination tables for

(a) the symmetries of a rectangle which is not a square,

(b) addition mod 4 for the set $\{0, 1, 2, 3\}$,

(c) multiplication mod 10 for the set $\{2, 4, 6, 8\}$.

Show that (a), (b) and (c) are groups and that (b) and (c) are isomorphic.

For (c) solve the equations $x^2 = x$ and $x^3 = x$.

18　Define a group.

Show that the set $\{1, -1, i, -i\}$ under multiplication is a group and that the set of matrices $\{\mathbf{A}, \mathbf{B}, \mathbf{C}, \mathbf{D}\}$, where

$$\mathbf{A} = \begin{pmatrix} 0 & 0 \\ 1 & 1 \end{pmatrix}, \qquad\qquad \mathbf{B} = \begin{pmatrix} 0 & 0 \\ -1 & -1 \end{pmatrix},$$

$$\mathbf{C} = \begin{pmatrix} 0 & 0 \\ i & i \end{pmatrix}, \qquad\qquad \mathbf{D} = \begin{pmatrix} 0 & 0 \\ -i & -i \end{pmatrix},$$

under matrix multiplication is also a group. Are the two groups isomorphic? Give reasons to justify your answer.

(The associativity of multiplication may be assumed in both cases.)　[L]

19　The matrices (with complex elements)

$$\mathbf{E} = \begin{pmatrix} 1 & 0 \\ 0 & 0 \end{pmatrix}, \quad \mathbf{A} = \begin{pmatrix} -1 & 0 \\ 0 & 0 \end{pmatrix}, \quad \mathbf{B} = \begin{pmatrix} i & 0 \\ 0 & 0 \end{pmatrix}, \quad \mathbf{C} = \begin{pmatrix} x & y \\ z & t \end{pmatrix}$$

are given to form a group under the operation of matrix multiplication. Find x, y, z and t, confirm that $\mathbf{A}(\mathbf{BC})$ is then equal to $(\mathbf{AB})\mathbf{C}$, and exhibit the group operation table.

(The associative law for matrix multiplication may be assumed except for the relation quoted.)　　　　　　　　　　　　　　[C]

20　The group G with operation $*$ is a cyclic group of order n with generator g. Given that $g^5 = g^{17}$, determine, with reasons, the set of possible values of n.

In the case $n = 6$, determine the order of each element of G and list all the generators of G.　　　　　　　　　　　　　　[C]

21　(a) A group H has fifteen elements,

$$1, \omega, \omega^2, \ldots, \omega^{14},$$

where ω is a complex 15th root of unity

$$(\omega^{15} = 1, \omega \neq 1, \omega^3 \neq 1, \omega^5 \neq 1),$$

the law of combination of elements being multiplication. Find a sub-group of H having five elements 1, a, b, c, d such that

$$a^2 = b, \quad b^2 = c, \quad c^2 = d, \quad d^2 = a,$$

and prove your result.

(b) A group G has distinct elements e, a, b, c, ..., where e is the identity element. The law of combination is denoted by the symbol $*$.

(i) Prove that, if

$$a * a = b, \quad b * b = a,$$

then the elements e, a, b form a sub-group of G.

(ii) Prove that, if

$$a * a = b, \quad b * b = c, \quad c * c = a,$$

then the elements e, a, b, c do not form a sub-group of G. [C]

22 Using any notation with which you are familiar, list all the one-to-one mappings of the set $\{a, b, c, d\}$ into itself which either map a and b into themselves or transform them into each other. Prove that, with respect to the operation of composition, these mappings form a group and that this group is not isomorphic with the additive group of integers modulo 4. [L]

23 The mappings f_1, f_2, f_3, f_4 of $D \rightarrow D$, under composition of mappings, are

$$f_1(x) = \frac{x - 1}{x + 1}, \quad f_2(x) = f_1[f_1(x)], \quad f_3(x) = f_1[f_2(x)], \quad f_4(x) = f_1[f_3(x)],$$

where

$$D = \{x : x \in \mathbb{R}, x \neq 0, x \neq 1, x \neq -1\}.$$

Determine $f_2(x)$, $f_3(x)$ and show that $f_4(x) = I(x)$, where I is the identity mapping.

Prove that the four mappings I, f_1, f_2, f_3 form a group with composition as the group operation, and show that this group is isomorphic with the group consisting of $\{1, i, -1, -i\}$, with multiplication as the group operation. [L]

24 Show that the set of rational numbers, with the operations of addition and multiplication, constitute a field. Show also that the set of integers does not form a field. [L]

25 A set S consists of ordered pairs of real numbers, i.e.

$$S \equiv \{(x, y) : x \text{ and } y \text{ are real}\}.$$

Two binary operations, $+$ and $\times$, are defined on the elements of S by the relations

$$(x, y) + (x', y') = (x + x', y + y'),$$
$$(x, y) \times (x', y') = (xx', yy'),$$

the operations on the right being addition and multiplication, in the ordinary sense, for real numbers. Prove that S is a commutative ring with unity, and determine whether it is a field. [C]

26 In each case below, determine whether the given set with the given operations forms (a) a ring, (b) a field, (c) neither a ring nor a field. Give your reasons, *briefly*.

(a) The integers (positive, negative and zero) under the usual arithmetic operations (addition and multiplication).

(b) The integers $\{0, 1, 2, 3, 4, 5\}$ under the usual arithmetic operations modulo 6.

(c) The integers $\{0, 1, 2, 3, 4\}$ under the usual arithmetic operations modulo 5.

(d) The set of vectors in three dimensions under vector addition and the scalar (or dot) product.

(e) The set of invertible (non-singular) 2×2 matrices under the usual matrix operations of addition and multiplication. [C]

27 Consider the set S consisting of the functions $f_1, f_2, \ldots, f_6$, where

$$f_1(t) = t, \quad f_3(t) = 1 - t, \quad f_5(t) = \frac{(t - 1)}{t},$$

$$f_2(t) = \frac{1}{t}, \quad f_4(t) = \frac{1}{(1 - t)}, \quad f_6(t) = \frac{t}{(t - 1)},$$

the domain of the f_j being the set of all real numbers except 0 and 1. Show that S is a group under the binary operation $\circ$ where $\circ$ is defined as the operation of composition, i.e. $(f_i \circ f_j)(t)$ is defined as $f_i[f_j(t)]$.

(The associativity of composition of functions may be assumed.) [L]

28 An algebraic structure involves two operations on a set X. The set is closed under both operations, both of which are commutative and associative. One operation is denoted by '$+$', the other by '$\cdot$', and there are identity elements '0' for '$+$' and '1' for '$\cdot$' respectively.

What additional properties should the structure possess in order to be a field?

Write out the tables for the operations $+$ and $\cdot$ when X is the set of residue classes modulo 5, denoted by $\{0, 1, 2, 3, 4\}$.

29 Define a field. (The term 'group' may be used without explanation.)

Let the set of residue classes of integers modulo n, with the operations of addition and multiplication modulo n, be denoted by $(R_n, +, .)$. You may assume that both operations are associative. If $\mathbf{a}, \mathbf{b}, \mathbf{c}$ are elements of R_n, prove that $\mathbf{a}.(\mathbf{b} + \mathbf{c}) = \mathbf{a}.\mathbf{b} + \mathbf{a}.\mathbf{c}$.

Show, by forming appropriate tables or otherwise, that $(R_5, +, .)$ is a field. Find which elements in R_5 have

(a) square roots,   (b) cube roots.

Solve the equation $x^2 + 3x + 2 = 0$, in which unknown and constants are elements of R_5.   [L]

30   (a) Define a group and prove from your definition that if a, b, c are elements of a group such that $a \circ b = a \circ c$, where $\circ$ denotes the group operation, then $b = c$.

(b) Prove that the set of numbers $\{x + y\sqrt{2}\}$, where x and y are rational, subject to the normal rules of addition and multiplication, form a field, stating the zero and unit elements. Give also the multiplicative inverse of $x + y\sqrt{2}$ in the form $x' + y'\sqrt{2}$.   [L]

31   (a) Consider (i) the set of 2×2 real matrices with the operations of matrix addition and multiplication, (ii) the set of integers with the normal operations of addition and multiplication. Show that neither is a field, in each case illustrating by argument or counter-example one field axiom which is not satisfied.

(b) Two operations $*$ and $\circ$ are defined on the set $\mathbb{R}$ of real numbers by $p * q = p + q + 1$, $p \circ q = pq + p + q$. Consider the function $f: x \mapsto x + 1$. If the images of $p, q, p + q + 1, pq + p + q$ under this mapping are a, b, c, d respectively, express a, b, c, d in terms of p and q, and show that $a + b = c$, $ab = d$. Deduce that $(\mathbb{R}, *, \circ)$ is a field. State the identity element for each operation and the inverse of p under each.   [L]

12 Vector spaces

12.1 Linear dependence

The n vectors $\mathbf{u}_1, \mathbf{u}_2, \ldots, \mathbf{u}_n$ are said to be *linearly dependent* if there exists a relation between them of the form

$$k_1\mathbf{u}_1 + k_2\mathbf{u}_2 + k_3\mathbf{u}_3 + \ldots + k_n\mathbf{u}_n = \mathbf{0},$$

where the scalars $k_1, k_2, k_3, \ldots, k_n$ are not all zero.

When such a relation exists, the vector $\mathbf{u}_i$ can be expressed in terms of the remaining $(n - 1)$ vectors, provided that k_i is not zero.

When no such relation exists, the n vectors are said to be *linearly independent*.

In two dimensions, the vectors $\begin{pmatrix} a_1 \\ a_2 \end{pmatrix}, \begin{pmatrix} b_1 \\ b_2 \end{pmatrix}$ are linearly dependent if non-zero scalars k_1, k_2 can be found such that

$$k_1\begin{pmatrix} a_1 \\ a_2 \end{pmatrix} + k_2\begin{pmatrix} b_1 \\ b_2 \end{pmatrix} = \begin{pmatrix} 0 \\ 0 \end{pmatrix}.$$

In this case, $k_1 a_1 + k_2 b_1 = 0$, $k_1 a_2 + k_2 b_2 = 0$

$$\Rightarrow \qquad a_1 : a_2 = b_1 : b_2.$$

Then the two vectors are parallel, and $\begin{vmatrix} a_1 & b_1 \\ a_2 & b_2 \end{vmatrix} = 0$.

Let $\mathbf{u}$ and $\mathbf{v}$ be two vectors in the plane of the vectors $\mathbf{i}$ and $\mathbf{j}$. When $\mathbf{u}$ and $\mathbf{v}$ are linearly independent, they define two distinct directions. Any other vector $\mathbf{w}$ in the same plane can be expressed as the sum of scalar multiples of $\mathbf{u}$ and $\mathbf{v}$. It follows that any three coplanar vectors $\mathbf{u}, \mathbf{v}$ and $\mathbf{w}$ will be linearly dependent.

Example 1

Find scalars p, q, r such that $p\begin{pmatrix} 1 \\ 4 \end{pmatrix} + q\begin{pmatrix} 2 \\ 2 \end{pmatrix} + r\begin{pmatrix} 4 \\ 6 \end{pmatrix} = \mathbf{0}$.

This vector equation gives

$$p + 2q + 4r = 0, \qquad \ldots (1)$$
$$4p + 2q + 6r = 0. \qquad \ldots (2)$$

By subtraction,
$$3p + 2r = 0$$
$$\Rightarrow \qquad r = -3p/2.$$

By substitution in equation (1),
$$q = 5p/2.$$

This shows that the scalars p, q, r are such that

$$p:q:r = 2:5:-3$$

$$\Rightarrow \qquad 2\begin{pmatrix}1\\4\end{pmatrix} + 5\begin{pmatrix}2\\2\end{pmatrix} - 3\begin{pmatrix}4\\6\end{pmatrix} = \mathbf{0}.$$

In three dimensions, the vectors $\mathbf{a}$, $\mathbf{b}$, $\mathbf{c}$ are linearly dependent if scalars p, q, r (not all zero) can be found such that

$$p\mathbf{a} + q\mathbf{b} + r\mathbf{c} = \mathbf{0}.$$

Let $\qquad \mathbf{a} = \begin{pmatrix}a_1\\a_2\\a_3\end{pmatrix}, \qquad \mathbf{b} = \begin{pmatrix}b_1\\b_2\\b_3\end{pmatrix}, \qquad \mathbf{c} = \begin{pmatrix}c_1\\c_2\\c_3\end{pmatrix}.$

When none of the three scalars p, q, r is zero, the vector equation

$$p\mathbf{a} + q\mathbf{b} + r\mathbf{c} = 0$$

gives three equations

$$pa_1 + qb_1 + rc_1 = 0, \qquad \ldots (1)$$

$$pa_2 + qb_2 + rc_2 = 0, \qquad \ldots (2)$$

$$pa_3 + qb_3 + rc_3 = 0. \qquad \ldots (3)$$

From equations (2) and (3),

$$\frac{p}{b_2 c_3 - b_3 c_2} = \frac{-q}{a_2 c_3 - a_3 c_2} = \frac{r}{a_2 b_3 - a_3 b_2}.$$

Substitution for p, q and r in equation (1) gives

$$a_1(b_2 c_3 - b_3 c_2) - b_1(a_2 c_3 - a_3 c_2) + c_1(a_2 b_3 - a_3 b_2) = 0$$

$$\Rightarrow \qquad \begin{vmatrix} a_1 & b_1 & c_1 \\ a_2 & b_2 & c_2 \\ a_3 & b_3 & c_3 \end{vmatrix} = 0.$$

Conversely, when this determinant is zero, any three scalars p, q, r in the ratios

$$(b_2 c_3 - b_3 c_2):(c_2 a_3 - c_3 a_2):(a_2 b_3 - a_3 b_2)$$

will be such that $p\mathbf{a} + q\mathbf{b} + r\mathbf{c} = \mathbf{0}$, showing that the vectors $\mathbf{a}$, $\mathbf{b}$, $\mathbf{c}$ are linearly dependent.

This shows that a necessary and sufficient condition for the vectors $\mathbf{a}$, $\mathbf{b}$, $\mathbf{c}$ to be linearly independent is that the above determinant is not zero.

In the special case when one vector $\mathbf{b}$ is a scalar multiple of another vector $\mathbf{c}$, i.e. when $q\mathbf{b} + r\mathbf{c} = 0$, the three vectors $\mathbf{a}$, $\mathbf{b}$, $\mathbf{c}$ are linearly dependent with $p = 0$.

In this case, the determinant above is clearly equal to zero, since the second column is a multiple of the third column.

Example 2

Show that the vectors $\begin{pmatrix} 3 \\ 4 \\ 1 \end{pmatrix}$, $\begin{pmatrix} 3 \\ -1 \\ -1 \end{pmatrix}$, $\begin{pmatrix} -1 \\ 2 \\ 1 \end{pmatrix}$ are linearly dependent, and find a linear relation between them.

To show that the given vectors are linearly dependent, it is sufficient to show that the determinant with columns formed by the elements of the vectors is zero.

$$\begin{vmatrix} 3 & 3 & -1 \\ 4 & -1 & 2 \\ 1 & -1 & 1 \end{vmatrix} = 3(-1+2) - 3(4-2) - (-4+1) = 0.$$

Since the determinant is 0, the vectors are linearly dependent, and scalars p, q, r exist, not all zero, such that

$$p\begin{pmatrix} 3 \\ 4 \\ 1 \end{pmatrix} + q\begin{pmatrix} 3 \\ -1 \\ -1 \end{pmatrix} + r\begin{pmatrix} -1 \\ 2 \\ 1 \end{pmatrix} = \begin{pmatrix} 0 \\ 0 \\ 0 \end{pmatrix}.$$

This gives the three equations

$$3p + 3q - r = 0, \qquad \ldots (1)$$
$$4p - q + 2r = 0, \qquad \ldots (2)$$
$$p - q + r = 0. \qquad \ldots (3)$$

From equations (1) and (3),

$$4p + 2q = 0 \quad \Rightarrow \quad q = -2p.$$

Substitution in equation (2) gives $r = -3p$.
Hence $p:q:r = 1:-2:-3$.

The required relation between the vectors is

$$\begin{pmatrix} 3 \\ 4 \\ 1 \end{pmatrix} - 2\begin{pmatrix} 3 \\ -1 \\ -1 \end{pmatrix} - 3\begin{pmatrix} -1 \\ 2 \\ 1 \end{pmatrix} = \begin{pmatrix} 0 \\ 0 \\ 0 \end{pmatrix}.$$

Exercises 12.1

1   Express the vector **w** in terms of **u** and **v** when
   (a) $\mathbf{u} = 3\mathbf{i} + 2\mathbf{j}$, $\mathbf{v} = 2\mathbf{i} + \mathbf{j}$, $\mathbf{w} = \mathbf{i} + 2\mathbf{j}$;
   (b) $\mathbf{u} = 2\mathbf{i} - \mathbf{j}$, $\mathbf{v} = -3\mathbf{i} + 4\mathbf{j}$, $\mathbf{w} = 4\mathbf{i} + 3\mathbf{j}$;
   (c) $\mathbf{u} = 4\mathbf{i} - 3\mathbf{j}$, $\mathbf{v} = 3\mathbf{i} + 4\mathbf{j}$, $\mathbf{w} = 5\mathbf{i}$;
   (d) $\mathbf{u} = 5\mathbf{i} + 7\mathbf{j}$, $\mathbf{v} = 6\mathbf{i} + 5\mathbf{j}$, $\mathbf{w} = 9\mathbf{i} - \mathbf{j}$.

2   Find scalars p, q, r such that $p\mathbf{a} + q\mathbf{b} + r\mathbf{c} = \mathbf{0}$ when

   (a) $\mathbf{a} = \begin{pmatrix} 1 \\ 2 \\ 3 \end{pmatrix}$, $\mathbf{b} = \begin{pmatrix} 2 \\ -1 \\ 1 \end{pmatrix}$, $\mathbf{c} = \begin{pmatrix} -1 \\ 2 \\ 1 \end{pmatrix}$;

(b) $\mathbf{a} = \begin{pmatrix} 4 \\ -6 \\ 2 \end{pmatrix}$, $\mathbf{b} = \begin{pmatrix} -6 \\ 9 \\ -3 \end{pmatrix}$, $\mathbf{c} = \begin{pmatrix} 5 \\ 7 \\ 1 \end{pmatrix}$;

(c) $\mathbf{a} = \begin{pmatrix} 3 \\ 4 \\ -5 \end{pmatrix}$, $\mathbf{b} = \begin{pmatrix} 0 \\ 1 \\ 1 \end{pmatrix}$, $\mathbf{c} = \begin{pmatrix} 4 \\ 5 \\ -7 \end{pmatrix}$;

(d) $\mathbf{a} = \begin{pmatrix} 2 \\ 5 \\ 6 \end{pmatrix}$, $\mathbf{b} = \begin{pmatrix} 1 \\ -1 \\ 3 \end{pmatrix}$, $\mathbf{c} = \begin{pmatrix} 1 \\ 4 \\ 3 \end{pmatrix}$.

3 Find the values of λ for which the vectors $\mathbf{a}$, $\mathbf{b}$, $\mathbf{c}$ are linearly dependent, where

$$\mathbf{a} = 2\mathbf{i} + \lambda\mathbf{j} + 3\mathbf{k}, \quad \mathbf{b} = 3\mathbf{i} + 3\mathbf{j} + 5\mathbf{k}, \quad \mathbf{c} = \lambda\mathbf{i} + 2\mathbf{j} + 2\mathbf{k}.$$

For each value of λ, express $\mathbf{a}$ in terms of $\mathbf{b}$ and $\mathbf{c}$.

4 Show that the vectors $2\mathbf{i} + \mathbf{j}$, $2\mathbf{j} + \mathbf{k}$, $p\mathbf{i} - 2\mathbf{k}$ are linearly independent, provided that p does not equal 8.

12.2 Vector spaces

A vector space (or linear space) over a field F is a set of elements V with the operations of addition $(+)$ and of multiplication by scalars from the field F such that $(V, +)$ is a commutative group and that for all $a, b \in V$ and all $\lambda, \mu \in F$ the following conditions hold:

(1) $$\lambda a \in V,$$

(2) $$\lambda(a + b) = \lambda a + \lambda b,$$

(3) $$(\lambda + \mu)a = \lambda a + \mu a,$$

(4) $$\lambda(\mu a) = (\lambda\mu)a,$$

(5) $$(1)a = a,$$

where 1 is the identity element in F for multiplication. It is customary to use the symbol V to denote the vector space generated by the set V and the field F with the given operations.

It should be noted that addition in F and addition in V are distinct operations, but that the same symbol $+$ is used to denote either operation. Similarly, multiplication of an element of V by an element of F and multiplication of one element of F by another are distinct operations.

Consider the set S of all vectors in the x-y plane. It has been shown in Chapter 11 that the set S with the operation of vector addition forms a group $(S, +)$. This group is commutative, since $\mathbf{a} + \mathbf{b} = \mathbf{b} + \mathbf{a}$ for all $\mathbf{a}, \mathbf{b} \in S$. The multiplication of elements of the set S by any real numbers p and q satisfies the following rules:

(1) $$p\mathbf{a} \in S,$$

(2) $$p(\mathbf{a} + \mathbf{b}) = p\mathbf{a} + p\mathbf{b},$$

(3) $$(p + q)\mathbf{a} = p\mathbf{a} + q\mathbf{a},$$

(4) $$p(q\mathbf{a}) = (pq)\mathbf{a},$$

(5) $$(1)\mathbf{a} = \mathbf{a}.$$

Hence with the stated operations the set S is a vector space over the field of real numbers.

Example 1
The sets P and Q are subsets of S, the set of vectors in the x-y plane.

Each member of P is of the form $2a\mathbf{i} + a\mathbf{j}$, and each member of Q is of the form $a\mathbf{i} + (1 - a)\mathbf{j}$, where $a \in \mathbb{R}$. Show that under the operation of vector addition, P is a vector space over the field of real numbers, but Q is not.

Under the operation of vector addition, the set P forms a commutative group $(P, +)$.

Multiplication of members of P by real numbers satisfies the rules set out above.

Hence P forms a vector space over the field of real numbers.

The vector space P is said to be a *subspace* of S.

Consider the vectors $p\mathbf{i} + (1 - p)\mathbf{j}$, $q\mathbf{i} + (1 - q)\mathbf{j}$ which are members of Q. The sum of these two vectors is

$$(p + q)\mathbf{i} + (2 - p - q)\mathbf{j},$$

which is not a member of Q, i.e. Q is not closed under the operation of vector addition. Since the set Q does not form a group under the operation of vector addition, the conditions for a vector space are not satisfied.

Dimension
The set of vectors $k\mathbf{a}$, where k is real, forms a vector space A over the field of real numbers. This is one-dimensional vector space.

If the vector $\overrightarrow{OP}$ represents $\mathbf{a}$ and the vector $\overrightarrow{OQ}$ represents $k\mathbf{a}$, then the points O, P and Q are in a straight line. The vector $\mathbf{a}$ is said to *span* the vector space A.

The set of vectors $p\mathbf{i} + q\mathbf{j}$, where p and q are real, forms a vector space in two dimensions. Each member of the set is a linear combination of the two vectors $\mathbf{i}$ and $\mathbf{j}$, which are said to span the vector space.

A vector space V is said to be spanned by a set of n members of V if each member of V can be expressed as a linear combination of these n members.

The smallest possible value of n is called the *dimension* of V. It follows that the dimension of V is the greatest number of linearly independent vectors in the set V.

Basis
Let the dimension of a vector space V be denoted by d. Then any set of d linearly

independent members of V forms a *basis* for V. Each member of V can be expressed as a linear combination of the members of a basis. The members of a basis must be linearly independent, and each basis must contain d members.

A set S is said to span the vector space V if each member of V can be expressed as a linear combination of members of S. The set S may have more than d members, but then the members of S will not be linearly independent.

Subspaces

Let T be a subset of the set V. Then if V is a vector space over a field F and if T is also a vector space over the same field, T is called a *subspace* of V.

Consider the set V of vectors of the form $a\mathbf{i} + b\mathbf{j} + c\mathbf{k}$. This set forms a commutative group $(V, +)$ under the operation of vector addition. Since the five conditions stated earlier in this section are satisfied, V forms a vector space over the field of real numbers.

The three vectors $\mathbf{i}, \mathbf{j}, \mathbf{k}$ form a basis for V, and the dimension of this space is 3.

Let T be the set of vectors of the form $s\mathbf{i} + 2s\mathbf{j} + t\mathbf{k}$, where s and t are parameters.

Then T forms a commutative group $(T, +)$ under the operation of vector addition. The same five conditions are satisfied, and T forms a vector space over the field of real numbers. This means that T is a subspace of V. A basis for T is given by the vectors $\{\mathbf{i} + 2\mathbf{j}, \mathbf{k}\}$, and the dimension of T is 2.

Every point with position vector of the form $s\mathbf{i} + 2s\mathbf{j} + t\mathbf{k}$ lies in the plane with vector equation

$$\mathbf{r} = s\mathbf{i} + 2s\mathbf{j} + t\mathbf{k}.$$

Example 1

Find a basis for the vector space V of all vectors of the form $x\mathbf{i} + y\mathbf{j} + z\mathbf{k}$, where $4x - 2y + 3z = 0$.

When
$$4x - 2y + 3z = 0,$$
$$x\mathbf{i} + y\mathbf{j} + z\mathbf{k} = (\tfrac{1}{2}y - \tfrac{3}{4}z)\mathbf{i} + y\mathbf{j} + z\mathbf{k}$$
$$= y(\tfrac{1}{2}\mathbf{i} + \mathbf{j}) + z(-\tfrac{3}{4}\mathbf{i} + \mathbf{k}).$$

This shows that any member of V can be expressed as a linear combination of the vectors $\mathbf{a}$ and $\mathbf{b}$, where

$$\mathbf{a} = \tfrac{1}{2}\mathbf{i} + \mathbf{j}, \qquad \mathbf{b} = -\tfrac{3}{4}\mathbf{i} + \mathbf{k}.$$

Hence $\{\mathbf{a}, \mathbf{b}\}$ is a basis for V, and the dimension of the space is 2.

Note that there is no unique basis. For this example, a basis is also given by $\{\mathbf{i} + 2\mathbf{j}, 3\mathbf{i} - 4\mathbf{k}\}$ or by $\{3\mathbf{i} - 4\mathbf{k}, 3\mathbf{j} + 2\mathbf{k}\}$.

Example 2

Find the dimension of the vector space over the field of real numbers spanned by the vectors $\mathbf{a}, \mathbf{b}, \mathbf{c}, \mathbf{d}$ where

$$\mathbf{a} = \begin{pmatrix} 1 \\ 2 \\ 3 \\ 4 \end{pmatrix}, \quad \mathbf{b} = \begin{pmatrix} 2 \\ 3 \\ 4 \\ 5 \end{pmatrix}, \quad \mathbf{c} = \begin{pmatrix} 2 \\ 4 \\ 6 \\ 8 \end{pmatrix}, \quad \mathbf{d} = \begin{pmatrix} 3 \\ 5 \\ 7 \\ 9 \end{pmatrix}.$$

Any basis must contain only linearly independent vectors.

Since $\mathbf{c} = 2\mathbf{a}$, $\mathbf{c}$ is not independent of $\mathbf{a}$.

Since $\mathbf{d} = \mathbf{a} + \mathbf{b}$, $\mathbf{d}$ is not independent of $\mathbf{a}$ and $\mathbf{b}$.

This leaves $\mathbf{a}$ and $\mathbf{b}$, which are linearly independent.

Hence the dimension of the space is 2, and $\{\mathbf{a}, \mathbf{b}\}$ is a basis.

Example 3

The members of the set S are all numbers of the form $a + b\sqrt{2}$, where a and b are rational numbers. Show that S forms a vector space over the field of real numbers.

First it must be proved that under the operation of addition S forms a commutative group $(S, +)$.

Let $a_1 + b_1\sqrt{2}$ and $a_2 + b_2\sqrt{2}$ be members of S.

Then their sum $(a_1 + a_2) + (b_1 + b_2)\sqrt{2}$ belongs to S.

This shows that S is closed under the operation of addition. Addition is clearly associative.

There exists an identity element $0 + 0\sqrt{2}$. Each element $a + b\sqrt{2}$ has its inverse element $-a - b\sqrt{2}$. Finally, addition is commutative, since

$$(a_1 + b_1\sqrt{2}) + (a_2 + b_2\sqrt{2}) = (a_2 + b_2\sqrt{2}) + (a_1 + b_1\sqrt{2}).$$

It follows that $(S, +)$ is a commutative group.

Secondly, let λ and μ be real numbers. Then

(1) $$\lambda(a + b\sqrt{2}) \in S,$$

(2) $$\lambda[(a_1 + b_1\sqrt{2}) + (a_2 + b_2\sqrt{2})] = \lambda(a_1 + b_1\sqrt{2}) + \lambda(a_2 + b_2\sqrt{2}),$$

(3) $$(\lambda + \mu)(a + b\sqrt{2}) = \lambda(a + b\sqrt{2}) + \mu(a + b\sqrt{2}),$$

(4) $$\lambda(\mu a + \mu b\sqrt{2}) = \lambda\mu(a + b\sqrt{2}),$$

(5) $$(1)(a + b\sqrt{2}) = a + b\sqrt{2}.$$

Since these five conditions are satisfied, the set S forms a vector space over the field of real numbers.

The two elements $1 + 0\sqrt{2}$ and $0 + 1\sqrt{2}$ form a basis for this vector space.

Note that in this example the elements of the space are not vectors.

Exercises 12.2

1   The members of the set S_1 are of the form $s(\mathbf{i} + \mathbf{j})$, and the members of the set S_2 are of the form $t(\mathbf{i} - \mathbf{j})$, where s and t are parameters. Show that S_1

and S_2 are vector spaces over the field of real numbers.
Find whether this is also true for the union $S_1 \cup S_2$ of the two sets.

2 The members of the set X are quadratic expressions of the form $ax^2 + bx + c$, where a, b and c are real. Show that X is a vector space over the field of real numbers.
Find a basis for X and state its dimension.

3 Show that the set of complex numbers is a vector space
(a) over the field of real numbers,
(b) over the field of rational numbers.

4 Show that the set of solutions of the differential equation $\dfrac{dy}{dx} = y$ is a vector space over the field of real numbers.
Find a basis for this space and state its dimension.

5 Show that the vectors $\mathbf{i} + \mathbf{j} + \mathbf{k}$, $2\mathbf{i} - 3\mathbf{j} + 4\mathbf{k}$ span the same space as the vectors $\mathbf{i} - 9\mathbf{j} + 5\mathbf{k}$, $3\mathbf{i} + 8\mathbf{j} + \mathbf{k}$.
Show also that the vector $3\mathbf{i} - 2\mathbf{j} + \mathbf{k}$ does not lie in this space.

6 Show that the set of solution vectors of the simultaneous equations

$$x + 2y - z = 0,$$
$$2x - y + z = 0$$

forms a vector space over the field of real numbers.
Find a basis for this space.

7 Show that the set of solution vectors of the equation $x + y + z = 0$ forms a vector space over the field of real numbers, and find a basis for this space.

8 Show that the three vectors

$$\mathbf{i} + 3\mathbf{j} + \mathbf{k}, \quad \mathbf{i} + 4\mathbf{j} - \mathbf{k}, \quad \mathbf{i} + \mathbf{j} + 5\mathbf{k}$$

are not linearly independent.
Show also that these vectors span the same space as that spanned by the vectors

$$\mathbf{i} + 2\mathbf{j} + 3\mathbf{k}, \quad \mathbf{j} - 2\mathbf{k}.$$

Prove that the vector $7\mathbf{i} - 2\mathbf{j} - \mathbf{k}$ does not lie in this space.

9 Find a basis for the vector space over the field of real numbers spanned by each of the following sets of vectors.
(a) $4\mathbf{i} + 6\mathbf{j}$, $2\mathbf{i} + 3\mathbf{j}$, $8\mathbf{i} + 12\mathbf{j}$,
(b) $\mathbf{i} + \mathbf{j} + \mathbf{k}$, $3\mathbf{i} + \mathbf{j} - \mathbf{k}$, $3\mathbf{i} + 2\mathbf{j} + \mathbf{k}$,
(c) $\mathbf{i} + \mathbf{j}$, $2\mathbf{j} - \mathbf{k}$, $2\mathbf{i} + \mathbf{k}$,
(d) $\mathbf{i} - 2\mathbf{j}$, $2\mathbf{i} - \mathbf{j}$, $4\mathbf{i} + 3\mathbf{j}$.

10 The members of the set V are vectors of the form $p\mathbf{i} + q\mathbf{j} + r\mathbf{k}$, where $p, q, r \in \mathbb{R}$. Show that V is a vector space over the field of real numbers. Determine which of the following sets are subspaces of V.
(a) All vectors with $p = q = r$.
(b) All vectors with $r = 1$.
(c) All vectors with $q = 0$.
(d) All vectors with $p, q, r \in \mathbb{Z}^+$.
(e) All vectors with $p = 3r$, $q = 2r$.
(f) All vectors with $p^2 + q^2 + r^2 = 1$.

Miscellaneous exercises 12

1 (a) Show that S_1, the set of symmetric 2×2 matrices, is a vector space over the field of real numbers.
Find a basis for S_1 and state its dimension.
(b) Show that S_2, the set of skew-symmetric 2×2 matrices, is a vector space over the field of real numbers. Find a basis for S_2 and state its dimension.
(c) Show also that the union $S_1 \cup S_2$ and the intersection $S_1 \cap S_2$ are vector spaces over the field of real numbers, and state their dimensions.

2 Show that the set of symmetric 3×3 matrices is a vector space over the field of real numbers. State its dimension.

3 Express the fourth column of the matrix $\mathbf{A}$, where

$$\mathbf{A} = \begin{pmatrix} 1 & 1 & 1 & 1 \\ 1 & 2 & 3 & 4 \\ 3 & 2 & 1 & 0 \end{pmatrix},$$

as a linear combination of the first and second columns.
Show that the set of vectors $\{\mathbf{x}\}$ such that $\mathbf{Ax} = \mathbf{0}$ forms a vector space. Find a basis for this space and state its dimension. [L]

4 Of the vectors $\mathbf{a}, \mathbf{b}, \mathbf{c}, \mathbf{d}$ defined by

$$\mathbf{a} = \begin{pmatrix} 5 \\ 6 \\ -11 \end{pmatrix}, \quad \mathbf{b} = \begin{pmatrix} 2 \\ -2 \\ 5 \end{pmatrix}, \quad \mathbf{c} = \begin{pmatrix} 9 \\ 2 \\ -1 \end{pmatrix}, \quad \mathbf{d} = \begin{pmatrix} -5 \\ -14 \\ 24 \end{pmatrix},$$

show that $\mathbf{a}, \mathbf{b}, \mathbf{d}$ form a set of basis vectors. Express $\mathbf{c}$ in terms of this basis.
If $\mathbf{a}, \mathbf{b}, \mathbf{c}\,\mathbf{d}$ are the position vectors of points A, B, C, D respectively, show that the point $P(1, -2, 3)$ lies on AD, find the ratio $AP:AD$ and show that BP is perpendicular to PC. [L]

5 F is the set of all 2×2 matrices of the form

$$\begin{pmatrix} a + ib & c + id \\ -c + id & a - ib \end{pmatrix},$$

where a, b, c, d are real numbers and $i^2 = -1$. Determine whether or not each of the following statements is true, giving reasons for your answers.

(a) F is a vector space over the field of complex numbers.

(b) Every non-zero element of F possesses a multiplicative inverse belonging to F. [L]

6 Given that

$$\begin{pmatrix} 4 & 1 & 2 & -7 \\ 1 & 0 & -1 & 1 \\ 3 & 1 & p & q \end{pmatrix} \begin{pmatrix} x \\ y \\ z \\ w \end{pmatrix} = \begin{pmatrix} 0 \\ 0 \\ 0 \end{pmatrix},$$

write down a set of simultaneous equations in x, y, z, w.

(a) If $p = 2$, $q = -5$, show by solving for x, y, z in terms of w that the equations have an infinite number of solutions, each of which may be expressed as $w\mathbf{v}$, where $\mathbf{v}$ is a fixed vector.

(b) If $p = 3$, $q = -8$, show that the set of solutions of the equation forms a two-dimensional vector space, and find a basis for this space. [L]

7 Show that the vectors $\mathbf{X}_1$, $\mathbf{X}_2$ and $\mathbf{X}_3$, where

$$\mathbf{X}_1 = \begin{pmatrix} 1 \\ 2 \\ 3 \end{pmatrix}, \quad \mathbf{X}_2 = \begin{pmatrix} 3 \\ 2 \\ 2 \end{pmatrix}, \quad \mathbf{X}_3 = \begin{pmatrix} 3 \\ 2 \\ 1 \end{pmatrix},$$

are linearly independent. Express $\mathbf{X}_4$, where $\mathbf{X}_4 = \begin{pmatrix} 4 \\ 4 \\ 3 \end{pmatrix}$, as a linear combination of $\mathbf{X}_1$, $\mathbf{X}_2$ and $\mathbf{X}_3$.

Find the value of a such that the vector $\begin{pmatrix} a \\ 3 \\ 4 \end{pmatrix}$ is orthogonal to $\mathbf{X}_4$. [L]

8 Explain what is meant by the statement that the m column vectors $\mathbf{x}_1$, $\mathbf{x}_2$, $\ldots$, $\mathbf{x}_m$, where each vector consists of n elements, are linearly independent.

Find the matrix $\mathbf{A}$, where $\mathbf{y} = \mathbf{Ax}$ represents a linear transformation for which

$$\begin{pmatrix} 3 \\ 2 \\ 1 \end{pmatrix} = \mathbf{A}\begin{pmatrix} 1 \\ 0 \\ 0 \end{pmatrix}, \quad \begin{pmatrix} 2 \\ 3 \\ 1 \end{pmatrix} = \mathbf{A}\begin{pmatrix} 0 \\ 1 \\ 0 \end{pmatrix}, \quad \begin{pmatrix} 3 \\ 2 \\ 1 \end{pmatrix} = \mathbf{A}\begin{pmatrix} 0 \\ 0 \\ 1 \end{pmatrix}.$$

Find the images under this transformation of $\mathbf{x}_1$, $\mathbf{x}_2$ and $\mathbf{x}_3$, where

$$\mathbf{x}_1 = \begin{pmatrix} 1 \\ 1 \\ 1 \end{pmatrix}, \quad \mathbf{x}_2 = \begin{pmatrix} 4 \\ 3 \\ -1 \end{pmatrix}, \quad \mathbf{x}_3 = \begin{pmatrix} 5 \\ 4 \\ 0 \end{pmatrix}.$$

Show that (a) $\mathbf{x}_1$ and $\mathbf{x}_2$ are linearly independent,

(b) $\mathbf{Ax}_1$ and $\mathbf{Ax}_2$ are linearly independent,

(c) $\mathbf{x}_1$, $\mathbf{x}_2$ and $\mathbf{x}_3$ are linearly dependent. [L]

9 Let V be the vector space of triples of real numbers over the field of real numbers. Show that the vectors

$$\mathbf{a} = \begin{pmatrix} 1 \\ 0 \\ 0 \end{pmatrix}, \quad \mathbf{b} = \begin{pmatrix} 1 \\ 1 \\ 0 \end{pmatrix}, \quad \mathbf{c} = \begin{pmatrix} 1 \\ 1 \\ 1 \end{pmatrix}$$

are linearly independent. Given any vector $\mathbf{x}$, where $\mathbf{x} = \begin{pmatrix} x_1 \\ x_2 \\ x_3 \end{pmatrix}$, find

real numbers α, β, γ such that

$$\alpha\mathbf{a} + \beta\mathbf{b} + \gamma\mathbf{c} = \mathbf{x},$$

giving α, β, γ in terms of x_1, x_2, x_3. Given that $\mathbf{d} = \begin{pmatrix} 0 \\ 0 \\ 1 \end{pmatrix}$, explain why

$\mathbf{b}$, $\mathbf{c}$, $\mathbf{d}$ is not a basis for V.

Show that $\begin{pmatrix} 2 \\ 2 \\ 1 \end{pmatrix}$ can be expressed as a linear combination of $\mathbf{b}$, $\mathbf{c}$ and $\mathbf{d}$,

but that $\begin{pmatrix} 2 \\ 3 \\ 1 \end{pmatrix}$ cannot be so expressed. [L]

10 If $\mathbf{A} = \begin{pmatrix} 1 & 0 & 2 \\ 2 & 1 & 0 \\ 0 & 2 & 1 \end{pmatrix}$, $\mathbf{B} = \begin{pmatrix} 2 & 0 & 1 \\ 0 & 1 & 2 \\ 1 & 2 & 0 \end{pmatrix}$, $\mathbf{C} = \begin{pmatrix} 1 & 2 & 0 \\ 0 & 1 & 2 \\ 2 & 0 & 1 \end{pmatrix}$,

$\mathbf{D} = \begin{pmatrix} 0 & 2 & 1 \\ 2 & 1 & 0 \\ 1 & 0 & 2 \end{pmatrix}$, consider the set V of all matrices expressible in the form

$k\mathbf{A} + l\mathbf{B} + m\mathbf{C} + n\mathbf{D}$, where k, l, m, n are any real numbers. Prove that $\mathbf{A}$, $\mathbf{B}$, $\mathbf{C}$, $\mathbf{D}$ are not linearly independent, and find a set which is linearly independent and forms a basis for V. [AEB]

11 (a) Show that the vectors $\mathbf{u}$ and $\mathbf{v}$, where $\mathbf{u} = \begin{pmatrix} 1 \\ 2 \end{pmatrix}$ and $\mathbf{v} = \begin{pmatrix} -2 \\ 1 \end{pmatrix}$, form

a basis for a two-dimensional space and express the vectors $\begin{pmatrix} 0 \\ 1 \end{pmatrix}$ and $\begin{pmatrix} 1 \\ 0 \end{pmatrix}$

in terms of this basis.

(b) A mapping of the plane to itself is represented by the matrix $\begin{pmatrix} 1 & 0 \\ 2 & 0 \end{pmatrix}$

in terms of the basis $\begin{pmatrix} 1 \\ 0 \end{pmatrix}, \begin{pmatrix} 0 \\ 1 \end{pmatrix}$.

(i) Let **u** and **v**, the vectors in (a), map to **u**′ and **v**′ respectively. Show that **u**′ and **v**′ are linearly dependent, but that nevertheless every vector in the image space can be represented in terms of them.

(ii) Show further that the set of points which map to the origin under this mapping lie on a straight line.

(iii) Show also that the set of points which map to the point (2, 4) lie on another straight line.

(iv) How many solutions are there to the equation

$$\begin{pmatrix} 1 & 0 \\ 2 & 0 \end{pmatrix} \begin{pmatrix} x \\ y \end{pmatrix} = \begin{pmatrix} 3 \\ 4 \end{pmatrix}?$$

Give a geometric reason for your answer. [AEB]

12 A vector space V_1 is spanned by the three vectors X_1, X_2 and X_3, where

$$X_1 = \begin{pmatrix} 1 \\ 1 \\ 0 \\ 1 \end{pmatrix}, \quad X_2 = \begin{pmatrix} 2 \\ 1 \\ -1 \\ 1 \end{pmatrix}, \quad X_3 = \begin{pmatrix} 5 \\ 3 \\ 4 \\ 2 \end{pmatrix}.$$

Find the value of a such that the vector $\begin{pmatrix} 2 \\ 4 \\ -1 \\ a \end{pmatrix}$ lies in the space V_1.

The vector space V_2 is spanned by the four vectors X_1, X_2, X_3 and X_4, where

$$X_4 = \begin{pmatrix} 2 \\ 4 \\ -1 \\ 3 \end{pmatrix}.$$

Express the vector $\begin{pmatrix} 1 \\ 0 \\ -1 \\ 1 \end{pmatrix}$ as a linear combination of X_1, X_2, X_3 and X_4. [C]

13 A vector space V_1 is spanned by the vectors **u** and **v**, where $\mathbf{u} = \begin{pmatrix} 1 \\ 1 \\ 1 \end{pmatrix}$ and $\mathbf{v} = \begin{pmatrix} 1 \\ 2 \\ 4 \end{pmatrix}$. Given that the vector $\begin{pmatrix} x \\ 7 \\ 5 \end{pmatrix}$ lies in this space, find the value of x.

Obtain a basis for the vector space V_2 onto which V_1 is mapped by the linear mapping given by the matrix $\mathbf{A}$, where

$$\mathbf{A} = \begin{pmatrix} 1 & 0 & 2 \\ 1 & 1 & 3 \\ 1 & 0 & 2 \end{pmatrix}.$$

Find a non-zero vector in V_1 which is mapped onto the zero vector in V_2.

[L]

14 Show that, if $\mathbf{a} = (3, 1, -2)$, $\mathbf{b} = (-2, -1, 1)$, $\mathbf{c} = (1, 5, 5)$, and $\mathbf{d} = (1, -1, -2)$, then $\{\mathbf{a}, \mathbf{b}, \mathbf{c}\}$ forms a basis for $\mathbb{R}^3$, and $\{\mathbf{a}, \mathbf{b}, \mathbf{d}\}$ does not.

Express the vector $(3, 5, 3)$ in the form $\lambda\mathbf{a} + \mu\mathbf{b} + \eta\mathbf{c}$, where λ, μ, η are constants to be determined.

Show that, if S is the subspace spanned by $\{\mathbf{a}, \mathbf{b}, \mathbf{d}\}$, then the vector $(-1, 1, 1) \notin S$.

[L]

15 The set of all real 2×2 matrices forms a vector space V under addition and multiplication by real scalars. Find the dimension of this space, and state a basis for V.

Show that the set of all symmetric 2×2 matrices is a vector subspace of V.

Prove by a counter-example that the set of all singular 2×2 matrices is not a vector subspace of V.

13 Determinants and matrices

13.1 Determinants of order 2

The value of the determinant $\begin{vmatrix} a & b \\ c & d \end{vmatrix}$ is defined to be $ad - bc$.

(1) If rows and columns are interchanged, the value of the determinant is unchanged, since

$$\begin{vmatrix} a & c \\ b & d \end{vmatrix} = ad - bc.$$

In consequence, any statement about rows applies equally to columns.

(2) If two rows are identical, the determinant equals 0, for

$$\begin{vmatrix} a & b \\ a & b \end{vmatrix} = ab - ab = 0.$$

(3) If two rows are interchanged, the sign of the determinant is changed, since

$$\begin{vmatrix} c & d \\ a & b \end{vmatrix} = cb - ad = -(ad - bc).$$

(4) If the elements in one row are multiplied by k, the value of the determinant is multiplied by k.

$$\begin{vmatrix} ka & kb \\ c & d \end{vmatrix} = k(ad - bc).$$

Note that
$$\begin{vmatrix} ka & kb \\ kc & kd \end{vmatrix} = k^2(ad - bc).$$

(5) If k times each element in one row is added to the corresponding element in the other row, the value of the determinant is unchanged.

$$\begin{vmatrix} a + kc & b + kd \\ c & d \end{vmatrix} = (a + kc)d - c(b + kd) = ad - bc.$$

(6) The simultaneous equations

$$ax + by = 0, \qquad cx + dy = 0,$$

where a, b, c, d are not zero, will have a solution other than $x = y = 0$ provided that $a/c = b/d$, i.e. provided that

$$\begin{vmatrix} a & b \\ c & d \end{vmatrix} = 0.$$

(7) Let $\mathbf{A}$ be the matrix $\begin{pmatrix} a & b \\ c & d \end{pmatrix}$ and let $\mathbf{B}$ be the matrix $\begin{pmatrix} p & q \\ r & s \end{pmatrix}$. The determinants of these matrices are given by

$$\det \mathbf{A} = ad - bc, \qquad \det \mathbf{B} = ps - qr.$$

Since
$$\mathbf{AB} = \begin{pmatrix} ap + br & aq + bs \\ cp + dr & cq + ds \end{pmatrix},$$

$$\det(\mathbf{AB}) = (ap + br)(cq + ds) - (aq + bs)(cp + dr)$$

$$= adps + bcqr - adqr - bcps$$

$$= (ad - bc)(ps - qr)$$

$$\Rightarrow \qquad \det(\mathbf{AB}) = (\det \mathbf{A})(\det \mathbf{B}).$$

Thus the determinant of $\mathbf{AB}$ equals the product of the determinants of $\mathbf{A}$ and $\mathbf{B}$. This result is also true for $n \times n$ matrices.

Example 1 Evaluate $\begin{vmatrix} 36 & 45 \\ 21 & 28 \end{vmatrix}$.

$$\begin{vmatrix} 36 & 45 \\ 21 & 28 \end{vmatrix} = 9 \begin{vmatrix} 4 & 5 \\ 21 & 28 \end{vmatrix}$$

$$= (9 \times 7) \begin{vmatrix} 4 & 5 \\ 3 & 4 \end{vmatrix}$$

$$= 63(4 \times 4 - 3 \times 5) = 63.$$

Example 2 Evaluate $\begin{vmatrix} 42 & 40 \\ 19 & 18 \end{vmatrix}$.

Subtract twice the second row from the first row.

$$\begin{vmatrix} 42 & 40 \\ 19 & 18 \end{vmatrix} = \begin{vmatrix} 4 & 4 \\ 19 & 18 \end{vmatrix}.$$

Subtract the second column from the first column.

$$\begin{vmatrix} 4 & 4 \\ 19 & 18 \end{vmatrix} = \begin{vmatrix} 0 & 4 \\ 1 & 18 \end{vmatrix} = -4.$$

Exercises 13.1

1 Evaluate the determinants

(a) $\begin{vmatrix} 10 & 11 \\ 9 & 10 \end{vmatrix}$, (b) $\begin{vmatrix} 4 & 5 \\ 3 & 4 \end{vmatrix}$, (c) $\begin{vmatrix} 40 & 50 \\ 30 & 40 \end{vmatrix}$, (d) $\begin{vmatrix} 48 & 60 \\ 44 & 55 \end{vmatrix}$.

2 Given that $\mathbf{A} = \begin{pmatrix} 7 & 5 \\ 9 & 6 \end{pmatrix}$, evaluate the determinants of the matrices

(a) **A**, (b) **2A**, (c) **A²**, (d) **3A**.

3 Given that $\mathbf{P} = \begin{pmatrix} 7 & -2 \\ 4 & 3 \end{pmatrix}$ and that $\mathbf{Q} = \begin{pmatrix} 8 & 9 \\ 4 & 5 \end{pmatrix}$, find the matrix **PQ**.

Verify that $\det \mathbf{PQ} = (\det \mathbf{P})(\det \mathbf{Q})$.

13.2 Determinants of order 3

Let **A** be the square matrix $\begin{pmatrix} a_1 & b_1 & c_1 \\ a_2 & b_2 & c_2 \\ a_3 & b_3 & c_3 \end{pmatrix}$.

By definition, det **A** is given by

$$\begin{vmatrix} a_1 & b_1 & c_1 \\ a_2 & b_2 & c_2 \\ a_3 & b_3 & c_3 \end{vmatrix} = a_1 \begin{vmatrix} b_2 & c_2 \\ b_3 & c_3 \end{vmatrix} - b_1 \begin{vmatrix} a_2 & c_2 \\ a_3 & c_3 \end{vmatrix} + c_1 \begin{vmatrix} a_2 & b_2 \\ a_3 & b_3 \end{vmatrix}$$

$$= a_1 b_2 c_3 - a_1 b_3 c_2 - b_1 a_2 c_3 + b_1 a_3 c_2 + c_1 a_2 b_3 - c_1 a_3 b_2.$$

The coefficient of any element in this expansion is known as its cofactor. The cofactors of a_1, a_2, a_3 are denoted by A_1, A_2, A_3 respectively, so that

$$\det A = a_1 A_1 + a_2 A_2 + a_3 A_3,$$

where $\quad A_1 = b_2 c_3 - b_3 c_2, \quad A_2 = b_3 c_1 - b_1 c_3, \quad A_3 = b_1 c_2 - b_2 c_1.$

Similarly, $\qquad\qquad \det A = b_1 B_1 + b_2 B_2 + b_3 B_3,$

where $\quad B_1 = c_2 a_3 - c_3 a_2, \quad B_2 = c_3 a_1 - c_1 a_3, \quad B_3 = c_1 a_2 - c_2 a_1.$

The value of det **A** is unchanged if rows and columns are interchanged.

For $\qquad \begin{vmatrix} a_1 & a_2 & a_3 \\ b_1 & b_2 & b_3 \\ c_1 & c_2 & c_3 \end{vmatrix} = a_1 \begin{vmatrix} b_2 & b_3 \\ c_2 & c_3 \end{vmatrix} - a_2 \begin{vmatrix} b_1 & b_3 \\ c_1 & c_3 \end{vmatrix} + a_3 \begin{vmatrix} b_1 & b_2 \\ c_1 & c_2 \end{vmatrix}$

$$= a_1 A_1 + a_2 A_2 + a_3 A_3 = \det A.$$

The interchange of two rows changes the sign of det **A**.

For example, consider the row transformation which interchanges row 2 with row 3. Let **P** be the matrix

$$\begin{pmatrix} 1 & 0 & 0 \\ 0 & 0 & 1 \\ 0 & 1 & 0 \end{pmatrix},$$

and let $\mathbf{PA} = \mathbf{B}$, so that

$$\begin{pmatrix} 1 & 0 & 0 \\ 0 & 0 & 1 \\ 0 & 1 & 0 \end{pmatrix} \begin{pmatrix} a_1 & b_1 & c_1 \\ a_2 & b_2 & c_2 \\ a_3 & b_3 & c_3 \end{pmatrix} = \begin{pmatrix} a_1 & b_1 & c_1 \\ a_3 & b_3 & c_3 \\ a_2 & b_2 & c_2 \end{pmatrix} = \mathbf{B}.$$

Then det $\mathbf{B}$ = (det $\mathbf{P}$)(det $\mathbf{A}$).

But det $\mathbf{P}$ = -1, and so det $\mathbf{B}$ = $-$det $\mathbf{A}$.

When two rows of the matrix $\mathbf{A}$ are identical, det $\mathbf{A}$ = 0. For the interchange of these two rows makes no difference to the value of det $\mathbf{A}$.

The value of a determinant is unchanged when a constant multiple of each element in one row is added to the corresponding element of another row.

For example, let k times row 3 of the matrix $\mathbf{A}$ be added to row 2. Let $\mathbf{Q}$ be the matrix

$$\begin{pmatrix} 1 & 0 & 0 \\ 0 & 1 & k \\ 0 & 0 & 1 \end{pmatrix},$$

and let $\mathbf{QA}$ = $\mathbf{C}$, so that

$$\begin{pmatrix} 1 & 0 & 0 \\ 0 & 1 & k \\ 0 & 0 & 1 \end{pmatrix} \begin{pmatrix} a_1 & b_1 & c_1 \\ a_2 & b_2 & c_2 \\ a_3 & b_3 & c_3 \end{pmatrix} = \begin{pmatrix} a_1 & b_1 & c_1 \\ a_2 + ka_3 & b_2 + kb_3 & c_2 + kc_3 \\ a_3 & b_3 & c_3 \end{pmatrix} = \mathbf{C}.$$

Then det $\mathbf{C}$ = (det $\mathbf{Q}$)(det $\mathbf{A}$).

But det $\mathbf{Q}$ = 1, and so det $\mathbf{C}$ = det $\mathbf{A}$.

If the elements of a row of the matrix $\mathbf{A}$ have a common factor k then k is a factor of det $\mathbf{A}$. This follows at once from the fact that each term in the expansion of det $\mathbf{A}$ contains one element from each row.

Example 1

Evaluate det $\mathbf{A}$, where

$$\mathbf{A} = \begin{pmatrix} 3 & 1 & 2 \\ 9 & 5 & 8 \\ 7 & 4 & 6 \end{pmatrix},$$

by converting the matrix $\mathbf{A}$ to echelon form, i.e. to a matrix in which each element beneath the principal diagonal is zero. (The principal diagonal of a matrix runs from the first element in the first row to the last element in the last row.)

The matrix can be converted to echelon form in the three steps, each equivalent to a row operation.

(1) Multiply row 1 by 3 and subtract from row 2. In matrix form, this operation is given by

$$\begin{pmatrix} 1 & 0 & 0 \\ -3 & 1 & 0 \\ 0 & 0 & 1 \end{pmatrix} \begin{pmatrix} 3 & 1 & 2 \\ 9 & 5 & 8 \\ 7 & 4 & 6 \end{pmatrix} = \begin{pmatrix} 3 & 1 & 2 \\ 0 & 2 & 2 \\ 7 & 4 & 6 \end{pmatrix}.$$

(2) Multiply row 1 by 7/3 and subtract from row 3. In matrix form, this operation is given by

$$\begin{pmatrix} 1 & 0 & 0 \\ 0 & 1 & 0 \\ -7/3 & 0 & 1 \end{pmatrix} \begin{pmatrix} 3 & 1 & 2 \\ 0 & 2 & 2 \\ 7 & 4 & 6 \end{pmatrix} = \begin{pmatrix} 3 & 1 & 2 \\ 0 & 2 & 2 \\ 0 & 5/3 & 4/3 \end{pmatrix}.$$

(3) Multiply the new row 2 by 5/6 and subtract from row 3. In matrix form,

$$\begin{pmatrix} 1 & 0 & 0 \\ 0 & 1 & 0 \\ 0 & -5/6 & 1 \end{pmatrix} \begin{pmatrix} 3 & 1 & 2 \\ 0 & 2 & 2 \\ 0 & 5/3 & 4/3 \end{pmatrix} = \begin{pmatrix} 3 & 1 & 2 \\ 0 & 2 & 2 \\ 0 & 0 & -1/3 \end{pmatrix}.$$

The determinant of the resulting matrix in echelon form is $3 \times 2 \times (-1/3)$, i.e. -2.

At each step, the determinant of the matrix used to produce the desired row transformation is equal to 1.

It follows that det $\mathbf{A}$ also is equal to -2.

Example 2

Given that det $\mathbf{A} = 0$,

$$\text{where } \mathbf{A} = \begin{pmatrix} a_1 & b_1 & c_1 \\ a_2 & b_2 & c_2 \\ a_3 & b_3 & c_3 \end{pmatrix},$$

show that the columns of the matrix $\mathbf{A}$, regarded as vectors, are linearly dependent.

Let A_1, B_1, C_1 be the cofactors of a_1, b_1, c_1 respectively in det $\mathbf{A}$.

Then $\qquad\qquad A_1 a_1 + B_1 b_1 + C_1 c_1 = \text{det } \mathbf{A}.$ $\qquad$ ... (1)

This result follows directly from the definition of det $\mathbf{A}$ at the beginning of this section.

Also $\qquad\qquad A_1 a_2 + B_1 b_2 + C_1 c_2 = 0,$ $\qquad\qquad$ ... (2)
$\qquad\qquad\qquad A_1 a_3 + B_1 b_3 + C_1 c_3 = 0.$ $\qquad\qquad$ ... (3)

These results can be verified by replacing A_1 by $(b_2 c_3 - b_3 c_2)$, B_1 by $(c_2 a_3 - c_3 a_2)$ and C_1 by $(a_2 b_3 - a_3 b_2)$.

These three equations give the vector equation

$$A_1 \begin{pmatrix} a_1 \\ a_2 \\ a_3 \end{pmatrix} + B_1 \begin{pmatrix} b_1 \\ b_2 \\ b_3 \end{pmatrix} + C_1 \begin{pmatrix} c_1 \\ c_2 \\ c_3 \end{pmatrix} = \begin{pmatrix} \text{det } \mathbf{A} \\ 0 \\ 0 \end{pmatrix}. \qquad \text{... (4)}$$

But det $\mathbf{A} = 0$, and so, assuming that A_1, B_1, C_1 are not all zero, it follows that the three column vectors are linearly dependent.

By the method used to obtain equation (4), it can be shown that

$$A_2 \begin{pmatrix} a_1 \\ a_2 \\ a_3 \end{pmatrix} + B_2 \begin{pmatrix} b_1 \\ b_2 \\ b_3 \end{pmatrix} + C_2 \begin{pmatrix} c_1 \\ c_2 \\ c_3 \end{pmatrix} = \begin{pmatrix} 0 \\ \text{det } \mathbf{A} \\ 0 \end{pmatrix}.$$

But det $\mathbf{A} = 0$, and so, assuming that A_2, B_2, C_2 are not all zero, it follows that the three column vectors are linearly dependent.

This leaves the case when A_1, B_1, C_1, A_2, B_2, C_2 are all zero.

Now
$$A_1 = 0 \Rightarrow b_2 c_3 - b_3 c_2 = 0 \Rightarrow b_2 : b_3 = c_2 : c_3,$$
$$B_1 = 0 \Rightarrow c_2 a_3 - c_3 a_2 = 0 \Rightarrow c_2 : c_3 = a_2 : a_3.$$

Hence
$$a_2 : a_3 = b_2 : b_3 = c_2 : c_3.$$

Similarly, from $A_2 = 0$ and $B_2 = 0$,
$$a_1 : a_3 = b_1 : b_3 = c_1 : c_3.$$

This shows that when A_1, A_2, B_1, B_2 are all zero, each column vector is a multiple of any other column vector.

Hence, in all cases, det $\mathbf{A} = 0$ is a sufficient condition for the column vectors of the matrix $\mathbf{A}$ to be linearly dependent.

Example 3
The system of equations

$$a_1 x + b_1 y + c_1 z = 0, \qquad \ldots \text{(1)}$$
$$a_2 x + b_2 y + c_2 z = 0, \qquad \ldots \text{(2)}$$
$$a_3 x + b_3 y + c_3 z = 0, \qquad \ldots \text{(3)}$$

has a solution other than $x = y = z = 0$. Show that det $\mathbf{A} = 0$, where

$$\mathbf{A} = \begin{pmatrix} a_1 & b_1 & c_1 \\ a_2 & b_2 & c_2 \\ a_3 & b_3 & c_3 \end{pmatrix}.$$

Let A_1, A_2, A_3 be the cofactors of a_1, a_2, a_3 respectively in det $\mathbf{A}$. Multiply equation (1) by A_1, equation (2) by A_2, equation (3) by A_3 and then add the three equations together.

The coefficients of y and z vanish identically, and the coefficient of x is $(a_1 A_1 + a_2 A_2 + a_3 A_3)$, which equals det $\mathbf{A}$. This shows that $(\text{det } \mathbf{A})x = 0$.

Similarly, it can be shown that $(\text{det } \mathbf{A})y = 0$ and $(\text{det } \mathbf{A})z = 0$. It follows that, for any solution other than $x = y = z = 0$ to exist, det $\mathbf{A}$ must be zero.

Intersection of three planes
Let the equations of three distinct planes be

$$a_1 x + b_1 y + c_1 z = d_1, \qquad \ldots \text{(1)}$$
$$a_2 x + b_2 y + c_2 z = d_2, \qquad \ldots \text{(2)}$$
$$a_3 x + b_3 y + c_3 z = d_3, \qquad \ldots \text{(3)}$$

When this system of three simultaneous equations has a unique solution or has more than one solution, the equations are said to be *consistent*.

When the system has no solution, the equations are said to be *inconsistent*.

When there is a unique solution, the three planes meet in a single point. For example, the planes

$$3x + y - 2z = 7,$$
$$x + 2y + 3z = 1,$$
$$2x + 3y + 4z = 3$$

have only the point $(4, -3, 1)$ in common.

When there is more than one solution, the three planes meet in a common line. For example, for all values of the parameter t, the equations

$$x + y - z = 2,$$
$$2x - y - z = 3,$$
$$x + 4y - 2z = 3$$

are satisfied by

$$x = 1 + 2t, \ y = t, \quad z = 3t - 1.$$

These parametric equations give the straight line through which each plane passes.

When there is no solution, there are three possibilities:
(1) The three planes are parallel to one another.
(2) Two planes only are parallel to one another.
(3) The three planes meet in three parallel lines, forming a prism.

Let the equations (1), (2), and (3) above be multiplied by A_1, A_2, A_3 respectively, where A_1, A_2, A_3 are the cofactors of a_1, a_2, a_3 in the determinant of the matrix $\mathbf{A}$, where

$$\mathbf{A} = \begin{pmatrix} a_1 & b_1 & c_1 \\ a_2 & b_2 & c_2 \\ a_3 & b_3 & c_3 \end{pmatrix}.$$

Since $b_1 A_1 + b_2 A_2 + b_3 A_3 = 0$ and $c_1 A_1 + c_2 A_2 + c_3 A_3 = 0$, the sum of the resulting equations gives the equation

$$(a_1 A_1 + a_2 A_2 + a_3 A_3)x = (d_1 A_1 + d_2 A_2 + d_3 A_3)$$

$$\Rightarrow \qquad (\det \mathbf{A})x = \begin{vmatrix} d_1 & b_1 & c_1 \\ d_2 & b_2 & c_1 \\ d_3 & b_3 & c_3 \end{vmatrix}.$$

Similarly,

$$(\det \mathbf{A})y = \begin{vmatrix} a_1 & d_1 & c_1 \\ a_2 & d_2 & c_2 \\ a_3 & d_3 & c_3 \end{vmatrix}$$

and

$$(\det \mathbf{A})z = \begin{vmatrix} a_1 & b_1 & d_1 \\ a_2 & b_2 & d_2 \\ a_3 & b_3 & d_3 \end{vmatrix}.$$

Provided that det $\mathbf{A}$ is not zero, these three equations give the values of x, y and z at the point of intersection of the three planes.

This shows that when det $\mathbf{A}$ is not zero, the equations (1), (2) and (3) have a unique solution.

Example 4

Show that the three equations

$$ax + 2y + 3z = 13, \qquad \ldots (1)$$

$$-x + 3y + z = 11, \qquad \ldots (2)$$

$$3x + y + 2z = 12 \qquad \ldots (3)$$

will have a unique solution provided that a is not equal to 4.
(a) Solve the equations when $a = 5$.
(b) Show that when $a = 4$ the equations represent three planes which meet in three parallel lines.

The equations will have a unique solution provided that

$$\begin{vmatrix} a & 2 & 3 \\ -1 & 3 & 1 \\ 3 & 1 & 2 \end{vmatrix} \neq 0$$

$$\Rightarrow \qquad a(5) - 2(-5) + 3(-10) \neq 0$$

$$\Rightarrow \qquad a \neq 4.$$

(a) Let $a = 5$.
Equation (1) $+ 5 \times$ equation (2) gives $17y + 8z = 68$.
Equation (3) $+ 3 \times$ equation (2) gives $10y + 5z = 45$.
These two equations give $y = -4$, $z = 17$.
Substitution in equation (2) gives $x = -6$.

(b) Let $a = 4$.
Consider the two planes given by

$$4x + 2y + 3z = 13,$$

$$-x + 3y + z = 11.$$

Elimination of x gives $\qquad 14y + 7z = 57.$
Elimination of z gives $\qquad 7x - 7y = -20.$
Hence the line of intersection of these planes is given by

$$x = t - 20/7, \quad y = t, \quad z = 57/7 - 2t.$$

The direction ratios of this line are $1:1:-2$.

Consider next the two planes given by

$$-x + 3y + z = 11,$$

$$3x + y + 2z = 12.$$

Elimination of x gives

$$10y + 5z = 45$$

$\Rightarrow$

$$2y + z = 9.$$

Elimination of z gives

$$5x - 5y = -10$$

$\Rightarrow$

$$x - y = -2.$$

Hence the line of intersection of these planes is given by

$$x = t - 2, \quad y = t, \quad z = 9 - 2t.$$

The direction ratios of this line are also $1:1:-2$.

Similarly, the line of intersection of the other pair of planes is given by

$$x = t + 10, \quad y = t, \quad z = -9 - 2t.$$

This line has direction ratios $1:1:-2$.

Thus, when $a = 4$, the planes meet in three parallel lines.

Example 5

The cartesian equations of three planes are

$$kx - y - 8z = 0,$$

$$x + y - z = 0,$$

$$3x + ky - 5z = 0.$$

Find the values of the constant k for which the planes meet in a common line, and in each case find the equations of the line.

By Example 3 above, the equations will have a solution other than $x = y = z = 0$ provided that

$$\begin{vmatrix} k & -1 & -8 \\ 1 & 1 & -1 \\ 3 & k & -5 \end{vmatrix} = 0$$

$\Rightarrow$

$$k(-5 + k) + (-5 + 3) - 8(k - 3) = 0$$

$\Rightarrow$

$$k^2 - 13k + 22 = 0$$

$\Rightarrow$

$$k = 2 \text{ or } k = 11.$$

(a) When $k = 2$, the equations become

$$2x - y - 8z = 0,$$

$$x + y - z = 0,$$

$$3x + 2y - 5z = 0.$$

From the first two equations, $3x - 9z = 0$, i.e. $x = 3z$.
By substitution in the first equation, $y = -2z$.
Hence $\qquad\qquad\qquad x:y:z = 3:-2:1$.
These ratios satisfy the third equation.
The straight line common to the three planes is given by the parametric equations

$$x = 3t, \qquad y = -2t, \qquad z = t.$$

(b) When $k = 11$, the equations become

$$11x - y - 8z = 0,$$

$$x + y - z = 0,$$

$$3x + 11y - 5z = 0.$$

From the first two equations, $12x - 9z = 0$, i.e. $4x = 3z$.
By substitution in the second equation, $4y = z$.
Hence $\qquad\qquad\qquad x:y:z = 3:1:4$.
These ratios satisfy the third equation.
This shows that the planes intersect in the straight line with parametric equations

$$x = 3t, \qquad y = t, \qquad z = 4t.$$

Exercises 13.2

1 Evaluate the following determinants:

(a) $\begin{vmatrix} 37 & 18 & 27 \\ 31 & 17 & 21 \\ 13 & 7 & 9 \end{vmatrix}$,

(b) $\begin{vmatrix} 13 & 8 & 12 \\ 10 & 6 & 9 \\ 12 & 7 & 11 \end{vmatrix}$,

(c) $\begin{vmatrix} 1 & 2 & 3 \\ 2 & 4 & 0 \\ -1 & 3 & 1 \end{vmatrix}$,

(d) $\begin{vmatrix} 9 & 11 & 7 \\ 5 & -4 & 3 \\ 7 & 3 & 5 \end{vmatrix}$,

(e) $\begin{vmatrix} 17 & 10 & 15 \\ 9 & 5 & 10 \\ 13 & 8 & 9 \end{vmatrix}$,

(f) $\begin{vmatrix} 61 & 83 & 75 \\ 29 & 37 & 33 \\ 43 & 53 & 47 \end{vmatrix}$.

2 Solve the equation

$$\begin{vmatrix} x & 2 & 3 \\ 3 & x+1 & 1 \\ 1 & 2 & x+2 \end{vmatrix} = 0.$$

3 Find the values of k for which the three planes given by the equations

$$2x - ky + 2z = 0,$$

$$6x + ky + 2z = 0,$$

$$4x - 3y + kz = 0,$$

meet in a straight line.

Find the direction ratios of the line in each case.

4 Show that

$$\begin{vmatrix} -2 & 2 & -1 \\ 2 & 6 & 5 \\ 2 & 4 & 4 \end{vmatrix} = 0.$$

Find the cofactors of the elements of the first row, and use them to show that the three column vectors are linearly dependent.

Find also the cofactors of the elements of the first column, and use them to show that the three row vectors are linearly dependent.

5 Use row transformations to convert the following matrices to echelon form, and hence evaluate their determinants.

(a) $\begin{pmatrix} 3 & 2 & 1 \\ 9 & 7 & 7 \\ 6 & 2 & 1 \end{pmatrix}$,

(b) $\begin{pmatrix} 4 & 8 & 12 \\ -1 & 1 & -1 \\ 2 & -2 & 4 \end{pmatrix}$,

(c) $\begin{pmatrix} 6 & 12 & 3 \\ 6 & 14 & 4 \\ 2 & -2 & 1 \end{pmatrix}$,

(d) $\begin{pmatrix} 12 & 24 & 2 \\ 6 & 3 & 1 \\ 4 & 11 & 1 \end{pmatrix}$.

6 Show that the three planes given by the equations

$$2x - 2y + z = 5,$$

$$5x + 7y + 3z = 1,$$

$$3x - 4y - 2z = 5,$$

intersect in a single point. Find the coordinates of this point.

7 Show that the three planes

$$5x - 2y - 3z = 1,$$

$$4x + 2y - 6z = 8,$$

$$x + 3y - 4z = 7,$$

meet in a common line.

Express the equations of this line in parametric form.

8 Show that the planes

$$x - 2y + 3z = 2,$$
$$x - y + z = 14,$$
$$2x - 3y + 4z = 6,$$

meet in three lines which are parallel to the line $2x = y = 2z$.

9 Show that the planes

$$3x - 2y - 2z = 1,$$
$$8x - 7y - 4z = 4,$$
$$6x - 4y - 4z = 2,$$

have a common line of intersection, and find its direction ratios.

10 Show that the equations

$$4x + 6y + 2z = 5,$$
$$3x - 2y + 4z = 6,$$
$$6x + 9y + 3z = 7,$$

are inconsistent, and give a geometrical interpretation.

13.3 Properties of matrices

A matrix with m rows and n columns is said to be of order $m \times n$. Two matrices **A** and **B** of the same order can be added together to form a third matrix in which each element is the sum of the corresponding elements in **A** and **B**.

Matrix addition is commutative, i.e.

$$\mathbf{A} + \mathbf{B} = \mathbf{B} + \mathbf{A}.$$

It is also associative, since

$$\mathbf{A} + (\mathbf{B} + \mathbf{C}) = (\mathbf{A} + \mathbf{B}) + \mathbf{C}.$$

In the matrix $k\mathbf{A}$, where k is a scalar, each element is k times the corresponding element in the matrix **A**.

The product of two matrices

When the matrix **P** is of order $m \times p$ and the matrix **Q** is of order $p \times n$, **Q** can be multiplied by **P** to form the product **PQ**. In this product **PQ**, the matrix **Q** is said to be premultiplied by **P**, while **P** is said to be postmultiplied by **Q**.

The ith row of **P** and the jth column of **Q** each contain p elements. The sum of the p products of corresponding elements in the ith row of **P** and the jth column of **Q** gives the value of the element in the ith row and jth column of the product **PQ**.

The matrix **PQ** will be of order $m \times n$.

For example, let the matrices **P** and **Q** be given by

$$\mathbf{P} = \begin{pmatrix} 4 & 1 & 2 \\ 3 & 2 & 5 \end{pmatrix}, \quad \mathbf{Q} = \begin{pmatrix} 0 & 1 & 0 & 1 \\ 2 & 1 & 6 & 0 \\ 0 & 1 & 0 & -2 \end{pmatrix}.$$

P is of order 2×3, and **Q** is of order 3×4.

The product **PQ** is of order 2×4, and is given by

$$\mathbf{PQ} = \begin{pmatrix} 2 & 7 & 6 & 0 \\ 4 & 10 & 12 & -7 \end{pmatrix}.$$

When **P** is of order $m \times p$ and **Q** is of order $p \times m$, each of the products **PQ** and **QP** can be formed.

The product **PQ**, in which **Q** is premultiplied by **P**, will be of order $m \times m$, i.e. a square matrix of order m.

The product **QP**, in which **P** is premultiplied by **Q**, will be of order $p \times p$, i.e. a square matrix of order p.

In the case when **P** and **Q** are square matrices of the same order, both products **PQ** and **QP** will exist, but in general they will not be equal. For example,

$$\begin{pmatrix} 3 & 0 \\ 2 & 1 \end{pmatrix}\begin{pmatrix} 1 & 1 \\ 0 & 2 \end{pmatrix} = \begin{pmatrix} 3 & 3 \\ 2 & 4 \end{pmatrix} \quad \text{but} \quad \begin{pmatrix} 1 & 1 \\ 0 & 2 \end{pmatrix}\begin{pmatrix} 3 & 0 \\ 2 & 1 \end{pmatrix} = \begin{pmatrix} 5 & 1 \\ 4 & 2 \end{pmatrix}.$$

In general, matrix multiplication is **not** commutative.

It can be shown that matrix multiplication is associative, i.e.

$$\mathbf{A(BC)} = \mathbf{(AB)C}.$$

Also, for three matrices of suitable orders, the distributive laws hold good, i.e.

$$\mathbf{A(B + C)} = \mathbf{AB} + \mathbf{AC},$$

$$\mathbf{(B + C)A} = \mathbf{BA} + \mathbf{CA}.$$

Transpose of a matrix

The transpose $\mathbf{A}^\mathrm{T}$ of the matrix **A** is formed by interchanging the rows of **A** with its columns.

For example, when **A** is the matrix

$$\begin{pmatrix} 4 & 2 & 0 \\ 3 & 1 & 2 \end{pmatrix}, \quad \mathbf{A}^\mathrm{T} = \begin{pmatrix} 4 & 3 \\ 2 & 1 \\ 0 & 2 \end{pmatrix}.$$

Clearly, the transpose of the matrix $\mathbf{A}^\mathrm{T}$ is **A**, i.e. $(\mathbf{A}^\mathrm{T})^\mathrm{T} = \mathbf{A}$.

Let **P** be a square matrix. If **P** and its transpose $\mathbf{P}^\mathrm{T}$ are equal, the matrix **P** is said to be *symmetric*.

Examples of symmetric matrices are

$$\begin{pmatrix} 1 & 3 \\ 3 & 2 \end{pmatrix} \quad \text{and} \quad \begin{pmatrix} 0 & 4 & 2 \\ 4 & -1 & 3 \\ 2 & 3 & 1 \end{pmatrix}.$$

Let $\mathbf{Q}$ be a square matrix such that $\mathbf{Q}^{\mathrm{T}} = -\mathbf{Q}$. Then $\mathbf{Q}$ is said to be *skew-symmetric*. For example,

$$\begin{pmatrix} 0 & 3 \\ -3 & 0 \end{pmatrix} \quad \text{and} \quad \begin{pmatrix} 0 & 4 & 2 \\ -4 & 0 & -3 \\ -2 & 3 & 0 \end{pmatrix} \text{ are skew-symmetric}.$$

Note that the elements on the principal diagonal of a skew-symmetric matrix must be zero.

Example 1

Express the matrix $\begin{pmatrix} 2 & 5 \\ 1 & 4 \end{pmatrix}$ in the form $\mathbf{P} + \mathbf{Q}$, where $\mathbf{P}$ is a symmetric matrix and $\mathbf{Q}$ is a skew-symmetric matrix.

Since the elements on the principal diagonal of $\mathbf{Q}$ must be zero, $\mathbf{P}$ and $\mathbf{Q}$ must be of the form

$$\mathbf{P} = \begin{pmatrix} 2 & a \\ a & 4 \end{pmatrix}, \quad \mathbf{Q} = \begin{pmatrix} 0 & b \\ -b & 0 \end{pmatrix}.$$

Then $\qquad\qquad a + b = 5, \ a - b = 1$

$\Rightarrow \qquad\qquad\qquad a = 3, b = 2$

$\Rightarrow \qquad \begin{pmatrix} 2 & 5 \\ 1 & 4 \end{pmatrix} = \begin{pmatrix} 2 & 3 \\ 3 & 4 \end{pmatrix} + \begin{pmatrix} 0 & 2 \\ -2 & 0 \end{pmatrix}.$

Transpose of a product

Consider first the product $\mathbf{AB}$, where $\mathbf{A} = \begin{pmatrix} a & b \\ c & d \end{pmatrix}$ and $\mathbf{B} = \begin{pmatrix} p & q \\ r & s \end{pmatrix}$.

$$\mathbf{AB} = \begin{pmatrix} ap + br & aq + bs \\ cp + dr & cq + ds \end{pmatrix}.$$

Also $\qquad \mathbf{A}^{\mathrm{T}} = \begin{pmatrix} a & c \\ b & d \end{pmatrix} \quad \text{and} \quad \mathbf{B}^{\mathrm{T}} = \begin{pmatrix} p & r \\ q & s \end{pmatrix}$

$\Rightarrow \qquad \mathbf{B}^{\mathrm{T}}\mathbf{A}^{\mathrm{T}} = \begin{pmatrix} ap + br & cp + dr \\ aq + bs & cq + ds \end{pmatrix}.$

This shows that for 2×2 matrices $(\mathbf{AB})^{\mathrm{T}} = \mathbf{B}^{\mathrm{T}}\mathbf{A}^{\mathrm{T}}$.

In the general case, the element in the ith row and jth column of the product $\mathbf{AB}$ is found from the ith row of $\mathbf{A}$ and the jth column of $\mathbf{B}$.

But the ith row of $\mathbf{A}$ is the ith column of $\mathbf{A}^{\mathrm{T}}$, and the jth column of $\mathbf{B}$ is the jth row of $\mathbf{B}^{\mathrm{T}}$.

Hence the element in the jth row and ith column of $\mathbf{B}^T\mathbf{A}^T$ is the same as the element in the ith row and the jth column of $\mathbf{AB}$. It follows that

$$(\mathbf{AB})^T = \mathbf{B}^T\mathbf{A}^T.$$

Note the reversal of the order of the matrices.

There is an important consequence when $\mathbf{B}$ is the transpose of $\mathbf{A}$. When $\mathbf{B} = \mathbf{A}^T$, $\mathbf{B}^T = \mathbf{A}$. This gives

$$(\mathbf{AA}^T)^T = \mathbf{AA}^T,$$

showing that the matrix $\mathbf{AA}^T$ is symmetric. For example,

$$\begin{pmatrix} 3 & -2 \\ 1 & 4 \end{pmatrix}\begin{pmatrix} 3 & 1 \\ -2 & 4 \end{pmatrix} = \begin{pmatrix} 13 & -5 \\ -5 & 17 \end{pmatrix}$$

Example 2

Express the transpose of the product $\mathbf{PQR}$ in terms of $\mathbf{P}^T$, $\mathbf{Q}^T$ and $\mathbf{R}^T$.

Let $\mathbf{QR} = \mathbf{S}$. Then $(\mathbf{PS})^T = \mathbf{S}^T\mathbf{P}^T$.

But
$$\mathbf{S}^T = (\mathbf{QR})^T = \mathbf{R}^T\mathbf{Q}^T$$

$\Rightarrow$
$$(\mathbf{PS})^T = \mathbf{R}^T\mathbf{Q}^T\mathbf{P}^T$$

$\Rightarrow$
$$(\mathbf{PQR})^T = \mathbf{R}^T\mathbf{Q}^T\mathbf{P}^T.$$

Example 3

Find the matrix $\mathbf{A}$ of the form $\begin{pmatrix} 2 & -1 \\ c & d \end{pmatrix}$ such that $\mathbf{A}^2 = \mathbf{A}$.

$$\begin{pmatrix} 2 & -1 \\ c & d \end{pmatrix}\begin{pmatrix} 2 & -1 \\ c & d \end{pmatrix} = \begin{pmatrix} 4 - c & -2 - d \\ 2c + cd & -c + d^2 \end{pmatrix}.$$

By equating each element in this product to the corresponding element in the matrix $\mathbf{A}$,

$$4 - c = 2, \quad -2 - d = -1, \quad 2c + cd = c, \quad -c + d^2 = d.$$

From the first two equations, $c = 2$ and $d = -1$.
These values satisfy the other two equations. Hence

$$\mathbf{A} = \begin{pmatrix} 2 & -1 \\ 2 & -1 \end{pmatrix}.$$

Exercises 13.3

1   Given that $\mathbf{A} = \begin{pmatrix} 4 & 1 \\ 5 & 2 \end{pmatrix}$ and $\mathbf{B} = \begin{pmatrix} 3 & -1 \\ -4 & 2 \end{pmatrix}$, find the products (a) $\mathbf{AB}$,

(b) $\mathbf{BA}$, (c)$\mathbf{B}^T\mathbf{A}^T$, (d) $\mathbf{A}^T\mathbf{B}^T$.

2 Find the products **PQ** and **QP** given that

$$\mathbf{P} = \begin{pmatrix} 3 & 1 & 2 \\ 2 & 0 & 4 \end{pmatrix}, \qquad \mathbf{Q} = \begin{pmatrix} 1 & -2 \\ -2 & 3 \\ -1 & 1 \end{pmatrix}.$$

3 Find the matrix **A** such that

$$2\mathbf{A} - 3\begin{pmatrix} -2 & 3 \\ 2 & 1 \\ 2 & 1 \end{pmatrix} = 2\begin{pmatrix} 4 & 0 \\ 2 & 2 \\ 3 & 4 \end{pmatrix} - 3\begin{pmatrix} 0 & 1 \\ 2 & 1 \\ 2 & 3 \end{pmatrix}.$$

4 Form the products **PQ** and **QP**

(a) when $\mathbf{P} = \begin{pmatrix} 2 & 1 \\ 1 & 1 \end{pmatrix}$ and $\mathbf{Q} = \begin{pmatrix} 1 & -1 \\ -1 & 2 \end{pmatrix}$,

(b) when $\mathbf{P} = \begin{pmatrix} 1 & 0 \\ 2 & 0 \end{pmatrix}$ and $\mathbf{Q} = \begin{pmatrix} 0 & 0 \\ 5 & 7 \end{pmatrix}$.

5 Given that A is a square matrix, show that $\mathbf{A} + \mathbf{A}^{\mathrm{T}}$ is symmetric, and that $\mathbf{A} - \mathbf{A}^{\mathrm{T}}$ is skew-symmetric.

Express as the sum of a symmetric and a skew-symmetric matrix

(a) $\begin{pmatrix} 4 & 4 \\ 0 & 1 \end{pmatrix}$, (b) $\begin{pmatrix} 1 & 1 & 0 \\ -1 & 2 & -2 \\ 2 & 2 & 3 \end{pmatrix}$.

6 Show that, for the set of diagonal matrices of order 3, multipliation is commutative. (A diagonal matrix is a square matrix in which all elements not on the principal diagonal are zero.)

7 A zero matrix (or null matrix) is one in which every element is zero. Such a matrix is denoted by **0**.

Show that, if $\mathbf{A} = \begin{pmatrix} 1 & 2 \\ -\frac{1}{2} & -1 \end{pmatrix}$, $\mathbf{A}^2 = \mathbf{0}$.

Show that the most general matrix of the form $\begin{pmatrix} 1 & b \\ c & d \end{pmatrix}$ such that its square is a zero matrix is $\begin{pmatrix} 1 & b \\ -1/b & -1 \end{pmatrix}$.

8 Find the sum of **PQ** and **QP** when

$$\mathbf{P} = \begin{pmatrix} 1 & -1 \\ 3 & -1 \end{pmatrix} \quad \text{and} \quad \mathbf{Q} = \begin{pmatrix} 1 & 1 \\ 5 & -1 \end{pmatrix}.$$

9 Find (a) $\mathbf{A}^2$, (b) det **A**, given that A is the matrix $\begin{pmatrix} 1 & -2 & -3 \\ -1 & 2 & 3 \\ 1 & -2 & -3 \end{pmatrix}$.

10 The matrices **A** and **B** are symmetric. Prove that the product **AB** is symmetric if and only if **AB** equals **BA**.

13.4 The inverse of a matrix

It was shown in *Advanced Mathematics 1*, Section 13.5, that when $ad - bc$ is not zero, the matrix **A** given by $\mathbf{A} = \begin{pmatrix} a & b \\ c & d \end{pmatrix}$ will possess an inverse matrix.

This inverse is denoted by $\mathbf{A}^{-1}$ and is given by

$$\mathbf{A}^{-1} = \frac{1}{(ad - bc)} \begin{pmatrix} d & -b \\ -c & a \end{pmatrix}.$$

Then $\mathbf{AA}^{-1} = \mathbf{I}$ and $\mathbf{A}^{-1}\mathbf{A} = \mathbf{I}$, where $\mathbf{I} = \begin{pmatrix} 1 & 0 \\ 0 & 1 \end{pmatrix}$.

Let **P** be a square matrix of order n, and let **I** denote the unit matrix of order n in which each element on the principal diagonal is 1.

If there exists a matrix **Q** such that

$$\mathbf{PQ} = \mathbf{QP} = \mathbf{I},$$

the matrix **Q** is known as the inverse of the matrix **P**.

A matrix whose determinant is zero is said to be singular. Such a matrix cannot possess an inverse. For if

$$\mathbf{PQ} = \mathbf{I},$$

$$\det(\mathbf{PQ}) = \det(\mathbf{I}) = 1$$

$\Rightarrow$ $$(\det \mathbf{P})(\det \mathbf{Q}) = 1$$

$\Rightarrow$ $$\det \mathbf{P} \neq 0.$$

A non-singular matrix cannot have two inverses.

For suppose that $\mathbf{AP} = \mathbf{PA} = \mathbf{I}$ and $\mathbf{AQ} = \mathbf{QA} = \mathbf{I}$.

Then $$\mathbf{PAQ} = \mathbf{P(AQ)} = \mathbf{PI} = \mathbf{P},$$

and $$\mathbf{PAQ} = \mathbf{(PA)Q} = \mathbf{IQ} = \mathbf{Q},$$

$\Rightarrow$ $$\mathbf{P} = \mathbf{Q}.$$

Inverse of a product

Let **A** and **B** be non-singular square matrices of the same order, and let

$$(\mathbf{AB})^{-1} = \mathbf{C}.$$

Then $$(\mathbf{AB})\mathbf{C} = \mathbf{I}.$$

Since matrix multiplication is associative, this gives

$$\mathbf{A(BC) = I.}$$

Premultiply by $\mathbf{A}^{-1}$: $\qquad\qquad\qquad \mathbf{BC = A^{-1}I = A^{-1}}.$

Premultiply by $\mathbf{B}^{-1}$: $\qquad\qquad\qquad \mathbf{C = B^{-1}A^{-1}}$

$\Rightarrow$ $\qquad\qquad\qquad\qquad\qquad \mathbf{(AB)^{-1} = B^{-1}A^{-1}}.$

Example 1

Given that the matrix $\mathbf{A}$ possesses an inverse, show that

$$\mathbf{(AB = AC) \Rightarrow (B = C)}.$$

From $\qquad\qquad\qquad\qquad\qquad \mathbf{AB = AC}$

$$\mathbf{A^{-1}(AB) = A^{-1}(AC)}$$

$\Rightarrow$ $\qquad\qquad\qquad \mathbf{(A^{-1}A)B = (A^{-1}A)C}$

$\Rightarrow$ $\qquad\qquad\qquad\qquad\qquad \mathbf{B = C}.$

The adjoint matrix

The adjoint (or adjugate) matrix of a square matrix $\mathbf{A}$ is the transpose of the matrix formed by replacing each element of $\mathbf{A}$ by its cofactor in det $\mathbf{A}$.

Let $\mathbf{A}$ be the matrix $\begin{pmatrix} a_{11} & a_{12} & a_{13} \\ a_{21} & a_{22} & a_{23} \\ a_{31} & a_{32} & a_{33} \end{pmatrix}$

in which a_{ij} denotes the element in the ith row and the jth column.

Let A_{ij} denote the cofactor of a_{ij}. Then the adjoint of $\mathbf{A}$ is given by

$$\text{adj } \mathbf{A} = \begin{pmatrix} A_{11} & A_{21} & A_{31} \\ A_{12} & A_{22} & A_{32} \\ A_{13} & A_{23} & A_{33} \end{pmatrix}.$$

Consider the product $\mathbf{A} \times (\text{adj } \mathbf{A})$.

The sum of the products of the elements of any row of $\mathbf{A}$ with their own cofactors equals det $\mathbf{A}$. It follows that each diagonal element in the product $\mathbf{A} \times (\text{adj } \mathbf{A})$ equals det $\mathbf{A}$. For

$$a_{11}A_{11} + a_{12}A_{12} + a_{13}A_{13} = \det \mathbf{A},$$

$$a_{21}A_{21} + a_{22}A_{22} + a_{23}A_{23} = \det \mathbf{A},$$

$$a_{31}A_{31} + a_{32}A_{32} + a_{33}A_{33} = \det \mathbf{A}.$$

The sum of the products of the elements of any row of $\mathbf{A}$ with the corresponding cofactors of another row is zero. For example,

$$a_{21}A_{11} + a_{22}A_{12} + a_{23}A_{13} = 0.$$

It follows that, except for the diagonal elements, every element in the product $\mathbf{A} \times (\text{adj } \mathbf{A})$ is zero. This gives

$$\mathbf{A} \times (\text{adj } \mathbf{A}) = (\det \mathbf{A})\mathbf{I},$$

where $\mathbf{I}$ is the unit matrix of order 3.

A similar argument shows that

$$(\text{adj } \mathbf{A}) \times \mathbf{A} = (\det \mathbf{A})\mathbf{I}.$$

Example 2

Form the adjoint of the matrix $\mathbf{A}$ when $\mathbf{A} = \begin{pmatrix} 4 & 2 & 1 \\ 3 & 1 & 2 \\ 2 & -1 & 1 \end{pmatrix}$.

$$A_{11} = \begin{vmatrix} 1 & 2 \\ -1 & 1 \end{vmatrix} = 3, \qquad A_{12} = -\begin{vmatrix} 3 & 2 \\ 2 & 1 \end{vmatrix} = 1, \qquad A_{13} = \begin{vmatrix} 3 & 1 \\ 2 & -1 \end{vmatrix} = -5,$$

$$A_{21} = -\begin{vmatrix} 2 & 1 \\ -1 & 1 \end{vmatrix} = -3, \quad A_{22} = \begin{vmatrix} 4 & 1 \\ 2 & 1 \end{vmatrix} = 2, \qquad A_{23} = -\begin{vmatrix} 4 & 2 \\ 2 & -1 \end{vmatrix} = 8,$$

$$A_{31} = \begin{vmatrix} 2 & 1 \\ 1 & 2 \end{vmatrix} = 3, \qquad A_{32} = -\begin{vmatrix} 4 & 1 \\ 3 & 2 \end{vmatrix} = -5, \; A_{33} = \begin{vmatrix} 4 & 2 \\ 3 & 1 \end{vmatrix} = -2.$$

$$\Rightarrow \qquad \text{adj } \mathbf{A} = \begin{pmatrix} 3 & -3 & 3 \\ 1 & 2 & -5 \\ -5 & 8 & -2 \end{pmatrix}.$$

It can be verified that in this example

$$\mathbf{A} \times (\text{adj } \mathbf{A}) = 9\mathbf{I} = (\det \mathbf{A})\mathbf{I}.$$

Inverse by means of the adjoint matrix

Provided that $\det \mathbf{A}$ is not zero, the relation between $\mathbf{A}$ and adj $\mathbf{A}$ gives a method for calculating the inverse of $\mathbf{A}$.

Let $\det \mathbf{A} = D$. Then
$$\mathbf{A}(\text{adj } \mathbf{A}) = (\text{adj } \mathbf{A})\mathbf{A} = D\mathbf{I}$$

$$\Rightarrow \qquad \mathbf{A}\left(\frac{1}{D} \text{adj } \mathbf{A}\right) = \left(\frac{1}{D} \text{adj } \mathbf{A}\right)\mathbf{A} = \mathbf{I},$$

by dividing by D.

This shows that $\mathbf{A}^{-1}$, the inverse of $\mathbf{A}$, is given by

$$\mathbf{A}^{-1} = \frac{1}{D} \text{adj } \mathbf{A} = \frac{1}{\det \mathbf{A}} \text{adj } \mathbf{A}.$$

Example 3

Find the inverse of the matrix $\mathbf{A}$, where $\mathbf{A} = \begin{pmatrix} 7 & 4 & 6 \\ 9 & 5 & 8 \\ 3 & 1 & 2 \end{pmatrix}$.

Step 1 Evaluate $\det \mathbf{A}$:

$$\det \mathbf{A} = 7 \times 2 - 4(-6) + 6(-6) = 2.$$

Step 2 Find the cofactors and replace each element by its own cofactor. This gives the matrix

$$\begin{pmatrix} 2 & 6 & -6 \\ -2 & -4 & 5 \\ 2 & -2 & -1 \end{pmatrix}.$$

Step 3 Transpose to obtain the adjoint matrix:

$$\text{adj } \mathbf{A} = \begin{pmatrix} 2 & -2 & 2 \\ 6 & -4 & -2 \\ -6 & 5 & -1 \end{pmatrix}.$$

Step 4 Divide adj **A** by det **A**, i.e. by 2:

$$\mathbf{A}^{-1} = \begin{pmatrix} 1 & -1 & 1 \\ 3 & -2 & -1 \\ -3 & 2\frac{1}{2} & -\frac{1}{2} \end{pmatrix}.$$

Step 5 Check that the product of the final matrix with **A** equals **I**.

Inverse by row transformations

There are three types of row transformations of a matrix, each of which is equivalent to premultiplication by a suitable matrix.

(1) *Interchange of rows*

The first and second rows of a 3 × 3 matrix will be interchanged when the matrix is premultiplied by the matrix

$$\begin{pmatrix} 0 & 1 & 0 \\ 1 & 0 & 0 \\ 0 & 0 & 1 \end{pmatrix}.$$

(2) *Multiplication of a row by a non-zero constant.*

(3) *Addition to one row of a multiple of another row.*

Let $\mathbf{A} = \begin{pmatrix} -2 & 2 & -1 \\ 4 & 0 & 2 \\ 3 & -2 & 1 \end{pmatrix}$. This matrix can be converted to the unit matrix by four row operations.

(a) Let $\mathbf{P} = \begin{pmatrix} 1 & 0 & 0 \\ 2 & 1 & 0 \\ 0 & 0 & 1 \end{pmatrix}$. Then $\mathbf{PA} = \begin{pmatrix} -2 & 2 & -1 \\ 0 & 4 & 0 \\ 3 & -2 & 1 \end{pmatrix}$.

This operation replaces row 2 of **A** by row 2 + 2(row 1).

(b) Let $\mathbf{Q} = \begin{pmatrix} 1 & 0 & 1 \\ 0 & 1 & 0 \\ 0 & 0 & 1 \end{pmatrix}$. Then $\mathbf{QPA} = \begin{pmatrix} 1 & 0 & 0 \\ 0 & 4 & 0 \\ 3 & -2 & 1 \end{pmatrix}$.

This operation replaces row 1 of **PA** by row 1 + row 3.

(c) Let $\mathbf{R} = \begin{pmatrix} 1 & 0 & 0 \\ 0 & 1 & 0 \\ -3 & 1/2 & 1 \end{pmatrix}$. Then $\mathbf{RQPA} = \begin{pmatrix} 1 & 0 & 0 \\ 0 & 4 & 0 \\ 0 & 0 & 1 \end{pmatrix}$.

This operation replaces row 3 of **QPA** by row 3 − 3(row 1) + $\frac{1}{2}$(row 2).

(d) Let $\mathbf{S} = \begin{pmatrix} 1 & 0 & 0 \\ 0 & 1/4 & 0 \\ 0 & 0 & 1 \end{pmatrix}$. Then $\mathbf{SRQPA} = \begin{pmatrix} 1 & 0 & 0 \\ 0 & 1 & 0 \\ 0 & 0 & 1 \end{pmatrix}$.

This operation multiplies row 2 by $\frac{1}{4}$.

Since **SRQPA** equals the unit matrix **I**, the product **SRQP** equals the inverse of **A**, i.e. $\mathbf{A}^{-1}$ = **SRQP**. This product can easily be calculated, and gives

$$\mathbf{A}^{-1} = \begin{pmatrix} 1 & 0 & 1 \\ 1/2 & 1/4 & 0 \\ -2 & 1/2 & -2 \end{pmatrix}.$$

Note that it is not necessary to write down the individual matrices **P, Q, R, S**, as the layout of the following example shows.

Example 4

Find the inverse of the matrix **A**, where $\mathbf{A} = \begin{pmatrix} 1 & 3 & 3 \\ 4 & 13 & 12 \\ 2 & 7 & 5 \end{pmatrix}$.

Set the matrix **A** side by side with the unit matrix **I**.

$$\mathbf{A} = \begin{pmatrix} 1 & 3 & 3 \\ 4 & 13 & 12 \\ 2 & 7 & 5 \end{pmatrix} \qquad \mathbf{I} = \begin{pmatrix} 1 & 0 & 0 \\ 0 & 1 & 0 \\ 0 & 0 & 1 \end{pmatrix}$$

Each row operation is applied to both matrices.
Let these row operations be equivalent to premultiplication by the matrices **P, Q, R, S**.

Step 1 Perform the operation row 2 − 4(row 1):

$$\mathbf{PA} = \begin{pmatrix} 1 & 3 & 3 \\ 0 & 1 & 0 \\ 2 & 7 & 5 \end{pmatrix} \qquad \mathbf{P} = \mathbf{PI} = \begin{pmatrix} 1 & 0 & 0 \\ -4 & 1 & 0 \\ 0 & 0 & 1 \end{pmatrix}$$

Step 2 Row 3 − 2(row 1) − row 2:

$$\mathbf{QPA} = \begin{pmatrix} 1 & 3 & 3 \\ 0 & 1 & 0 \\ 0 & 0 & -1 \end{pmatrix} \qquad \mathbf{QP} = \begin{pmatrix} 1 & 0 & 0 \\ -4 & 1 & 0 \\ 2 & -1 & 1 \end{pmatrix}$$

Step 3 Row 1 − 3(row 2) + 3(row 3):

$$\mathbf{RQPA} = \begin{pmatrix} 1 & 0 & 0 \\ 0 & 1 & 0 \\ 0 & 0 & -1 \end{pmatrix} \qquad \mathbf{RQP} = \begin{pmatrix} 19 & -6 & 3 \\ -4 & 1 & 0 \\ 2 & -1 & 1 \end{pmatrix}$$

Step 4 (Row 3) × (−1):

$$\mathbf{SRQPA} = \begin{pmatrix} 1 & 0 & 0 \\ 0 & 1 & 0 \\ 0 & 0 & 1 \end{pmatrix} \qquad \mathbf{SRQP} = \begin{pmatrix} 19 & -6 & 3 \\ -4 & 1 & 0 \\ -2 & 1 & -1 \end{pmatrix}$$

Since (**SRQP**)**A** = **I**, the product **SRQP** equals the inverse of the matrix **A**, i.e.

$$\mathbf{A}^{-1} = \begin{pmatrix} 19 & -6 & 3 \\ -4 & 1 & 0 \\ -2 & 1 & -1 \end{pmatrix}.$$

Exercises 13.4

1 Find the inverse of each of the following matrices.

(a) $\begin{pmatrix} 5 & 7 \\ 2 & 3 \end{pmatrix}$,

(b) $\begin{pmatrix} 7 & 4 \\ 5 & 3 \end{pmatrix}$,

(c) $\begin{pmatrix} 9 & 4 \\ 4 & 2 \end{pmatrix}$,

(d) $\begin{pmatrix} 4 & -5 \\ 2 & -2 \end{pmatrix}$,

(e) $\begin{pmatrix} 5 & 9 \\ 3 & 6 \end{pmatrix}$,

(f) $\begin{pmatrix} 16 & -12 \\ -10 & 8 \end{pmatrix}$,

(g) $\begin{pmatrix} 0 & -1/2 \\ 1/4 & 1/8 \end{pmatrix}$,

(h) $\begin{pmatrix} 1 & 1/2 \\ 1/5 & 1/5 \end{pmatrix}$.

2 Verify that $(\mathbf{AB})^{-1} = \mathbf{B}^{-1}\mathbf{A}^{-1}$ when $\mathbf{A} = \begin{pmatrix} 6 & 2 \\ 10 & 4 \end{pmatrix}$, $\mathbf{B} = \begin{pmatrix} 3 & 4 \\ 1 & 2 \end{pmatrix}$.

3 Describe the effect on a 3 × 3 matrix of premultiplication by each of the following matrices. Find the inverse of each of the given matrices.

(a) $\begin{pmatrix} 0 & 0 & 1 \\ 0 & 1 & 0 \\ 1 & 0 & 0 \end{pmatrix}$,

(b) $\begin{pmatrix} 0 & 1 & 0 \\ 0 & 0 & 1 \\ 1 & 0 & 0 \end{pmatrix}$,

(c) $\begin{pmatrix} 1 & 0 & 0 \\ 0 & 1 & 2 \\ 0 & 0 & 1 \end{pmatrix}$,

(d) $\begin{pmatrix} 1 & 3 & 5 \\ 0 & 1 & 2 \\ 0 & 0 & 1 \end{pmatrix}$.

4 Use the method of row transformations to find the inverse of each of the following matrices.

(a) $\begin{pmatrix} 1 & 2 & 3 \\ 2 & 4 & 5 \\ 3 & 5 & 6 \end{pmatrix}$,

(b) $\begin{pmatrix} 1 & 3 & 3 \\ 1 & 3 & 4 \\ 1 & 4 & 3 \end{pmatrix}$,

(c) $\begin{pmatrix} 1 & 3 & 2 \\ -1 & 2 & 1 \\ 2 & 1 & -1 \end{pmatrix}$,   (d) $\begin{pmatrix} 4 & 3 & 3 \\ -1 & 0 & -1 \\ -4 & -4 & -3 \end{pmatrix}$.

5 Use the adjoint method to find the inverse of each of the following matrices.

(a) $\begin{pmatrix} -5 & 7 & 1 \\ 1 & -5 & 7 \\ 7 & 1 & -5 \end{pmatrix}$,   (b) $\begin{pmatrix} 1 & 4 & 16 \\ 0 & 1 & 4 \\ 0 & 0 & 1 \end{pmatrix}$,

(c) $\begin{pmatrix} 4 & 1 & 2 \\ -1 & 3 & 4 \\ -1 & 2 & 3 \end{pmatrix}$,   (d) $\dfrac{1}{4}\begin{pmatrix} 2 & 5 & 4 \\ 2 & 3 & -4 \\ 1 & 2 & -3 \end{pmatrix}$.

6 Given that

$$\mathbf{P} = \begin{pmatrix} 1 & 0 & 0 \\ -2 & 1 & 0 \\ 6 & -3 & 1 \end{pmatrix}, \qquad \mathbf{Q} = \begin{pmatrix} 11 & 4 & 0 \\ 4 & 11 & 3 \\ 0 & 3 & 1 \end{pmatrix},$$

show that the product $\mathbf{P}^{\mathrm{T}}\mathbf{Q}\mathbf{P}$ is a diagonal matrix.
   Use the relation $(\mathbf{ABC})^{-1} = \mathbf{C}^{-1}\mathbf{B}^{-1}\mathbf{A}^{-1}$ to find the inverse of $\mathbf{Q}$.

7 Given that $\mathbf{P}$ and $\mathbf{Q}$ are 3×3 matrices, show the adj $(\mathbf{PQ})$ equals (adj $\mathbf{Q}$) (adj $\mathbf{P}$). Prove by a counter-example that in general adj $(\mathbf{PQ})$ is not equal to (adj $\mathbf{P}$) (adj $\mathbf{Q}$).

13.5 Linear transformations in three dimensions

In *Advanced Mathematics 1*, Section 13.6, linear transformations in two dimensions by means of 2×2 matrices were considered.
   In three dimensions, the equation

$$\begin{pmatrix} X \\ Y \\ Z \end{pmatrix} = \mathbf{T}\begin{pmatrix} x \\ y \\ z \end{pmatrix},$$

where $\mathbf{T}$ is a 3×3 matrix, defines a linear transformation which maps the vector $\begin{pmatrix} x \\ y \\ z \end{pmatrix}$ to the vector $\begin{pmatrix} X \\ Y \\ Z \end{pmatrix}$.

Note that the origin is mapped to itself.
   Consider the transformation which maps the unit vectors

$$\begin{pmatrix} 1 \\ 0 \\ 0 \end{pmatrix}, \quad \begin{pmatrix} 0 \\ 1 \\ 0 \end{pmatrix}, \quad \begin{pmatrix} 0 \\ 0 \\ 1 \end{pmatrix} \quad \text{to the vectors} \quad \begin{pmatrix} a_1 \\ a_2 \\ a_3 \end{pmatrix}, \quad \begin{pmatrix} b_1 \\ b_2 \\ b_3 \end{pmatrix}, \quad \begin{pmatrix} c_1 \\ c_2 \\ c_3 \end{pmatrix}$$

respectively.

The matrix $\mathbf{T}$ of this transformation is given by

$$\mathbf{T} = \begin{pmatrix} a_1 & b_1 & c_1 \\ a_2 & b_2 & c_2 \\ a_3 & b_3 & c_3 \end{pmatrix}.$$

Let the vectors $\mathbf{p}$ and $\mathbf{q}$ be mapped to the vectors $\mathbf{u}$ and $\mathbf{v}$,

i.e. $\qquad\qquad \mathbf{Tp} = \mathbf{u}, \qquad \mathbf{Tq} = \mathbf{v}.$

Then $\qquad\qquad \mathbf{T(p + q)} = \mathbf{u} + \mathbf{v},$

and for any scalar k,

$$\mathbf{T}(k\mathbf{p}) = k\mathbf{u}.$$

These are the essential properties of a linear transformation.

Example 1

The transformation defined by the equation

$$\begin{pmatrix} X \\ Y \\ Z \end{pmatrix} = \begin{pmatrix} 2 & -3 & 1 \\ 1 & 4 & -2 \\ -3 & 1 & 3 \end{pmatrix} \begin{pmatrix} x \\ y \\ z \end{pmatrix}$$

maps the points A(1, 2, 3), B(5, 0, 5) to the points C, D respectively. Verify that the mid-point of AB is mapped to the mid-point of CD.

Let $\mathbf{T}$ denote the matrix of the transformation.
Let M be the mid-point of AB. Then M is the point (3, 1, 4).

$$\mathbf{T}\begin{pmatrix} 1 \\ 2 \\ 3 \end{pmatrix} = \begin{pmatrix} -1 \\ 3 \\ 8 \end{pmatrix}, \qquad \mathbf{T}\begin{pmatrix} 5 \\ 0 \\ 5 \end{pmatrix} = \begin{pmatrix} 15 \\ -5 \\ 0 \end{pmatrix}, \qquad \mathbf{T}\begin{pmatrix} 3 \\ 1 \\ 4 \end{pmatrix} = \begin{pmatrix} 7 \\ -1 \\ 4 \end{pmatrix}.$$

The point A(1, 2, 3) is mapped to the point C(-1, 3, 8).
The point B(5, 0, 5) is mapped to the point D(15, -5, 0).
The point M(3, 1, 4) is mapped to the point (7, -1, 4), which is the mid-point of CD.

Example 2

The unit vectors $\begin{pmatrix} 1 \\ 0 \\ 0 \end{pmatrix}, \begin{pmatrix} 0 \\ 1 \\ 0 \end{pmatrix}, \begin{pmatrix} 0 \\ 0 \\ 1 \end{pmatrix}$ are mapped to the vectors $\begin{pmatrix} 1 \\ 2 \\ 1 \end{pmatrix}, \begin{pmatrix} 3 \\ 1 \\ 4 \end{pmatrix}, \begin{pmatrix} 4 \\ 3 \\ 2 \end{pmatrix}$

respectively in a linear transformation.

Find the image under this transformation of the straight line with equation

$$\mathbf{r} = \begin{pmatrix} 4 \\ -2 \\ 1 \end{pmatrix} + t\begin{pmatrix} 3 \\ 1 \\ -1 \end{pmatrix}.$$

The matrix **T** of the transformation is given by

$$\mathbf{T} = \begin{pmatrix} 1 & 3 & 4 \\ 2 & 1 & 3 \\ 1 & 4 & 2 \end{pmatrix}.$$

Since $\mathbf{T}\begin{pmatrix} 4 \\ -2 \\ 1 \end{pmatrix} = \begin{pmatrix} 2 \\ 9 \\ -2 \end{pmatrix}$ and $\mathbf{T}\begin{pmatrix} 3 \\ 1 \\ -1 \end{pmatrix} = \begin{pmatrix} 2 \\ 4 \\ 5 \end{pmatrix}$, the given line is

mapped to the line with vector equation

$$\mathbf{r} = \begin{pmatrix} 2 \\ 9 \\ -2 \end{pmatrix} + t\begin{pmatrix} 2 \\ 4 \\ 5 \end{pmatrix}.$$

Clearly, the set of lines parallel to the vector $\begin{pmatrix} 3 \\ 1 \\ -1 \end{pmatrix}$ will be mapped to the

set of lines parallel to the vector $\begin{pmatrix} 2 \\ 4 \\ 5 \end{pmatrix}$.

Also, the set of lines which pass through the point $(4, -2, 1)$ will be mapped to the set of lines which pass through the point $(2, 9, -2)$.

Example 3

Show that the set S of vectors of the form $\begin{pmatrix} s + t \\ -t \\ -s + t \end{pmatrix}$ is a vector space of

dimension 2.

Show also that the transformation with matrix **T** given by

$$\mathbf{T} = \begin{pmatrix} 2 & 2 & 1 \\ 1 & 3 & 1 \\ -1 & -2 & 0 \end{pmatrix}$$

maps each member of S to itself.

Let the vectors $\begin{pmatrix} a + b \\ -b \\ -a + b \end{pmatrix}$, $\begin{pmatrix} c + d \\ -d \\ -c + d \end{pmatrix}$ belong to the set S.

Then for any scalar k, the vector $\begin{pmatrix} ka + kb \\ -kb \\ -ka + kb \end{pmatrix}$ belongs to S, and the vector

$\begin{pmatrix} a + c + b + d \\ -b - d \\ -a - c + b + d \end{pmatrix}$ belongs to S.

It follows that the set S forms a vector space.

Let $\mathbf{u}$ and $\mathbf{v}$ be the vectors given by the values ($s = 1, t = 0$) and ($s = 0, t = 1$) respectively, i.e.

$$\mathbf{u} = \begin{pmatrix} 1 \\ 0 \\ -1 \end{pmatrix}, \quad \mathbf{v} = \begin{pmatrix} 1 \\ -1 \\ 1 \end{pmatrix}.$$

The vectors $\mathbf{u}$ and $\mathbf{v}$ are linearly independent, and any member of the set S can be expressed in terms of $\mathbf{u}$ and $\mathbf{v}$, for

$$\begin{pmatrix} s + t \\ -t \\ -s + t \end{pmatrix} = s\mathbf{u} + t\mathbf{v}.$$

Thus the vector space is spanned by $\mathbf{u}$ and $\mathbf{v}$, and the dimension of the space is 2.

The equation

$$\begin{pmatrix} 2 & 2 & 1 \\ 1 & 3 & 1 \\ -1 & -2 & 0 \end{pmatrix} \begin{pmatrix} s + t \\ -t \\ -s + t \end{pmatrix} = \begin{pmatrix} 2s + 2t - 2t - s + t \\ s + t - 3t - s + t \\ -s - t + 2t \end{pmatrix} = \begin{pmatrix} s + t \\ -t \\ -s + t \end{pmatrix}$$

shows that each member of the set S is mapped to itself.

Example 4
Find the matrix $\mathbf{T}$ of the transformation which maps the points $(1, 1, 0)$, $(2, 1, 0)$, $(1, 2, 2)$ to the points $(3, 4, 4)$, $(5, 7, 6)$, $(8, 9, 8)$ respectively.

Since
$$\mathbf{T}\begin{pmatrix} 1 \\ 1 \\ 0 \end{pmatrix} = \begin{pmatrix} 3 \\ 4 \\ 4 \end{pmatrix} \quad \text{and} \quad \mathbf{T}\begin{pmatrix} 2 \\ 1 \\ 0 \end{pmatrix} = \begin{pmatrix} 5 \\ 7 \\ 6 \end{pmatrix},$$

$$\mathbf{T}\begin{pmatrix} 1 \\ 0 \\ 0 \end{pmatrix} = \mathbf{T}\begin{pmatrix} 2 \\ 1 \\ 0 \end{pmatrix} - \mathbf{T}\begin{pmatrix} 1 \\ 1 \\ 0 \end{pmatrix} = \begin{pmatrix} 5 \\ 7 \\ 6 \end{pmatrix} - \begin{pmatrix} 3 \\ 4 \\ 4 \end{pmatrix} = \begin{pmatrix} 2 \\ 3 \\ 2 \end{pmatrix}.$$

This vector is the first column of the matrix $\mathbf{T}$.

$$\mathbf{T}\begin{pmatrix} 0 \\ 1 \\ 0 \end{pmatrix} = \mathbf{T}\begin{pmatrix} 1 \\ 1 \\ 0 \end{pmatrix} - \mathbf{T}\begin{pmatrix} 1 \\ 0 \\ 0 \end{pmatrix} = \begin{pmatrix} 3 \\ 4 \\ 4 \end{pmatrix} - \begin{pmatrix} 2 \\ 3 \\ 2 \end{pmatrix} = \begin{pmatrix} 1 \\ 1 \\ 2 \end{pmatrix}.$$

This vector is the second column of the matrix $\mathbf{T}$.

$$\mathbf{T}\begin{pmatrix} 0 \\ 0 \\ 1 \end{pmatrix} = \tfrac{1}{2}\mathbf{T}\begin{pmatrix} 1 \\ 2 \\ 2 \end{pmatrix} - \mathbf{T}\begin{pmatrix} 0 \\ 1 \\ 0 \end{pmatrix} - \tfrac{1}{2}\mathbf{T}\begin{pmatrix} 1 \\ 0 \\ 0 \end{pmatrix} = \tfrac{1}{2}\begin{pmatrix} 8 \\ 9 \\ 8 \end{pmatrix} - \begin{pmatrix} 1 \\ 1 \\ 2 \end{pmatrix} - \tfrac{1}{2}\begin{pmatrix} 2 \\ 3 \\ 2 \end{pmatrix} = \begin{pmatrix} 2 \\ 2 \\ 1 \end{pmatrix},$$

giving the third column. Hence

$$\mathbf{T} = \begin{pmatrix} 2 & 1 & 2 \\ 3 & 1 & 2 \\ 2 & 2 & 1 \end{pmatrix}.$$

Example 5

The matrix **T** of a linear transformation is given by $\mathbf{T} = \begin{pmatrix} 1 & 2 & 3 \\ 2 & 3 & 4 \\ 3 & 5 & 7 \end{pmatrix}$.

(a) Show that **T** is a singular matrix.

(b) Show that the linearly independent vectors $\begin{pmatrix} 1 \\ 0 \\ 0 \end{pmatrix}, \begin{pmatrix} 1 \\ 1 \\ 0 \end{pmatrix}, \begin{pmatrix} 1 \\ 1 \\ 1 \end{pmatrix}$ are mapped

to linearly dependent vectors.

(c) Find the set of vectors which are mapped to the zero vector.

(d) Show that the image of every vector lies in a vector space of dimension 2.

(a) It is easily shown that det **T** = 0, so that **T** is singular. This is also follows from the fact that row 3 is the sum of rows 1 and 2, so that the rows are not linearly independent.

(b) $\mathbf{T}\begin{pmatrix} 1 \\ 0 \\ 0 \end{pmatrix} = \begin{pmatrix} 1 \\ 2 \\ 3 \end{pmatrix} = \mathbf{u}, \qquad \mathbf{T}\begin{pmatrix} 1 \\ 1 \\ 0 \end{pmatrix} = \begin{pmatrix} 3 \\ 5 \\ 8 \end{pmatrix} = \mathbf{v}, \qquad \mathbf{T}\begin{pmatrix} 1 \\ 1 \\ 1 \end{pmatrix} = \begin{pmatrix} 6 \\ 9 \\ 15 \end{pmatrix} = \mathbf{w}.$

To find the relation between **u**, **v** and **w**, form a vector with zero for its first element (i) from **u** and **v**, (ii) from **v** and **w**.

$$\text{(i) } 3\mathbf{u} - \mathbf{v} = \begin{pmatrix} 0 \\ 1 \\ 1 \end{pmatrix}, \qquad \text{(ii) } 2\mathbf{v} - \mathbf{w} = \begin{pmatrix} 0 \\ 1 \\ 1 \end{pmatrix}.$$

It follows that $3\mathbf{u} - 3\mathbf{v} + \mathbf{w} = \mathbf{0}$, showing that **u**, **v** and **w** are linearly dependent vectors.

(c) The matrix equation $\mathbf{T}\begin{pmatrix} x \\ y \\ z \end{pmatrix} = \begin{pmatrix} 0 \\ 0 \\ 0 \end{pmatrix}$

is equivalent to the two independent equations

$$x + 2y + 3z = 0, \qquad 2x + 3y + 4z = 0,$$

which gives $x = t, y = -2t, z = t$.

Hence each vector of the form $t\begin{pmatrix} 1 \\ -2 \\ 1 \end{pmatrix}$ is mapped to the zero vector.

(d) Let $\mathbf{T}\begin{pmatrix} x \\ y \\ z \end{pmatrix} = \begin{pmatrix} a \\ b \\ c \end{pmatrix}$.

Then
$$a = x + 2y + 3z,$$
$$b = 2x + 3y + 4z,$$
$$c = 3x + 5y + 7z.$$

Since $c = a + b$,
$$\begin{pmatrix} a \\ b \\ c \end{pmatrix} = \begin{pmatrix} a \\ b \\ a + b \end{pmatrix} = a\begin{pmatrix} 1 \\ 0 \\ 1 \end{pmatrix} + b\begin{pmatrix} 0 \\ 1 \\ 1 \end{pmatrix}.$$

Hence the image of any vector lies in the vector space spanned by the vectors $\begin{pmatrix} 1 \\ 0 \\ 1 \end{pmatrix}$ and $\begin{pmatrix} 0 \\ 1 \\ 1 \end{pmatrix}$. In terms of cartesian coordinates, this space is the plane $x + y - z = 0$, which contains the lines $x = z$, $y = 0$ and $y = z$, $x = 0$ corresponding to these base vectors.

Exercises 13.5

1 Find the matrix of the linear transformation in three dimensions which maps the unit vectors **i**, **j**, **k** to the vectors $\begin{pmatrix} 2 \\ 0 \\ 3 \end{pmatrix}$, $\begin{pmatrix} -3 \\ 4 \\ 1 \end{pmatrix}$, $\begin{pmatrix} -1 \\ 2 \\ 2 \end{pmatrix}$ respectively.

Find the image under this transformation
(a) of the points $(1, 1, 1)$ and $(-1, -1, -1)$,
(b) of the vector $4\mathbf{i} + \mathbf{j} - 2\mathbf{k}$,
(c) of the straight line $\mathbf{r} = t\mathbf{i} + (1 - t)\mathbf{j} + (2 + t)\mathbf{k}$,

(d) of the plane $\mathbf{r} = s\begin{pmatrix} 1 \\ 0 \\ 1 \end{pmatrix} + t\begin{pmatrix} 1 \\ 1 \\ 0 \end{pmatrix}$.

2 Show that the transformation with matrix $\begin{pmatrix} 1 & 2 & 3 \\ 2 & 4 & 6 \\ 3 & 6 & 9 \end{pmatrix}$ maps each point (x, y, z) to a point on the line given by
$$x = t, \quad y = 2t, \quad z = 3t.$$

Find (a) the set of points which are mapped to the origin,
(b) the set of points which are mapped to the point $(1, 2, 3)$.

3 The matrix of a linear transformation is $\begin{pmatrix} 1 & 2 & 0 \\ 0 & 1 & 3 \\ 0 & 0 & 1 \end{pmatrix}$.

Find (a) the image of the vector $2\mathbf{i} - 3\mathbf{j} + \mathbf{k}$,
(b) the vector whose image is the vector $\mathbf{i} + \mathbf{j} + \mathbf{k}$,
(c) the set of vectors which are mapped to themselves.

4 The matrix of a linear transformation in three dimensions is non-singular. Show that the images of two linearly independent vectors will be linearly independent vectors.

5 A linear transformation maps the three vectors $\mathbf{i} + \mathbf{j} - \mathbf{k}$, $2\mathbf{i} + \mathbf{j} - \mathbf{k}$, $\mathbf{i} - 2\mathbf{j} + \mathbf{k}$ to the vectors $2\mathbf{i} + 2\mathbf{j} + 2\mathbf{k}$, $3\mathbf{i} + 4\mathbf{j} + 5\mathbf{k}$, $-3\mathbf{i} + \mathbf{j} + 2\mathbf{k}$ respectively. Find the matrix of the transformation.

6 In a linear transformation, the images of the unit vectors $\mathbf{i}, \mathbf{j}, \mathbf{k}$ are the vectors $\mathbf{k}, \mathbf{i} - 3\mathbf{k}, \mathbf{j} + 3\mathbf{k}$ respectively. Find the vector space of dimension 1 such that each member is mapped to itself.

7 Find the image under the transformation $\begin{pmatrix} X \\ Y \\ Z \end{pmatrix} = \begin{pmatrix} 1 & 3 & 2 \\ -1 & 1 & 2 \\ 2 & -2 & 1 \end{pmatrix} \begin{pmatrix} x \\ y \\ z \end{pmatrix}$

of

(a) the line $\mathbf{r} = (1 + 3t)\mathbf{i} - \mathbf{j} + (1 - t)\mathbf{k}$,
(b) the line $\mathbf{r} = s\mathbf{i} - (3 + s)\mathbf{j} - s\mathbf{k}$.
These lines intersect at the point P. Find the position vector of P and of its image.

8 Show that each column vector of the matrix $\begin{pmatrix} 2 & 0 & -5 \\ 2 & 1 & 4 \\ 1 & -2 & 2 \end{pmatrix}$ is at right

angles to each of the other column vectors.

Deduce that the images of the unit vectors $\mathbf{i}, \mathbf{j}, \mathbf{k}$ in the transformation given by this matrix will be mutually perpendicular.

13.6 Orthogonal matrices

A non-singular matrix $\mathbf{A}$ is said to be *orthogonal* if its transpose is equal to its inverse, i.e. if $\mathbf{AA}^T = \mathbf{I}$.

An orthogonal matrix has the following properties:
(1) Its determinant equals either 1 or -1.
(2) Each column vector and each row vector is a unit vector.
(3) The scalar product of any two column vectors is zero.
(4) The scalar product of any two row vectors is zero.
(5) In any linear transformation given by an orthogonal matrix, any vector is mapped to a vector of equal length.
Let the matrix $\mathbf{A}$ be orthogonal, i.e. $\mathbf{A}^T = \mathbf{A}^{-1}$.

Then
$$\mathbf{AA}^T = \mathbf{AA}^{-1} = \mathbf{I}$$
$\Rightarrow$
$$(\det \mathbf{A})(\det \mathbf{A}^T) = 1$$
$\Rightarrow$
$$(\det \mathbf{A})^2 = 1$$
$\Rightarrow$
$$\det \mathbf{A} = \pm 1.$$

Orthogonal matrices in two dimensions

The matrix $\mathbf{P}$ given by $\mathbf{P} = \begin{pmatrix} \cos\theta & -\sin\theta \\ \sin\theta & \cos\theta \end{pmatrix}$ is orthogonal, since $\mathbf{PP}^T = \mathbf{I}$.

The column vectors of $\mathbf{P}$ are unit vectors, and are mutually orthogonal, i.e. at right angles to one another.

Also
$$\det \mathbf{P} = \cos^2 \theta + \sin^2 \theta = 1.$$

It was seen in *Advanced Mathematics 1*, page 343, that the transformation

$$\begin{pmatrix} X \\ Y \end{pmatrix} = \begin{pmatrix} \cos \theta & -\sin \theta \\ \sin \theta & \cos \theta \end{pmatrix} \begin{pmatrix} x \\ y \end{pmatrix}$$

represents an anticlockwise rotation about the origin through an angle θ.

The matrix $\mathbf{Q}$ given by $\mathbf{Q} = \begin{pmatrix} \cos 2\alpha & \sin 2\alpha \\ \sin 2\alpha & -\cos 2\alpha \end{pmatrix}$ is orthogonal, since $\mathbf{QQ}^\mathsf{T} = \mathbf{I}$.

The column vectors of $\mathbf{Q}$ are unit vectors, and are mutually orthogonal.

Also
$$\det \mathbf{Q} = -\cos^2 2\alpha - \sin^2 2\alpha = -1.$$

In *Advanced Mathematics 1*, page 346, it was shown that the transformation

$$\begin{pmatrix} X \\ Y \end{pmatrix} = \begin{pmatrix} \cos 2\alpha & \sin 2\alpha \\ \sin 2\alpha & -\cos 2\alpha \end{pmatrix} \begin{pmatrix} x \\ y \end{pmatrix}$$

represents a reflection in the straight line $y = x \tan \alpha$. Clearly, lengths are unchanged by a rotation or by a reflection.

Consider the effect of a reflection in the line $y = x \tan \alpha$, followed by a reflection in the line $y = x \tan \beta$. This transformation is given by the equation

$$\begin{pmatrix} X \\ Y \end{pmatrix} = \begin{pmatrix} \cos 2\beta & \sin 2\beta \\ \sin 2\beta & -\cos 2\beta \end{pmatrix} \begin{pmatrix} \cos 2\alpha & \sin 2\alpha \\ \sin 2\alpha & -\cos 2\alpha \end{pmatrix} \begin{pmatrix} x \\ y \end{pmatrix}$$

$$\Rightarrow \quad \begin{pmatrix} X \\ Y \end{pmatrix} = \begin{pmatrix} \cos 2(\beta - \alpha) & -\sin 2(\beta - \alpha) \\ \sin 2(\beta - \alpha) & \cos 2(\beta - \alpha) \end{pmatrix} \begin{pmatrix} x \\ y \end{pmatrix}.$$

Comparison of this matrix with the matrix $\mathbf{P}$ above shows that the resulting transformation is an anticlockwise rotation about the origin through an angle $2\beta - 2\alpha$.

Thus two reflections in succession are equivalent to a rotation.

Note that the product of two orthogonal matrices of the same order is always an orthogonal matrix. For when $\mathbf{A}$ and $\mathbf{B}$ are orthogonal matrices of order n,

$$(\mathbf{AB})^\mathsf{T} = \mathbf{B}^\mathsf{T}\mathbf{A}^\mathsf{T} = \mathbf{B}^{-1}\mathbf{A}^{-1} = (\mathbf{AB})^{-1}.$$

Orthogonal matrices in three dimensions

Let the matrix $\mathbf{A}$ given by

$$\mathbf{A} = \begin{pmatrix} a_1 & b_1 & c_1 \\ a_2 & b_2 & c_2 \\ a_3 & b_3 & c_3 \end{pmatrix}$$

be orthogonal. Since $\mathbf{A}^{-1} = \mathbf{A}^\mathsf{T}$, i.e. the inverse of $\mathbf{A}$ equals the transpose of $\mathbf{A}$,

$$\mathbf{A}^\mathsf{T}\mathbf{A} = \mathbf{I}$$

$$\Rightarrow \qquad \begin{pmatrix} a_1 & a_2 & a_3 \\ b_1 & b_2 & b_3 \\ c_1 & c_2 & c_3 \end{pmatrix} \begin{pmatrix} a_1 & b_1 & c_1 \\ a_2 & b_2 & c_2 \\ a_3 & b_3 & c_3 \end{pmatrix} = \begin{pmatrix} 1 & 0 & 0 \\ 0 & 1 & 0 \\ 0 & 0 & 1 \end{pmatrix}.$$

This gives nine equations satisfied by the elements of the matrix $\mathbf{A}$.
Three equations arise from the diagonal elements of $\mathbf{I}$. They are

$$a_1^2 + a_2^2 + a_3^2 = 1,$$

$$b_1^2 + b_2^2 + b_3^2 = 1,$$

$$c_1^2 + c_2^2 + c_3^2 = 1.$$

It follows that the columns of $\mathbf{A}$ are unit vectors.
The remaining six equations are of the form

$$a_1 b_1 + a_2 b_2 + a_3 b_3 = 0,$$

showing that the columns are mutually orthogonal vectors.
The linear transformation

$$\begin{pmatrix} X \\ Y \\ Z \end{pmatrix} = \begin{pmatrix} a_1 & b_1 & c_1 \\ a_2 & b_2 & c_2 \\ a_3 & b_3 & c_3 \end{pmatrix} \begin{pmatrix} x \\ y \\ z \end{pmatrix}$$

gives
$$X = a_1 x + b_1 y + c_1 z,$$

$$Y = a_2 x + b_2 y + c_2 z,$$

$$Z = a_3 x + b_3 y + c_3 z.$$

Since $a_1^2 + a_2^2 + a_3^2 = 1$, the coefficient of x^2 in the expression $X^2 + Y^2 + Z^2$
is 1. The same is true of y^2 and z^2.

The coefficient of $2xy$ in the same expression is $a_1 b_1 + a_2 b_2 + a_3 b_3$, which is
zero. The same is true of coefficients of $2yz$ and $2zx$. This shows that

$$X^2 + Y^2 + Z^2 = x^2 + y^2 + z^2.$$

It follows that the length of any vector is unchanged in a linear transformation
using an orthogonal matrix.

Example 1
Describe the transformations given by the orthogonal matrices $\mathbf{P}$, $\mathbf{Q}$, $\mathbf{PQ}$ and
$\mathbf{QP}$, where

$$\mathbf{P} = \begin{pmatrix} 0 & 0 & -1 \\ 0 & 1 & 0 \\ 1 & 0 & 0 \end{pmatrix}, \qquad \mathbf{Q} = \begin{pmatrix} 1 & 0 & 0 \\ 0 & 0 & 1 \\ 0 & 1 & 0 \end{pmatrix}.$$

(a) The first column of the matrix $\mathbf{P}$ shows that the vector $\mathbf{i}$ is mapped to the
vector $\mathbf{k}$. The second column shows that the vector $\mathbf{j}$ is mapped to itself, while the
third column shows that the vector $\mathbf{k}$ is mapped to the vector $-\mathbf{i}$.

The transformation is a rotation through 90° about the y-axis, in the sense that takes the vector $\mathbf{i}$ to the vector $\mathbf{k}$.

Note that det $\mathbf{P} = 1$.

(b) The matrix $\mathbf{Q}$ gives a transformation which maps the vector $\mathbf{i}$ to itself, and interchanges the vectors $\mathbf{j}$ and $\mathbf{k}$. The point (x, y, z) is mapped to the point (x, z, y), so that any point in the plane $y = z$ is mapped to itself. Any other point is mapped to its reflection in the plane $y = z$.

Note that det $\mathbf{Q} = -1$.

(c) $$\mathbf{PQ} = \begin{pmatrix} 0 & -1 & 0 \\ 0 & 0 & 1 \\ 1 & 0 & 0 \end{pmatrix}.$$

In this case, the vectors $\mathbf{i}, \mathbf{j}, \mathbf{k}$ are mapped to the vectors $\mathbf{k}, -\mathbf{i}, \mathbf{j}$ respectively. The transformation consists of a reflection in the plane $y = z$, followed by a rotation through a right angle about the y-axis. The determinant of $\mathbf{PQ}$ is -1.

(d) $$\mathbf{QP} = \begin{pmatrix} 0 & 0 & -1 \\ 1 & 0 & 0 \\ 0 & 1 & 0 \end{pmatrix}.$$

The corresponding transformation consists of a rotation through 90° about the y-axis, followed by a reflection in the plane $y = z$. $\text{Det}(\mathbf{QP}) = -1$.

Example 2
Show that a reflection in the plane $y = z$ followed by a reflection in the plane $x = y$ is equivalent to a rotation through 120° about the line $x = y = z$.

The matrix of the first transformation, giving a reflection in the plane $y = z$, is

$$\begin{pmatrix} 1 & 0 & 0 \\ 0 & 0 & 1 \\ 0 & 1 & 0 \end{pmatrix}.$$

The matrix of the second transformation, giving a reflection in the plane $x = y$, is

$$\begin{pmatrix} 0 & 1 & 0 \\ 1 & 0 & 0 \\ 0 & 0 & 1 \end{pmatrix}.$$

The matrix of the resulting transformation is the product

$$\begin{pmatrix} 0 & 1 & 0 \\ 1 & 0 & 0 \\ 0 & 0 & 1 \end{pmatrix} \begin{pmatrix} 1 & 0 & 0 \\ 0 & 0 & 1 \\ 0 & 1 & 0 \end{pmatrix},$$

i.e. $$\begin{pmatrix} 0 & 0 & 1 \\ 1 & 0 & 0 \\ 0 & 1 & 0 \end{pmatrix}.$$

The vector **i** will be mapped to the vector **j**, the vector **j** will be mapped to **k** and the vector **k** will be mapped to **i**.

The point with coordinates (t, t, t) will be mapped to itself, so that each point on the line $x = y = z$ is mapped to itself. The angle of rotation can be found by considering the effect of the transformation on a vector at right angles to the line $x = y = z$.

Consider for example the point $(1, -1, 0)$. This is mapped to the point $(0, 1, -1)$, showing that the vector $\mathbf{i} - \mathbf{j}$ is mapped to the vector $\mathbf{j} - \mathbf{k}$. Let θ be the angle between these two vectors.

Then
$$\cos \theta = \frac{(\mathbf{i} - \mathbf{j}) \cdot (\mathbf{j} - \mathbf{k})}{(\sqrt{2})(\sqrt{2})} = -\tfrac{1}{2}$$

$\Rightarrow$
$$\theta = 120°.$$

Since an orthogonal transformation preserves lengths unchanged, this shows that every vector at right angles to the line $x = y = z$ will be rotated through the same angle of 120°.

Exercises 13.6

1 Find the 2×2 matrices which give the following transformations:
   (a) a reflection in the line $y = x \tan 15°$,
   (b) an anticlockwise rotation through 30°,
   (c) a reflection in the line $y + 2x = 0$,
   (d) a clockwise rotation through the angle $\tan^{-1}(4/3)$.

2 Describe the transformations given by the following matrices:

(a) $\begin{pmatrix} 1 & 0 & 0 \\ 0 & -1 & 0 \\ 0 & 0 & -1 \end{pmatrix}$,
(b) $\begin{pmatrix} 0 & 0 & 1 \\ 0 & 1 & 0 \\ 1 & 0 & 0 \end{pmatrix}$,

(c) $\begin{pmatrix} -1 & 0 & 0 \\ 0 & 0 & 1 \\ 0 & 1 & 0 \end{pmatrix}$,
(d) $\dfrac{1}{\sqrt{2}}\begin{pmatrix} 1 & 0 & -1 \\ 0 & 1 & 0 \\ 1 & 0 & 1 \end{pmatrix}$.

3 Find the matrix of each of the following transformations:
   (a) a reflection in the plane $y = 0$,
   (b) a reflection in the plane $x + z = 0$,
   (c) a rotation through 180° about the line $x + y = 0$, $z = 0$,
   (d) the vectors **i**, **j**, **k** are mapped to the vectors $(\mathbf{j} + \mathbf{k})/(\sqrt{2})$, **i**, $(\mathbf{j} - \mathbf{k})/(\sqrt{2})$ respectively.

4 Given that $\mathbf{A} = \tfrac{1}{7}\begin{pmatrix} 3 & -6 & 2 \\ -2 & -3 & -6 \\ 6 & 2 & -3 \end{pmatrix}$,

verify the following properties of the matrix **A**:

(a) Each column vector and each row vector is of unit length.

(b) Any two column vectors are at right angles to one another.

(c) Any two row vectors are at right angles to one another.

(d) Det $\mathbf{A} = 1$.

(e) The transformation given by the matrix $\mathbf{A}$ maps the vector $(2\mathbf{i} - \mathbf{j} + \mathbf{k})$ to itself, and maps the equilateral triangle with vertices at the points $(1, 0, 0)$, $(0, 1, 0)$, $(0, 0, 1)$ to an equilateral triangle.

5 Calculate the inverse of the matrix $\mathbf{A}$, where $\mathbf{A} = \begin{pmatrix} 16 & -15 & 12 \\ -12 & -20 & -9 \\ 15 & 0 & -20 \end{pmatrix}$.

Find the values of the constant k such that $k\mathbf{A}$ is an orthogonal matrix.

6 An orthogonal transformation maps the vector $\mathbf{i}$ to the vector $(\mathbf{i} + \mathbf{j} + \mathbf{k})/(\sqrt{3})$ and the vector $\mathbf{j}$ to the vector $(-\mathbf{i} + \mathbf{k})/(\sqrt{2})$. Given that the determinant of the matrix of the transformation is 1, find the vector to which the vector $\mathbf{k}$ is mapped.

13.7 Eigenvectors of a matrix

An *eigenvector* (or *latent vector*) of a matrix is a vector $\mathbf{u}$ which is mapped to $\lambda\mathbf{u}$, where λ is a scalar constant, by the linear transformation associated with the matrix.

Let $\mathbf{A}$ be the matrix $\begin{pmatrix} 3 & 1 \\ 2 & 4 \end{pmatrix}$ and let $\mathbf{u} = \begin{pmatrix} 1 \\ 2 \end{pmatrix}$.

Then $$\mathbf{Au} = \begin{pmatrix} 3 & 1 \\ 2 & 4 \end{pmatrix}\begin{pmatrix} 1 \\ 2 \end{pmatrix} = \begin{pmatrix} 5 \\ 10 \end{pmatrix} = 5\mathbf{u}.$$

This shows that the vector $\begin{pmatrix} 1 \\ 2 \end{pmatrix}$ is an eigenvector of the matrix $\mathbf{A}$. Note that any multiple of this vector is also an eigenvector of $\mathbf{A}$. The numerical factor 5 in the equation $\mathbf{A} = 5\mathbf{u}$ is known as the eigenvalue corresponding to the eigenvector $\mathbf{u}$.

The linear transformation given by

$$\begin{pmatrix} X \\ Y \end{pmatrix} = \begin{pmatrix} 3 & 1 \\ 2 & 4 \end{pmatrix}\begin{pmatrix} x \\ y \end{pmatrix}$$

maps any point on the line $y = 2x$ to a point on the same line. For example, the point $(3, 6)$ is mapped to the point $(15, 30)$. To find whether any other vector is mapped to a scalar multiple of itself in this transformation, consider the equation

$$\begin{pmatrix} 3 & 1 \\ 2 & 4 \end{pmatrix}\begin{pmatrix} x \\ y \end{pmatrix} = \lambda\begin{pmatrix} x \\ y \end{pmatrix},$$

where λ is a scalar. This gives

$$3x + y = \lambda x, \qquad\qquad 2x + 4y = \lambda y,$$
$$\Rightarrow \qquad (3 - \lambda)x + y = 0, \qquad 2x + (4 - \lambda)y = 0.$$

Elimination of x and y gives

$$(3 - \lambda)(4 - \lambda) = 2$$
$$\Rightarrow \qquad \lambda^2 - 7\lambda + 10 = 0$$
$$\Rightarrow \qquad\qquad \lambda = 5 \text{ or } 2.$$

These two values of λ are the *eigenvalues* (or *latent roots*) of the matrix **A**, the value 5 corresponding to the eigenvector **u** above. To find an eigenvector with eigenvalue 2, consider the equation

$$\begin{pmatrix} 3 & 1 \\ 2 & 4 \end{pmatrix}\begin{pmatrix} x \\ y \end{pmatrix} = 2\begin{pmatrix} x \\ y \end{pmatrix}.$$

This gives two equations

$$3x + y = 2x, \qquad 2x + 4y = 2y,$$

which are both satisfied by any pair of values of x and y such that $y = -x$.

Hence any scalar multiple of the vector $\begin{pmatrix} 1 \\ -1 \end{pmatrix}$ is an eigenvector of the matrix **A**, with eigenvalue 2.

Example 1

Find the eigenvalues of the matrix $\begin{pmatrix} 7 & 6 \\ 6 & 2 \end{pmatrix}$ and the corresponding eigenvectors of unit length.

Let $\begin{pmatrix} x \\ y \end{pmatrix}$ be an eigenvector with eigenvalue λ. Then

$$\begin{pmatrix} 7 & 6 \\ 6 & 2 \end{pmatrix}\begin{pmatrix} x \\ y \end{pmatrix} = \lambda\begin{pmatrix} x \\ y \end{pmatrix}$$
$$\Rightarrow \qquad 7x + 6y = \lambda x, \qquad 6x + 2y = \lambda y \qquad\qquad \dots (1)$$
$$\Rightarrow \qquad (7 - \lambda)x = -6y, \qquad 6x = (\lambda - 2)y.$$

Elimination of x and y gives

$$(7 - \lambda)(\lambda - 2) = -36$$
$$\Rightarrow \qquad \lambda^2 - 9\lambda - 22 = 0$$
$$\Rightarrow \qquad\qquad \lambda = -2 \text{ or } 11.$$

When $\lambda = -2$, equations (1) become

$$7x + 6y = -2x, \qquad 6x + 2y = -2y$$
$$\Rightarrow \qquad 9x = -6y, \qquad\qquad 6x = -4y.$$

These equations are satisfied by any pair of values of x and y such that the ratio of x to y is $-2:3$.

Hence any multiple of the vector $\begin{pmatrix} -2 \\ 3 \end{pmatrix}$ is an eigenvector. The unit eigenvectors corresponding to the eigenvalue -2 are

$$\frac{1}{\sqrt{13}}\begin{pmatrix} -2 \\ 3 \end{pmatrix} \quad \text{and} \quad \frac{1}{\sqrt{13}}\begin{pmatrix} 2 \\ -3 \end{pmatrix}.$$

When $\lambda = 11$, equations (1) become

$$7x + 6y = 11x, \qquad 6x + 2y = 11y$$
$$\Rightarrow \qquad 4x = 6y, \qquad\qquad 6x = 9y,$$

These equations are satisfied by any pair of values of x and y such that the ratio of x to y is $3:2$.

Any multiple of the vector $\begin{pmatrix} 3 \\ 2 \end{pmatrix}$ is an eigenvector, and the unit eigenvectors corresponding to the eigenvalue 11 are

$$\frac{1}{\sqrt{13}}\begin{pmatrix} 3 \\ 2 \end{pmatrix} \quad \text{and} \quad -\frac{1}{\sqrt{13}}\begin{pmatrix} 3 \\ 2 \end{pmatrix}.$$

Note that in this example the matrix is symmetric, and the eigenvectors $\begin{pmatrix} -2 \\ 3 \end{pmatrix}$ and $\begin{pmatrix} 3 \\ 2 \end{pmatrix}$ are at right angles to one another.

Example 2

The matrix $\begin{pmatrix} a & h \\ h & b \end{pmatrix}$ is such that $a \neq b$ and $ab - h^2$ is not zero. Show that
(a) the eigenvalues of this matrix are real and not zero,
(b) the eigenvectors are orthogonal.

The equation

$$\begin{pmatrix} a & h \\ h & b \end{pmatrix}\begin{pmatrix} x \\ y \end{pmatrix} = \lambda\begin{pmatrix} x \\ y \end{pmatrix}$$

gives $\qquad\quad ax + hy = \lambda x, \qquad hx + by = \lambda y$
$\Rightarrow \qquad\quad (\lambda - a)x = hy, \qquad hx = (\lambda - b)y.$ ... (1)

Elimination of x and y gives

$$(\lambda - a)(\lambda - b) = h^2$$
$$\Rightarrow \qquad \lambda^2 - (a + b)\lambda + ab - h^2 = 0. \qquad\qquad \text{... (2)}$$

Let λ_1 and λ_2 be the roots of this equation.
Neither root is zero, since $ab - h^2$ is not zero.
The roots are real, since

$$(a + b)^2 - 4(ab - h^2) = (a^2 - 2ab + b^2) + 4h^2$$
$$= (a - b)^2 + 4h^2,$$

which is positive.

When $\lambda = \lambda_1$, the equations (1) above are satisfied by any pair of values of x and y such that the ratio of x to y is $h:(\lambda_1 - a)$.

Hence the vector $\begin{pmatrix} h \\ \lambda_1 - a \end{pmatrix}$ is an eigenvector for the eigenvalue λ_1.

Similarly, the vector $\begin{pmatrix} h \\ \lambda_2 - a \end{pmatrix}$ is an eigenvector for the eigenvalue λ_2.

These two eigenvectors will be orthogonal, i.e. at right angles, if

$$h^2 + (\lambda_1 - a)(\lambda_2 - a) = 0.$$

From equation (2) above,

$$\lambda_1 \lambda_2 = ab - h^2, \qquad \lambda_1 + \lambda_2 = a + b.$$

Hence
$$\begin{aligned} h^2 + (\lambda_1 - a)(\lambda_2 - a) &= h^2 + \lambda_1 \lambda_2 - a(\lambda_1 + \lambda_2) + a^2 \\ &= h^2 + (ab - h^2) - a(a + b) + a^2 \\ &= 0. \end{aligned}$$

This shows that the eigenvectors of the symmetric matrix $\begin{pmatrix} a & h \\ h & b \end{pmatrix}$ are orthogonal.

Example 3
Find the eigenvalues and eigenvectors of the matrix

$$\begin{pmatrix} 2 & -2 & 3 \\ 1 & 1 & 1 \\ 1 & 3 & -1 \end{pmatrix}.$$

Let $\begin{pmatrix} x \\ y \\ z \end{pmatrix}$ be an eigenvector and let λ be the corresponding eigenvalue.

Then
$$\begin{pmatrix} 2 & -2 & 3 \\ 1 & 1 & 1 \\ 1 & 3 & -1 \end{pmatrix} \begin{pmatrix} x \\ y \\ z \end{pmatrix} = \lambda \begin{pmatrix} x \\ y \\ z \end{pmatrix}$$

$\Rightarrow$
$$\begin{cases} 2x - 2y + 3z = \lambda x, \\ x + y + z = \lambda y, \\ x + 3y - z = \lambda z. \end{cases}$$

$\Rightarrow$
$$\begin{cases} (2 - \lambda)x - 2y + 3z = 0, \\ x + (1 - \lambda)y + z = 0, \\ x + 3y - (1 + \lambda)z = 0. \end{cases} \qquad \dots \text{(A)}$$

A necessary and sufficient condition for this system of equations to have a solution other than $x = y = z = 0$ is that the determinant of the coefficients should equal zero. This gives

$$\begin{vmatrix} (2 - \lambda) & -2 & 3 \\ 1 & (1 - \lambda) & 1 \\ 1 & 3 & -(1 + \lambda) \end{vmatrix} = 0$$

$\Rightarrow \qquad\qquad -\lambda^3 + 2\lambda^2 + 5\lambda - 6 = 0$

$\Rightarrow \qquad\qquad -(\lambda - 1)(\lambda + 2)(\lambda - 3) = 0$

$\Rightarrow \qquad\qquad\qquad \lambda = 1, \quad -2 \quad \text{or} \quad 3.$

These are the three eigenvalues of the matrix.

When $\lambda = 1$, the system of equations (A) above becomes

$$x - 2y + 3z = 0, \qquad\qquad \ldots (1)$$
$$x + z = 0, \qquad\qquad \ldots (2)$$
$$x + 3y - 2z = 0. \qquad\qquad \ldots (3)$$

Equation (2) is satisfied by $x = t, z = -t$.

Then equations (1) and (3) each give $y = -t$.

This shows that when $\lambda = 1$ the equations (A) are satisfied when

$$x : y : z = 1 : -1 : -1.$$

It follows that the vector $\begin{pmatrix} 1 \\ -1 \\ -1 \end{pmatrix}$ is an eigenvector corresponding to the eigenvalue 1.

When $\lambda = -2$, the system of equations (A) becomes

$$4x - 2y + 3z = 0, \qquad\qquad \ldots (1)$$
$$x + 3y + z = 0, \qquad\qquad \ldots (2)$$
$$x + 3y + z = 0. \qquad\qquad \ldots (3)$$

Equation (1) $- 3 \times$ equation (2) gives $x - 11y = 0$.

Let $y = t, x = 11t$. Substitution in equation (2) gives $z = -14t$.

Hence $x : y : z = 11 : 1 : -14$, and so the vector $\begin{pmatrix} 11 \\ 1 \\ -14 \end{pmatrix}$ is an eigenvector corresponding to the eigenvalue -2.

When $\lambda = 3$, the system of equations (A) becomes

$$-x - 2y + 3z = 0,$$
$$x - 2y + z = 0,$$
$$x + 3y - 4z = 0.$$

These equations are satisfied by $x = t$, $y = t$, $z = t$, so that $\begin{pmatrix} 1 \\ 1 \\ 1 \end{pmatrix}$ is an eigen-vector corresponding to the eigenvalue 3.

The characteristic equation

The characteristic equation of a matrix is the equation whose roots are the eigenvalues of the matrix.

Let
$$\mathbf{A} = \begin{pmatrix} a_1 & b_1 & c_1 \\ a_2 & b_2 & c_2 \\ a_3 & b_3 & c_3 \end{pmatrix},$$

and let $\mathbf{u}$ be an eigenvector of this matrix corresponding to an eigenvalue λ. Then
$$\mathbf{Au} = \lambda\mathbf{u}.$$

This equation can be written in the form

$$\mathbf{Au} = \lambda\mathbf{Iu}$$
$$\Rightarrow \qquad (\mathbf{A} - \lambda\mathbf{I})\mathbf{u} = \mathbf{0}.$$

For a solution other than $\mathbf{u} = \mathbf{0}$ to exist, the determinant of the matrix $(\mathbf{A} - \lambda\mathbf{I})$ must be zero. This gives

$$\begin{vmatrix} a_1 - \lambda & b_1 & c_1 \\ a_2 & b_2 - \lambda & c_2 \\ a_3 & b_3 & c_3 - \lambda \end{vmatrix} = 0.$$

In the expansion of this determinant, the only terms in λ^3 and λ^2 come from the product $(a_1 - \lambda)(b_2 - \lambda)(c_3 - \lambda)$.

Hence the coefficient of λ^3 is -1 and the coefficient of λ^2 is $(a_1 + b_2 + c_3)$.

This shows that the sum of the eigenvalues of the matrix $\mathbf{A}$ equals the sum of the elements on the principal diagonal of $\mathbf{A}$.

The constant term in the expansion of the determinant is det $\mathbf{A}$. Since the coefficient of λ^3 is -1, the product of the eigenvalues of the matrix $\mathbf{A}$ must equal det $\mathbf{A}$.

Let the 3×3 matrix $\mathbf{P}$ be orthogonal.

The characteristic equation of $\mathbf{P}$ will be a cubic equation in λ, which must have at least one real root.

It was shown in Section 13.6 that the linear transformation associated with $\mathbf{P}$ leaves lengths unchanged. It follows that any real eigenvalue of the matrix $\mathbf{P}$ must equal either 1 or -1.

Since $\mathbf{P}$ is an orthogonal matrix, det $\mathbf{P} = \pm 1$, and so the product of the eigenvalues is either 1 or -1.

There are two possibilities:

(1) All three eigenvalues are real, each being 1 or -1,
(2) One eigenvalue is real and equals 1 or -1, while the other eigenvalues are conjugate complex numbers with modulus 1.

Example 4
Find the characteristic equation and the eigenvalues of the matrix $\mathbf{A}$, where

$$\mathbf{A} = \tfrac{1}{3}\begin{pmatrix} 1 & -2 & -2 \\ -2 & 1 & -2 \\ -2 & -2 & 1 \end{pmatrix}.$$

Show that the associated transformation maps
(a) any point (t, t, t) to the point $(-t, -t, -t)$,
(b) any point in the plane $x + y + z = 0$ to itself.
Hence show that this transformation is a reflection in the plane $x + y + z = 0$. Note that the matrix $\mathbf{A}$ is orthogonal, since $\mathbf{AA}^\mathrm{T} = \mathbf{I}$.

The characteristic equation is given by

$$|\mathbf{A} - \lambda\mathbf{I}| = 0,$$

i.e.
$$\begin{vmatrix} \tfrac{1}{3} - \lambda & -\tfrac{2}{3} & -\tfrac{2}{3} \\ -\tfrac{2}{3} & \tfrac{1}{3} - \lambda & -\tfrac{2}{3} \\ -\tfrac{2}{3} & -\tfrac{2}{3} & \tfrac{1}{3} - \lambda \end{vmatrix} = 0$$

$$\Rightarrow \quad \begin{vmatrix} 1 - 3\lambda & -2 & -2 \\ -2 & 1 - 3\lambda & -2 \\ -2 & -2 & 1 - 3\lambda \end{vmatrix} = 0$$

$$\Rightarrow \quad (1 - 3\lambda)(9\lambda^2 - 6\lambda - 3) + 2(6\lambda - 6) - 2(6 - 6\lambda) = 0$$
$$\Rightarrow \quad -27\lambda^3 + 27\lambda^2 + 27\lambda - 27 = 0$$
$$\Rightarrow \quad \lambda^3 - \lambda^2 - \lambda + 1 = 0$$
$$\Rightarrow \quad (\lambda + 1)(\lambda - 1)^2 = 0$$
$$\Rightarrow \quad \lambda = 1, 1, -1.$$

(a) $\mathbf{A}\begin{pmatrix} t \\ t \\ t \end{pmatrix} = \tfrac{1}{3}\begin{pmatrix} t - 2t - 2t \\ -2t + t - 2t \\ -2t - 2t + t \end{pmatrix} = \begin{pmatrix} -t \\ -t \\ -t \end{pmatrix}.$

This shows that the point (t, t, t) is mapped to the point $(-t, -t, -t)$.
(b) The coordinates of a point P in the plane $x + y + z = 0$ can be put in the form $(p, q, -p - q)$.

$$\mathbf{A}\begin{pmatrix} p \\ q \\ -p - q \end{pmatrix} = \tfrac{1}{3}\begin{pmatrix} p - 2q + 2p + 2q \\ -2p + q + 2p + 2q \\ -2p - 2q - p - q \end{pmatrix} = \begin{pmatrix} p \\ q \\ -p - q \end{pmatrix}$$

This shows that the point P is mapped to itself.
Let the point Q lie on the straight line through P perpendicular to the plane $x + y + z = 0$. The coordinates of Q can be taken to be

$$(p + t, q + t, p - q + t).$$

$$A\begin{pmatrix} p + t \\ q + t \\ -p - q + t \end{pmatrix} = A\begin{pmatrix} p \\ q \\ -p - q \end{pmatrix} + A\begin{pmatrix} t \\ t \\ t \end{pmatrix} = \begin{pmatrix} p \\ q \\ -p - q \end{pmatrix} - \begin{pmatrix} t \\ t \\ t \end{pmatrix}$$

$$= \begin{pmatrix} p - t \\ q - t \\ -p - q - t \end{pmatrix}.$$

Hence the point Q is mapped to a point R with coordinates

$$(p - t, q - t, -p - q - t).$$

The mid-point of QR is the point $(p, q, -p, -q)$, i.e. the point P.

It follows that R is the reflection of Q in the plane $x + y + z = 0$, since QR is at right angles to this plane.

Hence the transformation is a reflection in the plane $x + y + z = 0$.

Exercises 13.7

1   Find the eigenvalues and eigenvectors of the following matrices. Verify that the symmetric matrices have orthogonal eigenvectors.

(a) $\begin{pmatrix} 1 & 3 \\ 6 & 4 \end{pmatrix}$,　　(b) $\begin{pmatrix} 6 & 0 \\ 4 & 4 \end{pmatrix}$,　　(c) $\begin{pmatrix} 5 & 1 \\ 2 & 4 \end{pmatrix}$,　　(d) $\begin{pmatrix} 7 & 2 \\ 2 & 4 \end{pmatrix}$,

(e) $\begin{pmatrix} 5 & 3 \\ 3 & 13 \end{pmatrix}$,　　(f) $\begin{pmatrix} 2 & 3 \\ 6 & 5 \end{pmatrix}$,　　(g) $\begin{pmatrix} 6 & 4 \\ 4 & 0 \end{pmatrix}$,　　(h) $\begin{pmatrix} -2 & 2 \\ 2 & -5 \end{pmatrix}$.

2   Find the characteristic equations and the real eigenvalues of the following orthogonal matrices:

(a) $\begin{pmatrix} 1 & 0 & 0 \\ 0 & 0 & 1 \\ 0 & 1 & 0 \end{pmatrix}$,　　(b) $\begin{pmatrix} 0 & 0 & 1 \\ -1 & 0 & 0 \\ 0 & 1 & 0 \end{pmatrix}$,　　(c) $\begin{pmatrix} 0 & 1 & 0 \\ -1 & 0 & 0 \\ 0 & 0 & 1 \end{pmatrix}$.

3   Find the eigenvalues and eigenvectors of the following symmetric matrices. Verify in each case that the eigenvectors are mutually orthogonal.

(a) $\begin{pmatrix} 3 & -4 & 2 \\ -4 & -1 & 6 \\ 2 & 6 & -2 \end{pmatrix}$,　　(b) $\begin{pmatrix} 3 & -1 & 1 \\ -1 & 5 & -1 \\ 1 & -1 & 3 \end{pmatrix}$.

4   Find the characteristic equation of the orthogonal matrix

$$\tfrac{1}{7}\begin{pmatrix} 3 & 6 & 2 \\ -2 & 3 & -6 \\ 6 & -2 & -3 \end{pmatrix},$$

and show that only one eigenvalue is real.

5 Find the eigenvalues and eigenvectors of the following matrices:

(a) $\begin{pmatrix} 2 & 0 & -1 \\ 0 & 2 & 0 \\ -1 & 0 & 2 \end{pmatrix}$, (b) $\begin{pmatrix} 2 & 0 & -1 \\ 1 & 3 & 1 \\ 2 & 2 & 4 \end{pmatrix}$, (c) $\begin{pmatrix} 5 & 1 & -2 \\ -1 & 6 & 1 \\ 0 & 1 & 3 \end{pmatrix}$,

6 In Example 4 above the characteristic equation of the matrix **A** was found to be $\lambda^3 - \lambda^2 - \lambda + 1 = 0$. Verify that $\mathbf{A}^3 - \mathbf{A}^2 - \mathbf{A} + \mathbf{I} = \mathbf{0}$. (The Cayley–Hamilton theorem states that every matrix satisfies its own characteristic equation.)

13.8 Diagonalisation of a matrix

Let λ_1 and λ_2 be the eigenvalues of the 2×2 matrix **A**, and let $\begin{pmatrix} u_1 \\ u_2 \end{pmatrix}$ and $\begin{pmatrix} v_1 \\ v_2 \end{pmatrix}$ be the corresponding eigenvectors.

Let **P** be the non-singular matrix $\begin{pmatrix} u_1 & v_1 \\ u_2 & v_2 \end{pmatrix}$.

Then
$$\mathbf{AP} = \begin{pmatrix} \lambda_1 u_1 & \lambda_2 v_1 \\ \lambda_1 u_2 & \lambda_2 v_2 \end{pmatrix}$$

$$= \begin{pmatrix} u_1 & v_1 \\ u_2 & v_2 \end{pmatrix}\begin{pmatrix} \lambda_1 & 0 \\ 0 & \lambda_2 \end{pmatrix}$$

$\Rightarrow \qquad \mathbf{AP} = \mathbf{P}\begin{pmatrix} \lambda_1 & 0 \\ 0 & \lambda_2 \end{pmatrix}$.

Premultiply each side of this equation by $\mathbf{P}^{-1}$.

$$\mathbf{P}^{-1}\mathbf{AP} = \begin{pmatrix} \lambda_1 & 0 \\ 0 & \lambda_2 \end{pmatrix}.$$

Thus the matrix **A** is transformed into a diagonal matrix in which the diagonal elements are the eigenvalues of **A**.

Example 1

Transform the matrix $\begin{pmatrix} 3 & 1 \\ 2 & 4 \end{pmatrix}$ into a diagonal matrix.

It was shown in the previous section that the eigenvalues of this matrix are 2, with eigenvector $\begin{pmatrix} 1 \\ -1 \end{pmatrix}$, and 5, with eigenvector $\begin{pmatrix} 1 \\ 2 \end{pmatrix}$.

Step 1 Form a matrix **P** in which the columns are the eigenvectors of the given matrix **A**:

$$\mathbf{P} = \begin{pmatrix} 1 & 1 \\ -1 & 2 \end{pmatrix}.$$

Step 2 Form the product **AP**:

$$\mathbf{AP} = \begin{pmatrix} 3 & 1 \\ 2 & 4 \end{pmatrix}\begin{pmatrix} 1 & 1 \\ -1 & 2 \end{pmatrix} = \begin{pmatrix} 2 & 5 \\ -2 & 10 \end{pmatrix}.$$

Step 3 Find the inverse of **P**:

$$\mathbf{P}^{-1} = \tfrac{1}{3}\begin{pmatrix} 2 & -1 \\ 1 & 1 \end{pmatrix}.$$

Step 4 Form the product $\mathbf{P}^{-1}\mathbf{AP}$:

$$\mathbf{P}^{-1}\mathbf{AP} = \tfrac{1}{3}\begin{pmatrix} 2 & -1 \\ 1 & 1 \end{pmatrix}\begin{pmatrix} 2 & 5 \\ -2 & 10 \end{pmatrix} = \begin{pmatrix} 2 & 0 \\ 0 & 5 \end{pmatrix}.$$

This method applies also to 3×3 matrices, as in the next example.

Example 2

Transform to a diagonal matrix the matrix $\begin{pmatrix} 2 & -2 & 3 \\ 1 & 1 & 1 \\ 1 & 3 & -1 \end{pmatrix}$.

Let the matrix be denoted by **A**. The eigenvalues of **A** were shown in the previous section to be 1, -2, 3 with eigenvectors $\begin{pmatrix} 1 \\ -1 \\ -1 \end{pmatrix}$, $\begin{pmatrix} 11 \\ 1 \\ -14 \end{pmatrix}$, $\begin{pmatrix} 1 \\ 1 \\ 1 \end{pmatrix}$ respectively.

Let the matrix **P** be given by

$$\mathbf{P} = \begin{pmatrix} 1 & 11 & 1 \\ -1 & 1 & 1 \\ -1 & -14 & 1 \end{pmatrix}.$$

Then
$$\mathbf{AP} = \begin{pmatrix} 1 & -22 & 3 \\ -1 & -2 & 3 \\ -1 & 28 & 3 \end{pmatrix}.$$

The inverse of the matrix **P** is given by

$$\mathbf{P}^{-1} = \tfrac{1}{30}\begin{pmatrix} 15 & -25 & 10 \\ 0 & 2 & -2 \\ 15 & 3 & 12 \end{pmatrix}.$$

The product $\mathbf{P}^{-1}\mathbf{AP}$ is then a diagonal matrix.

$$\mathbf{P}^{-1}\mathbf{AP} = \tfrac{1}{30}\begin{pmatrix} 30 & 0 & 0 \\ 0 & -60 & 0 \\ 0 & 0 & 90 \end{pmatrix} = \begin{pmatrix} 1 & 0 & 0 \\ 0 & -2 & 0 \\ 0 & 0 & 3 \end{pmatrix}.$$

Quadratic forms

An expression of the form $ax^2 + 2hxy + by^2$ is known as a *quadratic* form in the two variables x and y. This expression is equal to the product of a row vector (x, y), a symmetric matrix and a column vector $\begin{pmatrix} x \\ y \end{pmatrix}$. Thus

$$ax^2 + 2hxy + by^2 = (x, y)\begin{pmatrix} a & h \\ h & b \end{pmatrix}\begin{pmatrix} x \\ y \end{pmatrix}.$$

Let $\mathbf{A} = \begin{pmatrix} a & h \\ h & b \end{pmatrix}$, and let the eigenvalues of $\mathbf{A}$ be λ_1 and λ_2 with eigenvectors $\mathbf{u}$ and $\mathbf{v}$ respectively. These eigenvectors can be chosen to be unit vectors, and since $\mathbf{A}$ is symmetric, they will be orthogonal.

Let $\mathbf{P}$ be the 2×2 matrix in which the first column is the vector $\mathbf{u}$ and the second column is the vector $\mathbf{v}$. Then $\mathbf{P}$ is an orthogonal matrix such that

$$\mathbf{P}^{-1}\mathbf{A}\mathbf{P} = \begin{pmatrix} \lambda_1 & 0 \\ 0 & \lambda_2 \end{pmatrix}.$$

Consider the linear transformation given by

$$\begin{pmatrix} X \\ Y \end{pmatrix} = \mathbf{P}^{-1}\begin{pmatrix} x \\ y \end{pmatrix}$$

$\Rightarrow \qquad\qquad \begin{pmatrix} x \\ y \end{pmatrix} = \mathbf{P}\begin{pmatrix} X \\ Y \end{pmatrix}.$

Consider the transpose of this equation.

The transpose of the column vector $\begin{pmatrix} x \\ y \end{pmatrix}$ is the row vector (x, y).

The transpose of $\mathbf{P}\begin{pmatrix} X \\ Y \end{pmatrix}$ is $(X, Y)\mathbf{P}^{\mathrm{T}}$, which equals $(X, Y)\mathbf{P}^{-1}$ since $\mathbf{P}$ is an orthogonal matrix.

Hence the transpose of the equation $\begin{pmatrix} x \\ y \end{pmatrix} = \mathbf{P}\begin{pmatrix} X \\ Y \end{pmatrix}$ is the equation

$$(x, y) = (X, Y)\mathbf{P}^{-1}.$$

When (x, y) and $\begin{pmatrix} x \\ y \end{pmatrix}$ are expressed in terms of (X, Y) and $\begin{pmatrix} X \\ Y \end{pmatrix}$,

$$(x, y)\mathbf{A}\begin{pmatrix} x \\ y \end{pmatrix} = [(X, Y)\mathbf{P}^{-1}]\mathbf{A}\left[\mathbf{P}\begin{pmatrix} X \\ Y \end{pmatrix}\right]$$

$$= (X, Y)[\mathbf{P}^{-1}\mathbf{A}\mathbf{P}]\begin{pmatrix} X \\ Y \end{pmatrix}$$

$$= (X, Y)\begin{pmatrix} \lambda_1 & 0 \\ 0 & \lambda_2 \end{pmatrix}\begin{pmatrix} X \\ Y \end{pmatrix}.$$

$\Rightarrow \qquad\qquad ax^2 + 2hxy + by^2 = \lambda_1 X^2 + \lambda_2 Y^2.$

By this linear transformation, the curve

$$ax^2 + 2hxy + by^2 = c,$$

where c is a constant, will be mapped to the curve

$$\lambda_1 X^2 + \lambda_2 Y^2 = c.$$

Since the transformation is orthogonal, the shape and size of the curve will be unchanged.

Example 1
Show that the curve with equation

$$8x^2 + 12xy + 17y^2 = 100$$

is an ellipse with axes of lengths $4\sqrt{5}$ and $2\sqrt{5}$.

Let
$$8x^2 + 12xy + 17y^2 = (x, y)\mathbf{A}\begin{pmatrix} x \\ y \end{pmatrix}.$$

Then the matrix $\mathbf{A}$ is given by $\mathbf{A} = \begin{pmatrix} 8 & 6 \\ 6 & 17 \end{pmatrix}$.

The eigenvalues of the symmetric matrix $\mathbf{A}$ are 5 and 20, with eigenvectors $\begin{pmatrix} 2 \\ -1 \end{pmatrix}$ and $\begin{pmatrix} 1 \\ 2 \end{pmatrix}$ respectively.

The unit eigenvectors can be taken to be $\dfrac{1}{\sqrt{5}}\begin{pmatrix} 2 \\ -1 \end{pmatrix}$ and $\dfrac{1}{\sqrt{5}}\begin{pmatrix} 1 \\ 2 \end{pmatrix}$.

Let
$$\mathbf{P} = \frac{1}{\sqrt{5}}\begin{pmatrix} 2 & 1 \\ -1 & 2 \end{pmatrix}.$$

Then $\mathbf{P}$ is an orthogonal matrix, with $\mathbf{P}^{-1} = \mathbf{P}^{\mathrm{T}}$.

$$\mathbf{P}^{-1} = \frac{1}{\sqrt{5}}\begin{pmatrix} 2 & -1 \\ 1 & 2 \end{pmatrix}.$$

Let
$$\begin{pmatrix} X \\ Y \end{pmatrix} = \mathbf{P}^{-1}\begin{pmatrix} x \\ y \end{pmatrix}.$$

Then
$$\begin{pmatrix} x \\ y \end{pmatrix} = \mathbf{P}\begin{pmatrix} X \\ Y \end{pmatrix} \quad \text{and} \quad (x, y) = (X, Y)\mathbf{P}^{-1}.$$

$$(x, y)\mathbf{A}\begin{pmatrix} x \\ y \end{pmatrix} = [(X, Y)\mathbf{P}^{-1}]\mathbf{A}\left[\mathbf{P}\begin{pmatrix} X \\ Y \end{pmatrix}\right]$$

$$= (X, Y)[\mathbf{P}^{-1}\mathbf{A}\mathbf{P}]\begin{pmatrix} X \\ Y \end{pmatrix}.$$

This gives
$$P^{-1}AP = \tfrac{1}{5}\begin{pmatrix} 2 & -1 \\ 1 & 2 \end{pmatrix}\begin{pmatrix} 8 & 6 \\ 6 & 17 \end{pmatrix}\begin{pmatrix} 2 & 1 \\ -1 & 2 \end{pmatrix}$$

$$= \begin{pmatrix} 5 & 0 \\ 0 & 20 \end{pmatrix}$$

$$\Rightarrow \qquad 8x^2 + 12xy + 17y^2 = (X, Y)\begin{pmatrix} 5 & 0 \\ 0 & 20 \end{pmatrix}\begin{pmatrix} X \\ Y \end{pmatrix}$$

$$= 5X^2 + 20Y^2.$$

Hence the curve $8x^2 + 12xy + 17y^2 = 100$ is mapped to the curve

$$5X^2 + 20Y^2 = 100,$$

i.e.
$$\frac{X^2}{20} + \frac{Y^2}{5} = 1.$$

This is the equation of an ellipse with axes of lengths $4\sqrt{5}$ and $2\sqrt{5}$. Since the transformation is orthogonal, the original curve is an ellipse of the same shape and size. The two curves are shown in Fig. 13.1.

The transformation

$$\begin{pmatrix} X \\ Y \end{pmatrix} = \frac{1}{\sqrt{5}}\begin{pmatrix} 2 & -1 \\ 1 & 2 \end{pmatrix}\begin{pmatrix} x \\ y \end{pmatrix}$$

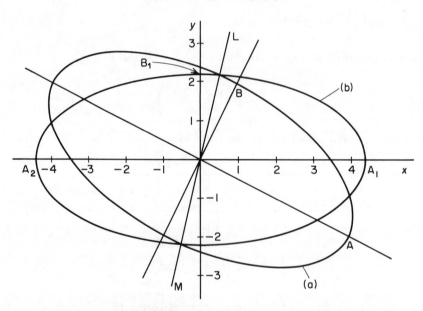

(a) $8x^2 + 12xy + 17y^2 = 100$
(b) $5x^2 + 20y^2 = 100$

Fig. 13.1

is an anticlockwise rotation about the origin through the angle $\tan^{-1}\frac{1}{2}$, and maps the point $A(4, -2)$ to the point $A_1(2\sqrt{5}, 0)$ and the point $B(1, 2)$ to the point $B_1(0, \sqrt{5})$.

An alternative is provided by the transformation

$$\binom{X}{Y} = \frac{1}{\sqrt{5}}\begin{pmatrix} -2 & 1 \\ 1 & 2 \end{pmatrix}\binom{x}{y},$$

which maps the point $A(4, -2)$ to the point $A_2(-2\sqrt{5}, 0)$ and the point B to the point B_1.

This transformation is a reflection in the line $y = (2 + \sqrt{5})x$, which is the line LM in Fig. 13.1

Example 2

Given that $f(x, y, z) = 4x^2 + y^2 + z^2 + 2xy + 4yz + 2zx$, express the quadratic form $f(x, y, z)$ as the product of a row vector, a symmetric matrix $\mathbf{A}$ and a column vector.

Find (a) the characteristic equation of the matrix $\mathbf{A}$, (b) an orthogonal matrix $\mathbf{P}$ such that $\mathbf{P}^{-1}\mathbf{AP}$ is a diagonal matrix. Hence determine the nature of the surface $f(x, y, z) = 24$.

The required product is given by

$$f(x, y, z) = (x, y, z)\begin{pmatrix} 4 & 1 & 1 \\ 1 & 1 & 2 \\ 1 & 2 & 1 \end{pmatrix}\begin{pmatrix} x \\ y \\ z \end{pmatrix}.$$

(a) The characteristic equation of the matrix $\mathbf{A}$ in this product is

$$\det(\mathbf{A} - \lambda\mathbf{I}) = 0$$

$$\Rightarrow \qquad \begin{vmatrix} 4 - \lambda & 1 & 1 \\ 1 & 1 - \lambda & 2 \\ 1 & 2 & 1 - \lambda \end{vmatrix} = 0$$

$$\Rightarrow \qquad \lambda^3 - 6\lambda^2 + 3\lambda + 10 = 0.$$

$$\Rightarrow \qquad (\lambda - 5)(\lambda - 2)(\lambda + 1) = 0.$$

Hence the eigenvalues of the matrix $\mathbf{A}$ are $5, 2$ and -1. By the method explained in Section 13.7, unit eigenvectors corresponding to these eigenvalues are found to be

$$\lambda = 5, \frac{1}{\sqrt{6}}\begin{pmatrix} 2 \\ 1 \\ 1 \end{pmatrix}; \quad \lambda = 2, \frac{1}{\sqrt{3}}\begin{pmatrix} 1 \\ -1 \\ -1 \end{pmatrix}; \quad \lambda = -1, \frac{1}{\sqrt{2}}\begin{pmatrix} 0 \\ 1 \\ -1 \end{pmatrix}.$$

Let $\mathbf{P}$ be the matrix in which these unit vectors form the columns,

i.e. $$\mathbf{P} = \frac{1}{\sqrt{6}}\begin{pmatrix} 2 & \sqrt{2} & 0 \\ 1 & -\sqrt{2} & \sqrt{3} \\ 1 & -\sqrt{2} & -\sqrt{3} \end{pmatrix}.$$

Note that $\mathbf{P}$ is an orthogonal matrix, so that $\mathbf{P}^{-1} = \mathbf{P}^{\mathrm{T}}$. Then

$$\mathbf{P}^{-1}\mathbf{A} = \frac{1}{\sqrt{6}}\begin{pmatrix} 2 & 1 & 1 \\ \sqrt{2} & -\sqrt{2} & -\sqrt{2} \\ 0 & \sqrt{3} & -\sqrt{3} \end{pmatrix}\begin{pmatrix} 4 & 1 & 1 \\ 1 & 1 & 2 \\ 1 & 2 & 1 \end{pmatrix}$$

$$= \frac{1}{\sqrt{6}}\begin{pmatrix} 10 & 5 & 5 \\ 2\sqrt{2} & -2\sqrt{2} & -2\sqrt{2} \\ 0 & -\sqrt{3} & \sqrt{3} \end{pmatrix}$$

$$\Rightarrow \quad \mathbf{P}^{-1}\mathbf{A}\mathbf{P} = \frac{1}{6}\begin{pmatrix} 10 & 5 & 5 \\ 2\sqrt{2} & -2\sqrt{2} & -2\sqrt{2} \\ 0 & -\sqrt{3} & \sqrt{3} \end{pmatrix}\begin{pmatrix} 2 & \sqrt{2} & 0 \\ 1 & -\sqrt{2} & \sqrt{3} \\ 1 & -\sqrt{2} & -\sqrt{3} \end{pmatrix}$$

$$= \begin{pmatrix} 5 & 0 & 0 \\ 0 & 2 & 0 \\ 0 & 0 & -1 \end{pmatrix}.$$

Thus $\mathbf{P}^{-1}\mathbf{A}\mathbf{P}$ is a diagonal matrix, its elements on the principal diagonal being the eigenvalues of $\mathbf{A}$.

Consider the linear transformation

$$\begin{pmatrix} X \\ Y \\ Z \end{pmatrix} = \mathbf{P}^{-1}\begin{pmatrix} x \\ y \\ z \end{pmatrix}.$$

This gives $\begin{pmatrix} x \\ y \\ z \end{pmatrix} = \mathbf{P}\begin{pmatrix} X \\ Y \\ Z \end{pmatrix}$ and $(x, y, z) = (X, Y, Z)\mathbf{P}^{-1}$.

It follows that

$$(x, y, z)\mathbf{A}\begin{pmatrix} x \\ y \\ z \end{pmatrix} = (X, Y, Z)\mathbf{P}^{-1}\mathbf{A}\mathbf{P}\begin{pmatrix} X \\ Y \\ Z \end{pmatrix}$$

$$= (X, Y, Z)\begin{pmatrix} 5 & 0 & 0 \\ 0 & 2 & 0 \\ 0 & 0 & -1 \end{pmatrix}\begin{pmatrix} X \\ Y \\ Z \end{pmatrix}$$

$$= 5X^2 + 2Y^2 - Z^2.$$

This shows that the surface given by $f(x, y, z) = 24$ is mapped to the surface given by

$$5X^2 + 2Y^2 - Z^2 = 24.$$

Since the transformation is orthogonal, the two surfaces are identical in size and shape. Consequently, the nature of the first surface can be deduced from the nature of the second surface.

The surface with equation

$$5X^2 + 2Y^2 - Z^2 = 24$$

is symmetrical about each of the planes $X = 0$, $Y = 0$ and $Z = 0$.
The plane $X = 0$ cuts the surface in the hyperbola $2Y^2 - Z^2 = 24$.
The plane $Y = 0$ cuts the surface in the hyperbola $5X^2 - Z^2 = 24$.
The plane $Z = 0$ cuts the surface in the ellipse $5X^2 + 2Y^2 = 24$, of eccentricity $\sqrt{(3/5)}$.

Any parallel plane $Z = k$, where k is a constant, cuts the surface in a larger ellipse of the same eccentricity.

Exercises 13.8

1 Express each of the following quadratic forms as a product of a row vector, a symmetric matrix and a column vector:
(a) $7x^2 - 6xy + 5y^2$,
(b) $2x^2 + 8xy + 3y^2$,
(c) $x^2 + 4xy - 6y^2$,
(d) $3x^2 - xy - 2y^2$.

2 Find the symmetric matrix $\mathbf{A}$ such that

$$f(x, y, z) \equiv (x, y, z)\mathbf{A}\begin{pmatrix} x \\ y \\ z \end{pmatrix},$$

(a) when $f(x, y, z) \equiv x^2 + 2y^2 + 3z^2 + 4xy + 6yz$,
(b) when $f(x, y, z) \equiv 2x^2 - y^2 + 2xy - 4yz + 6zx$.

3 Apply to each of the following equations the transformation given:

(a) $5x^2 + 6xy + 5y^2 = 8;$ $\begin{pmatrix} X \\ Y \end{pmatrix} = \frac{1}{\sqrt{2}}\begin{pmatrix} 1 & 1 \\ 1 & -1 \end{pmatrix}\begin{pmatrix} x \\ y \end{pmatrix},$

(b) $2x^2 + 4xy - y^2 = 6;$ $\begin{pmatrix} X \\ Y \end{pmatrix} = \frac{1}{\sqrt{5}}\begin{pmatrix} 2 & 1 \\ 1 & -2 \end{pmatrix}\begin{pmatrix} x \\ y \end{pmatrix},$

(c) $7x^2 + 48xy - 7y^2 = 25;$ $\begin{pmatrix} X \\ Y \end{pmatrix} = \frac{1}{5}\begin{pmatrix} 3 & -4 \\ 4 & 3 \end{pmatrix}\begin{pmatrix} x \\ y \end{pmatrix},$

(d) $7x^2 - 2\sqrt{6}xy + 6y^2 = 36;$ $\begin{pmatrix} X \\ Y \end{pmatrix} = \frac{1}{\sqrt{5}}\begin{pmatrix} \sqrt{3} & -\sqrt{2} \\ \sqrt{2} & \sqrt{3} \end{pmatrix}\begin{pmatrix} x \\ y \end{pmatrix}.$

4 Find the linear transformation which maps the curve $5x^2 - 6xy + 5y^2 = 2$ to the curve $4X^2 + Y^2 = 1$. Sketch both curves.

5 Find the linear transformation which maps the curve $2x^2 - 12xy + 7y^2 = 22$ to the curve $11X^2 - 2Y^2 = 22$. Sketch both curves.

6 The equation of a curve is $17x^2 - 18xy - 7y^2 = 20$. Apply to this equation

(a) the transformation $\begin{pmatrix} X \\ Y \end{pmatrix} = \frac{1}{\sqrt{10}}\begin{pmatrix} 3 & -1 \\ 1 & 3 \end{pmatrix}\begin{pmatrix} x \\ y \end{pmatrix},$

(b) the transformation $\begin{pmatrix} X \\ Y \end{pmatrix} = \dfrac{1}{\sqrt{10}} \begin{pmatrix} -3 & 1 \\ 1 & 3 \end{pmatrix} \begin{pmatrix} x \\ y \end{pmatrix}$.

Describe each transformation in geometrical terms.

7 By means of an orthogonal transformation, show that the curve

$$2x^2 - 12xy - 3y^2 = 42$$

is a hyperbola with eccentricity $\sqrt{(13/7)}$.

8 By means of an orthogonal transformation, show that the curve

$$10x^2 - 12xy + 5y^2 = 14$$

is an ellipse with eccentricity $\sqrt{(13/14)}$.

9 Express the quadratic form

$$3x^2 + 2xy + 2xz + 4yz$$

in the form $\qquad\qquad (x, y, z)\mathbf{A} \begin{pmatrix} x \\ y \\ z \end{pmatrix}$,

where $\mathbf{A}$ is a symmetric matrix. Show that the eigenvalues of the matrix $\mathbf{A}$ are 4, 1, -2.
 Find an orthogonal matrix $\mathbf{P}$ such that $\mathbf{P}^{-1}\mathbf{A}\mathbf{P}$ is a diagonal matrix.

10 Express f(x, y, z), where

$$f(x, y, z) \equiv 7x^2 + 6y^2 + 5z^2 + 4xy + 4yz$$

as the product of a row vector, a symmetric matrix $\mathbf{A}$ and a column vector.
 Find the eigenvalues of the matrix $\mathbf{A}$.
 Use an orthogonal transformation to reduce the quadratic form f(x, y, z) to the form $9X^2 + 6Y^2 + 3Z^2$.

Miscellaneous exercises 13

1 Show that when $k = 4$ the matrix $\begin{pmatrix} 2 & k & 4 \\ 3 & 12 & 8 \\ 1 & 8 & k \end{pmatrix}$ is singular.
 Find another value of k for which this matrix is singular.

2 Find the eigenvalues and eigenvectors of the following matrices:

(a) $\begin{pmatrix} -2 & 2 & 2 \\ 2 & 1 & 2 \\ -3 & -6 & -7 \end{pmatrix}$, (b) $\begin{pmatrix} 4 & 2 & 2 \\ 1 & 5 & 1 \\ -2 & -4 & 0 \end{pmatrix}$, (c) $\begin{pmatrix} 3 & 9 & 0 \\ 0 & -3 & 8 \\ 0 & 0 & 6 \end{pmatrix}$.

3 Find the matrix $\mathbf{P}$ such that $\mathbf{AP} = \mathbf{BP} = \mathbf{C}$, where

$$\mathbf{A} = \begin{pmatrix} 3 & 1 & -2 \\ 0 & 4 & -3 \end{pmatrix}, \quad \mathbf{B} = \begin{pmatrix} 2 & -1 & 0 \\ 2 & 0 & -1 \end{pmatrix}, \quad \mathbf{C} = \begin{pmatrix} 1 & 2 \\ 0 & 0 \end{pmatrix}.$$

4 Find the values of λ for which a non-zero vector $\mathbf{u}$ can be found such that

$$\begin{pmatrix} 2 & -2 & 3 \\ 1 & 1 & 1 \\ 1 & 3 & -1 \end{pmatrix} \mathbf{u} = \lambda \mathbf{u},$$

and find a vector of unit modulus corresponding to each value of λ.

5 Given that $f(\lambda) = \begin{vmatrix} \lambda & 0 & 2\lambda \\ 0 & \lambda & 1 \\ 1 & -1 & \lambda \end{vmatrix}$, find $f(2)$.

Find also the two values of λ, λ_1 and λ_2, for which $\mathbf{M}$ has no inverse, where

$$\mathbf{M} = \begin{pmatrix} \lambda & 0 & 2\lambda \\ 0 & \lambda & 1 \\ 1 & -1 & \lambda \end{pmatrix}.$$

If $\lambda \neq \lambda_1$ and $\lambda \neq \lambda_2$, find $\mathbf{M}^{-1}$ in terms of λ and write out $\mathbf{M}$ and $\mathbf{M}^{-1}$ when $\lambda = 2$. Check your result by evaluating $\mathbf{MM}^{-1}$ in this case. [L]

6 Given that $\mathbf{A} = \begin{pmatrix} 1 & 0 & 0 \\ 8 & -3 & 2 \\ 20 & -10 & 6 \end{pmatrix},$

evaluate the determinant of $\mathbf{A}$. Verify that $\mathbf{A}^2 - 3\mathbf{A} + 2\mathbf{I} = \mathbf{0}$, where $\mathbf{I}$ is the unit matrix of order 3.
(a) By multiplying this equation by $\mathbf{A}^{-1}$, obtain the matrix $\mathbf{A}^{-1}$.
(b) Prove by induction that $\mathbf{A}^n = \mathbf{I} + (2^n - 1)(\mathbf{A} - \mathbf{I})$, where n is a positive integer. [L]

7 Without expanding the following determinants, show that each is zero:

(a) $\begin{vmatrix} 4 & -5 & 1 \\ 10 & -2 & -8 \\ 1 & 4 & -5 \end{vmatrix}$, (b) $\begin{vmatrix} 5 & 9 & -2 \\ 9 & 11 & 8 \\ 11 & 12 & 13 \end{vmatrix}$, (c) $\begin{vmatrix} 1 & 2 & 3 \\ 4 & 5 & 6 \\ 7 & 8 & 9 \end{vmatrix}.$

8 Show that the matrices $\mathbf{A}$ and $\mathbf{B}$, where

$$\mathbf{A} = \begin{pmatrix} 0 & -1 & 0 \\ 1 & 0 & 0 \\ 0 & 0 & 1 \end{pmatrix}, \quad \mathbf{B} = \begin{pmatrix} 1 & 0 & 0 \\ 0 & 0 & -1 \\ 0 & 1 & 0 \end{pmatrix},$$

are orthogonal matrices, and that each has one real eigenvalue.
Show also that the linear transformation associated with each matrix is a rotation through a right angle.
Form the products $\mathbf{AB}$ and $\mathbf{BA}$, and show that the linear transformation

associated with (a) the matrix **AB**, (b) the matrix **BA**, is a rotation. Find in each case the equation of the axis of rotation.

9 (a) Find the set of all matrices $\begin{pmatrix} x \\ y \\ z \end{pmatrix}$ which map to $\begin{pmatrix} 0 \\ 0 \end{pmatrix}$ under the

transformation $\begin{pmatrix} x \\ y \\ z \end{pmatrix} \mapsto \begin{pmatrix} 7 & 1 & -1 \\ 8 & -1 & -2 \end{pmatrix} \begin{pmatrix} x \\ y \\ z \end{pmatrix}$.

(b) Hence, or otherwise, find all the solutions of

$$7x + y - z = 7,$$
$$8x - y - 2z = 5.$$

(c) Using your result, or otherwise, find a relation connecting k, l, and m if the equations

$$kx + ly + mz = n,$$
$$7x + y - z = 7,$$
$$8x - y - 2z = 5$$

have a unique solution. [AEB]

10 (a) If **A** is a 2×2 matrix representing a one–one transformation of the plane to itself, and $\begin{pmatrix} x \\ y \end{pmatrix}$ represents the vector $x\mathbf{i} + y\mathbf{j}$, interpret geometrically the equation

$$\mathbf{A} \begin{pmatrix} x \\ y \end{pmatrix} = \lambda \begin{pmatrix} x \\ y \end{pmatrix},$$

where λ is a real number.

(b) If $\mathbf{B} = \begin{pmatrix} \cos 2\theta & \sin 2\theta \\ \sin 2\theta & -\cos 2\theta \end{pmatrix}$, find two solutions $\begin{pmatrix} x_1 \\ y_1 \end{pmatrix}$ and $\begin{pmatrix} x_2 \\ y_2 \end{pmatrix}$ of the equation $\mathbf{B} \begin{pmatrix} x \\ y \end{pmatrix} = -1 \begin{pmatrix} x \\ y \end{pmatrix}$. [AEB]

11 The mapping f where $f : \begin{pmatrix} x \\ y \\ z \end{pmatrix} \mapsto \begin{pmatrix} 1 & 1 & 1 \\ 1 & 2 & 3 \end{pmatrix} \begin{pmatrix} x \\ y \\ z \end{pmatrix}$ maps the set of triples of

real numbers to the set of pairs of numbers.

(a) Find the set of all triples which map to $\begin{pmatrix} 0 \\ 0 \end{pmatrix}$ under f.

(b) Find by trial, or otherwise, one solution to the simultaneous equations

$$x + y + z = 31,$$
$$x + 2y + 3z = 41,$$

and find all positive integral solutions of these last two equations.   [AEB]

12   Find the eigenvalues of the matrix

$$A = \begin{pmatrix} 2 & -2 & 3 \\ 1 & 1 & 1 \\ 1 & 3 & -1 \end{pmatrix}.$$

Describe the effect on points on the line $x = y = z$ of the linear transformation defined by

$$\begin{pmatrix} x_2 \\ y_2 \\ z_2 \end{pmatrix} = A \begin{pmatrix} x_1 \\ y_1 \\ z_1 \end{pmatrix},$$

and find the matrix of the inverse transformation. Explain the nature of the transformation with matrix $(A - 3I)$, where I is the unit matrix of order 3.

13   Find the inverse A^{-1} and the characteristic equation of the matrix A, where

$$A = \begin{pmatrix} 3 & -2 & 3 \\ 1 & 2 & 1 \\ 1 & 3 & 0 \end{pmatrix}.$$

Find a constant λ and a non-zero vector x such that

$$Ax = A^{-1}x = \lambda x.$$ [L]

14   (a) If 2-dimensional space is transformed by

$$\begin{pmatrix} x \\ y \end{pmatrix} \mapsto \begin{pmatrix} 4 & -1 \\ 6 & -3 \end{pmatrix} \begin{pmatrix} x \\ y \end{pmatrix},$$

find the equations of the straight lines which are mapped onto themselves.
(b) Show that 4 is one of the roots of the characteristic equation of the matrix

$$\begin{pmatrix} 5 & 2 & -3 \\ 3 & 3 & -2 \\ -5 & 4 & 5 \end{pmatrix}.$$

Given an interpretation of such a root when the matrix is used to define a linear transformation of geometrical space.   [L]

15   Show that the equations

$$\begin{aligned} x - y + 2z &= -3, \\ 2x + 3y - 4z &= 5, \\ 5x + az &= b \end{aligned}$$

have a unique solution unless $a = 2$.

Show also that if $a = 2$, $b = 4$ the equations are inconsistent, but that if $a = 2$, $b = -4$ the equations have an infinite number of solutions. State in each case how the planes represented by these equations meet. [L]

16 Calculate the inverse $\mathbf{A}^{-1}$ of the matrix $\mathbf{A}$, where

$$\mathbf{A} = \begin{pmatrix} 2 & 2 & 1 \\ 2 & 4 & 1 \\ 3 & 2 & 0 \end{pmatrix}.$$

Find the values of λ for which the determinant of the matrix $(\mathbf{A} - \lambda\mathbf{I})$ equals 0, where $\mathbf{I}$ is the unit matrix of order 3.
Show that $\mathbf{A}^2 - 6\mathbf{A} - \mathbf{I}$ is a multiple of $\mathbf{A}^{-1}$. [L]

17 Obtain the matrix $\mathbf{T}_1$ such that

$$\mathbf{T}_1 \begin{pmatrix} x \\ y \end{pmatrix} = \begin{pmatrix} x_1 \\ y_1 \end{pmatrix},$$

where $\begin{pmatrix} x_1 \\ y_1 \end{pmatrix}$ is the vector obtained by rotating $\begin{pmatrix} x \\ y \end{pmatrix}$ about the origin O through a positive angle α.

Obtain the matrix $\mathbf{T}_2$ such that $\mathbf{T}_2 \begin{pmatrix} x \\ y \end{pmatrix} = \begin{pmatrix} x_2 \\ y_2 \end{pmatrix}$, where (x_2, y_2) is the reflection of the point (x, y) in the x-axis. Hence, or otherwise, find the matrix of the transformation which represents a reflection in the line $4y = 3x$. [L]

18 Evaluate the determinant of the coefficients in the simultaneous equations

$$x - (4k - 3)y + 2z = 0,$$
$$kx - (2k - 1)y + (k + 1)z = 0,$$
$$(2k + 2)x + 3ky + (k + 2)z = 0.$$

Find all the solutions of the equations when $k = 1$. Show that if $k \neq 1$ the only real solution is $x = y = z = 0$. [L]

19 (a) Show that the transformation given by

$$\begin{pmatrix} x_2 \\ y_2 \end{pmatrix} = \begin{pmatrix} 1/2 & \sqrt{3}/2 \\ \sqrt{3}/2 & -1/2 \end{pmatrix} \begin{pmatrix} x_1 \\ y_1 \end{pmatrix}$$

is equivalent to a reflection in a straight line through the origin.
(b) Calculate det $\mathbf{A}$, when $\mathbf{A}$ is the matrix

$$\begin{pmatrix} -2 & 1 & 1 \\ 1 & -2 & 1 \\ 1 & 1 & -2 \end{pmatrix}.$$

Show that when $k \neq 1$ the system of equations

$$A\begin{pmatrix} x \\ y \\ z \end{pmatrix} = \begin{pmatrix} k \\ -2 \\ 1 \end{pmatrix}$$

has no solution. Find the solution when $k = 1$.   [L]

20 Explain why only one of the transformations defined below is linear:

$$T_1(x, y, z) = (x + 2y - 2z, 3x + 3, 2x - 2y + 5z),$$
$$T_2(x, y, z) = (x + 2y - 2z, 3x + 3z, 2x - 2y + 5z),$$
$$T_3(x, y, z) = (x + 2y - 2z, 3x + 3z^2, 2x - 2y + 5z).$$

Write down the matrix of the linear transformation and show that under this transformation
(a) all points are mapped onto a particular plane, whose equation should be given,
(b) all points on the line $\dfrac{x - 1}{-1} = \dfrac{2y + 2}{3} = \dfrac{z}{1}$ are mapped onto the same point, whose coordinates should be given,
(c) all points which are mapped onto the origin lie on a certain straight line, whose equations should be given.   [L]

21 The transformation T_1 operates on the vector $\begin{pmatrix} x \\ y \end{pmatrix}$ according to the equation

$$T_1\begin{pmatrix} x \\ y \end{pmatrix} = \begin{pmatrix} 0 \\ 2 \end{pmatrix} + \begin{pmatrix} 1 & 0 \\ 0 & -1 \end{pmatrix}\begin{pmatrix} x \\ y \end{pmatrix}.$$

Show that T_1 corresponds to a reflection in the line $y = 1$.
    A second transformation is given by

$$T_2\begin{pmatrix} x \\ y \end{pmatrix} = \frac{1}{25}\begin{pmatrix} -7 & 24 \\ 24 & 7 \end{pmatrix}\begin{pmatrix} x \\ y \end{pmatrix}.$$

By considering $T_2\begin{pmatrix} 3t + 4\lambda \\ 4t - 3\lambda \end{pmatrix}$,

or otherwise, show that T_2 corresponds to a reflection in the line $3y = 4x$ and find the point Q which is invariant under both T_1 and T_2.   [L]

22 Linear transformations T_1 and T_2 are reflections of the x-y plane in the lines $y = x$ and $y = x \tan (\pi/3)$ respectively. Write down the matrix representations of T_1 and T_2, find the matrix representation of the combined transformation $T_2 T_1$ and interpret the combined transformation geometrically.

A transformation T_3 is a magnification from the origin by a factor 2; T_4 is a translation with vector $\begin{pmatrix} -\sqrt{3} \\ -1 \end{pmatrix}$. Find the matrix of the transformation $T_3 T_2 T_1$ and the image of the point $(0, 1)$ under the transformation $T_4 T_3 T_2 T_1$. [L]

23 Find the eigenvalues and corresponding eigenvectors for the matrix

$$\begin{pmatrix} 1 & 1 & -2 \\ -1 & 2 & 1 \\ 0 & 1 & -1 \end{pmatrix}.$$

Show that under the transformation associated with this matrix, any point in the plane $x = z$ is mapped to a point in the same plane. Find the locus of points which are mapped to themselves.

24 The three planes

$$2x + y + z = 4$$
$$x + 2y + z = 2$$
$$x + y + 2z = 6$$

meet only in the point $(1, -1, 3)$. The x, y, z coordinate system is transformed by the linear transformation

$$\begin{pmatrix} x \\ y \\ z \end{pmatrix} = \frac{1}{3} \begin{pmatrix} 1 & 2 & 2 \\ 2 & -2 & 1 \\ 2 & 1 & -2 \end{pmatrix} \begin{pmatrix} X \\ Y \\ Z \end{pmatrix}.$$

In the X, Y, Z system, obtain the equations of the planes and the coordinates of the point(s) in which they meet. [L]

25 Find the inverse of the matrix $\mathbf{M}$, where

$$\mathbf{M} = \begin{pmatrix} 2 & 1 & 1 \\ 1 & -1 & 2 \\ 3 & 2 & -1 \end{pmatrix}.$$

If $\mathbf{MX} = \mathbf{N}$, where $\mathbf{X} = \begin{pmatrix} x \\ y \\ z \end{pmatrix}$ and $\mathbf{N} = \begin{pmatrix} a \\ 2a \\ b \end{pmatrix}$, find x, y and z in terms of a and b. [L]

26 For two non-singular $n \times n$ matrices $\mathbf{A}$ and $\mathbf{B}$, prove that

$$(\mathbf{AB})^{-1} = \mathbf{B}^{-1}\mathbf{A}^{-1}.$$

Find the inverse of the product

$$\begin{pmatrix} 1 & 3 & 5 \\ 0 & 1 & 2 \\ 0 & 0 & 1 \end{pmatrix} \begin{pmatrix} 1 & 0 & 0 \\ 0 & 2 & 0 \\ 0 & 0 & 3 \end{pmatrix} \begin{pmatrix} 1 & 0 & 0 \\ 0 & 0 & 1 \\ 0 & 1 & 0 \end{pmatrix}.$$

27 Show that the characteristic equation of the matrix $\mathbf{A}$, where

$$\mathbf{A} = \begin{pmatrix} 1 & -4 & 1 \\ -4 & 1 & 1 \\ 4 & 4 & 4 \end{pmatrix},$$

has two equal roots.

In a linear transformation, the point with position vector $\begin{pmatrix} x \\ y \\ z \end{pmatrix}$ becomes

the point with position vector $\mathbf{A} \begin{pmatrix} x \\ y \\ z \end{pmatrix}$. Show that the straight line $x = y = -z$ is mapped to itself. Find the equation of the plane which is transformed into itself, but does not contain this line.

28 Show that if $k = 10$ the value of the determinant

$$\begin{vmatrix} 5 - k & 4 & -1 \\ 4 & 5 - k & -1 \\ -4 & -4 & 2 - k \end{vmatrix}$$

is zero, and show that this is so for only one other value of k. Verify that for each of these values of k, the equations

$$(5 - k)x + 4y - z = 1,$$
$$4x + (5 - k)y - z = 4,$$
$$-4x - 4y + (2 - k)z = 5$$

are inconsistent, and give a geometrical interpretation of these equations in each case. [L]

29 Find the eigenvalues of the symmetric matrix $\mathbf{A}$ given by

$$(x, y, z)\mathbf{A} \begin{pmatrix} x \\ y \\ z \end{pmatrix} = x^2 + 4y^2 + z^2 + 4xz.$$

Using the eigenvectors of $\mathbf{A}$, or otherwise, obtain an orthogonal matrix $\mathbf{P}$ such that $\mathbf{P}^{-1}\mathbf{AP}$ is a diagonal matrix. Verify that the diagonal elements of the matrix $\mathbf{P}^{-1}\mathbf{AP}$ are the eigenvalues of $\mathbf{A}$. Show that the transformation

$$\begin{pmatrix} X \\ Y \\ Z \end{pmatrix} = \mathbf{P}^{-1} \begin{pmatrix} x \\ y \\ z \end{pmatrix}$$

reduces to the form $x^2 + 4y^2 + z^2 + 4xz$ to the form $aX^2 + bY^2 + cZ^2$. [L]

30 Show that the matrix $\mathbf{A}$ given by $\mathbf{A} = \begin{pmatrix} 3 & 2 & 2 \\ 2 & 2 & 0 \\ 2 & 0 & 4 \end{pmatrix}$ has an eigenvector

$\begin{pmatrix} 2 \\ -2 \\ -1 \end{pmatrix}$ with eigenvalue 0.

Find the other eigenvectors of $\mathbf{A}$. Find also an orthogonal matrix $\mathbf{P}$ such that $\mathbf{P}^{-1}\mathbf{AP}$ is a diagonal matrix.

Show that the transformation with matrix $\mathbf{P}^{-1}$ converts the expression

$$3x^2 + 4xy + 4xz + 2y^2 + 4z^2$$

to a form containing only two terms. [L]

31 Given that $a = 1 - \sqrt{3}$ and $b = 1 + \sqrt{3}$, show that the column vectors of the matrix $\mathbf{P}$, where

$$\mathbf{P} = \frac{1}{3}\begin{pmatrix} 1 & a & b \\ b & 1 & a \\ a & b & 1 \end{pmatrix},$$

are mutually orthogonal.

Show that the linear transformation given by

$$\begin{pmatrix} X \\ Y \\ Z \end{pmatrix} = \mathbf{P}\begin{pmatrix} x \\ y \\ z \end{pmatrix}$$

maps any point (t, t, t) to itself, and maps the vector $\mathbf{i} - 2\mathbf{j} + \mathbf{k}$ to the vector $(\sqrt{3})(\mathbf{i} - \mathbf{k})$.

Deduce that this transformation is a rotation about the line $x = y = z$ through a right angle.

32 The transformation T of the points of a three-dimensional space is given by $\mathbf{r}' = \mathbf{Mr}$, where

$$\mathbf{r} = \begin{pmatrix} x \\ y \\ z \end{pmatrix}, \qquad \mathbf{r}' = \begin{pmatrix} x' \\ y' \\ z' \end{pmatrix} \qquad \text{and } \mathbf{M} \text{ is a } 3 \times 3 \text{ matrix,}$$

and

$$\mathbf{e}_1 = \begin{pmatrix} 1 \\ 0 \\ 0 \end{pmatrix}, \quad \mathbf{e}_2 = \begin{pmatrix} 0 \\ 1 \\ 0 \end{pmatrix} \quad \text{and} \quad \mathbf{e}_3 = \begin{pmatrix} 0 \\ 0 \\ 1 \end{pmatrix}$$

form a right-handed set of mutually perpendicular vectors.

Given that

$$\mathbf{f}_1 = \begin{pmatrix} \frac{2}{3} \\ \frac{2}{3} \\ \frac{1}{3} \end{pmatrix} \quad \text{and} \quad \mathbf{f}_2 = \begin{pmatrix} -\frac{2}{3} \\ \frac{1}{3} \\ \frac{2}{3} \end{pmatrix},$$

obtain a unit vector $\mathbf{f}_3$ such that $\mathbf{f}_1$, $\mathbf{f}_2$ and $\mathbf{f}_3$ form a right-handed set of mutually perpendicular unit vectors. Write down the matrix $\mathbf{M}_1$ of the transformation T_1 which takes $\mathbf{e}_1$, $\mathbf{e}_2$ and $\mathbf{e}_3$ into $\mathbf{f}_1$, $\mathbf{f}_2$ and $\mathbf{f}_3$ respectively.

Find the image of the vector $\begin{pmatrix} 1 \\ 0 \\ 1 \end{pmatrix}$ under T_1.

A second transformation T_2 has matrix

$$\mathbf{M}_2 = \begin{pmatrix} 0 & 1 & 1 \\ 1 & 1 & 0 \\ -1 & 0 & 1 \end{pmatrix}.$$

Show that T_2 maps the whole space onto the plane $x - y - z = 0$.

Find the image under T_2 of (a) the line $x/2 = y/3 = z/5$,

(b) the plane $2x + y - z = 0$. [JMB]

33 The orthogonal matrix $\mathbf{P}$, where

$$\mathbf{P} = \tfrac{1}{2}\begin{pmatrix} \sqrt{2} & a & \sqrt{2} \\ -1 & b & 1 \\ -1 & c & 1 \end{pmatrix},$$

is such that $\det \mathbf{P} = 1$. Find the values of a, b and c.

Show that the vector $\begin{pmatrix} 1 \\ -1 \\ \sqrt{2} - 1 \end{pmatrix}$ is an eigenvector of $\mathbf{P}$.

Show also that under the transformation given by

$$\begin{pmatrix} X \\ Y \\ Z \end{pmatrix} = \mathbf{P}\begin{pmatrix} x \\ y \\ z \end{pmatrix}$$

any vector at right angles to this eigenvector is rotated through an angle θ where $4 \cos \theta = 2\sqrt{2} - 1$.

34 Given that the matrix $\begin{pmatrix} 1/2 & \sqrt{3}/2 & 0 \\ 0 & 0 & 1 \\ \sqrt{3}/2 & -1/2 & 0 \end{pmatrix}$ corresponds to a rotation about a fixed axis through the origin, determine this axis and the angle of rotation. [JMB]

35 Show that the eigenvalues of the matrix $\mathbf{A}$, where

$$\mathbf{A} = \begin{pmatrix} -2 & -3 & -1 \\ 1 & 2 & 1 \\ 3 & 3 & 2 \end{pmatrix},$$

are 1, -1, 2 and find the associated eigenvectors.

Verify that there is a matrix $\mathbf{P}$, formed from the eigenvectors, such that $\mathbf{P}^{-1}\mathbf{AP}$ is a diagonal matrix in which the elements in the leading diagonal are the eigenvalues of $\mathbf{A}$. [L]

36 The three planes π_1, π_2, π_3 have the equations

$$\pi_1: \qquad 2x - z = 0,$$
$$\pi_2: x + y + 2z = 0,$$
$$\pi_3: x - 5y + 2z = 0.$$

Show that each of these three planes is perpendicular to the other two planes.

Verify that, for all values of the parameters s and t, the point P $(2s, 2t, -s - t)$ lies in the plane π_2. Find a vector equation of the line L which is perpendicular to π_2 and contains the point P.

The transformation $T: \mathbb{R}^3 \to \mathbb{R}^3$ is given by $T(\mathbf{r}) = \mathbf{r}'$, where $\mathbf{r}' = \mathbf{Mr}$, and

$$\mathbf{r} = \begin{pmatrix} x \\ y \\ z \end{pmatrix}, \qquad \mathbf{r}' = \begin{pmatrix} x' \\ y' \\ z' \end{pmatrix}, \qquad \mathbf{M} = \tfrac{1}{6}\begin{pmatrix} 5 & -1 & -2 \\ -1 & 5 & -2 \\ -2 & -2 & 2 \end{pmatrix}.$$

Show that each point of the line L is transformed by T to the point P. Show also that the image of the plane π_1 under the transformation T is a line and find a vector equation of this line. [JMB]

37 Given that $\mathbf{X} = \begin{pmatrix} x \\ y \end{pmatrix}$, $\mathbf{X}_1 = \begin{pmatrix} x_1 \\ y_1 \end{pmatrix}$, $\mathbf{A} = \begin{pmatrix} -5 & 0 \\ 0 & 5 \end{pmatrix}$, find the image of the curve $y = x^2 + 2x + 1$ under the transformation $\mathbf{X}_1 = \mathbf{AX}$.

Find also the image of the curve $x^2 - y^2 = 1$ under the transformation $\mathbf{X}_1 = \mathbf{BX}$, where $\mathbf{B}$ is the matrix

$$\begin{pmatrix} \tfrac{1}{2} & \tfrac{1}{2} \\ \tfrac{1}{2} & -\tfrac{1}{2} \end{pmatrix}.$$

Describe each of these transformations in geometrical terms.

38 Write down the symmetric matrix $\mathbf{A}$ of the quadratic form Q, where

$$Q \equiv 2x^2 + y^2 - 4xy - 4yz,$$

and find the eigenvalues of $\mathbf{A}$.

Find also an orthogonal matrix $\mathbf{H}$ which is such that the transformation

$$\begin{pmatrix} x \\ y \\ z \end{pmatrix} = \mathbf{H}\begin{pmatrix} u \\ v \\ w \end{pmatrix}$$

transforms Q to a sum of squares of the form $au^2 + bv^2 + cw^2$, and find the set $\{a, b, c\}$. [L]

39 Prove that the equations

$$x - 5y + 2z = 1,$$
$$x + ky + 4z = 2,$$
$$kx + y + (3k + 1)z = 5$$

have a unique solution except when $k = 1$ or $k = 3$.

Show that when $k = 1$ the equations have no solution.

Show also that when $k = 3$ the equations are not independent, and in this case express each of x and z in terms of y.

Interpret these three cases geometrically. [L]

40 A square matrix $\mathbf{A}$ of order n is said to be orthogonal if $\mathbf{AA}^{\mathrm{T}} = \mathbf{I}$, where $\mathbf{I}$ is the unit matrix of order n.

(a) If $\mathbf{A}$ is an orthogonal matrix of order 2, prove that the determinant of $\mathbf{A}$ is plus or minus one.

(b) Prove that, if $\mathbf{A}$ and $\mathbf{B}$ are orthogonal matrices of order n, then so also is $\mathbf{AB}$.

(c) Given that the matrix $\mathbf{Q}$, where $\mathbf{Q} = \begin{pmatrix} a & 0 & b \\ 0 & c & 0 \\ b & 0 & -a \end{pmatrix}$, is orthogonal, prove that the matrix $\mathbf{R}$, where

$$\mathbf{R} = \frac{1}{\sqrt{2}} \begin{pmatrix} a & -c & b \\ a & c & b \\ b\sqrt{2} & 0 & -a\sqrt{2} \end{pmatrix},$$

is also orthogonal.

Show that there is a matrix $\mathbf{P}$, with elements independent of a, b and c, such that $\mathbf{PQ} = \mathbf{R}$. [L]

14 The vector product

14.1 The vector product

In *Advanced Mathematics 1* we considered the product of a scalar and a vector, i.e. $\lambda\mathbf{a}$, where λ is a scalar and $\mathbf{a}$ is a vector. This product was defined as a vector in the same direction as $\mathbf{a}$ and having magnitude $\lambda|\mathbf{a}|$.

We also considered the *scalar product* (or *dot* product) of two vectors, i.e. $\mathbf{a} \cdot \mathbf{b}$. The scalar product $\mathbf{a} \cdot \mathbf{b}$ was defined as $ab \cos \theta$, where a and b are $|\mathbf{a}|$ and $|\mathbf{b}|$ respectively and θ is the angle between the vectors $\mathbf{a}$ and $\mathbf{b}$.

We now consider another product, the *vector product* (or *cross* product) of two vectors, $\mathbf{a} \times \mathbf{b}$.

The vector product $\mathbf{a} \times \mathbf{b}$ is defined as the vector with
(1) magnitude $ab|\sin \theta|$, where θ is the angle between the vectors $\mathbf{a}$ and $\mathbf{b}$,
(2) direction perpendicular to both the vector $\mathbf{a}$ and the vector $\mathbf{b}$,
(3) sense given by the sense in which a right-handed screw would advance when rotated from the direction of $\mathbf{a}$ to the direction of $\mathbf{b}$.

Thus the vector product $\mathbf{a} \times \mathbf{b}$ can be defined by

$$\mathbf{a} \times \mathbf{b} = (ab|\sin \theta|)\mathbf{c},$$

where $\mathbf{c}$ is a unit vector perpendicular to both $\mathbf{a}$ and $\mathbf{b}$ in the sense given by the sense of advance of a right-handed screw rotated from the direction of $\mathbf{a}$ to the direction of $\mathbf{b}$.

It should be noted that the vector (or cross) product $\mathbf{a} \times \mathbf{b}$ is a vector, whereas the scalar (or dot) product $\mathbf{a} \cdot \mathbf{b}$ is a scalar.

In Fig. 14.1 the sense of the vector product $\mathbf{a} \times \mathbf{b}$ will be upwards out of the page towards the reader. The sense of the vector product $\mathbf{b} \times \mathbf{a}$ will be downwards into the book and away from the reader.

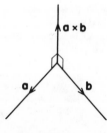

Fig. 14.1

The magnitude of the vector product $\mathbf{a} \times \mathbf{b}$ has been defined as $ab|\sin \theta|$ and this gives the area of the parallelogram formed by the vectors $\mathbf{a}$ and $\mathbf{b}$.

It will be seen that

$$\mathbf{b} \times \mathbf{a} = ab \, |\sin \theta|(-\mathbf{c}) = -ab|\sin \theta|\mathbf{c}.$$

Therefore $\mathbf{b} \times \mathbf{a} = -(\mathbf{a} \times \mathbf{b})$.

Thus it will be seen that the operation of forming the vector product of two vectors is not commutative. (The operation of forming the scalar product of two vectors is commutative since $\mathbf{a} \cdot \mathbf{b} = \mathbf{b} \cdot \mathbf{a}$.)

Also from the definitions of $\lambda\mathbf{a}$ and of $\mathbf{a} \times \mathbf{b}$, it is clear that $(\lambda\mathbf{a}) \times (\mu\mathbf{b}) = \lambda\mu(\mathbf{a} \times \mathbf{b})$.

The equation $\mathbf{a} \times \mathbf{b} = \mathbf{0}$ has the following possible solutions:
(1) $|\mathbf{a}| = 0$,
(2) $|\mathbf{b}| = 0$,
(3) $\theta = 0$, i.e. $\mathbf{a}$ and $\mathbf{b}$ are parallel and in the same sense,
(4) $\theta = \pi$, i.e. $\mathbf{a}$ and $\mathbf{b}$ are parallel but in opposite senses.
Thus $\mathbf{a} \times \mathbf{a} = \mathbf{0}$.

The unit vectors i, j, k
These three vectors (see Fig. 14.2) form what is known as an *orthogonal right-handed triad*, i.e $\mathbf{i}, \mathbf{j}, \mathbf{k}$ are mutually at right angles and in the cyclic order $\mathbf{i}, \mathbf{j}, \mathbf{k}, \mathbf{i}, \mathbf{j}, \mathbf{k}, \ldots$ and are such that the right-handed screw rule implies that

$$\mathbf{i} \times \mathbf{j} = \mathbf{k}, \qquad \mathbf{j} \times \mathbf{k} = \mathbf{i}, \qquad \mathbf{k} \times \mathbf{i} = \mathbf{j}$$

and consequently that

$$\mathbf{j} \times \mathbf{i} = -\mathbf{k}, \, \mathbf{k} \times \mathbf{j} = -\mathbf{i}, \, \mathbf{i} \times \mathbf{k} = -\mathbf{j}.$$

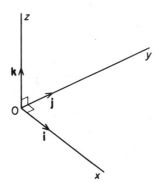

Fig. 14.2

Example
Evaluate (a) $(\lambda\mathbf{i}) \times (\mu\mathbf{j})$, (b) $(-\lambda\mathbf{i}) \times (\mu\mathbf{k})$, (c) $(\lambda\mathbf{a}) \times (-\mu\mathbf{a})$.

(a) $(\lambda \mathbf{i}) \times (\mu \mathbf{j}) = \lambda \mu (\mathbf{i} \times \mathbf{j}) = \lambda \mu \mathbf{k}$.

(b) $(-\lambda \mathbf{i}) \times (\mu \mathbf{k}) = -\lambda \mu (\mathbf{i} \times \mathbf{k}) = -\lambda \mu (-\mathbf{j}) = \lambda \mu \mathbf{j}$.

(c) $(\lambda \mathbf{a}) \times (-\mu \mathbf{a}) = -\lambda \mu (\mathbf{a} \times \mathbf{a}) = \mathbf{0}$.

Exercises 14.1

1 Evaluate   (a) $(2\mathbf{i}) \times (3\mathbf{j})$,   (b) $(3\mathbf{j}) \times (2\mathbf{i})$,   (c) $(-2\mathbf{i}) \times (3\mathbf{j})$,
              (d) $(2\mathbf{i}) \times (-3\mathbf{j})$,   (e) $(-2\mathbf{i}) \times (-3\mathbf{j})$.

2 Evaluate   (a) $(4\mathbf{j}) \times (5\mathbf{k})$,   (b) $(-4\mathbf{j}) \times (5\mathbf{k})$,   (c) $(4\mathbf{k}) \times (5\mathbf{j})$,
              (d) $(-5\mathbf{j}) \times (-4\mathbf{k})$, (e) $(4\mathbf{j}) \times (-5\mathbf{k})$.

3 Evaluate   (a) $(a\mathbf{i}) \times (b\mathbf{k})$,   (b) $(b\mathbf{i}) \times (a\mathbf{k})$,   (c) $(-a\mathbf{i}) \times (b\mathbf{k})$,
              (d) $(-a\mathbf{k}) \times (-b\mathbf{i})$, (e) $(a\mathbf{i}) \times (b\mathbf{i})$,   (f) $(-a\mathbf{k}) \times (-b\mathbf{k})$,
              (g) $(-a\mathbf{j}) \times (-b\mathbf{j})$.

14.2   The distributive law

It was shown in Section 12.7 of *Advanced Mathematics 1* that the operation of forming a scalar product of two vectors is distributive over vector addition, i.e.

$$\mathbf{a} \cdot (\mathbf{b} + \mathbf{c}) = \mathbf{a} \cdot \mathbf{b} + \mathbf{a} \cdot \mathbf{c}.$$

We will now show that the operation of forming a vector product of two vectors is also distributive over vector addition, i.e. that

$$\mathbf{a} \times (\mathbf{b} + \mathbf{c}) = \mathbf{a} \times \mathbf{b} + \mathbf{a} \times \mathbf{c}.$$

In Fig. 14.3, $\overrightarrow{OA} = \mathbf{a}$, $\overrightarrow{OB} = \mathbf{b}$, $\overrightarrow{OC} = \mathbf{c}$ and $\overrightarrow{OD} = \mathbf{b} + \mathbf{c} \; (= \mathbf{d}$, say).

The plane Π is perpendicular to $\mathbf{a}$ and contains O. The feet of the perpendiculars from B, C and D to the plane Π are B_1, C_1 and D_1 respectively.

Then $\overrightarrow{OD}_1 = \overrightarrow{OB}_1 + \overrightarrow{OC}_1$.

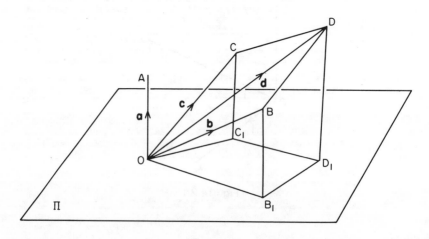

Fig. 14.3

The vector product, $\mathbf{a} \times \mathbf{b}$, is perpendicular to both $\mathbf{a}$ and $\mathbf{b}$, i.e. it is perpendicular to the plane AOB. Since BB_1 is perpendicular to the plane Π, it is parallel to AO and so $\mathbf{a} \times \mathbf{b}$ is perpendicular to OA and OB_1 and is obtained in direction by rotating OB_1 through a right angle about OA as shown in Fig. 14.4.

Similarly, the direction of $\mathbf{a} \times \mathbf{c}$ is obtained by rotating OC_1 through a right angle about OA and the direction of $\mathbf{a} \times \mathbf{d}$ is obtained by rotating OD_1 through a right angle about OA. The parallelogram $OB_1D_1C_1$ is rotated anticlockwise about O through an angle $\pi/2$ to the position $OB_2D_2C_2$.

Then
$$\overrightarrow{OD_2} = \overrightarrow{OB_2} + \overrightarrow{OC_2}. \qquad \ldots (1)$$

Now
$$OB_1 = OB \sin \beta,$$

where β is the angle between $\mathbf{a}$ and $\mathbf{b}$.

$\therefore$
$$OB_2 = OB \sin \beta.$$

Similarly $OC_2 = OC \sin \gamma$, where γ is the angle between $\mathbf{a}$ and $\mathbf{c}$, and $OD_2 = OD \sin \phi$, where ϕ is the angle between $\mathbf{a}$ and $(\mathbf{a} + \mathbf{c})$.

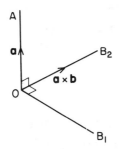

Fig. 14.4

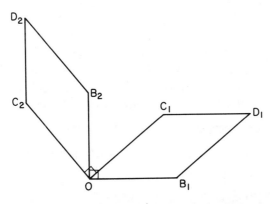

Fig. 14.5   (showing parallelogram $OB_1D_1C_1$ and $OB_2D_2C_2$ on plane Π)

Multiplying equation (1) by |**a**| gives

$$|\mathbf{a}| \ \overrightarrow{OD_2} = |\mathbf{a}|\overrightarrow{OB_2} + |\mathbf{a}|\overrightarrow{OC_2}.$$

The magnitude of $|\mathbf{a}|\overrightarrow{OB_2} = |\mathbf{a}|\,OB \sin \beta = ab \sin \beta$
$$= |\mathbf{a} \times \mathbf{b}|.$$

OB_2 lies in the plane Π and is therefore perpendicular to **a**. OB_2 is perpendicular to OB and to OB_1 and so OB_2 is perpendicular to **b**.
Thus $\overrightarrow{OB_2}$ is perpendicular to both **a** and **b**.
Also $\overrightarrow{OB_2}$ is in the sense of $\mathbf{a} \times \mathbf{b}$.

Therefore $|\mathbf{a}|\overrightarrow{OB_2} = \mathbf{a} \times \mathbf{b}.$
Similarly $|\mathbf{a}|\overrightarrow{OC_2} = \mathbf{a} \times \mathbf{c}$
and $|\mathbf{a}| \ \overrightarrow{OD_2} = \mathbf{a} \times \mathbf{d} = \mathbf{a} \times (\mathbf{b} + \mathbf{c}).$

Thus from equation (1)

$$\mathbf{a} \times (\mathbf{b} + \mathbf{c}) = \mathbf{a} \times \mathbf{b} + \mathbf{a} \times \mathbf{c}.$$

Exercises 14.2

1   Show that (a) $(\mathbf{a} + \mathbf{b}) \times (\mathbf{a} + \mathbf{b}) = \mathbf{0}$, (b) $(\mathbf{a} - \mathbf{b}) \times (\mathbf{a} + \mathbf{b}) = 2\mathbf{a} \times \mathbf{b}.$

2   Given that $\mathbf{a} + \mathbf{b} + \mathbf{c} = \mathbf{0}$, show that $\mathbf{a} \times \mathbf{b} = \mathbf{b} \times \mathbf{c} = \mathbf{c} \times \mathbf{a}.$

Deduce the sine rule $\dfrac{a}{\sin A} = \dfrac{b}{\sin B} = \dfrac{c}{\sin C}.$

14.3   The vector product in component form

We now obtain an expression for the vector product, $\mathbf{a} \times \mathbf{b}$, when **a** and **b** are in component form.

Let $\mathbf{a} = a_1\mathbf{i} + a_2\mathbf{j} + a_3\mathbf{k}$ and $\mathbf{b} = b_1\mathbf{i} + b_2\mathbf{j} + b_3\mathbf{k}.$

Then $\mathbf{a} \times \mathbf{b} = (a_1\mathbf{i} + a_2\mathbf{j} + a_3\mathbf{k}) \times (b_1\mathbf{i} + b_2\mathbf{j} + b_3\mathbf{k})$
$$= (a_1\mathbf{i}) \times (b_1\mathbf{i}) + (a_1\mathbf{i}) \times (b_2\mathbf{j}) + (a_1\mathbf{i}) \times (b_3\mathbf{k})$$
$$+ (a_2\mathbf{j}) \times (b_1\mathbf{i}) + (a_2\mathbf{j}) \times (b_2\mathbf{j}) + (a_2\mathbf{j}) \times (b_3\mathbf{k})$$
$$+ (a_3\mathbf{k}) \times (b_1\mathbf{i}) + (a_3\mathbf{k}) \times (b_2\mathbf{j}) + (a_3\mathbf{k}) \times (b_3\mathbf{k})$$
$$= a_1b_1(\mathbf{i} \times \mathbf{i}) + a_1b_2(\mathbf{i} \times \mathbf{j}) + a_1b_3(\mathbf{i} \times \mathbf{k})$$
$$+ a_2b_1(\mathbf{j} \times \mathbf{i}) + a_2b_2(\mathbf{j} \times \mathbf{j}) + a_2b_3(\mathbf{j} \times \mathbf{k})$$
$$+ a_3b_1(\mathbf{k} \times \mathbf{i}) + a_3b_2(\mathbf{k} \times \mathbf{j}) + a_3b_3(\mathbf{k} \times \mathbf{k}).$$

By the vector product properties of **i**, **j**, **k**, this gives

$$\mathbf{a} \times \mathbf{b} = (a_2b_3 - a_3b_2)\mathbf{i} - (a_1b_3 - a_3b_1)\mathbf{j} + (a_1b_2 - a_2b_1)\mathbf{k}$$

$$= \begin{vmatrix} \mathbf{i} & \mathbf{j} & \mathbf{k} \\ a_1 & a_2 & a_3 \\ b_1 & b_2 & b_3 \end{vmatrix}$$

(assuming that the 'determinant' is evaluated in terms of elements of the first row).

Instead of using the determinant notation, $\mathbf{a} \times \mathbf{b}$ can be obtained as follows (the full lines carry a $+$ sign and the dotted lines a $-$ sign):

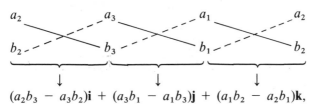

$$(a_2 b_3 - a_3 b_2)\mathbf{i} + (a_3 b_1 - a_1 b_3)\mathbf{j} + (a_1 b_2 - a_2 b_1)\mathbf{k},$$

i.e. $\mathbf{a} \times \mathbf{b}$.

The vector $a_1\mathbf{i} + a_2\mathbf{j} + a_3\mathbf{k}$ is often expressed in the form $\begin{pmatrix} a_1 \\ a_2 \\ a_3 \end{pmatrix}$ or as (a_1, a_2, a_3). It may be thought that the latter form could be confused with cartesian coordinates. However the context in which (a_1, a_2, a_3) appears will indicate whether it refers to a vector or to cartesian coordinates. In the same way we use the symbol $+$ for the addition of numbers and for the addition of vectors, e.g. $3 + 4 = 7$, $\mathbf{a} + \mathbf{b} = \mathbf{c}$, the context indicating the nature of the addition.

Example 1

Obtain $\mathbf{a} \times \mathbf{b}$ when $\mathbf{a} = 3\mathbf{i} - 2\mathbf{j} + 7\mathbf{k}$ and $\mathbf{b} = \mathbf{i} - 4\mathbf{k}$.

Using the procedure just outlined:

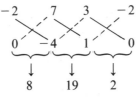

Thus
$$\mathbf{a} \times \mathbf{b} = 8\mathbf{i} + 19\mathbf{j} + 2\mathbf{k}.$$

It is most important that the student, having obtained a vector product $\mathbf{a} \times \mathbf{b}$, should check that the scalar product of $\mathbf{a} \times \mathbf{b}$ and the vector $\mathbf{a}$ equals 0 and the scalar product of $\mathbf{a} \times \mathbf{b}$ and the vector $\mathbf{b}$ equals 0.

Here we check that $\begin{pmatrix} 8 \\ 19 \\ 2 \end{pmatrix} \cdot \begin{pmatrix} 3 \\ -2 \\ 7 \end{pmatrix} = (8)(3) + (19)(-2) + (2)(7) = 0$

and that $\begin{pmatrix} 8 \\ 19 \\ 2 \end{pmatrix} \cdot \begin{pmatrix} 1 \\ 0 \\ -4 \end{pmatrix} = (8)(1) + (19)(0) + (2)(-4) = 0.$

Example 2

Find the area of the parallelogram formed by the vectors **a** and **b** where
a = (2, 3, 4) and **b** = (4, − 1, − 3).

$$\mathbf{a} \times \mathbf{b} = \begin{vmatrix} \mathbf{i} & \mathbf{j} & \mathbf{k} \\ 2 & 3 & 4 \\ 4 & -1 & -3 \end{vmatrix}$$

$$= (-9 + 4)\mathbf{i} - (-6 - 16)\mathbf{j} + (-2 - 12)\mathbf{k}$$
$$= -5\mathbf{i} + 22\mathbf{j} - 14\mathbf{k}.$$

By definition, the magnitude of the vector product **a** × **b** is $ab|\sin \theta|$, which equals the area of the parallelogram formed by the vectors **a** and **b**.

$$\text{Required area} = |\mathbf{a} \times \mathbf{b}|$$
$$= |-5\mathbf{i} + 22\mathbf{j} - 14\mathbf{k}|$$
$$= \sqrt{(5^2 + 22^2 + 14^2)}$$
$$= \sqrt{705}.$$

Example 3

The points A, B and C have position vectors **a**, **b** and **c** respectively, where

$$\mathbf{a} = \mathbf{i} + 2\mathbf{j} + 3\mathbf{k},$$
$$\mathbf{b} = 2\mathbf{i} \qquad - 4\mathbf{k},$$
$$\mathbf{c} = 5\mathbf{i} - \mathbf{j} - 3\mathbf{k}.$$

Find the area of the triangle ABC.

$$\overrightarrow{AB} = \mathbf{b} - \mathbf{a} = \mathbf{i} - 2\mathbf{j} - 7\mathbf{k},$$
$$\overrightarrow{AC} = \mathbf{c} - \mathbf{a} = 4\mathbf{i} - 3\mathbf{j} - 6\mathbf{k}$$

Area of $\triangle ABC = \frac{1}{2}|\overrightarrow{AB} \times \overrightarrow{AC}|$

and
$$\overrightarrow{AB} \times \overrightarrow{AC} = \begin{vmatrix} \mathbf{i} & \mathbf{j} & \mathbf{k} \\ 1 & -2 & -7 \\ 4 & -3 & -6 \end{vmatrix}$$

$$= -9\mathbf{i} - 22\mathbf{j} + 5\mathbf{k}.$$

∴ area of $\triangle ABC = \frac{1}{2}|-9\mathbf{i} - 22\mathbf{j} + 5\mathbf{k}|$
$$= \frac{1}{2}\sqrt{(81 + 484 + 25)} = \frac{1}{2}\sqrt{590}.$$

Example 4

Find a vector equation for the plane containing the points A. B and C with position vectors **a**, **b** and **c**, where

$$\mathbf{a} = \mathbf{i} + \mathbf{j} + \mathbf{k}, \quad \mathbf{b} = 2\mathbf{i} - \mathbf{j} + 3\mathbf{k}, \quad \mathbf{c} = 3\mathbf{i} - 4\mathbf{k}.$$
$$\overrightarrow{AB} = \mathbf{b} - \mathbf{a} = \mathbf{i} - 2\mathbf{j} + 2\mathbf{k}$$
$$\overrightarrow{BC} = \mathbf{c} - \mathbf{b} = \mathbf{i} + \mathbf{j} - 7\mathbf{k}$$

By definition $\overrightarrow{AB} \times \overrightarrow{BC}$ is perpendicular to both $\overrightarrow{AB}$ and $\overrightarrow{BC}$, i.e. $\overrightarrow{AB} \times \overrightarrow{BC}$

is normal to the plane containing both AB and BC and so the vector product $\overrightarrow{AB} \times \overrightarrow{BC}$ is normal to the plane ABC.

$$\overrightarrow{AB} \times \overrightarrow{BC} = (\mathbf{i} - 2\mathbf{j} + 2\mathbf{k}) \times (\mathbf{i} + \mathbf{j} - 7\mathbf{k})$$
$$= 12\mathbf{i} + 9\mathbf{j} + 3\mathbf{k}$$

Thus the vector $12\mathbf{i} + 9\mathbf{j} + 3\mathbf{k}$ is normal to the plane ABC and so too is the vector $4\mathbf{i} + 3\mathbf{j} + \mathbf{k}$.

$\therefore$ the equation of the plane ABC is of the form,

$$(4\mathbf{i} + 3\mathbf{j} + \mathbf{k}) \cdot \mathbf{r} = \lambda, \text{ a constant.}$$

The point A is on the plane and so $\mathbf{r} = \mathbf{i} + \mathbf{j} + \mathbf{k}$ satisfies this equation

i.e. $\qquad\qquad (4\mathbf{i} + 3\mathbf{j} + \mathbf{k}) \cdot (\mathbf{i} + \mathbf{j} + \mathbf{k}) = \lambda$

giving $\qquad\qquad\qquad\qquad \lambda = 8.$

$\therefore$ an equation for the plane ABC is

$$(4\mathbf{i} + 3\mathbf{j} + \mathbf{k}) \cdot \mathbf{r} = 8.$$

Example 5
Find the cartesian equation of the plane through the points D(1, 1, 1), E(4, -2, 3) and F(2, -3, -4).

$$\overrightarrow{DE} = \mathbf{e} - \mathbf{d} = (4, -2, 3) - (1, 1, 1) = (3, -3, 2),$$
$$\overrightarrow{EF} = \mathbf{f} - \mathbf{e} = (2, -3, -4) - (4, -2, 3) = (-2, -1, -7).$$

The vector product $\overrightarrow{DE} \times \overrightarrow{EF}$ is perpendicular to both DE and EF and so is normal to the plane DEF.

$$\overrightarrow{DE} \times \overrightarrow{EF} = \begin{vmatrix} \mathbf{i} & \mathbf{j} & \mathbf{k} \\ 3 & -3 & 2 \\ -2 & -1 & -7 \end{vmatrix}$$
$$= 23\mathbf{i} + 17\mathbf{j} - 9\mathbf{k} \qquad \text{or} \qquad (23, 17, -9).$$

Therefore the cartesian equation of the plane DEF is of the form

$$23x + 17y - 9z = \lambda.$$

Since D(1, 1, 1) lies in the plane, $\lambda = 23 + 17 - 9 = 31$.
Thus the cartesian equation of the plane DEF is

$$23x + 17y - 9z = 31.$$

Example 6
The point A has position vector $\mathbf{i} + 4\mathbf{j} - 3\mathbf{k}$ referred to the origin O. The line L has vector equation $\mathbf{r} = t\mathbf{i}$. The plane Π contains the line L and the point A. Find
(a) a vector which is normal to the plane Π,
(b) a vector equation for the plane Π.

The line L, with equation $\mathbf{r} = t\mathbf{i}$, goes through the origin O.
Thus the plane Π contains both O and A, i.e. it contains the line OA.

$$\overrightarrow{OA} = \mathbf{i} + 4\mathbf{j} - 3\mathbf{k}.$$

The direction vector of the line L is $t\mathbf{i}$.
The vector product $(\mathbf{i} + 4\mathbf{j} - 3\mathbf{k}) \times (t\mathbf{i})$ is perpendicular to both OA and the line L, i.e. it is normal to the plane Π.

$$(\mathbf{i} + 4\mathbf{j} - 3\mathbf{k}) \times (t\mathbf{i}) = -3t\mathbf{j} - 4t\mathbf{k}.$$

The vector $3\mathbf{j} + 4\mathbf{k}$ is also normal to the plane Π.
Thus an equation for the plane is $\mathbf{r} \cdot (3\mathbf{j} + 4\mathbf{k}) = p$.
The plane contains the origin and so $p = 0$.
Therefore an equation for the plane is

$$\mathbf{r} \cdot (3\mathbf{j} + 4\mathbf{k}) = 0.$$

An alternative solution is as follows:

The line L, with equation $\mathbf{r} = t\mathbf{i}$, contains both the origin and the point with position vector $\mathbf{i}$, i.e. the required plane contains the three points $(1, 4, -3)$, $(0, 0, 0)$ and $(1, 0, 0)$.

The method of the previous example could now be used or we can say that, to pass through the points $(0, 0, 0)$ and $(1, 0, 0)$, the cartesian equation of the plane must be of the form $m_2 y + m_3 z = 0$.

The plane contains the point $(1, 4, -3)$ and so

$$4m_2 - 3m_3 = 0, \quad \text{i.e.} \quad m_2 : m_3 = 3 : 4.$$

Thus the cartesian equation of the plane is $\quad 3y + 4z = 0$.
In vector form an equation is $\qquad \mathbf{r} \cdot (3\mathbf{j} + 4\mathbf{k}) = 0$.

Exercises 14.3

1   Find the vector product $\mathbf{a} \times \mathbf{b}$, where $\mathbf{a}$ and $\mathbf{b}$ are respectively
   (a) $3\mathbf{i} + 4\mathbf{j} - 5\mathbf{k}$ and $2\mathbf{i} - \mathbf{j} + \mathbf{k}$,
   (b) $3\mathbf{i} - \mathbf{j} + 4\mathbf{k}$ and $-\mathbf{i} + 3\mathbf{j} + 7\mathbf{k}$,
   (c) $4\mathbf{i} + 3\mathbf{k}$ and $-2\mathbf{j} - 3\mathbf{k}$,
   (d) $\mathbf{j} - 4\mathbf{k}$ and $2\mathbf{i} - \mathbf{j} - \mathbf{k}$,
   (e) $5\mathbf{k}$ and $4\mathbf{j}$,
   (f) $-\mathbf{i} + 3\mathbf{j} - 6\mathbf{k}$ and $4\mathbf{i} - \mathbf{j} + 5\mathbf{k}$.

2   Find the area of the parallelogram formed by the vectors
   (a) $\mathbf{i} + \mathbf{j} + \mathbf{k}$ and $2\mathbf{i} + 3\mathbf{j} + 4\mathbf{k}$,
   (b) $2\mathbf{i} - \mathbf{k}$ and $3\mathbf{i} + 5\mathbf{j} - 2\mathbf{k}$,
   (c) $3\mathbf{i} + 2\mathbf{j}$ and $\mathbf{i} - \mathbf{j} - 2\mathbf{k}$,
   (d) $4\mathbf{i} - 3\mathbf{j}$ and $-2\mathbf{i} - \mathbf{j} + \mathbf{k}$,

(e) $2\mathbf{i} - 5\mathbf{j} + 6\mathbf{k}$ and $\mathbf{i} - 2\mathbf{j} - 3\mathbf{k}$,
(f) $-\mathbf{i} + 5\mathbf{j} - 3\mathbf{k}$ and $2\mathbf{i} + 3\mathbf{j} - 7\mathbf{k}$.

3 The points A, B and C have position vectors $\mathbf{a}$, $\mathbf{b}$ and $\mathbf{c}$. Find the area of the triangle ABC when
(a) $\mathbf{a} = \mathbf{i} + \mathbf{j} + \mathbf{k}$, $\mathbf{b} = 2\mathbf{i} + 3\mathbf{j} + 4\mathbf{k}$, $\mathbf{c} = 5\mathbf{i} + 4\mathbf{j} + 3\mathbf{k}$;
(b) $\mathbf{a} = \mathbf{i} + 2\mathbf{k}$, $\mathbf{b} = -\mathbf{i} + 2\mathbf{j} - 3\mathbf{k}$, $\mathbf{c} = 2\mathbf{i} + 4\mathbf{k}$;
(c) $\mathbf{a} = 3\mathbf{i} - \mathbf{j} - 2\mathbf{k}$, $\mathbf{b} = 3\mathbf{j} + \mathbf{k}$, $\mathbf{c} = -\mathbf{i} - 2\mathbf{j} + 5\mathbf{k}$;
(d) $\mathbf{a} = 2\mathbf{i} + 5\mathbf{j}$, $\mathbf{b} = -3\mathbf{j} - \mathbf{k}$, $\mathbf{c} = -2\mathbf{i} - 3\mathbf{k}$;
(e) $\mathbf{a} = \mathbf{i} + 2\mathbf{j} - 3\mathbf{k}$, $\mathbf{b} = -\mathbf{i} - 4\mathbf{j} + 3\mathbf{k}$, $\mathbf{c} = 5\mathbf{i} + 3\mathbf{j} + \mathbf{k}$.

4 Find an equation, in the form $\mathbf{r} . \mathbf{n} = p$, for the plane which is parallel to the vectors $\mathbf{p}$ and $\mathbf{q}$ and which contains the point with position vector $\mathbf{a}$, given that
(a) $\mathbf{a} = \mathbf{0}$, $\mathbf{p} = \mathbf{i} + \mathbf{j} + \mathbf{k}$, $\mathbf{q} = 2\mathbf{i} + 3\mathbf{j} + 4\mathbf{k}$;
(b) $\mathbf{a} = \mathbf{0}$, $\mathbf{p} = 2\mathbf{j} + 3\mathbf{k}$, $\mathbf{q} = 2\mathbf{i} + \mathbf{k}$;
(c) $\mathbf{a} = \mathbf{i} + \mathbf{j} + \mathbf{k}$, $\mathbf{p} = \mathbf{i} + 2\mathbf{k}$, $\mathbf{q} = 5\mathbf{i} + 2\mathbf{j}$;
(d) $\mathbf{a} = 2\mathbf{i} - 3\mathbf{j} - 4\mathbf{k}$, $\mathbf{p} = \mathbf{j} + 3\mathbf{k}$, $\mathbf{q} = \mathbf{i} + 3\mathbf{j}$;
(e) $\mathbf{a} = 5\mathbf{i} - \mathbf{j} - 2\mathbf{k}$, $\mathbf{p} = 2\mathbf{i} - \mathbf{j}$, $\mathbf{q} = -\mathbf{i} + 2\mathbf{j} - 3\mathbf{k}$;
(f) $\mathbf{a} = -3\mathbf{i} - 2\mathbf{j} + \mathbf{k}$, $\mathbf{p} = -\mathbf{i} - 3\mathbf{j} + 4\mathbf{k}$, $\mathbf{q} = 4\mathbf{i}$.

5 Find an equation, in the form $\mathbf{r} . \mathbf{n} = p$, for the plane containing the points with position vectors $\mathbf{a}$, $\mathbf{b}$ and $\mathbf{c}$ when
(a) $\mathbf{a} = \mathbf{0}$, $\mathbf{b} = \mathbf{i} + \mathbf{j} + \mathbf{k}$, $\mathbf{c} = \mathbf{i} + 2\mathbf{j} + 3\mathbf{k}$;
(b) $\mathbf{a} = \mathbf{0}$, $\mathbf{b} = 2\mathbf{i} + \mathbf{k}$, $\mathbf{c} = 3\mathbf{j} + 2\mathbf{k}$;
(c) $\mathbf{a} = 2\mathbf{i}$, $\mathbf{b} = 4\mathbf{j}$, $\mathbf{c} = 5\mathbf{k}$;
(d) $\mathbf{a} = \mathbf{i} + \mathbf{j}$, $\mathbf{b} = \mathbf{j} + \mathbf{k}$, $\mathbf{c} = \mathbf{i} + \mathbf{k}$;
(e) $\mathbf{a} = \mathbf{i} + \mathbf{j} + \mathbf{k}$, $\mathbf{b} = 5\mathbf{i} + 4\mathbf{j} + 3\mathbf{k}$, $\mathbf{c} = 2\mathbf{i} + 3\mathbf{j} + 4\mathbf{k}$;
(f) $\mathbf{a} = 3\mathbf{i} - 4\mathbf{j} + 2\mathbf{k}$, $\mathbf{b} = -\mathbf{i} + 2\mathbf{j} - 3\mathbf{k}$, $\mathbf{c} = 5\mathbf{i} - \mathbf{j} + \mathbf{k}$.

6 Find the cartesian equation of the plane containing the points
(a) $(1, 0, 1)$, $(0, 2, 0)$, $(6, -4, 2)$;
(b) $(2, 3, 5)$, $(0, 4, 5)$, $(3, 4, 3)$;
(c) $(1, 2, 3)$, $(-1, 4, 5)$, $(6, -6, -3)$;
(d) $(1, -3, 3)$, $(-3, -8, 6)$, $(2, 3, 7)$;
(e) $(1, 2, 2)$, $(3, 5, 2)$, $(1, -1, 2)$;
(f) $(a, 0, 0)$, $(0, b, 0)$, $(0, 0, c)$.

7 The points A, B and C have position vectors $\mathbf{a}$, $\mathbf{b}$ and $\mathbf{c}$. Show that the area of the triangle ABC is $\frac{1}{2}|\mathbf{a} \times \mathbf{b} + \mathbf{b} \times \mathbf{c} + \mathbf{c} \times \mathbf{a}|$.

8 Given that $\mathbf{a} = \cos A\mathbf{i} + \sin A\mathbf{j}$ and $\mathbf{b} = \cos B\mathbf{i} + \sin B\mathbf{j}$, obtain $\mathbf{a} . \mathbf{b}$ and $\mathbf{a} \times \mathbf{b}$. Hence show that
(a) $\cos (A - B) = \cos A \cos B + \sin A \sin B$,
(b) $\sin (A - B) = \sin A \cos B - \cos A \sin B$.

14.4 The perpendicular distance from a point to a line

Figure 14.6 shows the line L with vector equation $r = a + tb$. This line passes through the point A with position vector a, and is parallel to the vector b.

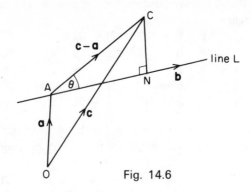

Fig. 14.6

The point C has position vector c and CN is the perpendicular from C to the line L. Let θ denote the angle CAN.

$$|(c - a) \times b| = AC\,|b|\sin\theta$$
$$= |b|CN$$
$$\Rightarrow \qquad CN = \frac{|(c - a) \times b|}{|b|},$$

which is the perpendicular distance from the point with position vector c to the line with vector equation $r = a + tb$.

Example 1

Find the perpendicular distance from the point C with position vector $2i - j + 2k$ to the line with vector equation

$$r = (i + j + k) + t(6i - 3j + 2k).$$

The line passes through the point A with position vector a where $a = i+j+k$. The direction vector of the line is $6i - 3j + 2k$.

$$\therefore \qquad c - a = (2i - j + 2k) - (i + j + k)$$
$$= i - 2j + k$$

$$\therefore \quad \text{the required distance} = \frac{|(i - 2j + k) \times (6i - 3j + 2k)|}{|6i - 3j + 2k|}$$

$$= \frac{|-i + 4j + 9k|}{7}$$

$$= \frac{\sqrt{98}}{7} = \sqrt{2}$$

Example 2
Find the perpendicular distance from the point C $(4, -1, 3)$ to the line with cartesian equations

$$\frac{x-1}{2} = \frac{y+1}{-3} = \frac{z-3}{-6}.$$

The point A $(1, -1, 3)$ lies on the line.

$$\overrightarrow{AC} = \mathbf{c} - \mathbf{a} = (4, -1, 3) - (1, -1, 3)$$
$$= (3, 0, 0)$$

The direction vector of the line is $(2, -3, -6)$.

$$\therefore \quad \text{the required distance} = \frac{|(3, 0, 0) \times (2, -3, -6)|}{|(2, -3, -6)|}$$

$$= \frac{|(0, 18, -9)|}{7}$$

$$= \frac{9\sqrt{5}}{7}.$$

Example 3
Find the perpendicular distance from the point A $(1, -2, 5)$ to the line with cartesian equations $x = 3$, $y - z = 2$.

The equations of the line represent two planes with normal vectors $(1, 0, 0)$ and $(0, 1, -1)$.
   The direction vector of the line is given by $(1, 0, 0) \times (0, 1, -1)$, i.e. $(0, 1, 1)$.
   When $x = 3$ and $y = 3$, we get $z = 1$, i.e. the point B $(3, 3, 1)$ lies on the line.

Then $\quad \overrightarrow{AB} = \mathbf{b} - \mathbf{a} = (3, 3, 1) - (1, -2, 5)$
$$= (2, 5, -4).$$

$$\therefore \quad \text{the required distance} = \frac{|(2, 5, -4) \times (0, 1, 1)|}{|(0, 1, 1)|}$$

$$= \frac{|(9, -2, 2)|}{\sqrt{2}}$$

$$= \sqrt{(89/2)}.$$

Exercises 14.4
1   Find the perpendicular distance from the point with position vector **a** to the
    line L when
    (a) $\mathbf{a} = 3\mathbf{i} + 2\mathbf{j} + \mathbf{k}$ and L has equation $\mathbf{r} = t(3\mathbf{j} + 4\mathbf{k})$;
    (b) $\mathbf{a} = 4\mathbf{j} + 5\mathbf{k}$ and L has equation $\mathbf{r} = (\mathbf{i} + \mathbf{j}) + t(\mathbf{i} + 2\mathbf{j} + 2\mathbf{k})$;
    (c) $\mathbf{a} = \mathbf{0}$ and L has equation $\mathbf{r} = (4\mathbf{i} - \mathbf{j} + \mathbf{k}) + t(3\mathbf{i} - 2\mathbf{j} + 6\mathbf{k})$;
    (d) $\mathbf{a} = \mathbf{i} + \mathbf{k}$ and L has equation $\mathbf{r} = (2\mathbf{i} - 3\mathbf{j} + 4\mathbf{k}) + t(4\mathbf{i} - 8\mathbf{j} + \mathbf{k})$;

2 Find the perpendicular distance from the point A to the line L when the coordinates of A and the cartesian equations of L are:

(a) A $(-2, 3, 7)$,   L: $\dfrac{x-4}{1} = \dfrac{y-2}{1} = \dfrac{z+1}{-1}$;

(b) A $(4, 0, 5)$,   L: $\dfrac{x+2}{-1} = \dfrac{y-3}{-2} = \dfrac{z}{2}$;

(c) A $(2, 3, 0)$,   L: $\dfrac{x}{6} = \dfrac{y}{3} = \dfrac{z}{2}$;

(d) A $(5, 6, 7)$,   L: $\dfrac{x+1}{1} = \dfrac{y-1}{8} = \dfrac{z+3}{-4}$;

3 Find the perpendicular distance from the point A to the line L when the coordinates of A and the parametric equations of L are:
(a) A $(0, 0, 0)$,   L: $x = 1 + t, y = 2 - 2t, z = 3 - 2t$;
(b) A $(5, 1, 1)$,   L: $x = 2t, y = 3t, z = 6t$;
(c) A $(2, 0, -3)$,   L: $x = 1, y = 2 + 3t, z = 5 - 4t$;
(d) A $(3, 4, -5)$,   L: $x = t, y = 3, z = -t$.

4 Find the perpendicular distance from the point A to the line L when the coordinates of A and the cartesian equations of L are:
(a) A $(1, 2, 3)$,   L: $3x - 2y - 5 = 0, 2y - z = 0$;
(b) A $(0, 0, 0)$,   L: $2x - y - 4 = 2x - z - 3 = 0$;
(c) A $(2, 0, -3)$,   L: $x + y = z = 1$;
(d) A $(3, 2, 0)$,   L: $x - 2y = -2, y - 4z = 5$.

14.5   The scalar triple product and the vector triple product

The scalar product of the vector **a** with the vector product of the vectors **b** and **c**, i.e. **a** . (**b** × **c**), is known as a *scalar triple* product. This product, **a** . (**b** × **c**), clearly exists since **b** × **c** is a vector (**d**, say) and **a** . **d** exists and is a scalar quantity – hence the name scalar triple product. There is also a *vector triple* product, which is the name given to the triple product **a** × (**b** × **c**). (Note that not all products exist. For instance **a** . (**b** . **c**) would be meaningless since **b** . **c** is not a vector.)

Let the points A, B, C have position vectors **a**, **b**, **c** respectively referred to the origin O. Fig. 14.7 shows the parallelepiped OBDCAEFG.

Volume of parallelepiped OBDCAEFG
$$= \text{(area of parallelogram OBDC)(AN)}$$
$$= |\mathbf{b} \times \mathbf{c}|\text{AN}$$
$$= |\mathbf{b} \times \mathbf{c}| \, |\mathbf{a}| \cos \theta$$
$$= |\mathbf{a}| \, |\mathbf{b} \times \mathbf{c}| \cos \theta$$
$$= \mathbf{a} \, . \, (\mathbf{b} \times \mathbf{c}).$$

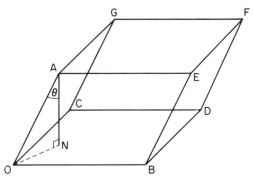

Fig. 14.7

Thus the scalar triple product $\mathbf{a} \cdot (\mathbf{b} \times \mathbf{c})$ gives the volume of the parallelepiped formed by the vectors $\mathbf{a}$, $\mathbf{b}$ and $\mathbf{c}$.

Also, the volume of the tetrahedron OABC $= \frac{1}{3}$(area of triangle OBC)(AN)

$$= \tfrac{1}{3}(\tfrac{1}{2} \text{ area of parallelogram OBDC})(\text{AN})$$
$$= \tfrac{1}{6}\mathbf{a} \cdot (\mathbf{b} \times \mathbf{c}).$$

Thus the scalar triple product $\mathbf{a} \cdot (\mathbf{b} \times \mathbf{c})$ also gives six times the volume of the tetrahedron OABC. The volume of the parallelepiped OBDCAEFG can similarly be expressed as $\mathbf{b} \cdot (\mathbf{c} \times \mathbf{a})$ and $\mathbf{c} \cdot (\mathbf{a} \times \mathbf{b})$.

Therefore $\qquad \mathbf{a} \cdot (\mathbf{b} \times \mathbf{c}) = \mathbf{b} \cdot (\mathbf{c} \times \mathbf{a}) = \mathbf{c} \cdot (\mathbf{a} \times \mathbf{b}).$

Also, the operation of forming a scalar product is commutative and

so $\qquad \mathbf{a} \cdot (\mathbf{b} \times \mathbf{c}) = (\mathbf{b} \times \mathbf{c}) \cdot \mathbf{a}, \mathbf{b} \cdot (\mathbf{c} \times \mathbf{a}) = (\mathbf{c} \times \mathbf{a}) \cdot \mathbf{b}$

and $\qquad \mathbf{c} \cdot (\mathbf{a} \times \mathbf{b}) = (\mathbf{a} \times \mathbf{b}) \cdot \mathbf{c}.$

Combining these results gives

$$\mathbf{a} \cdot (\mathbf{b} \times \mathbf{c}) = (\mathbf{a} \times \mathbf{b}) \cdot \mathbf{c}$$
$$= \mathbf{b} \cdot (\mathbf{c} \times \mathbf{a}) = (\mathbf{b} \times \mathbf{c}) \cdot \mathbf{a}$$
$$= \mathbf{c} \cdot (\mathbf{a} \times \mathbf{b}) = (\mathbf{c} \times \mathbf{a}) \cdot \mathbf{b}.$$

In view of these results, the notation $[\mathbf{a}, \mathbf{b}, \mathbf{c}]$ is often used to represent any of the above six scalar triple products.

Notice that the cyclic interchange of the three elements $\mathbf{a}$, $\mathbf{b}$ and $\mathbf{c}$ leaves the scalar triple product unchanged.

It can easily be shown that the effect of an interchange of two elements is to leave the magnitude of the scalar triple product unchanged but to change its sign.

Next we find the scalar triple product $\mathbf{a} \cdot (\mathbf{b} \times \mathbf{c})$ when the vectors $\mathbf{a}$, $\mathbf{b}$ and $\mathbf{c}$ are given in component form.

Let $\mathbf{a} = a_1\mathbf{i} + a_2\mathbf{j} + a_3\mathbf{k}$, $\mathbf{b} = b_1\mathbf{i} + b_2\mathbf{j} + b_3\mathbf{k}$ and $\mathbf{c} = c_1\mathbf{i} + c_2\mathbf{j} + c_3\mathbf{k}$.

Then $\quad \mathbf{a} \cdot (\mathbf{b} \times \mathbf{c}) = (a_1\mathbf{i} + a_2\mathbf{j} + a_3\mathbf{k}) \cdot \begin{vmatrix} \mathbf{i} & \mathbf{j} & \mathbf{k} \\ b_1 & b_2 & b_3 \\ c_1 & c_2 & c_3 \end{vmatrix}$

$$= a_1 \begin{vmatrix} b_2 & b_3 \\ c_2 & c_3 \end{vmatrix} - a_2 \begin{vmatrix} b_1 & b_3 \\ c_1 & c_3 \end{vmatrix} + a_3 \begin{vmatrix} b_1 & b_2 \\ c_1 & c_2 \end{vmatrix}$$

$$= \begin{vmatrix} a_1 & a_2 & a_3 \\ b_1 & b_2 & b_3 \\ c_1 & c_2 & c_3 \end{vmatrix}.$$

Example 1
Evaluate   (a) $(3\mathbf{i} + 4\mathbf{j} + 5\mathbf{k}) \cdot [(2\mathbf{i} - \mathbf{j} + 7\mathbf{k}) \times (-\mathbf{i} - 3\mathbf{j} + 4\mathbf{k})]$,
             (b) $(-\mathbf{i} - 3\mathbf{j} + 4\mathbf{k}) \cdot [(2\mathbf{i} - \mathbf{j} + 7\mathbf{k}) \times (3\mathbf{i} + 4\mathbf{j} + 5\mathbf{k})]$.

(a) $(2\mathbf{i} - \mathbf{j} + 7\mathbf{k}) \times (-\mathbf{i} - 3\mathbf{j} + 4\mathbf{k}) = \begin{vmatrix} \mathbf{i} & \mathbf{j} & \mathbf{k} \\ 2 & -1 & 7 \\ -1 & -3 & 4 \end{vmatrix}$

$$= 17\mathbf{i} - 15\mathbf{j} - 7\mathbf{k}.$$
$(3\mathbf{i} + 4\mathbf{j} + 5\mathbf{k}) \cdot (17\mathbf{i} - 15\mathbf{j} - 7\mathbf{k}) = 51 - 60 - 35 = -44.$
$\therefore \quad (3\mathbf{i} + 4\mathbf{j} + 5\mathbf{k}) \cdot [(2\mathbf{i} - \mathbf{j} + 7\mathbf{k}) \times (-\mathbf{i} - 3\mathbf{j} + 4\mathbf{k})] = -44.$

(b) $(2\mathbf{i} - \mathbf{j} + 7\mathbf{k}) \times (3\mathbf{i} + 4\mathbf{j} + 5\mathbf{k}) = \begin{vmatrix} \mathbf{i} & \mathbf{j} & \mathbf{k} \\ 2 & -1 & 7 \\ 3 & 4 & 5 \end{vmatrix}$

$$= -33\mathbf{i} + 11\mathbf{j} + 11\mathbf{k}.$$
$(-\mathbf{i} - 3\mathbf{j} + 4\mathbf{k}) \cdot (-33\mathbf{i} + 11\mathbf{j} + 11\mathbf{k}) = 33 - 33 + 44 = 44.$

This scalar triple product could have been obtained at once from (a) since two elements of the scalar triple product in (a) have been interchanged to give the scalar triple product in (b). This results only in a change of sign.

Example 2
The points A, B, C, D have position vectors $\mathbf{a}$, $\mathbf{b}$, $\mathbf{c}$, $\mathbf{d}$ respectively referred to the origin O. Given that $\mathbf{a} = (2, 1, 3)$, $\mathbf{b} = (1, 0, -2)$, $\mathbf{c} = (-3, 2, -1)$, $\mathbf{d} = (0, 2, -5)$, find the volumes of the tetrahedra (a) OABC, (b) ABCD.

(a) Volume of tetrahedron OABC $= \frac{1}{6}\mathbf{a} \cdot (\mathbf{b} \times \mathbf{c})$
$$= \frac{1}{6}(2, 1, 3) \cdot [(1, 0, -2) \times (-3, 2, -1)]$$

$$= \frac{1}{6}\begin{vmatrix} 2 & 1 & 3 \\ 1 & 0 & -2 \\ -3 & 2 & -1 \end{vmatrix}$$

$$= 7/2.$$

(b) The position vectors of A, B, C referred to D are, respectively, $(2, -1, 8)$, $(1, -2, 3)$, $(-3, 0, 4)$.
Since

$$\frac{1}{6}\begin{vmatrix} 2 & -1 & 8 \\ 1 & -2 & 3 \\ -3 & 0 & 4 \end{vmatrix} = -8\tfrac{1}{2},$$

volume of tetrahedron ABCD $= 8\tfrac{1}{2}$.

Example 3
Show that the points A $(1, -1, 3)$, B $(-1, 1, 5)$, C$(2, 4, 4)$ and D $(-3, 0, 6)$ are coplanar.

If the points are coplanar, the volume of the tetrahedron ABCD $= 0$.

$\overrightarrow{DA} = (4, -1, -3)$, $\overrightarrow{DB} = (2, 1, -1)$, $\overrightarrow{DC} = (5, 4, -2)$.
Volume of tetrahedron ABCD $= (4, -1, -3) \cdot [(2, 1, -1) \times (5, 4, -2)]$

$$= \begin{vmatrix} 4 & -1 & -3 \\ 2 & 1 & -1 \\ 5 & 4 & -2 \end{vmatrix}$$

$$= 0.$$

Therefore the points ABCD are coplanar.

The vector triple product $\mathbf{a} \times (\mathbf{b} \times \mathbf{c})$
The vector triple product $\mathbf{a} \times (\mathbf{b} \times \mathbf{c})$ is a vector which is perpendicular to both $\mathbf{a}$ and $(\mathbf{b} \times \mathbf{c})$. Let $\mathbf{a} \times (\mathbf{b} \times \mathbf{c}) = \mathbf{d}$. Then, because $\mathbf{d}$ is perpendicular to $\mathbf{b} \times \mathbf{c}$, it lies in the plane of $\mathbf{b}$ and $\mathbf{c}$. Therefore $\mathbf{d}$ can be expressed as $\lambda\mathbf{b} + \mu\mathbf{c}$, where λ, μ are scalars.

Taking $\mathbf{i}$, $\mathbf{j}$, $\mathbf{k}$ as perpendicular unit vectors with $\mathbf{j}$ and $\mathbf{k}$ in the plane of $\mathbf{b}$ and $\mathbf{c}$ and with $\mathbf{k}$ in the direction of $\mathbf{c}$, we have

$$\begin{aligned} \mathbf{c} &= \qquad\qquad c_3\mathbf{k} \\ \mathbf{b} &= \qquad b_2\mathbf{j} + b_3\mathbf{k} \\ \mathbf{a} &= a_1\mathbf{i} + a_2\mathbf{j} + a_3\mathbf{k} \end{aligned}$$

where a_1, a_2, a_3, b_2, b_3 and c_3 are scalars.

Then $\quad \mathbf{d} = (a_1\mathbf{i} + a_2\mathbf{j} + a_3\mathbf{k}) \times [(b_2\mathbf{j} + b_3\mathbf{k}) \times (c_3\mathbf{k})]$

$$= (a_1\mathbf{i} + a_2\mathbf{j} + a_3\mathbf{k}) \times \begin{vmatrix} \mathbf{i} & \mathbf{j} & \mathbf{k} \\ 0 & b_2 & b_3 \\ 0 & 0 & c_3 \end{vmatrix}$$

$$= (a_1\mathbf{i} + a_2\mathbf{j} + a_3\mathbf{k}) \times (b_2 c_3\mathbf{i})$$
$$= a_3 b_2 c_3\mathbf{j} - a_2 b_2 c_3\mathbf{k}$$

$$= a_3c_3(\mathbf{b} - b_3\mathbf{k}) - a_2b_2c_3\mathbf{k} \quad \text{since} \quad \mathbf{b} = b_2\mathbf{j} + b_3\mathbf{k}$$
$$= a_3c_3\mathbf{b} - (a_3b_3c_3 + a_2b_2c_3)\mathbf{k}$$
$$= a_3c_3\mathbf{b} - (a_3b_3 + a_2b_2)\mathbf{c} \quad \text{since} \quad \mathbf{c} = c_3\mathbf{k}.$$

But $\quad \mathbf{a} \cdot \mathbf{c} = (a_1, a_2, a_3) \cdot (0, 0, c_3) = a_3c_3$

and $\quad \mathbf{a} \cdot \mathbf{b} = (a_1, a_2, a_3) \cdot (0, b_2, b_3) = a_2b_2 + a_3b_3.$

$\therefore \qquad \mathbf{d} = (\mathbf{a} \cdot \mathbf{c})\mathbf{b} - (\mathbf{a} \cdot \mathbf{b})\mathbf{c},$

i.e. $\qquad\qquad \mathbf{a} \times (\mathbf{b} \times \mathbf{c}) = (\mathbf{a} \cdot \mathbf{c})\mathbf{b} - (\mathbf{a} \cdot \mathbf{b})\mathbf{c}.$

It should be noted that the operation of forming a vector product is not associative, i.e. $\mathbf{a} \times (\mathbf{b} \times \mathbf{c}) \neq (\mathbf{a} \times \mathbf{b}) \times \mathbf{c}$.

The vector triple product, $\mathbf{a} \times (\mathbf{b} \times \mathbf{c})$, is perpendicular to $\mathbf{b} \times \mathbf{c}$, i.e. lies in the plane containing $\mathbf{b}$ and $\mathbf{c}$, whereas the vector triple product, $(\mathbf{a} \times \mathbf{b}) \times \mathbf{c}$, is perpendicular to $\mathbf{a} \times \mathbf{b}$, i.e. lies in the plane containing $\mathbf{a}$ and $\mathbf{b}$.

Example 4

Evaluate (a) $\mathbf{a} \times (\mathbf{b} \times \mathbf{c})$, (b) $(\mathbf{a} \times \mathbf{b}) \times \mathbf{c}$, given that $\mathbf{a} = (3, -2, 1)$, $\mathbf{b} = (0, 4, -5)$, $\mathbf{c} = (5, 0, -3)$.

(a) $\mathbf{a} \times (\mathbf{b} \times \mathbf{c}) = (\mathbf{a} \cdot \mathbf{c})\mathbf{b} - (\mathbf{a} \cdot \mathbf{b})\mathbf{c}.$

$\qquad\qquad \mathbf{a} \cdot \mathbf{b} = (3, -2, 1) \cdot (0, 4, -5) = -8 - 5 = -13$

$\qquad\qquad \mathbf{a} \cdot \mathbf{c} = (3, -2, 1) \cdot (5, 0, -3) = 15 - 3 = 12$

$\therefore \quad \mathbf{a} \times (\mathbf{b} \times \mathbf{c}) = 12\mathbf{b} + 13\mathbf{c}$

$\qquad\qquad = 12(0, 4, -5) + 13(5, 0, -3)$

$\qquad\qquad = (65, 48, -99) \qquad \text{or} \qquad 65\mathbf{i} + 48\mathbf{j} - 99\mathbf{k}.$

(b) Since $\mathbf{p} \times \mathbf{q} = -\mathbf{q} \times \mathbf{p}$,

$\quad (\mathbf{a} \times \mathbf{b}) \times \mathbf{c} = -\mathbf{c} \times (\mathbf{a} \times \mathbf{b})$

$\qquad\qquad = -[(\mathbf{c} \cdot \mathbf{b})\mathbf{a} - (\mathbf{c} \cdot \mathbf{a})\mathbf{b}]$

$\qquad\qquad = -[(5, 0, -3) \cdot (0, 4, -5)\mathbf{a} - (5, 0, -3) \cdot (3, -2, 1)\mathbf{b}]$

$\qquad\qquad = -[15\mathbf{a} - 12\mathbf{b}]$

$\qquad\qquad = -15(3, -2, 1) + 12(0, 4, -5)$

$\qquad\qquad = (-45, 78, -75) \qquad \text{or} \qquad -45\mathbf{i} + 78\mathbf{j} - 75\mathbf{k}.$

Exercises 14.5

1 Evaluate the scalar triple products (a) $\mathbf{a} \cdot (\mathbf{b} \times \mathbf{c})$, (b) $\mathbf{c} \cdot (\mathbf{a} \times \mathbf{b})$, (c) $\mathbf{b} \cdot (\mathbf{a} \times \mathbf{c})$, (d) $(\mathbf{a} \times \mathbf{c}) \cdot \mathbf{b}$, given that

(a) $\mathbf{a} = \mathbf{i} + 2\mathbf{j} + 3\mathbf{k}$, $\mathbf{b} = -2\mathbf{i} + 3\mathbf{k}$, $\mathbf{c} = -4\mathbf{j} + 2\mathbf{k}$;

(b) $\mathbf{a} = 4\mathbf{i} + 3\mathbf{j} + 2\mathbf{k}$, $\mathbf{b} = -4\mathbf{i} + 2\mathbf{j} - \mathbf{k}$, $\mathbf{c} = 2\mathbf{i} - 2\mathbf{j}$;

(c) $\mathbf{a} = -2\mathbf{i} + 3\mathbf{j} - 5\mathbf{k}$, $\mathbf{b} = -\mathbf{i} + 7\mathbf{k}$, $\mathbf{c} = 3\mathbf{i} - \mathbf{j}$;

(d) $\mathbf{a} = 3\mathbf{j} + 5\mathbf{k}$, $\mathbf{b} = 5\mathbf{i} - 2\mathbf{j}$, $\mathbf{c} = 3\mathbf{i} - 4\mathbf{k}$;

(e) $\mathbf{a} = p\mathbf{i} + q\mathbf{j} + r\mathbf{k}$, $\mathbf{b} = -p\mathbf{i} + q\mathbf{j} - r\mathbf{k}$, $\mathbf{c} = q\mathbf{i} - p\mathbf{j} + r\mathbf{k}$.

2 Evaluate the vector triple products (a) $\mathbf{a} \times (\mathbf{b} \times \mathbf{c})$, (b) $(\mathbf{a} \times \mathbf{b}) \times \mathbf{c}$, given that

(a) $\mathbf{a} = (3, 2, 1)$, $\mathbf{b} = (4, -1, 0)$, $\mathbf{c} = (-1, 0, -5)$;
(b) $\mathbf{a} = (2, -1, 3)$, $\mathbf{b} = (-3, -2, 1)$, $\mathbf{c} = (3, -1, 0)$;
(c) $\mathbf{a} = (4, 3, 2)$, $\mathbf{b} = (-1, 2, -3)$, $\mathbf{c} = (3, -2, -1)$;
(d) $\mathbf{a} = (p, 0, 0)$, $\mathbf{b} = (0, p, 0)$, $\mathbf{c} = (0, 0, p)$.

3   Express, in terms of $p\mathbf{a}$, $q\mathbf{b}$ and $r\mathbf{c}$, where p, q and r are scalars, the vector triple products
   (a) $\mathbf{a} \times (\mathbf{c} \times \mathbf{b})$,   (b) $\mathbf{b} \times (\mathbf{a} \times \mathbf{c})$,   (c) $\mathbf{b} \times (\mathbf{c} \times \mathbf{a})$,
   (d) $\mathbf{c} \times (\mathbf{a} \times \mathbf{b})$,   (e) $\mathbf{c} \times (\mathbf{b} \times \mathbf{a})$,   (f) $(\mathbf{c} \times \mathbf{b}) \times \mathbf{a}$,
   (g) $(\mathbf{c} \times \mathbf{a}) \times \mathbf{b}$,   (h) $(\mathbf{b} \times \mathbf{c}) \times \mathbf{a}$,   (i) $(\mathbf{b} \times \mathbf{a}) \times \mathbf{c}$,
   (j) $(\mathbf{a} \times \mathbf{c}) \times \mathbf{b}$.

4   Find the volume of the parallelepiped formed by the vectors $\mathbf{a}$, $\mathbf{b}$ and $\mathbf{c}$, given that
   (a) $\mathbf{a} = (3, 2, 1)$, $\mathbf{b} = (4, 5, 6)$, $\mathbf{c} = (3, 7, 4)$;
   (b) $\mathbf{a} = (4, 6, -3)$, $\mathbf{b} = (0, 1, -3)$, $\mathbf{c} = (4, 0, -1)$;
   (c) $\mathbf{a} = (-1, -2, -3)$, $\mathbf{b} = (5, 6, 0)$, $\mathbf{c} = (3, -1, 2)$;
   (d) $\mathbf{a} = (6, 1, -3)$, $\mathbf{b} = (2, -4, 3)$, $\mathbf{c} = (-1, 2, -3)$.

5   Find the volume of the tetrahedron ABCD, where the coordinates of A, B, C and D are
   (a) A $(0, 0, 0)$,   B $(3, 0, 0)$,   C $(0, 4, 0)$,   D $(0, 0, 5)$;
   (b) A $(0, 0, 0)$,   B $(1, 2, 3)$,   C $(4, 2, 3)$,   D $(0, -2, 1)$;
   (c) A $(0, 0, 0)$,   B $(4, 3, 2)$,   C $(-1, 2, -3)$, D $(4, 2, 1)$;
   (d) A $(1, 4, 3)$,   B $(5, 3, 1)$,   C $(-4, -2, 3)$, D $(-1, -3, -5)$;
   (e) A $(-1, 2, 7)$, B $(-4, 2, -3)$, C $(2, -4, 5)$, D $(5, 2, 1)$;
   (f) A $(5, -1, 3)$, B $(-2, 4, 3)$,   C $(1, 4, 7)$ D $(-3, -2, 0)$.

6   Investigate whether or not the points P, Q, R, S are coplanar, given that the coordinates of P, Q, R, S are
   (a) P $(0, 0, 0)$,   Q $(1, 2, 3)$,   R $= (-3, 2, -1)$,   S $(4, 0, -3)$;
   (b) P $(1, 0, 0)$,   Q $(0, 2, 0)$,   R $= (0, 0, 3)$,   S $(-1, 4, 3)$;
   (c) P $(5, 2, 3)$,   Q $(-4, 1, -5)$, R $= (0, 1, -1)$,   S $(4, 1, 3)$;
   (d) P $(0, 11, 2)$, Q $(4, 0, 1)$,   R $= (5, 4, 3)$,   S $(-3, 8, -1)$.

7   Ascertain which of the following sets of vectors are linearly independent:
   (a) $\mathbf{i} + 2\mathbf{j} + 3\mathbf{k}$, $2\mathbf{i} - \mathbf{j} - 3\mathbf{k}$, $4\mathbf{i} - 4\mathbf{j} - 8\mathbf{k}$;
   (b) $3\mathbf{i} + \mathbf{j} + 3\mathbf{k}$, $9\mathbf{i} - 3\mathbf{j}$, $12\mathbf{i} - 2\mathbf{j} + 3\mathbf{k}$;
   (c) $2\mathbf{i} - \mathbf{j} + \mathbf{k}$, $\mathbf{i} - 5\mathbf{j} + 2\mathbf{k}$, $-3\mathbf{i} - 2\mathbf{j} + 3\mathbf{k}$;
   (d) $-3\mathbf{i} + \mathbf{j} - 11\mathbf{k}$, $\mathbf{i} - 7\mathbf{j} + 17\mathbf{k}$, $-2\mathbf{i} - 6\mathbf{j} + 6\mathbf{k}$.

8   Given that $\mathbf{p} \cdot (\mathbf{a} \times \mathbf{q}) = \mathbf{p} \cdot (\mathbf{b} \times \mathbf{q}) \neq 0$, show that the lines with vector equations $\mathbf{r} = \mathbf{a} + t\mathbf{p}$, $\mathbf{r} = \mathbf{b} + s\mathbf{q}$ intersect.
   Show also that the position vector of the point of intersection is

$$\mathbf{a} + \left[\frac{\mathbf{a} \cdot (\mathbf{b} \times \mathbf{q})}{\mathbf{q} \cdot (\mathbf{a} \times \mathbf{p})} \right] \mathbf{p}.$$

9 Show that the vector equation

$$\mathbf{r} \cdot (\mathbf{b} \times \mathbf{c}) + \mathbf{r} \cdot (\mathbf{c} \times \mathbf{a}) + \mathbf{r} \cdot (\mathbf{a} \times \mathbf{b}) = \mathbf{a} \cdot (\mathbf{b} \times \mathbf{c})$$

represents the plane passing through the points with position vectors $\mathbf{a}$, $\mathbf{b}$ and $\mathbf{c}$.

10 Show that the vector equations
(a) $\mathbf{r} = \mathbf{a} + \lambda\mathbf{b} + \mu\mathbf{c}$,
(b) $(\mathbf{r} - \mathbf{a}) \cdot [(\mathbf{b} - \mathbf{a}) \times (\mathbf{c} - \mathbf{a})] = 0$
represent the same plane.

Miscellaneous exercises

1 Find the magnitude of the vector $(\mathbf{i} + \mathbf{j} - \mathbf{k}) \times (\mathbf{i} - \mathbf{j} + \mathbf{k})$. [L]

2 Find a unit vector which is perpendicular to the vector $(4\mathbf{i} + 4\mathbf{j} - 7\mathbf{k})$ and to the vector $(2\mathbf{i} + 2\mathbf{j} + \mathbf{k})$. [L]

3 Show that the vectors $\mathbf{i} + \mathbf{j} - \mathbf{k}$ and $\mathbf{i} + \mathbf{j} + 2\mathbf{k}$ are at right angles to one another. Find a unit vector which is at right angles to each of these vectors. [L]

4 A plane passes through three points A, B and C whose position vectors $2\mathbf{i} - \mathbf{j} + \mathbf{k}$, $3\mathbf{i} + 2\mathbf{j} - \mathbf{k}$ and $-\mathbf{i} + 3\mathbf{j} + 2\mathbf{k}$ respectively. Show that a unit vector normal to the plane is

$$\frac{1}{3\sqrt{35}}(11\mathbf{i} + 5\mathbf{j} + 13\mathbf{k}).$$

Show that the equation of the plane is $11x + 5y + 13z = 30$. [JMB]

5 A plane passes through the three points A, B, C whose position vectors, referred to an origin O, are $(\mathbf{i} + 3\mathbf{j} + 3\mathbf{k})$, $(3\mathbf{i} + \mathbf{j} + 4\mathbf{k})$, $(2\mathbf{i} + 4\mathbf{j} + \mathbf{k})$ respectively. Find, in the form $(l\mathbf{i} + m\mathbf{j} + n\mathbf{k})$, a unit vector normal to this plane.

Find also the cartesian equation of the plane, and the perpendicular distance from the origin to this plane.

6 (a) If $\mathbf{r}$ is the position vector of a variable point P and A and B are fixed points with position vectors $\mathbf{a}$ and $\mathbf{b}$ respectively, describe the locus of P when

$$\mathbf{a} \times \mathbf{r} = \mathbf{a} \times \mathbf{b} \quad (\mathbf{a} \neq \mathbf{0}).$$

(b) The position vectors of points A, B, C are $\mathbf{a}$, $\mathbf{b}$, $\mathbf{c}$ respectively. By expressing the area of the triangle ABC in two ways, or otherwise, show that the shortest distance from B to AC is

$$\frac{|\mathbf{b} \times \mathbf{a} + \mathbf{a} \times \mathbf{c} + \mathbf{c} \times \mathbf{b}|}{|\mathbf{a} - \mathbf{c}|}.$$

Find the cartesian equation of the sphere, centre B $(4, 3, -2)$, which touches the line AC, where A and C have coordinates $(1, 2, 3)$ and $(-1, -2, 5)$ respectively. [L]

7 (a) Define the scalar and vector products of two vectors $\mathbf{a}, \mathbf{b}$ in terms of a, b and the angle θ between the vectors, explaining carefully any sign conventions used.

Deduce the formula

$$|\mathbf{a} \times \mathbf{b}|^2 = a^2 b^2 - (\mathbf{a} \cdot \mathbf{b})^2.$$

(b) Find the most general form for the vector $\mathbf{u}$ when

$$\mathbf{u} \times (\mathbf{i} + \mathbf{j} + 2\mathbf{k}) = \mathbf{i} - \mathbf{j}.$$ [JMB]

8 The position vectors with respect to the origin O of the points A, B and C are $(2\mathbf{i} + \mathbf{j} + \mathbf{k}), (\mathbf{i} + 2\mathbf{j} + \mathbf{k})$ and $(\mathbf{i} + \mathbf{j} + 2\mathbf{k})$ respectively. Derive a vector equation of the plane ABC.

The position vector of the point D is $(2\mathbf{i} + 2\mathbf{j} + 2\mathbf{k})$. Prove that the line OD is perpendicular to the plane ABC and find the position vector of the point P at which the line and the plane intersect. Find also the length of the shortest distance from P to the line BC. [L]

9 (a) The points A, B, C have coordinates $(1, 1, 1)$, $(2, 3, 4)$, $(3, 2, 2)$ respectively referred to a rectangular system of axes Ox, Oy, Oz. Find the equation of the plane ABC. If the vectors $\overrightarrow{OA}, \overrightarrow{OB}$ are denoted by $\mathbf{a}, \mathbf{b}$ respectively, find $\mathbf{a} \cdot \mathbf{b}$ and show that $\mathbf{a} \times \mathbf{b} = \mathbf{i} - 2\mathbf{j} + \mathbf{k}$, where $\mathbf{i}, \mathbf{j}, \mathbf{k}$ are unit vectors in the directions Ox, Oy, Oz.

Hence, or otherwise, find
(i) the cosine of the angle AOB,   (ii) the area of the triangle OAB.
(b) If $\mathbf{w}$ is any vector and $\mathbf{i}, \mathbf{j}, \mathbf{k}$ are unit vectors directed along rectangular axes Ox, Oy, Oz, show that

$$\mathbf{w} = (\mathbf{w} \cdot \mathbf{i})\mathbf{i} + (\mathbf{w} \cdot \mathbf{j})\mathbf{j} + (\mathbf{w} \cdot \mathbf{k})\mathbf{k}.$$ [L]

10 A tetrahedron OABC has its vertices at the points O $(0, 0, 0)$, A $(1, 2, -1)$, B $(-1, 1, 2)$ and C $(2, -1, 1)$. Write down expressions for $\overrightarrow{AB}$ and $\overrightarrow{AC}$ in terms of the mutually orthogonal unit vectors $\mathbf{i}, \mathbf{j}$ and $\mathbf{k}$ and find $\overrightarrow{AB} \times \overrightarrow{AC}$. Deduce the area of the triangle ABC. Find also the length of the perpendicular from O to the plane ABC. Hence, or otherwise, find the volume of the tetrahedron OABC. [L]

11 Given that

$$\overrightarrow{OA} = 3\mathbf{i} + 4\mathbf{j} + 5\mathbf{k},$$
$$\overrightarrow{OB} = 4\mathbf{i} + 6\mathbf{j} + 7\mathbf{k},$$
$$\overrightarrow{OC} = \mathbf{i} + 5\mathbf{j} + 3\mathbf{k},$$

find
(a) the cosine of the angle BAC,

(b) the area of the triangle ABC,

(c) the direction cosines of the normal to the plane ABC,

(d) the volume of the tetrahedron OABC. [L]

12 If the non-collinear points A, B, C have position vectors **a**, **b**, **c** respectively, write down the position vector of the point D such that ABCD is a parallelogram.

If A, B, C are the points (1, 0, 3), (2, 2, 3), (2, 4, 6) respectively, find

(a) the coordinates of D,

(b) the cartesian equation of the plane ABC, verifying that D does in fact lie in this plane,

(c) the area of the parallelogram ABCD,

(d) the volume of the pyramid VABCD, where V is the point (5, 1, 11). [L]

13 If $\overrightarrow{OA} = 2\mathbf{i} + 3\mathbf{j} - \mathbf{k}$, $\overrightarrow{OB} = \mathbf{i} - 2\mathbf{j} + 3\mathbf{k}$, where **i**, **j**, **k** are the cartesian unit vectors, calculate $\overrightarrow{OC} = \overrightarrow{OA} \times \overrightarrow{OB}$ in terms of **i**, **j** and **k**.

Calculate also

(a) the cosine of the acute angle between OA and BC,

(b) the area of the triangle ABC,

(c) the volume of the tetrahedron OABC,

(d) the angle between the planes OAB and OAC. [L]

14 The plane passing through the points P (3, 1, 5), Q (0, 5, 4) and R (6, 2, 4) cuts the coordinate axes O*x*, O*y*, O*z* at A, B, C respectively.

Find

(a) the equation of the plane PQR,

(b) the volume of the tetrahedron OABC,

(c) the coordinates of the foot of the perpendicular from O to the plane ABC,

(d) the area of the triangle ABC. [JMB]

15 The non-collinear points A, B and C have position vectors **a**, **b** and **c** respectively relative to an origin O. Show that the vector $(\mathbf{a} - \mathbf{b}) \times (\mathbf{a} - \mathbf{c})$ is normal to the plane through the points A, B and C.

Show that the points having position vectors

$$5\mathbf{j} - 4\mathbf{k}, \quad \mathbf{i} + \mathbf{j} - \mathbf{k}, \quad 2\mathbf{i} - \mathbf{k}, \quad 3\mathbf{j} - 2\mathbf{k}$$

are coplanar.

Find, in any form, the equation of the plane containing these four points. [C]

16 The points O, A, B, C have coordinates (0, 0, 0), (1, 0, 0), (0, 2, 0), (0, 0, 3) respectively. Find, in any form, equations for the plane ABC, and for the line through O which is normal to this plane.

Find the coordinates of the foot of the perpendicular from O to the plane, and the distance of O from the plane.

Obtain also the volume of the tetrahedron OABC, and the area of the triangle ABC. [L]

17 The position vectors of the four non-coplanar points A, B, C and D relative to an origin O are respectively **a**, **b**, **c** and **d**. Find the following in terms of **a**, **b**, **c** and **d**:
(a) the equation of the plane ABC,
(b) the shortest distance from D to the plane ABC,
(c) the equation of the plane through AB parallel to CD,
(d) the shortest distance between the lines AB and CD. [C]

18 For a tetrahedron OABC with O as origin, the vertices A, B, C have position vectors **a**, **b**, **c** respectively. Prove that the lines joining the vertices to the centroids of the opposite faces are concurrent.

Write down an expression in terms of the vectors **a**, **b**, **c** and their scalar and vector products that is equal to the volume of the tetrahedron.

Prove also that, if two pairs of opposite edges of this tetrahedron are perpendicular, then the third pair is also perpendicular. [L]

19 (a) Prove that
(i) $|\mathbf{a} \times \mathbf{b}| = \sqrt{[(\mathbf{a} \cdot \mathbf{a})(\mathbf{b} \cdot \mathbf{b}) - (\mathbf{a} \cdot \mathbf{b})^2]}$,
(ii) $|\mathbf{a} \times \mathbf{b}|^2 = -\mathbf{b} \cdot [\mathbf{a} \times (\mathbf{a} \times \mathbf{b})]$.
(b) Show that the shortest distance between the non-parallel lines

$$\mathbf{r} = \mathbf{a} + s\mathbf{u}, \qquad \mathbf{r} = \mathbf{b} + t\mathbf{v}$$

is given by

$$\left| \frac{(\mathbf{b} - \mathbf{a}) \cdot (\mathbf{u} \times \mathbf{v})}{|\mathbf{u} \times \mathbf{v}|} \right|.$$

(c) If $\mathbf{a} = \begin{pmatrix} 3 \\ 2 \\ -1 \end{pmatrix}$ and $\mathbf{b} = \begin{pmatrix} 1 \\ x \\ x^2 \end{pmatrix}$, show that there are just two vectors $\mathbf{b}_1, \mathbf{b}_2$ of the form **b** orthogonal to **a**. Explain why $\mathbf{b}_1 \times \mathbf{b}_2$ and **a** are linearly dependent, and obtain a relation between them. [L]

20 The position vectors of the points A, B, C with respect to the origin O are respectively

$$\mathbf{a} = 2\mathbf{i} + 3\mathbf{j} + \mathbf{k}, \mathbf{b} = 6\mathbf{i} - 7\mathbf{j} + 6\mathbf{k}, \mathbf{c} = 4\mathbf{i} + 2\mathbf{j} + 4\mathbf{k}.$$

Calculate the area of the triangle ABC and the volume of the tetrahedron OABC.

Evaluate the triple product $\mathbf{a} \times (\mathbf{b} \times \mathbf{c})$ of the given vectors, and find scalars λ and μ such that

$$\mathbf{a} \times (\mathbf{b} \times \mathbf{c}) = \lambda\mathbf{b} + \mu\mathbf{c}.$$

Find the position vector **d** of the point D on the line BC such that

$\mathbf{a} \cdot \mathbf{d} = 0$. Obtain a vector equation for the plane which is perpendicular to the plane OAD and which passes through the line BC.   [L]

21 Show that the planes

$$\mathbf{r} \cdot (4\mathbf{i} + 4\mathbf{j} + 7\mathbf{k}) = 7, \qquad \mathbf{r} \cdot (8\mathbf{i} - \mathbf{j} - 4\mathbf{k}) = 5$$

are at right angles. Show also that the line

$$\mathbf{r} = \mathbf{j} + 3\mathbf{k} + t(\mathbf{i} - 8\mathbf{j} + 4\mathbf{k})$$

is parallel to the line in which the planes meet, and is equidistant from the two planes.

Find unit vectors $\mathbf{a}$ and $\mathbf{b}$ such that the equation $\mathbf{r} \cdot (\mathbf{a} \times \mathbf{b}) = 0$ represents a plane perpendicular to the two given planes.   [L]

22 The corners of the fixed triangle ABC have non-zero position vectors $\mathbf{a}, \mathbf{b}, \mathbf{c}$ respectively, relative to the origin O, given by

$$\mathbf{a} = a_1\mathbf{i} + a_2\mathbf{j} + a_3\mathbf{k}, \quad \mathbf{b} = b_1\mathbf{i} + b_2\mathbf{j} + b_3\mathbf{k}, \quad \mathbf{c} = c_1\mathbf{i} + c_2\mathbf{j} + c_3\mathbf{k}.$$

The variable point P has position vector $\mathbf{r}$, where

$$\mathbf{r} = x\mathbf{i} + y\mathbf{j} + z\mathbf{k}.$$

Assuming that no three of the points O, A, B, C are collinear,
(a) interpret geometrically the simultaneous vector equations

$$\mathbf{a} \cdot (\mathbf{b} \times \mathbf{c}) = 0, (\mathbf{a} \times \mathbf{b}) \times (\mathbf{c} \times \mathbf{r}) = \mathbf{0},$$

(b) find, in any form, vector and cartesian equations of
(i) the line through C perpendicular to the plane OAB,
(ii) the plane through C perpendicular to AB.   [L]

Notation

Real numbers

$\mathbb{N}$ – the set of positive integers and zero $\{0, 1, 2, 3, \ldots\}$
$\mathbb{Z}$ – the set of integers $\{0, \pm 1, \pm 2, \ldots\}$
$\mathbb{Z}^+$ – the set of positive integers $\{1, 2, 3, \ldots\}$
$\mathbb{Q}$ – the set of rational numbers
$\mathbb{Q}^+$ – the set of positive rational numbers $\{x \in \mathbb{Q}; x > 0\}$
$\mathbb{R}$ – the set of real numbers
$\mathbb{R}^+$ – the set of positive real numbers $\{x \in \mathbb{R}; x > 0\}$

Complex numbers

$\mathbb{C}$ the set of complex numbers
$|z|$ the modulus of z
$\arg z$ the argument of z
$\operatorname{Re} z$ x, when $z = x + iy$
$\operatorname{Im} z$ y, when $z = x + iy$
z^* $x - iy$, when $z = x + iy$

Series

$$\sum_{r=1}^{n} u_r = u_1 + u_2 + u_3 + \ldots + u_n \;(= S_n)$$

$$\sum_{r=1}^{\infty} u_r = \lim_{n \to \infty} S_n$$

Vectors and matrices

$\mathbf{a} \cdot \mathbf{b}$ the scalar product of $\mathbf{a}$ and $\mathbf{b}$
$\mathbf{a} \times \mathbf{b}$ the vector product of $\mathbf{a}$ and $\mathbf{b}$
$\det \mathbf{A}$ the determinant of the matrix $\mathbf{A}$
$\mathbf{A}^{-1}$ the inverse of the non-singular matrix $\mathbf{A}$
$\mathbf{A}^{\mathrm{T}}$ the transpose of the matrix $\mathbf{A}$

Formulae

Case 3   Equal roots $m_1 = m_2 = m$ *Page*
General solution $y = (Ax + B)e^{mx}$

(b) $a\dfrac{d^2y}{dx^2} + 2h\dfrac{dy}{dx} + by = \phi(x)$ 127

General solution $y = G(x) + F(x)$,

where $y = G(x)$ is the general solution of the differential equation
(a) above, and $y = F(x)$ is any solution of (b) containing no arbitrary
constants.

Polar coordinates
The straight line through the point P (p, α) at right angles to OP 163

$$r = p \sec(\alpha - \theta)$$

The angle ϕ between tangent and radius vector 172

$$\tan \phi = r\frac{d\theta}{dr} = r\left/\frac{dr}{d\theta}\right.$$

Area of sector $\displaystyle\int_\alpha^\beta \tfrac{1}{2}r^2\,d\theta$ 177

Length of arc $\displaystyle\int_\alpha^\beta \sqrt{\left[r^2 + \left(\frac{dr}{d\theta}\right)^2\right]}\,d\theta$ 253

Coordinate geometry
Ellipse $\dfrac{x^2}{a^2} + \dfrac{y^2}{b^2} = 1$,   where $b^2 = a^2(1 - e^2)$ 362

Hyperbola $\dfrac{x^2}{a^2} - \dfrac{y^2}{b^2} = 1$,   where $b^2 = a^2(e^2 - 1)$ 349

Rectangular hyperbola   (i) $x^2 - y^2 = a^2$ (ii) $xy = c^2$ 353, 343
Pair of straight lines through the origin

$$ax^2 + 2hxy + by^2 = 0, \quad \text{where } h^2 \geqslant ab$$ 377

Matrices

Singular matrix	$\det \mathbf{A} = 0$	451
Inverse matrix $\mathbf{A}^{-1}$	$\mathbf{AA}^{-1} = \mathbf{A}^{-1}\mathbf{A} = \mathbf{I}$	451
Orthogonal matrix	$\mathbf{A}^T = \mathbf{A}^{-1}$	463
Symmetrical matrix	$\mathbf{A}^T = \mathbf{A}$	447
Skew-symmetric matrix	$\mathbf{A}^T = -\mathbf{A}$	448
Characteristic equation	$\det(\mathbf{A} - \lambda\mathbf{I}) = 0$	473
Eigenvalue λ with eigenvector $\mathbf{u}$,	$\mathbf{Au} = \lambda\mathbf{u}$	468

Solution of equations

$a_n x^n + a_{n-1} x^{n-1} + \ldots + a_1 x + a_0 = 0, \quad$ where $a_n \neq 0$

Sum of the roots $= -a_{n-1}/a_n$

Product of the roots $= (-1)^n \, a_0/a_n$

Triple vector product

$$\mathbf{a} \times (\mathbf{b} \times \mathbf{c}) = (\mathbf{a} \cdot \mathbf{c})\mathbf{b} - (\mathbf{a} \cdot \mathbf{b})\mathbf{c}$$

Differentiation

$f(x)$	$f'(x)$	
$\sinh x$	$\cosh x$	78
$\cosh x$	$\sinh x$	77
$\tanh x$	$\operatorname{sech}^2 x$	81
$\sinh^{-1} x$	$\dfrac{1}{\sqrt{(x^2 + 1)}}$	85
$\cosh^{-1} x$	$\dfrac{1}{\sqrt{(x^2 - 1)}}, \quad x > 1$	87
$\tanh^{-1} x$	$\dfrac{1}{1 - x^2}, \quad -1 < x < 1$	87

Leibnitz's theorem

$$\frac{d^n}{dx^n}(uv) = u\frac{d^n v}{dx^n} + \binom{n}{1}\frac{du}{dx}\frac{d^{n-1}v}{dx^{n-1}} + \ldots + \binom{n}{r}\frac{d^r u}{dx^r}\frac{d^{n-r}v}{dx^{n-r}} + \ldots + \frac{d^n u}{dx^n}v \qquad 221$$

Integration

$f(x)$	$\displaystyle\int f(x)\,dx$	
$\sinh x$	$\cosh x + c$	
$\cosh x$	$\sinh x + c$	83
$\tanh x$	$\ln \cosh x + c$	
$\dfrac{1}{\sqrt{(x^2 + a^2)}}$	$\sinh^{-1}(x/a) + c, \quad a > 0$	89
$\dfrac{1}{\sqrt{(x^2 - a^2)}}$	$\cosh^{-1}(x/a) + c, \quad x \geqslant a > 0$	90
Mean value	$\dfrac{1}{b - a}\displaystyle\int_a^b f(x)\,dx$	256

Root mean square value $\sqrt{\left[\dfrac{1}{b-a}\displaystyle\int_a^b [f(x)]^2\ \mathrm{d}x\right]}$ 257

Length of arc $\displaystyle\int_a^b \sqrt{\left[1+\left(\dfrac{\mathrm{d}y}{\mathrm{d}x}\right)^2\right]}\ \mathrm{d}x$ 251

Area of surface of revolution

$$2\pi\int_a^b y\ \sqrt{\left[1+\left(\dfrac{\mathrm{d}y}{\mathrm{d}x}\right)^2\right]}\ \mathrm{d}x$$ 254

Theorems of Pappus 260
Volume of revolution $=(\overline{y_A}\alpha)A$
Area of surface of revolution $=(\overline{y_c}\alpha)L$

Answers

Chapter 1

1.1   **1**   (a) $\sim p \lor q$ (b) $p \land \sim q$
        (c) $q \land \sim p$ (d) $\sim q \lor p$
     4   (a) $x = 2$; $x = 1, 2, 3$
        (b) $x = 2n\pi$; $x = n\pi, n \in \mathbf{Z}$
        (c) $1 < x < 2$; $0 < x < 3$
        (d) $0 < x < 1$; $-1 < x < 2$

1.2   **2**   (a) T;F   (b) F;T   (c) T;F   (d) F;T
     3   (a) F;T   (b) F;F   (c) T;T   (d) T;F
     4   (a) T, T, T, T   (b) T, F, F, T
        (c) T, F, F, T   (d) F, T, T, F

1.3   **1**   angle BAC $= 90°$
     7   (a) necessary   (b) necessary
        (c) sufficient   (d) neither
     8   (i) a   (ii) c   (iii) b

1.4   **1** E   **2** A   **3** C   **4** A   **5** D
     6 D   **7** A   **8** B   **9** E   **10** A
     11 C   **12** E   **13** B   **14** C   **15** D
     16 B   **17** E   **18** C   **19** B   **20** D
     21 E   **22** B   **23** E   **24** C   **25** D

1.5   **1** A   **2** C   **3** D   **4** A   **5** E
     6 B   **7** D   **8** B   **9** E   **10** C
     11 B   **12** A   **13** D   **14** E   **15** B
     16 C   **17** A   **18** C   **19** C   **20** E
     21 E   **22** D   **23** A   **24** B   **25** C

1.6   **1**   (a) 8   (b) 8/3   (c) $0, -4$ (d) 3
     2   0 or $-4/3$
     3 $0, \pi/6, 5\pi/6, \pi$

Chapter 2

2.1   **1**   (a) -64 (b) $32i$
     2   (a) $27i$ (b) $(-1 + i\sqrt{3})/16$
     3 $3 \sin \theta - 4 \sin^3 \theta$; $5 \sin \theta - 20 \sin^3 \theta$
        $+ 16 \sin^5 \theta$
     4 $8 \cos^3 \theta - 4 \cos \theta$;
        $32 \cos^5 \theta - 32 \cos^3 \theta + 6 \cos \theta$
     5 $16 \cos^5 \theta - 20 \cos^3 \theta + 5 \cos \theta$,
        $16 \cos^4 \theta - 12 \cos^2 \theta + 1$
     6 $64 \cos^7 \theta - 112 \cos^5 \theta + 56 \cos^3 \theta$
        $- 7 \cos \theta$;
        $64 \cos^6 \theta - 80 \cos^4 \theta + 24 \cos^2 \theta - 1$
     11   (a) $(\cos 4\theta + 4 \cos 2\theta + 3)/8$
        (b) $(\cos 5\theta + 5 \cos 3\theta + 10 \cos \theta)/16$
     12   16/35

13 $3\pi/8$
14 $2/15$

2.2   **1**   (a) $\pm(1 + i)$
        (b) $\pm(2 - i)$
        (c) $\pm(5 + 3i)$
        (d) $\pm(3 - 2i)$
     2   (a) $2^{1/4}$; $\pi/8$ or $-7\pi/8$
        (b) $12^{1/4}$; $\pi/12$ or $-11\pi/12$
        (c) 2; $3\pi/8$ or $-5\pi/8$
        (d) $\sqrt{2}$; $-5\pi/12$ or $7\pi/12$

2.3   **1**   (a) $2, -1 \pm i\sqrt{3}$
        (b) $-2i, \pm\sqrt{3} + i$
        (c) $-2, 1 \pm i\sqrt{3}$
     2   (a) 3; $-90°, 30°, 150°$
        (b) $\sqrt{2}$; $-135°, -15°, 105°$
        (c) 2; $-110°, 10°, 130°$
     3   (a) -1
        (b) $a^3 + b^3$
        (c) $i3\sqrt{3}$
        (d) 0
        (e) 8

2.4   **1**   (a) $0, \pm\pi/2, \pi$
        (b) $0, \pm 2\pi/7, \pm 4\pi/7, \pm 6\pi/7$
        (c) $\pm\pi/8, \pm 3\pi/8, \pm 5\pi/8, \pm 7\pi/8$
        (d) $\pm\pi/5, \pm 3\pi/5, \pi$
     2   (a) $\pm(1 \pm i)$
        (b) $\pm(\sqrt{3} + i)/\sqrt{2}, \pm(1 - i\sqrt{3})/\sqrt{2}$
     3   (a) $r = 1$; $\theta = k\pi/6, k = \pm 1, \pm 3,$
        ± 5
        (b) $r = \sqrt{2}$; $\theta = k\pi/4, k = 0, \pm 1,$
        $\pm 2, \pm 3, 4$
        (c) $r = 3$; $\theta = k\pi/5, k = \pm 1, \pm 3, 5$
     4 $\pm a(1 \pm i)/\sqrt{2}$; $(z^2 - \sqrt{2}az + a^2) \times$
        $(z^2 + \sqrt{2}az + a^2)$
     5   2; $-120°, -30°, 60°, 150°$
     6   2; $-156°, -84°, -12°, 60°, 132°$
     8 $\pm(1 \pm i)/\sqrt{2}$

2.5   **1**   (a) 13   (b) 2   (c) 9   (d) 25
     2   (a) $1 \pm i$ (b) $-2 \pm 2i$
     3 $3 + 4i$ or $3 - 4i$
     4 $3 - 3i, 2$
     5 $-i, 2 \pm i$
     6 $-2, 1 \pm 2i$
     7 $z = -4, \pm 2i$
     8 $1 - 2i, -1 \pm \sqrt{2}$; $(z^2 + 2z - 1) \times$

$(z^2 - 2z + 5)$

9 $z = 1 \pm 3i, -1 \pm 7i$

10 $z = 1 \pm 2i, -2 \pm i$

2.7 1 (a) $z = 1 + 3i + t(1 - 2i)$

(b) $z = 2 - 5i + t(3 + 7i)$

(c) $z = -2 + i + t(1 - i)$

(d) $z = i + 5t$

2 $|z - 2 - i| = |z - 1 - 2i|$

$|z - 4 + 3i| = |z - 5 + 2i|$

3 $4 - i$

4 1

5 $z = 6 + i + t(1 - 2i)$

8 $2\sqrt{5}$

9 $\sqrt{3}$

2.8 1 (a) $|z - 1 - i| = \sqrt{2}$

(b) $|z - 4 + 3i| = 5$

(c) $|z - 3i| = 3$

(d) $|z - 3 - 3i| = 3\sqrt{2}$

2 (a) $|z - 2 - i| = 5$

(b) $|z - 3i| = 2$

(c) $|z - 4| = 1$

(d) $|z - 1 - i| = 2\sqrt{2}$

3 (a) $x^2 + y^2 - 2x - 2y = 0$

(b) $x^2 + y^2 + 2y = 3$

(c) $x^2 + y^2 - 6x - 8y = 25$

(d) $5x^2 + 5y^2 - 8x + 6y = 75$

2.9 1 (a) $w = (-1 + i)z/\sqrt{2}$

(b) $w = (1 + i\sqrt{3})z/2$

(c) $w = (-1 - i\sqrt{3})z/2$

(d) $w = (\sqrt{3} - i)z/2$

3 (a) $Z = 4z, w = Z - 12i$

(b) $Z = z - 3i, w = 4Z$

4 all

6 (a) $Z = 2z, W = -iZ, w = W + 2$

(b) $Z = 4\sqrt{2}z, W = (1 + i)Z/\sqrt{2}$,
$w = W + i$

(c) $Z = 2z, W = (\sqrt{3} + i)Z/2$,
$w = W - 4$

(d) $Z = 3\sqrt{2}z, W = (1 + i)Z/\sqrt{2}$,
$w = W - 2$

2.10 1 $\frac{1}{2} \leqslant |w| \leqslant 1, -\pi/4 \leqslant \arg w \leqslant 0$

6 $|w| \geqslant \frac{1}{2}, -\pi/3 \leqslant \arg w \leqslant -\pi/6$

7 $u^2 + v^2 + u/3 + 2v/3 = 0$

9 four semicircles

10 $3\pi/16$

Miscellaneous exercises 2

1 $2e^{i\pi/3}, 2e^{-i\pi/3}; -1/2 - i\sqrt{3}/2$

3 $1; \pi/2 - \theta$

4 $2 - 3i, 2$

6 $(\cos 5\theta + 5 \cos 3\theta + 10 \cos \theta)/16$

7 (a) (i) $x^2 + y^2 - 2x - 2y = 7$ (ii) $x = 1$,
$y \geqslant 1$

(b) $8e^{i\pi/6}$ (ii) $r = 2, \theta = -11\pi/18, \pi/18$,
$13\pi/18$

8 $x + 3y = 14; (7\sqrt{10})/5$

10 $\theta = k\pi$ or $k\pi \pm \pi/6, k \in \mathbb{Z}$

11 $(0, 0), (2, 2). (1, 3). (-1, 1)$

12 $r = 1, \theta = 0, \pm 2\pi/7, \pm 4\pi/7, \pm 6\pi/7$;
$z^2 - 2z \cos \theta + 1$

13 (a) $\dfrac{5t - 10t^3 + t^5}{1 - 10t^2 + 5t^4}$, where $t = \tan \theta$

(b) $z = 5/3$, radius $4/3$

14 $\tan (k\pi/16), k = 1, 5, 9, 13$

15 (a) $(5 + 14i)/13$

18 $(x - e^{ik\pi/5}), k = 0, \pm 2, \pm 4; (\sqrt{5} - 1)/4$

19 $2 \cos (\pi/4 - \phi/2)$

20 (a) $2k\pi + 2\pi/3, k \in \mathbb{Z}$ (b) $x + y = 1$

21 the circle centre $z = 2$, radius 2
the line $x = 1$
the real axis excluding $0 \leqslant x \leqslant 2$

22 $z = (-1 \pm i\sqrt{3})/2$ or $(-1 \pm i\sqrt{15})/4$

23 $64c^7 - 112c^5 + 56c^3 - 7c$;
$A = 1/64, B = 7/64, C = 21/64$,
$D = 35/64$

25 $Z = z - 2, W = iZ, w = 2W$;
$|w + 4i| = 4$;
$0, \pm 2\sqrt{3} - 6i; 2, -1 \mp i\sqrt{3}$

26 (a) $(4\sqrt{10})/3$ (b) $(8 - 6i)/3; 4 + 2i$

27 $8e^{i\theta}, \theta = k\pi/8, k = -5, -1, 3, 7$

28 $Z = z + \sqrt{2}, W = e^{i\pi/4}Z, w = \sqrt{2}W$

29 $|z - \sec \theta| = \tan \theta$

30 (a) $1; r = 1, \theta = \pi/3; r = 2, \theta = 2\pi/3$;
$r = 1, \theta = \pi$

(b) $e^{\pm i\theta}, PQ = 2$

31 (a) $1 - i, -1 \pm 2i$ (b) $(3 + i\sqrt{3})/2; \pi/6$,
$2\pi/3$

32 $2 + i$

33 $r = 1, \theta = \pi/2, -\pi/6, -5\pi/6; -1 - i$,
$2 - i(1 \pm \sqrt{3})$

34 $|w - 5| = 8, |w - 2i| = |w - 6i|$

35 (a) $-1 + i\sqrt{3}, \frac{1}{2}(\sqrt{3} + i)$

(b) $9, 2\theta; \frac{1}{3}, -\theta; \frac{2}{3}\sqrt{(16 + 9 \cos^2 \theta)}$,
$\tan^{-1}(4 \tan \theta/5)$

36 $\sqrt{2}(\cos 45° - i \sin 45°); r = 2^{7/6}$,
$\theta = -135°, -15°, 105°$

37 (a) $(5 + 12i)/169; \pm(3 - 2i)$

(b) (i) $4x^2 + 4y^2 + 9x - y + 4 = 0$

(ii) $x + y + 1 = 0$

38 $3x^2 + 3y^2 + 10x + 3 = 0$

42 $a = -\frac{1}{2}, b = 0, \pm\frac{1}{2}\sqrt{3}, \pm 1/(2\sqrt{3})$

Chapter 3

3.1 1 (a) neither (b) odd (c) even
(d) odd

2 (a) $(12x^2 + 1) + (8x^3 + 6x)$

(b) $\dfrac{x^2 + 1}{x^2 - 1} + \dfrac{2x}{x^2 - 1}$

(c) $\cos 2x + \sin 2x$

(d) $\sec 2x + \tan 2x$

3 even (b) even (c) odd

4 yes; yes

5 (a) 0 (b) -2 (c) π and $-\pi$

(d) 0

6 (a) $\pi/2$ (b) 2π (c) π (d) $\pi/4$

10 domain $x \in \mathbb{R}$, $x > 2$, range $\mathbb{R}$

12 both continuous

3.2 1 $\frac{1}{2}(\cosh 2x + 1)$, $\frac{1}{2}(\cosh 2x - 1)$

2 $1; \sqrt{2}$

3 (a) $\sqrt{2}, \ln(\sqrt{2} - 1)$

(b) $13/12, \ln(3/2)$

4 $\pm\sqrt{3}; \pm\ln(2 + \sqrt{3})$

6 (a) $\sinh 2x$ (b) $\sinh 2x$

7 (a) $3 \sinh 3x$

(b) $\frac{1}{2}(1 + 1/x^2)$

(c) $-2 \tanh 2x \operatorname{sech} 2x$

9 (a) $0, \ln 5$ (b) $\ln 2$ (c) $\ln 3$

3.3 1 (a) $99/101$ (b) $-12/13$

2 (a) e^{2x} (b) -1

3 (a) $\ln 2$ (b) $\pm\ln 3$

4 (a) $2 \operatorname{sech}^2 2x$ (b) $\tanh x$

(c) $-\operatorname{cosech}^2 x$ (d) $\coth x$

(e) $-\tanh x \operatorname{sech} x$ (f) $2 \operatorname{cosech} 2x$

3.5 1 (a) $\frac{1}{2} \cosh 2x + c$

(b) $2 \ln \cosh(x/2) + c$

(c) $\frac{1}{2}x + \frac{1}{4} \sinh 2x + c$

(d) $\frac{1}{2} \ln \sinh 2x + c$

2 (a) $x \cosh x - \sinh x + c$

(b) $x \sinh x - \cosh x + c$

(c) $x \tanh x - \ln \cosh x + c$

4 (a) $1 \cdot 407$ (b) $0 \cdot 368$ (c) $0 \cdot 238$

(d) $0 \cdot 866$

6 $1 - e^{-1}$

7 $(\pi/2)(e - 5e^{-1})$

3.6 2 (a) $(x^2 + 1)^{-1/2}$

(b) $(x^2 - 1)^{-1/2}$

(c) $2/(1 - x^2)$ (d) $1/x$

3 (a) $2/\sqrt{(x^2 + 1)}$

(b) $2/\sqrt{(x^2 - 1)}$ (c) 0

(d) $(x^2 - 4)^{-1/2}$

4 (a) $0 \cdot 881$ (b) $1 \cdot 317$ (c) $1 \cdot 472$

(d) $-0 \cdot 693$ (e) $0 \cdot 693$ (f) $0 \cdot 549$

3.7 1 (a) $\sinh^{-1}(x/2) + c$

(b) $\frac{1}{2} \sinh^{-1}(2x/3) + c$

(c) $\cosh^{-1}(x/4) + c$

(d) $\frac{1}{2} \cosh^{-1}(2x) + c$

2 (a) $\ln(\sqrt{2} + 1)$

(b) $\ln[(4 + \sqrt{15})(2 - \sqrt{3})]$

(c) $\ln[(3 + 2\sqrt{2})(2 - \sqrt{3})]$

(d) $\ln(\sqrt{2} + 1)$

3 (a) $\cosh^{-1}(x - 2) + c$

(b) $\sinh^{-1}(x + 3) + c$

(c) $\sinh^{-1}(\frac{1}{2}x + 2) + c$

(d) $\cosh^{-1}[\frac{1}{2}(x - 1)] + c$

4 (a) $\frac{1}{2}\cosh^{-1}(x + \frac{1}{2}) + c$

(b) $\frac{1}{2} \sinh^{-1}(\frac{2}{3}x + \frac{1}{3}) + c$

(c) $\cosh^{-1}(\frac{1}{3}x + 1) + c$

(d) $(1/\sqrt{2}) \sinh^{-1}(2x - 1) + c$

5 (a) $\frac{1}{2}x \sqrt{(x^2 + 9)} + (9/2) \sinh^{-1}(x/3)$

$+ c$

(b) $\frac{1}{2}x \sqrt{(x^2 + 25)} + (25/2)$

$\sinh^{-1}(x/5) + c$

(c) $\frac{1}{2}x \sqrt{(x^2 - 9)} - (9/2)$

$\cosh^{-1}(x/3) + c$

(d) $\frac{1}{2}x \sqrt{(4x^2 - 1)} - \frac{1}{4}$

$\cosh^{-1}(2x) + c$

6 (a) $1 \cdot 710$ (b) $0 \cdot 415$

Miscellaneous exercises 3

1 $x = n\pi, n \in \mathbb{Z}$; (a) $\pi^2/2 - 2$

(b) $5\pi^2/8 - 3$

2 2

3 discontinuous, 0

4 f neither, g odd; $1 - \pi/3$; $k = -1, 12\pi$

6 $8/15$

7 (a) 4 (b) -3

9 $1; \frac{1}{2}; 1$

10 (a) $-\ln 2$ (b) $0, \ln 3$ (c) $\ln 3$

13 (a) $\frac{1}{3}x \sinh 3x - 1/9 \cosh 3x + c$

(b) (i) $\ln \frac{2}{3}$ (ii) $-\frac{1}{2} \ln 5$

14 (a) $1 + x^2/2 + x^4/24$, $0 \cdot 23$ per cent

15 $-\frac{2}{3}$

16 $x = -\ln 2$, $y = \ln(3/2)$

19 (a) $(\ln 2, \ln 3)$, $[\ln(17/6), \ln(17/9)]$

20 $\ln(1 + \sqrt{2})$, -1; minimum

21 $\frac{1}{3} \sinh^3 1 - (1/9) \cosh^3 1 + \frac{1}{3} \cosh 1 - 2/9$

22 $\ln \cosh x - \frac{1}{2} \tanh^2 x + c$

23 $\sinh 1$, $\cosh 1 - 1$, $1 - e^{-1}$

24 (a) $\{x : x \geqslant 1\}$ (b) $\mathbb{R}$;

(a) $\{x : x > \sqrt{(a^2 - b^2)}\}$ (b) $\mathbb{R}$

25 (a) $x^{2n-1}/(2n - 1)$

(b) $(e^4 - e^{-4})/64 - (e^2 - e^{-2})/8 + 3/8$

26 (a) $a^2 \sinh 2u$

27 $\ln(7 + 5\sqrt{2})$

30 $\pm\ln[u + \sqrt{(u^2 - 1)}]$; $3/5$; $x = 2n\pi + \pi/2$,

$n \in \mathbb{Z}$; $y = \pm \ln(3 + 2\sqrt{2})$

32 $\pi/4$; $(\pi/2)[3 \ln(1 + \sqrt{2}) - \sqrt{2}]$

34 (a) $\frac{1}{2} + \ln(4/3)$

(b) $\frac{1}{2} \ln[(x + 1)/(x - 1)]$, $|x| > 1$

35 (a) 0 or $\ln 3$ (b) $0 \cdot 562$

36 (a) $e^x = \cosh x + \sinh x$,

$e^{-x} = \cosh x - \sinh x$ (b) $-\frac{1}{2}$

37 (a) $x = -\ln 2$, $y = \ln(3/2)$

38 (b) $|\lambda| < 1$; $\sqrt{5}$

Chapter 4

4.2
1 $y = e^{Cx}$
2 $y = -\ln(1 + Cx)$
3 $y = Ce^{\sin x}$
4 $y + \ln(y - 1) = x^3/3 + C$
5 $y = [B\exp(2e^x) + 1]/[1 - B\exp(2e^x)]$
6 $y^2 = 1 + Cx^2/(1 - x^2)$
7 $\sin y = 1 - ce^{\sin x}$
8 $y^2 = (3x^2 - 1)/(1 + x^2)$
9 $y = 4x$
10 $y = 1/(e^x - x)$
11 $\ln(\sec y + \tan y) = 2\tan^{-1}(e^x) - \pi/2$
12 $\sin y = \sin x + \cos x$
13 $\sin y = 2\sin x - 1$
14 $\sqrt{y} = 3\sqrt{\sec x} - 1$
15 $y = (x^2 - 1)/(x^2 + 3)$
16 $y = \exp(C/x)$
17 $y = Cx^x e^{-x}$
18 $y = -\sqrt{[(x - \ln x)^2 - 1]}$

4.3
1 $y = (C + x)\operatorname{cosec} x - \cos x$
2 $y = 2\cos x - 2\cos^2 x$
3 $y = Ce^{-x} + 2x - 2$
4 $y = C\sin^2 x - 3\sin x \cos x$
5 $y = Ce^{-x}/(x - 1) + x - 1 + 1/(x - 1)$
6 $y = C\cos x + x^2 \sin x$
7 $y = x^4 - 2x$
8 $y = C\sin x + 2\sin^2 x$
9 $y = (x^5 - 1)/(5x^2)$
10 $y = C\operatorname{sech} x + \tfrac{2}{3}\cosh^2 x$
11 $y = Cx^2 e^{1/x} - 2x^2 - 2x - 1$
12 $y = C/x + (x - x^{-1})e^{x^2}$
13 $y = \sin x \cos x$
14 $y = x - \tfrac{1}{4}x^3 + \tfrac{1}{2}x^3 \ln x$
15 $y = \tfrac{1}{4}(2t - 1 + e^{-2t})$

4.4
1 $y = 8(x - 1)^3 - x^3$
2 $y = -x/(1 + \ln x)$
3 $x\dfrac{dv}{dx} = 1 + v^2;\ y = x\tan(\ln x)$
4 $\dfrac{dz}{dx} + z = 2\sin x;$
   $y^2 = e^{-x} + \sin x - \cos x$
5 $y = A + B/x$
6 $(1 + x^2)\dfrac{dz}{dx} - 2xz = -2x^3;$
   $y^{-2} = 1 + 2x^2 - (1 + x^2)\ln(1 + x^2)$
7 $\cos y = (2\sin x - 1 + e^{-2\sin x})/4$
8 $y = \tan(x + \pi/4) - x$
9 $\dfrac{dz}{dx} = \dfrac{2}{1 - 2z};$

$(x + y)^2 + x - y = C$
10 $2x^3 + 3x^2y + y^3 = C$
11 $2\ln(y - x) + 3x/(y - x) = C$
12 $e^{y/x} = \ln(Cx)$
13 $y + \sqrt{(x^2 + y^2)} = Cx^2$
14 $(x - y)^2 = x + y$
15 $y^{-2} = C\cos^2 x - \tfrac{1}{2}\sin^2 x \tan^2 x$
16 $x^2(1 - x^2)y^{-2} = C - 2x^5/5$
17 $y^{-2}\sec^2 x = C - 2\ln(\sec x + \tan x)$
18 $y^{-1} = 5e^{-2x}/4 + \tfrac{1}{2}x - \tfrac{1}{4}$
19 $y^{-2} = \tfrac{1}{2}e^{2x} + x + \tfrac{1}{2}$
20 $y^{-1} = Cx - x\ln x$

4.5(i)
1 $y = Ce^{3x}$
2 $y = Ce^{-2x}$
3 $y = C_1 e^x + C_2 e^{4x}$
4 $y = C_1 e^{-4x} + C_2 e^{3x}$
5 $y = C_1 e^{2x} + C_2 e^{-2x}$
   (or $A\cosh 2x + B\sinh 2x$)
6 $y = C_1 e^{-x} + C_2 e^{-2x}$
7 $y = C_1 e^x + C_2 e^{-5x}$
8 $y = C_1 e^{-2x} + C_2 e^{-x} + C_3 e^x$

4.5(ii)
1 $y = e^{-x}(C_1 \cos 2x + C_2 \sin 2x)$
2 $y = C_1 \cos 3x + C_2 \sin 3x$
3 $y = e^{-2x}(C_1 \cos x + C_2 \sin x)$
4 $y = e^{4x}(C_1 \cos 3x + C_2 \sin 3x)$
5 $y = C_1 \cos 4x + C_2 \sin 4x$
6 $y = e^{-kx}[C_1 \cos nx + C_2 \sin nx]$

4.5(iii)
1 $y = (C_1 x + C_2)e^{-2x}$
2 $y = (C_1 x + C_2)e^{5x}$
3 $y = (C_1 x + C_2)e^{-x/2}$
4 $y = (C_1 x + C_2)e^{-4x/3}$

4.5(iv)
1 $y = C_1 + xC_2 + C_3 e^{2x}$
2 $y = C_1 e^{-4x} + e^{2x}[C_2 \cos(2x\sqrt{3}) + C_3 \sin(2x\sqrt{3})]$
3 $y = C_1 e^{-x} + C_2 \cos(x\sqrt{3}) + C_3 \sin(x\sqrt{3})$
4 $y = C_1 \cos nx + C_2 \sin nx + C_3 \cosh nx + C_4 \sinh nx$
5 $y = C_1 e^{3x} + C_2 \cos 2x + C_3 \sin 2x$

4.5(v)
1 $k = \tfrac{1}{2};\ y = C_1 \cos 3x + C_2 \sin 3x + \tfrac{1}{2}e^{3x};$
   $y = \tfrac{1}{2}(e^{3x} + \cos 3x - \sin 3x);$
   neither
2 $k = \tfrac{1}{9};\ y = e^{2x}(C_1 \cos 3x + C_2 \sin 3x) + e^{2x}/9$
3 $c = -3/4;\ y = (19e^{4x} + 96e^{-x})/20 + x - 3/4$
4 $p = -1, q = 7;$
   $y = (C_1 + xC_2)e^{-2x} + 7x - 1;$
   $y = (1 - 5x)e^{-2x} + 7x - 1$
5 $k = 3;\ y = 3e^{-2x}(1 - \cos 5x)$

6 $a = 3, b = 0$;
$y = 7(e^{-3x/2} - e^{x/3})/11 + 3x$

8 $p = 1$; $y = C_1 \cos 3x + C_2 \sin 3x$
$+ x \sin 3x$; $y = \frac{1}{3} \cos 3x -$
$\sin 3x + x \sin 3x$; $x = 1$ or
$(2n + 1)\pi/6$, where $n \in \mathbb{Z}$

9 $a = 0, b = -1/(2n)$;
$y = (\sin nx - nx \cos nx)/(2n^2)$

10 $f(t) = C_1 + C_2 t + t^2$; $C_1 = 1$,
$C_2 = 2$

11 $p = 1, q = -1$; $y =$
$e^{-x/2}[-\cos (3x/2) + \frac{1}{3} \sin (3x/2)] +$
$\cos x - \sin x$

12 $a = 2; y = (3 + 2x)e^{-x} - 2e^{-2x}$

13 $a = 4, b = 3; y = e^{-x}(C_1 \cos 2x$
$+ C_2 \sin 2x) + 4 + 3x$

14 $y = x \cos x$

15 $y = (x^3 - 2x^2 + x)/2$

4.6 1 $y = \frac{1}{2}(e^{2x} - 1)$; $(e^2 - 3)/4$

2 $y = (x + 2)e^{-x}$; $y = 2 - x$

3 $y = -(1 - x)^2$; $y = -\ln (2 - x)$

4 $x = -1, x = 1, y = 1$;
$(\frac{1}{2}, -3), (2, 0)$

5 $y = (3x^2 - 2x - 1)/2$

6 $y = Ce^x - \frac{1}{2}e^{-x}$

7 $y = c, y = c \cos x$

8 $y = 5x + C \sin x$;
$y = 5(2x - \pi \sin x)/2$

9 $y = x^2/4, y = cx - c^2$; yes, both
solutions satisfy the differential
equation and dy/dx is continous at
$x = 1$

4.7 1 $a = x + y$; $(\ln 81)/(ak)$

3 $dN/dt = \frac{1}{2}N - \frac{1}{4}N_0$;
$N = N_0(e^{t/2} + 1)/2; 2 \ln 3$

4 $15 \ln 10 \approx 34.5$ days; 32 days

6 $dx/dt = x(b - x)$; $x = bpe^{bt}/(b - p$
$+ pe^{bt})$; the population tends to the
limit b

7 $dx/dt = \lambda y, (d/dt)(x + y) = \mu x$;
$d^2x/dt^2 + \lambda \, dx/dt - \lambda\mu x = 0$;
$\lambda = 2, \mu = 3/2$;
$y = (5e^t + 3e^{-3t})/2$

8 $30 \ln 2 \approx 20.8$ minutes; $100e^{-1} \approx$
36.8 kg

10 $\beta = (T_0 - T_1)/(T_0 h)$;
$dp/dx = -kp/[T_0(1 - \beta x)]$;
$p = p_0(1 - \beta x)^{k/(T_0\beta)}$; for $x \geqslant h$,
$dp/dx = -kp/T_1, p = p_1 e^{-k(x-h)/T_1}$

11 (a) $x = a/b - (a - bc)e^{-bt}/b$;
$a < bc$

4.8 1 $\mathbf{r} = \mathbf{U}(e^{-\lambda t} - e^{-4\lambda t})/(3\lambda)$

2 $\mathbf{r} = \mathbf{A} + \mathbf{B}e^{-\lambda t} + \mathbf{a}t/\lambda$

3 $\mathbf{r} = \left(\mathbf{a} - \dfrac{\mathbf{g}}{k^2 + n^2}\right)e^{-kt} \cos nt +$

$\dfrac{1}{n}\left(\mathbf{V} + k\mathbf{a} - \dfrac{k\mathbf{g}}{k^2 + n^2}\right)e^{-kt} \sin nt$

$+ \dfrac{\mathbf{g}}{k^2 + n^2}$;

$\mathbf{r} = \dfrac{\mathbf{g}}{k^2 + n^2} + \lambda\left(\mathbf{a} - \dfrac{\mathbf{g}}{k^2 + n^2}\right) +$

$\mu\left(\mathbf{V} + k\mathbf{a} - \dfrac{\mathbf{g}}{k^2 + n^2}\right)$,

where λ μ are parameters

5 (a) $\mathbf{r} = (\mathbf{i} - \mathbf{j} + \mathbf{k})e^{4t}$
(b) $\mathbf{r} = (\mathbf{i} + \mathbf{k})e^{-t}(\cos t - \sin t)$
(c) $\mathbf{r} = \mathbf{i} \sin t + \mathbf{j} \cos 2t$; $y = 1 - 2x^2$

Miscellaneous exercises 4

1 (a) $y = [x + \sqrt{(1 + x^2)}]^{3/2}$
(b) $y = x^4 - 2x$; (c) $y = Ce^{y/x}$

2 (a) $y = x - x \ln x$ (b) $k = \frac{1}{4}$;
$y = A \cos 2x + (B + \frac{1}{4}x) \sin 2x$

3 (a) $y = (C + e^{x^2})/x$
(b) $p = 1, q = -\frac{3}{4}, y = \frac{1}{4}e^{4x} + \frac{1}{2}e^{-x} + x$
$- \frac{3}{4}$

4 (a) $y = \frac{1}{3} \cot x + \frac{2}{3} \sin x \cos x + \operatorname{cosec} x$;
(b) $y = (4 + 3x)e^{2x} + 2x + 4$

5 (a) $y = A \cosh 2x + B \sinh 2x - 1$;
$f(0.02) = 1.001\ 600\ 2$

6 (a) $\tan y = C + 2 \sin^{-1}[(2x - 3)/3]$

(b) $y = \frac{1}{2}(1 + x)$

$- \frac{1}{4}(1 - x^2) \ln\left(\dfrac{1 + x}{1 - x}\right) - C(1 - x^2)$

7 (a) $y = \frac{1}{2} x^2 \cos x + A \cos x$,
(b) $k = \frac{1}{2}, y = \frac{1}{2}(1 + x + x^2)e^{-2x}$

8 (a) $y = x^2 - 1 + Ce^{-x^2}$
(b) $1/y^2 = 4x - 2x^2 - x^3$

9 (a) $y = Ax^2 + mx$
(b) $y = e^{-x} - e^{-2x} \cos x$

10 (a) $y = -2e^\pi \sin (\ln x)$
(b) $y = x^2 + C/x$

11 (a) $y = \frac{1}{2}x(x - 1) + Cx/(x - 1)$
(b) $a = 1, b = 8, y = e^{-2x}(A \cos x + B \sin x) + \cos 2x + 8 \sin 2x$

12 (a) $P = 1, Q = 1, y = Ae^{-x} + Be^{-5x/3} + \cos 2x + \sin 2x$

(b) $\dfrac{d^2y}{dx^2} - 9y = -9x^2 - 16$

13 (a) $y = (x + 1)(\frac{1}{3}x^3 - x)/(x - 1) + C(x + 1)/(x - 1)$
(b) $k = \frac{1}{2}, y = (1 - 2x + \frac{1}{2}x^2)e^x$

14 (a) $y^2 = x^2 - 2x$ (b) $y(d^2y/dx^2) = (dy/dx)^2$, $dy/dx = \pm 6$

16 (a) $x = Ae^{2\alpha t} + Be^{-2\alpha t}$, $x = N(e^{2\alpha t} + 9e^{-2\alpha t})/2$ $y = N(9e^{-2\alpha t} - e^{2\alpha t})/4$, $[\ln(\sqrt{13} - 2)]/(2\alpha)$

17 (a) $\dfrac{1}{\beta}\ln\left(\dfrac{\theta_1 - \theta_0}{\theta_2 - \theta_1}\right) + \dfrac{1}{\alpha}(\theta_1 - \theta_2)$

18 (a) $x = e^{-2t}$, $y = 2(e^{-2t} - e^{-3t})$, $z = 1 - 3e^{-2t} + 2e^{-3t}$
(b) $E(\omega \sin pt - p \sin \omega t)/[\omega(\omega^2 - p^2)]$

19 $u = A\cos 2x + B\sin 2x + 2$; $y = 1 + (x\sin 2x + \tfrac{1}{2}\cos 2x)/x^2 + (\tfrac{1}{2} - \pi^2/4)/x^2$

20 (a) $y = (\tfrac{1}{3}x^3 - x + A)/(1 + x^2)^2$
(b) $k = 2$, $y = (1 + x + 2x^2)e^{ax}$

21 (a) $y = 2\operatorname{cosec} x + \tfrac{1}{2}(1 - \cot x) + Ae^{-x}/\sin x$
(b) $k = 2$, $y = (A + Bx + 2x^2)e^{3x}$, $y = (x + 2x^2)e^{3x}$

23 (a) $2\dot{\mathbf{r}}.\ddot{\mathbf{r}}$, (b) $\dot{\mathbf{r}} \times \mathbf{H}$, (c) $\ddot{\mathbf{r}}.\mathbf{H}$
(i) $\dot{\mathbf{r}}^2 = 2\mathbf{E}.\mathbf{r} + \lambda^2\mathbf{H}^2$
(ii) $\dot{\mathbf{r}} = \mathbf{E}t + \mathbf{r} \times \mathbf{H} + \lambda\mathbf{H}$

24 (a) $y = 3e^{3x} + e^{-3x} - 4$, $z = 3e^{3x} - e^{-3x} - 2$

Chapter 5

5.1　1　$\pi/3$, $(4, \pi/2)$
2　$r = (1/\sqrt{2})\sec(\pi/4 - \theta)$
3　(a) $x + y = 5\sqrt{2}$
(b) $y = x$ (c) $y\sqrt{3} = x + 6$
(d) $x = 2$
4　$r = 2\sec(\pi/4 - \theta)$
5　$r = 2\sqrt{2}\sec\theta$; $r = 2\sqrt{2}\operatorname{cosec}\theta$
6　2
7　$r = \sqrt{2}\sec(3\pi/4 - \theta)$
8　$r = \sec(\pi/6 - \theta)$
9　$(4, 2\pi/3)$
10　$3 + 2\sqrt{2}$

5.2　1　(a) $r = 2\sec(\pi/3 - \theta)$
(b) $r = 5\sec(3\pi/4 - \theta)$
(c) $r = 4\sec(\pi/4 + \theta)$
2　(a) $(6, \pi/3)$, $(6, -\pi/3)$ (b) $(2\sqrt{2}, 0)$, $(2\sqrt{2}, \pi/2)$ (c) $(4, -\pi/6)$, $(4, \pi/2)$
3　(a) $r = \sec(2\pi/3 - \theta)$ (b) $r = 3\operatorname{cosec}\theta$ (c) $r = \tfrac{1}{2}\sec(\pi/6 + \theta)$
4　(a) $r^2 - 10r\cos(\theta - \pi/4) + 9 = 0$
(b) $r^2 - 6r\cos(\theta - \pi/3) - 16 = 0$
(c) $r = 4\cos(\theta - \pi/6)$
6　$(2, \pm\pi/3)$
7　(a) $(1, 0)$, $\sqrt{3}$ (b) $(2, 2\pi/3)$, 2
(c) $(\sqrt{2}, \pi/4)$, $\sqrt{2}$
8　$9\sqrt{3}$; $r^2 - 4\sqrt{3}r\cos\theta + 9 = 0$

9　$r^2 - 4r\sin\theta + 3 = 0$; $\theta = \pi/3$, $\theta = 2\pi/3$
10　$\theta = \pi/3$
5.3　9　$(3, \pm\pi/3)$
10　$(1, \pm\pi/3)$, $(4, \pi)$

Miscellaneous exercises 5

1　$(4\sqrt{2})/r = \cos\theta + 3\sin\theta$
2　$r = 2a\cos^2\alpha\sec(2\alpha - \theta)$
3　$r = a\sec(\pi/6 + \theta)$
5　$(\sqrt{2}, \pi/4)$, 1
8　$(\sqrt{2}, \pi/4)$; $r = \operatorname{cosec}\theta$, $r = \sec\theta$
9　$r = 2\sqrt{2}(\cos\theta + \sin\theta)$; $r = (\sqrt{2} \pm 2)\operatorname{cosec}\theta$
10　$(a, \pm\pi/3)$
13　$3\pi a^2/2$
14　$(4, \pm\pi/3)$
15　$\pi a^2/8$
16　$4/3$
17　$a^2(\tan\tfrac{1}{2}\alpha + \tfrac{1}{3}\tan^3\tfrac{1}{2}\alpha)$
18　$\pi/12$
20　$\sqrt{3}a^2/8$
22　$\pi a^2/12$
23　(a) $\pi/2 - 1$ (b) $r = a(1 - \cos\theta)$
24　$(a^2/16)(3\pi + 2 + 8\sqrt{2})$
25　$8/3$, $4/\sqrt{3}$; $8\pi(3\sqrt{3})$; $(4/3$; $\pm 2\pi/3)$
26　$(a, \pi/6)$, $(a, 5\pi/6)$
27　$\pi/4$
28　$r = a/\sqrt{2}$, $\theta = \pm\pi/6$ or $\pm 5\pi/6$; $a^2(1 - 1/\sqrt{2})$
29　$\tfrac{1}{4}b^2\tan\alpha(e^{4\pi\cot\alpha} - 1)$; $r = 2c\sin\theta$
30　$\pi/6$, $5\pi/6$; $5\pi/6$, $\pi/6$; $(16\pi - 21\sqrt{3})(a^2/6)$
31　$(1, \pi/6)$, $(1, 5\pi/6)$; $\sqrt{3}/7$; $(5\sqrt{3})/4 - 2\pi/3$
32　$9\pi + (3\sqrt{3})/2$, $18\pi - (3\sqrt{3})/2$
33　$(5 + 3\cos\theta)/(3\sin\theta)$; $r = 5$, $\theta = \pm\cos^{-1}(-3/5)$; $x^2 + y^2 + 12x + 11 = 0$; $\tan^{-1}(3/4)$
35　$2/3$

Chapter 6

6.1　1　0　　2　$\tfrac{3}{4}$　　3　0　　4　$\tfrac{3}{7}$　　5　0　　6　0
7　0　　8　1　　9　0　　10　0　　11　0　　12　0
13　1 if $x > 1$, 0 if $x = 1$, -1 if $0 < x < 1$　　14　0
15　diverges　　16　diverges
17　converges to 0　18　converges to 0
19　oscillates boundedly　20　diverges
21　converges to 0　22　converges to 0
23　converges to $\tfrac{1}{2}$　24　converges for all x
25　diverges if $|x| > 1$, converges to 0 if $|x| \leqslant 1$
26　diverges if $|x| \geqslant 1$, converges if $|x| < 1$

6.2 2 $A = 1, B = -1; \dfrac{1}{4} - \dfrac{n+1}{(n+2)^2}, \dfrac{1}{4}$

    3 $\arctan(2n+1) - \pi/4, \pi/4$

    4 (a) converges when $-2 < x < 0$
       (b) converges when $e^{-2} < x < 1$

6.3 1 (a) c (b) d (c) d (d) c (e) d
      (f) c (g) c (h) d (i) d (j) c
      (k) c (l) c

    2 (a) c for $0 < x < \frac{1}{2}$, d for $x \geqslant \frac{1}{2}$,
      (b) d for $0 < x \leqslant e^{-2}$,
        c for $e^{-2} < x < 1$, d for $x \geqslant 1$,
      (c) c for $0 < x < 1$, d for $x \geqslant 1$
      (d) c
      (e) c for $0 < x < \ln 2$, d for $x \geqslant \ln 2$
      (f) c for $x > 1$, d for $0 < x \leqslant 1$
      (g) c
      (h) c except when $x = 1$
        when it diverges
      (i) c for $x \neq (n + \frac{1}{2})\pi$,
        d for $x = (n + \frac{1}{2})\pi, n \in \mathbb{Z}$

6.4(i)   1 c to 1   2 d   3 d
      4 c to $\pi/4$ 5 c to 1
      6 c to $\ln(1 + \sqrt{2})$ 7 c to $\pi/2$
      8 c to 1   9 d   10 d
      11 c to 1   12 c to $\frac{1}{2}$

6.4(ii) 1 d   2 c to $2\sqrt{2}$ 3 d
      4 c to $\pi/2$ 5 c to $\ln(2 + \sqrt{3})$
      6 c to -1 7 c to 16/3 8 d
      9 c to $3\pi/2$ 10 c to 1

6.4(iii) 1 d   2 d   3 c to $\pi/2$
      4 c to $b/(a^2 + b^2)$ 5 c to $-1/9$
      6 c to $\frac{1}{2}\pi + 1$ 7 d
      8 c to $\pi/2$ 9 c to 4   10 d
      11 c to $\pi/8$ 12 d
      13 (b) (i) $\ln 2$ (ii) 20 14 $\pi/\sqrt{3}$

6.5 1 $x - x^3/3! + x^5/5!$,
      $(-1)^{n-1}x^{2n-1}/(2n-1)!$
    3 $-2x - 4x^2$
    4 (a) $ab(a^2 - b^2)/6$, (b) 1/12
    5 $y = 1 + (p - q)x - (p^2 + 2pq - 3q^2)x^2/2$, $z = 1 + (p - q)x/2 + (3p^2 - 2pq - q^2)x^2/8$;
      $p:q = 1:1$ or $-7:5$
    6 $\sec x; 1 + x^2/2 + 5x^4/24$
    7 $2\displaystyle\sum_{n=0}^{\infty} \dfrac{x^{4n}}{(4n)!}, 2\sum_{n=0}^{\infty} \dfrac{x^{4n}}{(4n+1)!}$;
      $\frac{1}{2}[(\cosh x + \cos x) - (\sinh x + \sin x)/x]$
    8 (a) $2 - 4x + 5x^2 - 14x^3/3$,
      $(-1)^n[3^n + 1]x^n/n!$
      (b) $x^2 - 2x^3 + 7x^4/2$,
      $(-1)^n(2^n - 2)x^n/n$,
      $\{x: -\frac{1}{2} < x \leqslant \frac{1}{2}\}$

9 $\ln 2 + x - 11x^2/2 + x^3/3$,
    $\{x:|x| < \frac{1}{2}\}$

11 $(-1)^n 2^n/[n(n-1)]$

12 $2\displaystyle\sum_{n=1}^{\infty} x^{2n-1}/(2n-1)$,
    $\{x: -1 < x < 1\}$;
    $2\displaystyle\sum_{r=1}^{\infty} \dfrac{1}{(2r-1)(2y+1)^{2r-1}}$; $1/y$

13 $1 + t^2, 2t + 2t^3, 2 + 8t^2 + 6t^4,$
    $16t + 40t^3 + 24t^5$;
    (a) $x + x^3/3 + 2x^5/15$
    (b) $1 + 2h + 2h^2 + 8h^3/3 + 10h^4/3$

14 (a) $y = \displaystyle\sum_{n=0}^{\infty} \dfrac{(-1)^n x^n}{n!}$;
    $e^{-2} (\approx 0\cdot13535)$

    (b) (a) $4x^2 - 16x^4/3 + 128x^6/45$
    (b) $-\sqrt{2}[(x - \pi/4) - (x - \pi/4)^3/6 + (x - \pi/4)^5/120]$

15 $y = 9x - 120x^3 + 432x^5 - 576x^7 + 256x^9$

6.6 2 $(64x^3 + 384x^2 + 288x - 480)e^{2x}$
    3 $x^2 + x^3 + 7x^4/12 + x^5/4$
    4 $(1 - x^2)y_{n+2} - 2(n + 1)xy_{n+1} - (n^2 + n - 20)y_n = 0$;
      $f(x) = 1 - 10x^2 + 35x^4/3$
    5 $x + x^5/20 + x^9/1440$

Miscellaneous exercises 6

1 (a) $1/\sqrt{3}$ (b) $5e - 1$ (c) $\mathbb{R}, x^2 \neq 1$
2 $a = 1, b = 11/6, c = 361/120$
3 (a) (i) convergent (ii) convergent
   (b) $\arctan(n + 1) - \pi/4, \pi/4$
4 (a) 1/10, 990 (b) $\frac{1}{2}$
5 (a)(i) a/b (ii) $\frac{1}{3}$ (c) convergent
6 (a) $1 + 2x + 2x^2 + 4x^3$ (b) both
   converge to $\ln 4$
8 $y = 1 - x^2/4 + x^4/64 - x^6/2304$;
   $y = \displaystyle\sum_{n=0}^{\infty} \dfrac{(-1)^n x^{2n}}{2^{2n} (n!)^2}$
9 (a) c (b) c (c) c (d) c (e) c (f) d
10 (a), (c) c (b) d
12 (b), (d) c (a), (c) d
13 (b) (i) 1 (ii) $(\sinh 2a)/(2a)$
14 (a) convergent (b) $3\frac{2}{3}$
15 (a) $\frac{2}{3}$ (b) $e^{1/2} - 1$ (c) $\ln 2$ (d) 32/49
16 $k = 1$; $\ln 2$
17 (a) converges to $\pi/2$ (b) diverges
18 (a) (b) (c) diverge
   (d) converges (e) converges when
   $|x| \leqslant 1$, diverges when $|x| > 1$;
   (f) converges for all x
19 (a) $\{x: 0 < x < 1\}$ (b) $\mathbb{R}^+$

20 (a) c (b) d (c) d (d) d (e) c (f) c
(g) c
22 (b) converges
23 (a) c (b) d (c) c (d) d (e) c (f) c

Chapter 7

7.1 **1** (a) π (b) $\pi[\cos(\pi/n) - \tfrac{1}{3}\cos^3(\pi/n)]$
4 (a) $\pi\phi/(\sin\phi)$ (b) $\pi^2/4$ (c) $-\pi^2/2$
5 $\tfrac{1}{2}\pi(\pi - 2)$

6 $\displaystyle\int_0^{na} f(x)\,dx = n\int_0^a f(x)\,dx$

7.2 **1** (a) $\pi/4$ (b) $8/15$ (c) $35\pi/256$
(d) $5\pi/32$ (e) $\pi/(4b^3)$
2 $I_0 = e - 1; I_3 = 6 - 2e$
3 $12(\pi^2 - 8)$
4 $-16e^{-1} + 6$
5 (a) $(9e - 65e^{-1})/2$
(b) $6 - e - 8e^{-1}$
6 $\pi^5 - 20\pi^3 + 120\pi$
7 $13/15 - \pi/4$
9 $3\pi/8$
10 $6 - 2e$
12 $2(\ln 2)^3 - 3(\ln 2)^2 + 3\ln 2 - 9/8$
14 $(3\pi + 8)/32$
15 $32/315$

16 $\dfrac{7}{32} + \left(\dfrac{3\sqrt{2}}{64}\right)\ln(1 + \sqrt{2})$

17 $a^x \ln a; 3(e^\pi + 1)/10$
20 (a) $1/40$ (b) $3\pi/512$ (c) $3\pi a^8/256$
(d) $b^{10}/1260$ (e) $\pi/(32a^3)$

7.3(i) **1** $4a$
2 $8a$
5 $9; 8(8\sqrt{5} - 1)\pi/3$
8 $12a; 6\pi a^2; 18\pi^2 a$
9 $4a[1 - \cos(\theta/2)]; \tfrac{4}{3}\pi a^2[8 - 9$
$\cos(\theta/2) + \cos(3\theta/2)]$
10 $12(2 - \sqrt{2})a$
11 $a(1 + k^2)^{\frac{1}{2}} (e^{2k\pi} - 1)/k$
12 $\tfrac{1}{2}a[\sinh^{-1}(2\pi) + 2\pi\sqrt{(1 + 4\pi^2)}]$

7.3(ii) **1** (a) $(\ln 4)/\pi$ (b) $\sqrt{[(4 - \pi)/\pi]}$
2 (a) 1 (b) $\sqrt{[(2\pi^2 - 3)/12]}$
3 (a) $\pi a/4$ (b) $a\sqrt{(2/3)}$
4 (a) 0 (b) $5/\sqrt{2}$
5 $2a/\pi$
6 $\sqrt{(k/2)}$

7.3(iii) **1** $\bar{x} = 0, \bar{y} = (4\pi + 3\sqrt{3})a/(16\pi)$
2 $(2a/5, a)$; (a) $(a/2, 0)$
(b) $(0, 5a/4)$
3 on bisecting radius, distant
$(2a\sin\alpha)/(3\alpha)$ from O
4 $\pi(1 - \ln 2)$,
$\bar{x} = (3 - 4\ln 2)/(2 - 2\ln 2)$

5 $4/15; \pi/12; 32\pi/105$
6 $16c^2/15; \bar{x} = 4c/7$
7 $a/(20 - 6\pi)$ from its vertex
8 on $Ox, \bar{x} = 1\,059/490$
9 $8(5\sqrt{5} - 1)\pi a^2/3$
10 $12(\pi^2 - 8)a/\pi^3$
12 $1, \bar{x} = 3 - 4\ln 2, \bar{y} = 9/4$
13 $\bar{x} = 9a/20, \bar{y} = 9a/10, 6\pi a^3/5$
15 $8a^2/15, \bar{x} = 4a/7, \bar{y} = 0$;
$(32\pi a^3\sqrt{2})/105$
16 (a) $\bar{x} = 2a/\pi = \bar{y}$
(b) $\bar{x} = 4a/(3\pi), \bar{y} = 4b/(3\pi)$
17 $\pi ab/2 - 4b^2/3$;
$4\pi a^2 b/3 - 16\pi b^3/15$;
$(20a^2 - 16b^2)/(15\pi a - 40b)$
18 $3\pi a^2; \bar{x} = \pi a, \bar{y} = 5a/6; 5\pi^2 a^3$
20 36π
21 $32\pi^2 a^3, 32\pi^2 a^2$

Miscellaneous exercises 7

1 (a) (i) $(\pi - 2\ln 2)/4$ (ii) $(5\ln 3)/3$
(b) $[\tan^{-1}(\tfrac{2}{3})]/6$
2 (a) $2 + \tfrac{1}{2}\ln(17/10)$ (b) $\pi^2/72$
3 (a) $\ln[p + 2)/(p + 1)]$
(b) and (c) $\ln\left[\dfrac{p + 2 + \sqrt{(4p + 4 + q)}}{p + 1 + \sqrt{(2p + 1 + q)}}\right]$
4 volume $256\pi r^3/3$, surface area $128\pi r^2$
5 (a) $\tfrac{1}{2}\pi - 1$ (b) $(e^x - e^{-x})/(e^x + e^{-x})$,
$e + \tfrac{1}{2}\pi - 1 - 2\tan^{-1}e$ (c) $3\pi/16$
6 (a) $e - 2$ (b) $(\sqrt{2})/3$
(c) $8/15 - (43\sqrt{2})/120$
7 (a) $\{x: -5 < x < -1\} \cup \{x: 2 < x < 3\}$
(b) $1 + \dfrac{2}{x - 3} + \dfrac{3}{x + 1}$
(e) $1 + 2\ln 2 + 3\ln(6/5)$
9 (a) (i) $1/6$ (ii) $\pi/(3\sqrt{3})$ (b) $\pi/32$
10 (a) $\tfrac{1}{2}\pi^3 - 12\pi + 24$ (b) $(2\pi - 3)a^2/6$
11 (a) $2\ln(2 + \sqrt{3}) - \sqrt{3}$ (b) $\tfrac{2}{3}\sqrt{(x^3 + 1)}$
$+ c, I_8 = (14\sqrt{2} - 16)/45$
12 $\tfrac{1}{2}\ln 2, \pi/4 - 2/3$
13 (b) $x\ln(x + \sqrt{x}) - \ln(1 + \sqrt{x}) - x +$
$\sqrt{x} + c$
14 (a) $\tfrac{1}{2}(\arctan x)^2 + c$
(b) $\tfrac{1}{4}\ln[(1 + 2\tan x)/(1 - 2\tan x)] + c$
15 (a) (i) $x\tan x - \ln\sec x + c$
(b) $-\tfrac{3}{4}\sqrt{(1 - 4x^2)} - \sin^{-1}(2x) + c$
16 (a) $120 - 44e$ (b) $3\pi a^2/16$
17 (a) $\pi/4$
18 (a) $\pi^3/2 - 12\pi + 24$
19 (a) $(3\pi + 8)/(32a^5)$ (b) $\tfrac{1}{2}\ln 2$
20 $6\pi a^2/5$
21 (arc) $2a/\pi$ from diameter, (area) $4a/(3\pi)$
from diameter

22 $(12a/7, 0)$

23 (a) $\frac{1}{3}$ (b) $\pi a^2/4$

24 $2(2 \ln 2 - 1)\pi a^2$

Chapter 8

8.1   **1** $25, 3 \pm \sqrt{2}$

    2 $-3, -2, 4$

    4 $\frac{1}{2}, \frac{3}{2}, 1 \pm \frac{1}{2}i$

    5 $2(p^2 - 3q)$

    6   (a) 14 (b) $\frac{3}{2}, \frac{3}{2}, -6$

    7 $-\alpha, -i\beta; \pm i, 1 \pm i$;
       $x^4 - 2x^3 + 3x^2 - 2x + 2$

    8 $p = 1, q = 0, r = 4$

  10 $\{k : 0 < k < 4\}; k = 2$

  11 $(0, 0), (\pm 1/\sqrt{3}, \frac{1}{4}); \{k : -\frac{1}{2} < k < \frac{1}{2},$
       $k \neq 0\}$

  12   (a) $[-7 \pm 3\sqrt{5}]/2, (1 \pm i\sqrt{3})/2$
      (b) $-\frac{1}{6}, \frac{1}{2}, -\frac{3}{2}$

  13 $-\frac{1}{3}, \frac{3}{2}, -3$

  14 $x^2 - 47x + 1 = 0$

  16   (b) $-a_1/(3a_0); \frac{3}{2}, (3 \pm i\sqrt{7})/4$

  17 $x^2 + 2$ (a) $\pm\sqrt{6}$ (b) $\pm\sqrt{6}, \pm 2i$

8.2   **2** $\{y : y \leqslant 1\} \cup \{y : y \geqslant 9\}$; min $(2, 9)$,
      max $(-2, 1)$

    3 $\{x : -2 < x < -1\} \cup \{x : x > 2\}$

    4 $\{x : 5/3 < x < 5\}$

    6 $\{x : |x| < \sqrt{(5/2)}\}$

    7 $\{x : |x| < 2\sqrt{2}\}$

    8   (a) $\{x : -3 < x < -1\}$
      $\cup \{x : 1 < x < 2\}$
      (b) $\{x : |x| < \sqrt{(3/2)}\}$

    9 $(0, \frac{1}{2})\{x : x > 3\} \cup \{x : x < 2\}$;
      $\frac{1}{2} - \dfrac{x}{12} + \dfrac{x^2}{72}, |x| < 2$

  10   (a) $\{x : x > \ln 3\}$
      (b) $\{x : \pi/6 < x < 5\pi/6\}$
      (c) $\{x : x > \frac{1}{2}\ln(2 + \sqrt{5})\}$

  11   (a) $\{x : 0 \leqslant x \leqslant \ln(5/3)\}$
      (b) $\{x : x > \ln 4\}$

8.3   **8**   (a) $\sqrt{2}, -1$ (b) $2, 0$

Miscellaneous exercises 8

1 $(x + 2)(x - 2)(4x - 9)$
    (a) ± 2 (b) $\pm 2, 9/4$;
    $(x + i\sqrt{2})(x - i\sqrt{2})(x + \sqrt{2})$
    $(x - \sqrt{2})(2x + 3)(2x - 3)$
    (c) $\pm 3/2$ (d) $\pm 3/2, \pm\sqrt{2}$ (e) $\pm 3/2,$
    $\pm\sqrt{2}, \pm i\sqrt{2}$

2 $a_1 = -n(n + 1)/2, a_n = (-1)^n n!$

3   (a) $x^3 - 3(2^{1/3} 4^{1/3})x - (2 + 4) = 0$;
    $2^{1/3} + 4^{1/3}$

4 $2(p^2 - 3q)$

5 $\alpha = 3, \beta = 2, A = 3, B = -4,$
    $u_n = 3^{n+1} - 2^{n+2}$

6 $f'(x) = 2(8x^3 + 6x^2 - 11x - 3)$
    $= 2(x - 1)(4x + 1)(2x + 3),$
    $f(x) = (x - 1)^2(2x + 3)^2$

8 $-2, \sqrt{3}, \sqrt{3}$

9 $z^2 - 4z + 13; f = 1, g = -2; 1, -2,$
    $2 \pm 3i$

10   (a) $c^3 = b^3 d, 2, -2, 2$ (b) $-1/2, 2, 9/2$

11 $\{x : -2 < x < 0\} \cup \{x : 2 < x < 4\}$

12 $\{x : x < 0\} \cup \{x : x > 2\}$

13 $(x - 2)^2(x^2 + 3); \dfrac{1}{(x - 2)^2} - \dfrac{1}{x^2 + 3}$;
    $\frac{1}{2} - \dfrac{\pi}{6\sqrt{3}}$

14 $(0, 4)$

15   min $(-1, -161)$, max $(2, 28)$,
    min $(4, -36); -3 < x < -2, 1 < x < 2,$
    $2 < x < 3, 4 < x < 5; 36\frac{5}{8}$

16 $e^2 y = 4x, y = 0$

17   (a) (i) $\{x : x < -6\} \cup \{x : -2 < x < \frac{2}{3}\}$
    (ii) $\{x : 1 \leqslant x \leqslant 3\}$;
    (b) (i) 1 (ii) $-\sqrt{2}$

20   (b) $\{a : a > 2\sqrt{6}\}$

21 $1 + \sqrt{5}$

24   (a) $\{\theta : \pi/3 < \theta < 2\pi/3\} \cup$
    $\{4\pi/3 < \theta < 5\pi/3\}$
    (b) $\{x : -2 < x < -1\} \cup \{x : 0 < x < 1\}$

Chapter 9

9.2   **1** $1 \cdot 02, 1 \cdot 0404, 1 \cdot 0612, 1 \cdot 0824, 1 \cdot 1041$

    2 $1 \cdot 01, 1 \cdot 0202, 1 \cdot 0306, 1 \cdot 0412$

    3 $1 \cdot 004, 1 \cdot 008, 1 \cdot 012, 1 \cdot 017, 1 \cdot 021$

    4 $1 \cdot 244$

    5 $1 \cdot 2245$

    6 $0 \cdot 4511; 0 \cdot 4512$

9.3   **1** $0 \cdot 995; 0 \cdot 983$

    2 $1 \cdot 226$

    4 $0 \cdot 55$

    5 $\frac{1}{2} + 2x - \frac{1}{2}x^2 - 4x^3/3 + x^4/4$

    6 $0 \cdot 424; 0 \cdot 52$

    7 $1 \cdot 725; 1 \cdot 29$

    8 $x + 2x^3/3 + 8x^5/15$

    9 $1 - x/2 + 5x^2/4 - 11x^3/24$

  10 $3(x - 2) + (x - 2)^2/2 - 3(x - 2)^3/2$
    $- 3(x - 2)^4/8$

9.4(i)   **1** $1 + x - x^4/12 - x^5/20$

    2 $x + x^4/12 + x^7/504$

    3   (a) $1 - x^2/2 + x^4/8$
      (b) $x - x^3/3 + x^5/15$

    5 $2x - x^3 - x^5/4$

    6 $1 - (x - 1) - (x - 1)^2/2 + (x - 1)^3/3$
      $+ (x - 1)^4/6$

    7 $1 - x + x^2/4 - x^3/36$

    8 $1 + (x - 1)^2 + (x - 1)^3 + (x - 1)^4/2$
      $+ 2(x - 1)^5/5$

9.4(ii)   **1** 0·4016; 0·6096
      2 0·811; 0·734
      3 0·1987
      4 1·29847
      5 1·1945
      6 0·185, 0·371, 0·564, 0·770
      7 0·810, 0·570, 0·294
      8 1·032, 1·128, 1·295
9.5   **3** (b) and (c)
    10 1·35321

Miscellaneous exercises 9
1 1·181
2 1·11237
3 0·825
4 0·7181
5 1·020; 1·072
6 (a) 1·211 (b) 1·224
7 0·2428
8 0·51, 0·5202, 0·5306, 0·5412, 0·5520
9 1·1353; 1·2636
11 1·16; 1·32
12 0·605; 0·956
13 $x + x^5/20 + x^6/30$
14 (a) $x + x^3/6 - x^4/12$
     (b) $1 - x^3/6 - x^5/40$
15 (a) $1 - x^2/2 - x^3/2 - 11x^4/24$
     (b) $1 + x + x^2/2 + x^3/3 + 7x^4/24$
16 $x + x^2/2 + x^4/12 + x^5/20 + x^6/120$;
     0·22013
17 $1 - x^4/6 + x^8/2520$
18 1·102; 1·215
19 0; 2·03717, $-0·77363$
20 3; 0·6694, 0·6691
21 2; 1·1573, 1·1757; 3·127
22 1·50524
23 3·592, 3·514
24 $|f'(x)| < 1$; 1·9294
25 8
26 4
27 1·52361
28 2·18032
29 $\sqrt{b}$
30 $2\ln y = \tan^{-1} x$; 1·2096; 1·2080

Chapter 10
10.2   **1** (a) $4x + y - 8 = 0, x - 4y + 15$
       $= 0; x + y - 4 = 0, x - y = 0$
      (b) $x + y - 6 = 0, x - y = 0$;
       $x + y + 6 = 0, x - y = 0$
      (c) $9x + 4y + 72 = 0,$
       $4x - 9y - 65 = 0;$
       $9x + y + 36 = 0, x - 9y - 160 =$
       0

2 (a) $x + y - 4 = 0, x - y = 0$;
    $x + 16y - 16 = 0,$
    $32x - 2y - 255 = 0$
    (b) $x + 4y - 12 = 0, 8x - 2y -$
    $45 = 0; x + y + 6 = 0, x - y = 0$
    (c) $x + 4y + 16 = 0, 4x - y + 30$
    $= 0; x + 25y + 40 = 0,$
    $125x - 5y + 2496 = 0$
3 (a) $(2, 0), (0, 8); (-15, 0), (0, 15/4)$;
    $(4, 0), (0, 4); (0, 0)$
    (b) $(6, 0), (0, 6); (0, 0); (-6, 0),$
    $(0, -6); (0, 0)$
    (c) $(-8, 0), (0, -18); (65/4, 0),$
    $(0, -65/9); (-4, 0), (0, -36);$
    $(160, 0), (0, -160/9)$
4 (a) $(4, 0), (0, 4); (0, 0); (16, 0), (0, 1)$;
    $(255/32, 0), (0, -255/2)$
    (b) $(12, 0), (0, 3); (45/8, 0),$
    $(0, -45/2); (-6, 0), (0, -6); (0, 0)$
    (c) $(-16, 0), (0, -4); (-15/2, 0),$
    $(0, 30); (-40, 0), (0, -8/5)$;
    $\left(\dfrac{-2496}{125}, 0\right), \left(0, \dfrac{2496}{5}\right)$
5 (a) $4\sqrt{17}$ (b) 32
6 $(2cp/3, 2c/3p), 9xy = 4c^2$
7 (a) $4xy = 27$
    (b) $18xy = 81 - 8x^4$
    (c) $18xy = 81 - x^4$
8 (a) 2 (b) $(p^4 - 1)^2/(2p^4)$
9 $(-2/p^3, -2p^3), (-2p, -2/p)$
10 ± 12
11 $y = -x + 12, y = -4x + 24$
12 $9x + y + 48 = 0, x + 25y + 80 =$
    0

10.3   **3** (a) $\sqrt{2}, \sqrt{2}, \sqrt{2}$ (b) $(\pm\sqrt{2}, 0),$
    $(\pm 2\sqrt{2}, 0), (\pm 2\sqrt{(2/3)}, 0)$
    (c) $x = \pm 1/\sqrt{2}, x = \pm\sqrt{2},$
    $x = \pm\frac{1}{3}\sqrt{6}$
    (d) $y = \pm x, y = \pm x, y = \pm x$
  4 (a) $\frac{1}{2}\sqrt{13}, \sqrt{(5/3)}, 3/2$ (b) $(\pm\sqrt{13}, 0),$
    $(\pm\sqrt{5}, 0), \left(\pm\dfrac{3}{2\sqrt{5}}, 0\right)$
    (c) $x = \pm 4/\sqrt{13}, x = \pm 3/\sqrt{5},$
    $x = \pm 2/(3\sqrt{5})$
    (d) $y = \pm 3x/2, y = \pm x\sqrt{(2/3)},$
    $y = \pm\frac{1}{2}x\sqrt{5}$
  5 (a) $x^2/9 - y^2/16 = 1$
    (b) $x^2/4 - y^2/4 = 1$
    (c) $x^2/4 - y^2/25 = 1$
    (d) $x^2/144 - y^2/25 = 1$
  6 $x^2/16 - y^2/48 = 1$
  7 $x^2/64 - y^2/36 = 1$

1 (a) $8x - 5y = 7, 5x + 8y = 60$
(b) $-3x + 2y = 5$,
$2x + 3y = -12$
(c) $-20x - 9y = 73, 9x - 20y = -105$ (d) $9x + 4y = 38$,
$4x - 9y = 60$

2 (a) $(7/8, 0), (0, -7/5)$
(b) $(-5/3, 0), (0, 5/2)$
(c) $(-73/20, 0), (0, -73/9)$
(d) $(38/9, 0), (0, 19/2)$

3 (a) $(12, 0), (0, 15/2)$ (b) $(-6, 0)$,
$(0, -4)$
(c) $(-35/3, 0), (0, 21/4)$
(d) $(15, 0), (0, -20/3)$

4 $c = \pm 3; y = 2x + 3, y = 2x - 3$

5 $y = x \pm a/\sqrt{2}$

6 $y = 2x - 3, y = -2x - 3$

7 $y = \pm x \sqrt{(3/2)} - 1$

8 $\pm 2, 4/3$

9 $[a(\sec\theta + \tan\theta), b(\sec\theta + \tan\theta)]$,
$[a(\sec\theta - \tan\theta), b(\tan\theta - \sec\theta)]$

11 $y^2(a^2 - 4x^2) = b^2 x^2$

12 (a) $4a^2 x^2 - 4b^2 y^2 = (a^2 + b^2)^2$
(b) $9a^2 x^2 - 9b^2 y^2 = (a^2 + b^2)^2$

15 both $\left[\dfrac{a^2 mc}{b^2 - a^2 m^2}, \dfrac{b^2 c}{b^2 - a^2 m^2} \right]$

3 (a) $\frac{1}{3}\sqrt{5}, \frac{1}{2}\sqrt{3}, \frac{1}{6}\sqrt{11}, \frac{1}{3}\sqrt{5}$
(b) $(\pm\sqrt{5}, 0), (\pm\sqrt{3}, 0), (\pm\sqrt{11}, 0)$,
$(0, \pm\sqrt{5})$
(c) $x = \pm 9/\sqrt{5}, x = \pm 4/\sqrt{3}$,
$x = \pm 36/\sqrt{11}, y = \pm 9/\sqrt{5}$

4 (a) $x^2/36 + y^2/16 = 1$ (b) $x^2/16 + y^2/4 = 1$ (c) $x^2/16 + y^2/25 = 1$

5 (a) $\frac{1}{3}\sqrt{5}, \frac{1}{2}\sqrt{3}, 3/5$
(b) $(\pm 2\sqrt{5}, 0), (\pm 2\sqrt{3}, 0), (0, \pm 3)$
(c) $x = \pm 18/\sqrt{5}, x = \pm 8/\sqrt{3}$,
$y = \pm 25/3$

1 (a) $2x + 3y = 25, 3x - 2y = 18$
(b) $3x + 4y = 17, 4x - 3y = 6$
(c) $-5x + 36y = 507, 36x + 5y = -480$ (d) $3x + 5y = -8$,
$5x - 3y = -2$

3 $y = x \pm 5$

4 $2y = x \pm 6$

5 $4y = x - 9, 7x + 8y = 27$

6 $-\frac{1}{3}, 7; 11/2$

7 (a) $b^2 x^2 + a^2 y^2 = 4x^2 y^2$
(b) $b^2 x^2 + a^2 y^2 = 9x^2 y^2$

8 $9a^2 x^2 + 9b^2 y^2 = (a^2 - b^2)^2$

12 $[a(\cos\theta - \sin\theta), b(\cos\theta + \sin\theta)]$
$x^2/a^2 + y^2/b^2 = 2$

14 $\left(\dfrac{-a^2 mc}{b^2 + a^2 m^2}, \dfrac{b^2 c}{b^2 + a^2 m^2} \right)$

1 (a) $Y = X + 2$ (b) $Y = X - 3$
(c) $Y = X - 1$ (d) $\dot{Y}^2 = 4(X + 3)$
(e) $(Y + 1)^2 = 4X$
(f) $(Y + 1)^2 = 4(X + 3)$
(g) $(X + 5)^2 + (Y + 4)^2 = 4$
(h) $(X - 5)^2 + (Y - 4)^2 = 9$
(i) $X^2 + Y^2 + 6X - 4Y + 9 = 0$
(j) $X^2 + Y^2 - 14X + 12Y + 76 = 0$
(k) $X^2 + Y^2 - 12X - 16Y + 50 = 0$

2 (a) $7Y = X$ (b) $2Y = X + 3$
(c) $16X^2 - 24XY + 9Y^2 - 15X - 20Y = 0$
(d) $16X^2 + 24XY + 9Y^2 + 15X - 20Y + 25 = 0$
(e) $9X^2 - 24XY + 16Y^2 - 40X - 30Y = 0$
(f) $XY = -2$ (g) $XY = 2$
(h) $X^2 + Y^2 = 1$

3 (a) $X - 2Y + 11 = 0$
(b) $9X^2 - 24XY + 16Y^2 - 20X - 15Y + 50 = 0$
(c) $119X^2 - 240XY - 119Y^2 + 754X + 52Y + 338 = 0$
(d) $13X^2 + 13Y^2 - 22X - 150Y + 390 = 0$

1 (b), (c); (e) represents a pair of
coincident straight lines.

2 (a) $\sqrt{89}$ (b) $\frac{1}{2}$ (c) $(2\sqrt{3})/5$
(d) $\sqrt{19}$ (e) $1/6$

4 $9Y^2 - X^2 = 0$

5 (a) $Y^2 - X^2 = 0$ (b) $X^2 - Y^2 = 0$

1 (a) hyperbola $2x - 3y = 0$,
$3x + 4y = 0$
(b) ellipse
(c) rectangular hyperbola
$2x - 3y = 0, 3x + 2y = 0$
(d) no real curve
(e) parallel lines $x + 3y = \pm 1$
(f) ellipse
(g) hyperbola $3x \pm 5y = 0$
(h) ellipse
(i) rectangular hyperbola
$2x - y = 0, x + 2y = 0$
(j) parallel lines $2x - 3y = \pm 2$

2 (a) $x = -4, y = 3$
(b) $x = -1, y = x - 3$
(c) $y = 1, x + y + 1 = 0$
(d) $x + 2y + 1 = 0, 2x - y = 0$
(e) $2x + 3y - 1 = 0$,
$4x - y + 1 = 0$

1 ellipse $x = 2y$
2 circle $5x = 2y$

3 rectangular hyperbola $x = 0$
4 rectangular hyperbola $y = 2x$
5 hyperbola $y = 0$
6 circle $y = 4x$
7 parabola $y = x$
8 ellipse $y = 0$

Miscellaneous exercises 10

1 $x + t^2 y = 2ct$
4 $x + yt_1 t_2 = t_1 + t_2$ (a) $y + 2x = 0$ (b) $2xy = x + y$
6 $x + p^2 y = 2cp, py - p^3 x = c(1 - p^4)$;
$$\frac{2cp}{\sqrt{(1 + p^4)}}, \frac{c(p^4 - 1)}{p\sqrt{(1 + p^4)}}$$
9 $c^4 - x^4 = 2c^2 xy$
12 $\dfrac{4c\sqrt{m}}{\sqrt{(1 + m^2)}}$
14 $(-ct, -c/t)$
15 $x + t^2 y = 2t$
16 $y = \pm x$
17 $(h, k), x = h, y = k$ $(1\frac{1}{2}, 3), (-\frac{1}{2}, -1)$
18 $(-c/t_1 t_2 t_3, -ct_1 t_2 t_3)$
19 $2, 2\sqrt{3}$
20 $y = \pm \frac{4}{3} x$
21 $\dfrac{x^2}{a^2} - \dfrac{y^2}{a^2 m^2} = 1$
25 $t(t^2 + 1)y = t(1 - t^2)x - 2a(1 - t^4)$
26 $\dfrac{x}{b} \tan \theta + \dfrac{y}{a} \sec \theta$
$$= \left(\frac{a^2 + b^2}{ab}\right) \sec \theta \tan \theta;$$
$$\left(\frac{2x}{a}\right)^2 - \left(\frac{2yb}{2b^2 + a^2}\right)^2 = 1$$
27 $x^2 - y^2 = 50$; (a) $5\sqrt{2}$ (b) $90°$ (c) $x = 5$
28 $x^2/25 + y^2/16 = 1$
29 $(\pm 4, 0)$
32 $a^2/x^2 + b^2/y^2 = 4$
36 $\left[0, \dfrac{3}{2}\left(\dfrac{1}{\sin \theta} - \sin \theta\right)\right]$,
$(y^2 + 2x^2)(18 - x^2) = 162$
37 $3x \cos \theta + 2y \sin \theta = 6; x^2/4 + y^2/9$
$= 4 - 2\sqrt{2}; 2\sqrt{10} \sqrt{(2 - \sqrt{2})}$
38 $\pm 1; [\pm\sqrt{(a^2 + b^2)}, 0]$ and $[0, \pm\sqrt{(a^2 + b^2)}]$
39 $x = 4, y = x + 5; (4, 0), (-16/5, 9/5)$
40 $x^2 + y^2 = a^2 + b^2$
41 $bx \cos \theta + ay \sin \theta = ab$,
$ax \sec \theta - by \cosec \theta = a^2 - b^2$;
$$\left(\frac{2ax}{a^2 - b^2}\right)^2 + \left(\frac{b}{2y}\right)^2 = 1$$
42 $3x + 4\sqrt{10}y = 28$

45 $\left(\dfrac{a^2 - 3b^2}{a} \cos \theta, -2b \sin \theta\right)$;
$$\frac{a^2 x^2}{(a^2 - 3b^2)^2} + \frac{y^2}{4b^2} = 1$$
46 $\left[\dfrac{a(a^2 - b^2) \cos \theta}{a^2 + b^2}, \dfrac{-b(a^2 - b^2) \sin \theta}{a^2 + b^2}\right]$,
$$\frac{x^2}{a^2} + \frac{y^2}{b^2} = \left(\frac{a^2 - b^2}{a^2 + b^2}\right)^2$$
47 $9a^2 x^2 + 9b^2 y^2 = (a^2 - b^2)^2$
48 $25x^2 + 9y^2 = 64, (0, \pm 32/15)$
49 $b^2 h(x - h) + a^2 k(y - k) = 0$;
(a) $b^2 h(a - h) + a^2 k(b - k) = 0$
(b) $a^2 k(a - h) = b^2 h(b - k)$
50 $(b^2 \cos^2 \theta + a^2 \sin^2 \theta)r^2 + 2(hb^2 \cos \theta + ka^2 \sin \theta)r + (b^2 h^2 + a^2 k^2 - a^2 b^2) = 0$,
$$\frac{xh}{a^2} + \frac{yk}{b^2} = \frac{h^2}{a^2} + \frac{k^2}{b^2}$$
51 $a + b = 0$
52 2α
54 $y = 2x$; an ellipse

Chapter 11

11.1 **1** $e = 0, 1^{-1} = 2, 2^{-1} = 1$
2 **0** has no inverse, not a group
3 $e = 1; 1, 2$ are self-inverses
4 $e = 1$; all self-inverses
5 $e = a$; all self-inverses
6 $e = p; q^{-1} = r, r^{-1} = q$
7 $e = 1$; all self-inverses

11.2 **1** (a)

	I	R
I	I	R
R	R	I

$R^{-1} = R.$

Abelian.

(b)

	I	R	M_1	M_2
I	I	R	M_1	M_2
R	R	I	M_2	M_1
M_1	M_1	M_2	I	R
M_2	M_2	M_1	R	I

R = rotation through π radians.
M_1, M_2 = reflection in axes of symmetry.
Each element is its own inverse.
Abelian.

(c)

	I	M
I	I	M
M	M	I

$M^{-1} = M.$

Abelian.

(d)

	I	R	M_1	M_2
I	I	R	M_1	M_2
R	R	I	M_2	M_1
M_1	M_1	M_2	I	R
M_2	M_2	M_1	R	I

Notation as for (b) above.
Each element is its own inverse.
Abelian.
(e) As for (d).

2 (a)

	I	R_1	R_2
I	I	R_1	R_2
R_1	R_1	R_2	I
R_2	R_2	I	R_1

(b)

	I	R_1	R_2	R_3	R_4
I	I	R_1	R_2	R_3	R_4
R_1	R_1	R_2	R_3	R_4	I
R_2	R_2	R_3	R_4	I	R_1
R_3	R_3	R_4	I	R_1	R_2
R_4	R_4	I	R_1	R_2	R_3

(c)

	I	R_1	R_2	R_3	R_4	R_5
I	I	R_1	R_2	R_3	R_4	R_5
R_1	R_1	R_2	R_3	R_4	R_5	I
R_2	R_2	R_3	R_4	R_5	I	R_1
R_3	R_3	R_4	R_5	I	R_1	R_2
R_4	R_4	R_5	I	R_1	R_2	R_3
R_5	R_5	I	R_1	R_2	R_3	R_4

11.3 1 (a) $\{I, H\}, \{I, V\}, \{I, R\}$ with the
notation of section 11.2
(b) $\{I, R\}, \{I, M_1\}, \{I, M_2\}$ when $R = \frac{1}{2}$ turn; M_1, M_2 are reflections in the
axes of symmetry
(c) $\{I, R\}, \{I, M_1\}, \{I, M_2\}$ with
notation of (b)
(d) $\{I, R\}, \{I, M_1\}, \{I, M_2\}$ with
notation of (b)
(e) $\{I, R_1, R_2\}, \{I, M_1\}, \{I, M_2\}, \{I, M_3\}$
with the notation of section 11.3
(f) $\{I, M_1\}, \{I, M_2\}, \{I, D_1\}, \{I, D_2\},$
$\{I, R_2\}, \{I, R_1, R_2, R_3\}, \{I, R_2, M_1, M_2\},$
$\{I, R_2, D_1, D_2\}$, where R_1, R_2, R_3 are
rotations through $\pi/2, \pi, 3\pi/2$
radians, M_1, M_2 are reflections in
axes of symmetry parallel to sides,
D_1, D_2 are reflections in the
diagonals.

2 (a) all sub-groups are cyclic;
generators H, V, R
(b) all sub-groups are cyclic;
generators R, M_1, M_2
(c) all sub-groups are cyclic;
generators R, M_1, M_2
(d) all sub-groups are cyclic;
generators R, M_1, M_2
(e) all sub-groups are cyclic;
generators R_1 or R_2; M_1, M_2, M_3
(f) all five sub-groups of order 2 are
cyclic with generators $M_1, M_2, D_1,$
D_2, R_2; $\{I, R_1, R_2, R_3\}$ is cyclic with
generator R_1
3 (a) all 2 (b) all 2 (c) all 2 (d) all
2 (e) R_1, R_2 both 3; M_1, M_2, M_3
all 2 (f) M_1, M_2, D_1, D_2, R_2 all 2;
R_1, R_3 both 4

11.4 1

	P_1	P_2
P_1	P_1	P_2
P_2	P_2	P_1

11.5 2 (b), (c), (d), (e), (f) are isomorphic
11.6 1 (a) a group (b) not a group
2 (a) not a group (b) a group
4 a group
11.7 1 field
2 ring
3 field
4 ring
5 ring
6 ring
7 field
8 field
9 ring

Miscellaneous exercises 11
2 (a) $1, 2 - g$
4 (b) $(\mathbb{Q}, *)$ does
5 $(S, *)$ is a group
6 yes; no
7 (a) possess closure, commutativity and
identity element. Not associative and no
unique invers for each element.
(b) possesses associativity and identity
element but not closure; no inverse for 2;
remainders 1, 2

8

n	1	2	3	4	5	6
inverse	1	4	5	2	3	6
order	1	3	6	3	6	2

$5 = 6 \times 2$
11 no

12 (a) $\frac{1}{2}$, $1/(4x)$

(b) $\begin{pmatrix} 1 & 0 \\ 0 & 1 \end{pmatrix}, \begin{pmatrix} -1 & 0 \\ 0 & -1 \end{pmatrix} = C$

	I	A	B	C
I	I	A	B	C
A	A	I	C	B
B	B	C	I	A
C	C	B	A	I

where $A = \begin{pmatrix} -1 & 0 \\ 0 & 1 \end{pmatrix}, B = \begin{pmatrix} 1 & 0 \\ 0 & -1 \end{pmatrix}$.

16 $\{1, z^3\}, \{1, z^2, z^4\}$.

17 (a)

	I	R	H	V
I	I	R	H	V
R	R	I	V	H
H	H	V	I	R
V	V	H	R	I

where $R = \frac{1}{2}$ turn and H, V are reflections in the axes of symmetry.

(b)

	0	1	2	3
0	0	1	2	3
1	1	2	3	0
2	2	3	0	1
3	3	0	1	2

$x = 6; x = 4, 6$

(c)

	2	4	6	8
2	4	8	2	6
4	8	6	4	2
6	2	4	6	8
8	6	2	8	4

18 yes

19 $x = -i, y = z = t = 0$

	E	A	B	C
E	E	A	B	C
A	A	E	C	B
B	B	C	A	E
C	C	B	E	A

20 $n \in \{1, 2, 3, 4, 6, 12\}$, g, g^5 are generators of G, each with order 6; g^2, g^4 have order 3; g^3 has order 2; $g^\circ = e$ has order 1.

21 (a) $1, \omega^3, \omega^6, \omega^{12}, \omega^9$

23 $-1/x, -(x+1)/(x-1)$

25 not a field

26 (a) ring (b) ring (c) field (d) neither (e) neither

28

+	0	1	2	3	4
0	0	1	2	3	4
1	1	2	3	4	0
2	2	3	4	0	1
3	3	4	0	1	2
4	4	0	1	2	3

	0	1	2	3	4
0	0	0	0	0	0
1	0	1	2	3	4
2	0	2	4	1	3
3	0	3	1	4	2
4	0	4	3	2	1

29 (a) 0, 1, 4 (b) 0, 1, 2, 3, 4; 3, 4

30 (b) $x' = \dfrac{x}{x^2 - 2y^2}, y' = \dfrac{-y}{x^2 - 2y^2}$

31 (b) $-1, 0; -2 - p, -p/(p+1)$

Chapter 12

12.1 **1** (a) $3\mathbf{u} - 4\mathbf{v}$ (b) $5\mathbf{u} + 2\mathbf{v}$
(c) $(4\mathbf{u} + 3\mathbf{v})/5$ (d) $-3\mathbf{u} + 4\mathbf{v}$

2 (a) $3, -4, -5$ (b) $3, 2, 0$
(c) $4, -1, -3$ (d) $5, -3, -7$

3 $\lambda = 1, \mathbf{a} = \mathbf{b} - \mathbf{c}$;
$\lambda = 2, \mathbf{a} = \frac{1}{2}\mathbf{b} + \frac{1}{4}\mathbf{c}$

12.2 **1** no

2 $\{1, x, x^2\}; 3$

4 $\{e^x\}; 1$

6 $\{(1, -3, -5)^T\}$

7 $\{(1, -1, 0)^T, (1, 0, -1)^T\}$

9 (a) $\{2\mathbf{i} + 3\mathbf{j}\}$ (b) $\{\mathbf{i} - \mathbf{k}, 2\mathbf{i} + \mathbf{j}\}$
(c) $\{\mathbf{i} + \mathbf{j}, 2\mathbf{j} - \mathbf{k}\}$ (d) $\{\mathbf{i}, \mathbf{j}\}$

10 (a), (c), (e)

Miscellaneous exercises 12

1 (a) $\begin{pmatrix} 1 & 0 \\ 0 & 0 \end{pmatrix}, \begin{pmatrix} 0 & 1 \\ 1 & 0 \end{pmatrix}, \begin{pmatrix} 0 & 0 \\ 0 & 1 \end{pmatrix}; 3$

(b) $\begin{pmatrix} 0 & -1 \\ 1 & 0 \end{pmatrix}; 1$ (c) $4; 0$

2 6

3 $C_4 = 3C_2 - 2C_1$;
$\{(1, 1, 1, 1)^T, (1, 2, 3, 4)^T\}; 2$

4 $\mathbf{c} = \mathbf{a} + 2\mathbf{b}; 2 : 3$

5 (a) yes (b) yes

6 (a) $\mathbf{v} = (2, -7, 3, 1)^T$ (b) $\{(1, -6, 1, 0)^T,$
$(-1, 11, 0, 1)^T\}$

7 $\mathbf{X}_4 = \mathbf{X}_1 - \mathbf{X}_2 + 2\mathbf{X}_3; -6$

8 $A = \begin{pmatrix} 3 & 2 & 3 \\ 2 & 3 & 2 \\ 1 & 1 & 1 \end{pmatrix}; \begin{pmatrix} 8 \\ 7 \\ 3 \end{pmatrix}, \begin{pmatrix} 15 \\ 15 \\ 6 \end{pmatrix}, \begin{pmatrix} 23 \\ 22 \\ 9 \end{pmatrix}$

9 $x_1 - x_2, x_2 - x_3, x_3$

10 A, B, C

11 (a) $(2\mathbf{u} + \mathbf{v})/5, (\mathbf{u} - 2\mathbf{v})/5$ (b) (iv) none

12 $4\frac{1}{2}; (10X_1 + X_2 - X_3 - 2X_4)/3$

13 $8; \begin{pmatrix} 3 \\ 5 \\ 3 \end{pmatrix}; \begin{pmatrix} 2 \\ 1 \\ -1 \end{pmatrix}$

14 2, 2, 1

15 $4; \begin{pmatrix} 1 & 0 \\ 0 & 0 \end{pmatrix}, \begin{pmatrix} 0 & 1 \\ 0 & 0 \end{pmatrix}, \begin{pmatrix} 0 & 0 \\ 1 & 0 \end{pmatrix}, \begin{pmatrix} 0 & 0 \\ 0 & 1 \end{pmatrix}$

Chapter 13

13.1
1 (a) 1 (b) 1 (c) 100 (d) 0
2 (a) -3 (b) -12 (c) 9
(d) -27
3 $\begin{pmatrix} 48 & 53 \\ 44 & 51 \end{pmatrix}$

13.2
1 (a) 6 (b) -1 (c) 30 (d) -4
(e) 0 (f) -12
2 $x = 1, 1, -5$
3 $k = 3, 3 : -2 : -6;$
$k = -1, 1 : 2 : -2$
4 $4, 2, -4; 4, -12, 16$
5 (a) 21 (b) 24 (c) 36 (d) -36
6 $(1, -1, 1)$
7 $x = t, y = t + 1, z = t - 1$
9 $6 : 4 : 5$
10 (a) and (c) represent parallel planes

13.3
1 (a) $\begin{pmatrix} 8 & -2 \\ 7 & -1 \end{pmatrix}$ (b) $\begin{pmatrix} 7 & 1 \\ -6 & 0 \end{pmatrix}$
(c) $\begin{pmatrix} 8 & 7 \\ -2 & -1 \end{pmatrix}$ (d) $\begin{pmatrix} 7 & -6 \\ 1 & 0 \end{pmatrix}$

2 $\begin{pmatrix} -1 & -1 \\ -2 & 0 \end{pmatrix}; \begin{pmatrix} -1 & 1 & -6 \\ 0 & -2 & 8 \\ -1 & -1 & 2 \end{pmatrix}$

3 $\begin{pmatrix} 1 & 3 \\ 2 & 2 \\ 3 & 1 \end{pmatrix}$

4 (a) $\begin{pmatrix} 1 & 0 \\ 0 & 1 \end{pmatrix}, \begin{pmatrix} 1 & 0 \\ 0 & 1 \end{pmatrix}$
(b) $\begin{pmatrix} 0 & 0 \\ 0 & 0 \end{pmatrix}, \begin{pmatrix} 0 & 0 \\ 19 & 0 \end{pmatrix}$

5 (a) $\begin{pmatrix} 4 & 2 \\ 2 & 1 \end{pmatrix} + \begin{pmatrix} 0 & 2 \\ -2 & 0 \end{pmatrix}$

(b) $\begin{pmatrix} 1 & 0 & 1 \\ 0 & 2 & 0 \\ 1 & 0 & 3 \end{pmatrix} + \begin{pmatrix} 0 & 1 & -1 \\ -1 & 0 & -2 \\ 1 & 2 & 0 \end{pmatrix}$

8 0 **9** (a) **0** (b) **0**

13.4
1 (a) $\begin{pmatrix} 3 & -7 \\ -2 & 5 \end{pmatrix}$ (b) $\begin{pmatrix} 3 & -4 \\ -5 & 7 \end{pmatrix}$
(c) $\frac{1}{2}\begin{pmatrix} 2 & -4 \\ -4 & 9 \end{pmatrix}$
(d) $\frac{1}{2}\begin{pmatrix} -2 & 5 \\ -2 & 4 \end{pmatrix}$
(e) $\frac{1}{3}\begin{pmatrix} 6 & -9 \\ -3 & 5 \end{pmatrix}$ (f) $\frac{1}{4}\begin{pmatrix} 4 & 6 \\ 5 & 8 \end{pmatrix}$
(g) $\begin{pmatrix} 1 & 4 \\ -2 & 0 \end{pmatrix}$ (h) $\begin{pmatrix} 2 & -5 \\ -2 & 10 \end{pmatrix}$

3 (a) $\begin{pmatrix} 0 & 0 & 1 \\ 0 & 1 & 0 \\ 1 & 0 & 0 \end{pmatrix}$ (b) $\begin{pmatrix} 0 & 0 & 1 \\ 1 & 0 & 0 \\ 0 & 1 & 0 \end{pmatrix}$
(c) $\begin{pmatrix} 1 & 0 & 0 \\ 0 & 1 & -2 \\ 0 & 0 & 1 \end{pmatrix}$
(d) $\begin{pmatrix} 1 & -3 & 1 \\ 0 & 1 & -2 \\ 0 & 0 & 1 \end{pmatrix}$

4 (a) $\begin{pmatrix} 1 & -3 & 2 \\ -3 & 3 & -1 \\ 2 & -1 & 0 \end{pmatrix}$
(b) $\begin{pmatrix} 7 & -3 & -3 \\ -1 & 0 & 1 \\ -1 & 1 & 0 \end{pmatrix}$
(c) $\frac{1}{10}\begin{pmatrix} 3 & -5 & 1 \\ -1 & 5 & 3 \\ 5 & -5 & -5 \end{pmatrix}$
(d) $\begin{pmatrix} 4 & 3 & 3 \\ -1 & 0 & -1 \\ -4 & -4 & -3 \end{pmatrix}$

5 (a) $\frac{1}{18}\begin{pmatrix} 1 & 2 & 3 \\ 3 & 1 & 2 \\ 2 & 3 & 1 \end{pmatrix}$
(b) $\begin{pmatrix} 1 & -4 & 0 \\ 0 & 1 & -4 \\ 0 & 0 & 1 \end{pmatrix}$
(c) $\frac{1}{5}\begin{pmatrix} 1 & 1 & -2 \\ -1 & 14 & -18 \\ 1 & -9 & 13 \end{pmatrix}$
(d) $\frac{1}{3}\begin{pmatrix} -1 & 23 & -32 \\ 2 & -10 & 16 \\ 1 & 1 & -4 \end{pmatrix}$

6 $\frac{1}{6}\begin{pmatrix} 2 & -4 & 12 \\ -4 & 11 & -33 \\ 12 & -33 & 105 \end{pmatrix}$

13.5 **1** $\begin{pmatrix} 2 & -3 & -1 \\ 0 & 4 & 2 \\ 3 & 1 & 2 \end{pmatrix}$;

(a) $(-2, 6, 6), (2, -6, -6)$

(b) $7\mathbf{i} + 9\mathbf{k}$

(c) $\mathbf{r} = (-5 + 4t)\mathbf{i} + (8 - 2t)\mathbf{j} + (5 + 4t)\mathbf{k}$

(d) $\mathbf{r} = s\begin{pmatrix} 1 \\ 2 \\ 5 \end{pmatrix} + t\begin{pmatrix} -1 \\ 4 \\ 4 \end{pmatrix}$

2 (a) the plane $x + 2y + 3z = 0$

(b) the plane $x + 2y + 3z = 1$

3 (a) $-4\mathbf{i} + \mathbf{k}$ (b) $5\mathbf{i} - 2\mathbf{j} + \mathbf{k}$

(c) $t\mathbf{i}$

5 $\begin{pmatrix} 1 & 3 & 2 \\ 2 & 1 & 1 \\ 3 & 2 & 3 \end{pmatrix}$

6 $t(\mathbf{i} + \mathbf{j} + \mathbf{k})$

7 (a) $\mathbf{r} = t\mathbf{i} - 5t\mathbf{j} + (5 + 5t)\mathbf{k}$

(b) $\mathbf{r} = (-9 - 4s)\mathbf{i} - (3 + 4s)\mathbf{j} + (6 + 3s)\mathbf{k}$

$(-2\mathbf{i} - \mathbf{j} + 2\mathbf{k}); (-\mathbf{i} + 5\mathbf{j})$

13.6 **1** (a) $\frac{1}{2}\begin{pmatrix} \sqrt{3} & 1 \\ 1 & -\sqrt{3} \end{pmatrix}$

(b) $\frac{1}{2}\begin{pmatrix} \sqrt{3} & -1 \\ 1 & \sqrt{3} \end{pmatrix}$

(c) $\frac{1}{5}\begin{pmatrix} -3 & -4 \\ -4 & 3 \end{pmatrix}$

(d) $\frac{1}{5}\begin{pmatrix} 3 & 4 \\ -4 & 3 \end{pmatrix}$

3 (a) $\begin{pmatrix} 1 & 0 & 0 \\ 0 & -1 & 0 \\ 0 & 0 & 1 \end{pmatrix}$

(b) $\begin{pmatrix} 0 & 0 & -1 \\ 0 & 1 & 0 \\ -1 & 0 & 0 \end{pmatrix}$

(c) $\begin{pmatrix} 0 & -1 & 0 \\ -1 & 0 & 0 \\ 0 & 0 & -1 \end{pmatrix}$

(d) $\frac{1}{\sqrt{2}}\begin{pmatrix} 0 & \sqrt{2} & 0 \\ 1 & 0 & 1 \\ 1 & 0 & -1 \end{pmatrix}$

5 $\frac{1}{625}\begin{pmatrix} 16 & -12 & 15 \\ -15 & -20 & 0 \\ 12 & -9 & -20 \end{pmatrix}$;

$k = \pm 1/25$

6 $(\mathbf{i} - 2\mathbf{j} + \mathbf{k})/\sqrt{6}$

13.7 **1** (a) $7, \begin{pmatrix} 1 \\ 2 \end{pmatrix}; -2, \begin{pmatrix} 1 \\ -1 \end{pmatrix}$

(b) $6, \begin{pmatrix} 1 \\ 2 \end{pmatrix}; 4, \begin{pmatrix} 0 \\ 1 \end{pmatrix}$

(c) $6, \begin{pmatrix} 1 \\ 1 \end{pmatrix}; 3, \begin{pmatrix} 1 \\ -2 \end{pmatrix}$

(d) $8, \begin{pmatrix} 2 \\ 1 \end{pmatrix}; 3, \begin{pmatrix} 1 \\ -2 \end{pmatrix}$

(e) $14, \begin{pmatrix} 1 \\ 3 \end{pmatrix}; 4, \begin{pmatrix} 3 \\ -1 \end{pmatrix}$

(f) $8, \begin{pmatrix} 1 \\ 2 \end{pmatrix}; -1, \begin{pmatrix} 1 \\ -1 \end{pmatrix}$

(g) $8, \begin{pmatrix} 2 \\ 1 \end{pmatrix}; -2, \begin{pmatrix} 1 \\ -2 \end{pmatrix}$

(h) $-1, \begin{pmatrix} 2 \\ 1 \end{pmatrix}; -6, \begin{pmatrix} 1 \\ -2 \end{pmatrix}$

2 (a) $(\lambda - 1)^2(\lambda + 1) = 0$,

$\lambda = 1, 1, -1$

(b) $\lambda^3 + 1 = 0, \lambda = -1$

(c) $(\lambda - 1)(\lambda^2 + 1) = 0, \lambda = 1$

3 (a) $6, (-2, 2, 1)^T; 3, (2, 1, 2)^T;$

$-9, (1, 2, -2)^T$

(b) $6, (1, -2, 1)^T; 3, (1, 1, 1)^T;$

$2, (1, 0, -1)^T$

4 $7\lambda^3 - 3\lambda^2 - 3\lambda + 7 = 0$

5 (a) $1, (1, 0, 1)^T; 2, (0, 1, 0)^T;$

$3, (1, 0, -1)^T$

(b) $2, (1, -1, 0)^T; 3, (2, -1, -2)^T;$

$4, (1, -1, -2)^T$

(c) $3, (1, 0, 1)^T; 5, (3, 2, 1)^T;$

$6, (1, 3, 1)^T$

13.8 **1** (a) $(x, y)\begin{pmatrix} 7 & -3 \\ -3 & 5 \end{pmatrix}\begin{pmatrix} x \\ y \end{pmatrix}$

(b) $(x, y)\begin{pmatrix} 2 & 4 \\ 4 & 3 \end{pmatrix}\begin{pmatrix} x \\ y \end{pmatrix}$

(c) $(x, y)\begin{pmatrix} 1 & 2 \\ 2 & -6 \end{pmatrix}\begin{pmatrix} x \\ y \end{pmatrix}$

(d) $(x, y)\begin{pmatrix} 3 & -\frac{1}{2} \\ -\frac{1}{2} & -2 \end{pmatrix}\begin{pmatrix} x \\ y \end{pmatrix}$

2 (a) $\begin{pmatrix} 1 & 2 & 0 \\ 2 & 2 & 3 \\ 0 & 3 & 3 \end{pmatrix}$

(b) $\begin{pmatrix} 2 & 1 & 3 \\ 1 & -1 & -2 \\ 3 & -2 & 0 \end{pmatrix}$

3 (a) $4X^2 + Y^2 = 4$

(b) $3X^2 - 2Y^2 = 6$

(c) $Y^2 - X^2 = 1$

(d) $9X^2 + 4Y^2 = 36$

4 $\begin{pmatrix} X \\ Y \end{pmatrix} = \frac{1}{\sqrt{2}}\begin{pmatrix} 1 & -1 \\ 1 & 1 \end{pmatrix}\begin{pmatrix} x \\ y \end{pmatrix}$

5 $\begin{pmatrix} X \\ Y \end{pmatrix} = \frac{1}{\sqrt{13}} \begin{pmatrix} 2 & -3 \\ 3 & 2 \end{pmatrix} \begin{pmatrix} x \\ y \end{pmatrix}$

6 (a) $2X^2 - Y^2 = 2$, a rotation
   (b) $2X^2 - Y^2 = 2$, a reflection

9 $A = \begin{pmatrix} 3 & 1 & 1 \\ 1 & 0 & 2 \\ 1 & 2 & 0 \end{pmatrix}$

$P = \frac{1}{\sqrt{6}} \begin{pmatrix} 2 & \sqrt{2} & 0 \\ 1 & -\sqrt{2} & \sqrt{3} \\ 1 & -\sqrt{2} & -\sqrt{3} \end{pmatrix}$

10 $A = \begin{pmatrix} 7 & 2 & 0 \\ 2 & 6 & 2 \\ 0 & 2 & 5 \end{pmatrix}$; 9, 6, 3

Miscellaneous exercises 13

1 20/3

2 (a) $-1, \begin{pmatrix} 0 \\ 1 \\ -1 \end{pmatrix}; -3, \begin{pmatrix} 2 \\ -1 \\ 0 \end{pmatrix};$
   $-4, \begin{pmatrix} 1 \\ 0 \\ -1 \end{pmatrix}$

(b) $4, \begin{pmatrix} 0 \\ 1 \\ -1 \end{pmatrix}; 3, \begin{pmatrix} 2 \\ -1 \\ 0 \end{pmatrix}; 2, \begin{pmatrix} 1 \\ 0 \\ -1 \end{pmatrix}$

(c) $6, \begin{pmatrix} 24 \\ 8 \\ 9 \end{pmatrix}; 3, \begin{pmatrix} 1 \\ 0 \\ 0 \end{pmatrix}; -3, \begin{pmatrix} -3 \\ 2 \\ 0 \end{pmatrix}$

3 $P = \begin{pmatrix} 2 & 4 \\ 3 & 6 \\ 4 & 8 \end{pmatrix}$

4 $\lambda = 3, \frac{1}{\sqrt{3}} \begin{pmatrix} 1 \\ 1 \\ 1 \end{pmatrix};$

$\lambda = 1, \frac{1}{\sqrt{3}} \begin{pmatrix} 1 \\ -1 \\ -1 \end{pmatrix};$

$\lambda = -2, \frac{1}{\sqrt{318}} \begin{pmatrix} 11 \\ 1 \\ -14 \end{pmatrix}$

5 2; 0 and 1

$\frac{1}{\lambda(\lambda - 1)^2} \begin{pmatrix} \lambda^2+1 & -2\lambda & -2\lambda^2 \\ 1 & \lambda^2-2\lambda & -\lambda \\ -\lambda & \lambda & \lambda^2 \end{pmatrix}$

6 2; (a) $\begin{pmatrix} 1 & 0 & 0 \\ -4 & 3 & -1 \\ -10 & 5 & -1\frac{1}{2} \end{pmatrix}$

8 (a) $x = t, y = t, z = t$
   (b) $x = t, y = -t, z = t$

9 (a) $t(1, -2, 5)^T$
   (b) $(1, 1, 1) + t(1, -2, 5)^T$
   (c) $k - 2l + 5m \neq 0$

10 (b) $\begin{pmatrix} -\sin\theta \\ \cos\theta \end{pmatrix}, \begin{pmatrix} \sin\theta \\ -\cos\theta \end{pmatrix}$

11 (a) $t(1, -2, 1)$
   (b) $(21, 10, 0) + t(1, -2, 1)$

12 $3, 1, -2; \frac{1}{6} \begin{pmatrix} 4 & -7 & 5 \\ -2 & 5 & -1 \\ -2 & 8 & -4 \end{pmatrix}$

13 $\frac{1}{8} \begin{pmatrix} 3 & -9 & 8 \\ -1 & 3 & 0 \\ -1 & 11 & -8 \end{pmatrix};$
   $\lambda^3 - 5\lambda^2 + 2\lambda + 8 = 0;$
   $\lambda = -1, x = (-11, -1, 14)^T$

14 (a) $y = x, y = 6x$

16 $\frac{1}{6} \begin{pmatrix} 2 & -2 & 2 \\ -3 & 3 & 0 \\ 8 & -2 & -4 \end{pmatrix}; 6, 1, -1$

17 $T_1 = \begin{pmatrix} \cos\alpha & -\sin\alpha \\ \sin\alpha & \cos\alpha \end{pmatrix}; T_2 = \begin{pmatrix} 1 & 0 \\ 0 & -1 \end{pmatrix};$
   $\frac{1}{25} \begin{pmatrix} 7 & 24 \\ 24 & -7 \end{pmatrix}$

18 $-4(k - 1)(k^2 + 1); x = 9t, y = -5t,$
   $z = -7t$

19 $0; x = t, y = 1 + t, z = t$

20 (a) $x - y + z = 0$ (b) $(-1, 3, 4)$
   (c) $x/(-2) = y/3 = z/2$

21 $x = 3/4, y = 1$

22 $\begin{pmatrix} 0 & 1 \\ 1 & 0 \end{pmatrix}, \frac{1}{2}\begin{pmatrix} -1 & \sqrt{3} \\ \sqrt{3} & 1 \end{pmatrix};$

$\frac{1}{2}\begin{pmatrix} \sqrt{3} & -1 \\ 1 & \sqrt{3} \end{pmatrix}; \begin{pmatrix} \sqrt{3} & -1 \\ 1 & \sqrt{3} \end{pmatrix};$

$(-\sqrt{3} - 1, \sqrt{3} - 1)$

23 $\lambda = 2, (1, 3, 1)^T; \lambda = 1, (3, 2, 1)^T;$
   $\lambda = -1, (1, 0, 1)^T$
   $x = 3t, y = 2t, z = t$

24 $2X + Y + Z = 4$
   $7X - Y + 2Z = 6$
   $7X + 2Y - Z = 18$
   $(5/3, 7/3, -5/3)$

25 $\frac{1}{6} \begin{pmatrix} -3 & 3 & 3 \\ 7 & -5 & -3 \\ 5 & -1 & -3 \end{pmatrix}; x = \frac{1}{2}(a + b),$
   $y = -\frac{1}{2}(a + b), z = \frac{1}{2}(a - b)$

26 $\begin{pmatrix} 1 & -3 & 1 \\ 0 & 0 & 1/3 \\ 0 & 1/2 & -1 \end{pmatrix}$

27 $4x + 4y - z = 0$

28 $k = 10$, planes meet in parallel lines,
   $k = 1$, planes are parallel

29 4, 3, −1;

$$\mathbf{P} = \frac{1}{\sqrt{2}}\begin{pmatrix} 0 & 1 & 1 \\ \sqrt{2} & 0 & 0 \\ 0 & 1 & -1 \end{pmatrix}$$

30 $\begin{pmatrix} 1 \\ 2 \\ -2 \end{pmatrix}; \begin{pmatrix} 2 \\ 1 \\ 2 \end{pmatrix}; \mathbf{P} = \frac{1}{3}\begin{pmatrix} 2 & 1 & 2 \\ -2 & 2 & 1 \\ -1 & -2 & 2 \end{pmatrix}$

32 $\frac{1}{3}\begin{pmatrix} 1 \\ -2 \\ 2 \end{pmatrix}; \frac{1}{3}\begin{pmatrix} 2 & -2 & 1 \\ 2 & 1 & -2 \\ 1 & 2 & 2 \end{pmatrix}; \begin{pmatrix} 1 \\ 0 \\ 1 \end{pmatrix}$

(a) $x/8 = y/5 = z/3$ (b) $x/2 = y = z$

33 $a = 0, b = \sqrt{2}, c = -\sqrt{2}$

34 $\mathbf{r} = t(\sqrt{3}\mathbf{i} + \mathbf{j} + \mathbf{k}); \cos^{-1}(-1/4)$

35 $\begin{pmatrix} 1 \\ -1 \\ 0 \end{pmatrix}, \begin{pmatrix} 1 \\ 0 \\ -1 \end{pmatrix}, \begin{pmatrix} 1 \\ -1 \\ -1 \end{pmatrix};$

$$\mathbf{P} = \begin{pmatrix} 1 & 1 & 1 \\ -1 & 0 & -1 \\ 0 & -1 & -1 \end{pmatrix}$$

36 $\mathbf{r} = (2s + \lambda)\mathbf{i} + (2t + \lambda)\mathbf{j} + (2\lambda - s - t)\mathbf{k}; \mathbf{r} = t(\mathbf{i} - 5\mathbf{j} + 2\mathbf{k})$

37 $5y_1 = (x_1 - 5)^2; 4x_1y_1 = 1$

38 $\mathbf{A} = \begin{pmatrix} 2 & -2 & 0 \\ -2 & 1 & -2 \\ 0 & -2 & 0 \end{pmatrix}; 4, 1, -2;$

$\mathbf{H} = \frac{1}{3}\begin{pmatrix} 2 & -2 & 1 \\ -2 & -1 & 2 \\ 1 & 2 & 2 \end{pmatrix}; \{4, 1, -2\}$

39 $k = 1$, two planes parallel
$k = 3$, three planes meet in a line;
$x = 13y, z = \frac{1}{2} - 4y$

40 $\mathbf{P} = \frac{1}{\sqrt{2}}\begin{pmatrix} 1 & -1 & 0 \\ 1 & 1 & 0 \\ 0 & 0 & \sqrt{2} \end{pmatrix}$

Chapter 14

14.1 **1** (a) $6\mathbf{k}$ (b) $-6\mathbf{k}$ (c) $-6\mathbf{k}$
  (d) $-6\mathbf{k}$ (e) $6\mathbf{k}$
2 (a) $20\mathbf{i}$ (b) $-20\mathbf{i}$ (c) $-20\mathbf{i}$
  (d) $20\mathbf{i}$ (e) $-20\mathbf{i}$
3 (a) $-ab\mathbf{j}$ (b) $-ab\mathbf{j}$ (c) $ab\mathbf{j}$
  (d) $ab\mathbf{j}$ (e) **0** (f) **0**
  (g) **0**

14.3 **1** (a) $-\mathbf{i} - 13\mathbf{j} - 11\mathbf{k}$
  (b) $-19\mathbf{i} - 25\mathbf{j} + 8\mathbf{k}$
  (c) $6\mathbf{i} + 12\mathbf{j} - 8\mathbf{k}$
  (d) $-5\mathbf{i} - 8\mathbf{j} - 2\mathbf{k}$
  (e) $-20\mathbf{i}$ (f) $9\mathbf{i} - 19\mathbf{j} - 11\mathbf{k}$
2 (a) $\sqrt{6}$ (b) $\sqrt{126}$ (c) $\sqrt{77}$
  (d) $5\sqrt{5}$ (e) $\sqrt{874}$, (f) $13\sqrt{6}$

3 (a) $\frac{5}{2}\sqrt{6}$ (b) $\frac{1}{2}\sqrt{21}$ (c) $\frac{1}{2}\sqrt{1403}$
  (d) $\frac{1}{2}\sqrt{849}$ (e) $\sqrt{602}$
4 (a) $\mathbf{r}.(\mathbf{i} - 2\mathbf{j} + \mathbf{k}) = 0$
  (b) $\mathbf{r}.(\mathbf{i} + 3\mathbf{j} - 2\mathbf{k}) = 0$
  (c) $\mathbf{r}.(2\mathbf{i} - 5\mathbf{j} - \mathbf{k}) = -4$
  (d) $\mathbf{r}.(9\mathbf{i} - 3\mathbf{j} + \mathbf{k}) = 23$
  (e) $\mathbf{r}.(\mathbf{i} + 2\mathbf{j} + \mathbf{k}) = 1$
  (f) $\mathbf{r}.(4\mathbf{j} + 3\mathbf{k}) = -5$
5 (a) $\mathbf{r}.(\mathbf{i} - 2\mathbf{j} + \mathbf{k}) = 0$
  (b) $\mathbf{r}.(3\mathbf{i} + 4\mathbf{j} - 6\mathbf{k}) = 0$
  (c) $\mathbf{r}.(10\mathbf{i} + 5\mathbf{j} + 4\mathbf{k}) = 20$
  (d) $\mathbf{r}.(\mathbf{i} + \mathbf{j} + \mathbf{k}) = 2$
  (e) $\mathbf{r}.(\mathbf{i} - 2\mathbf{j} + \mathbf{k}) = 0$
  (f) $\mathbf{r}.(9\mathbf{i} - 14\mathbf{j} - 24\mathbf{k}) = 35$
6 (a) $x + 2y + 3z = 4$
  (b) $2x + 4y + 3z = 31$
  (c) $2x - y + 3z = 9$
  (d) $2x - y + z = 8$
  (e) $z = 2$
  (f) $bcx + cay + abz = abc$

14.4 **1** (a) $\sqrt{10}$ (b) $\sqrt{10}$ (c) $\frac{1}{2}\sqrt{482}$
  (d) $17\sqrt{2/9}$
2 (a) $\sqrt{(134/3)}$ (b) $\frac{1}{3}\sqrt{530}$ (c) 2
  (d) $17\sqrt{5/3}$
3 (a) $\sqrt{5}$ (b) $\frac{1}{4}\sqrt{962}$ (c) $\frac{1}{5}\sqrt{1049}$
  (d) $\sqrt{3}$
4 (a) $\frac{1}{4}\sqrt{145}$ (b) $\frac{1}{3}\sqrt{29}$ (c) $\sqrt{(33/2)}$
  (d) $\frac{5}{9}\sqrt{2}$

14.5 **1** (a) (i) 44 (ii) 44 (iii) -44
  (iv) -44
  (b) (i) -6 (ii) -6 (iii) 6
  (iv) 6
  (c) (i) 44 (ii) 44 (iii) -44
  (iv) -44
  (d) (i) 90 (ii) 90 (iii) -90
  (iv) -90
  (e) (i) $2qr(p - q)$ (ii) $2qr(p - q)$
  (iii) $2qr(q - p)$
  (iv) $2qr(q - p)$
2 (a) (i) $(-22, 8, 50)$ (ii)
  $(-20, 16, 4)$ (b) (i) $(-18, -15, 7)$
  (ii) $(-7, -21, 28)$
  (c) (8, 0, −16) (ii) (12, 20, −4)
  (d) (i) (0, 0, 0) (ii) (0, 0, 0)
3 (a) $(\mathbf{a}.\mathbf{b})\mathbf{c} - (\mathbf{a}.\mathbf{c})\mathbf{b}$
  (b) $(\mathbf{b}.\mathbf{c})\mathbf{a} - (\mathbf{b}.\mathbf{a})\mathbf{c}$
  (c) $(\mathbf{b}.\mathbf{a})\mathbf{c} - (\mathbf{b}.\mathbf{c})\mathbf{a}$
  (d) $(\mathbf{c}.\mathbf{b})\mathbf{a} - (\mathbf{c}.\mathbf{a})\mathbf{b}$
  (e) $(\mathbf{c}.\mathbf{a})\mathbf{b} - (\mathbf{c}.\mathbf{b})\mathbf{a}$
  (f) $(\mathbf{a}.\mathbf{c})\mathbf{b} - (\mathbf{a}.\mathbf{b})\mathbf{c}$
  (g) $(\mathbf{b}.\mathbf{c})\mathbf{a} - (\mathbf{b}.\mathbf{a})\mathbf{c}$
  (h) $(\mathbf{a}.\mathbf{b})\mathbf{c} - (\mathbf{a}.\mathbf{c})\mathbf{b}$
  (i) $(\mathbf{c}.\mathbf{b})\mathbf{a} - (\mathbf{c}.\mathbf{a})\mathbf{b}$
  (j) $(\mathbf{b}.\mathbf{a})\mathbf{c} - (\mathbf{b}.\mathbf{c})\mathbf{a}$

4 (a) 49 (b) 64 (c) 77 (d) 39

5 (a) 10 (b) 4 (c) 7/2 (d) 31
(e) 78 (f) 143/6

6 (a) not coplanar (b) not coplanar
(c) coplanar (d) coplanar

7 (a) and (c) not coplanar
(b) and (d) coplanar

Miscellaneous exercises 14

1 $2\sqrt{2}$

2 $\dfrac{1}{\sqrt{2}}(\mathbf{i} - \mathbf{j})$

3 $\dfrac{1}{\sqrt{2}}(\mathbf{i} - \mathbf{j})$

5 $(3\mathbf{i} + 5\mathbf{j} + 4\mathbf{k})/(5\sqrt{2})$, $3x + 5y + 4z = 30$, $6/\sqrt{2}$

6 (a) a straight line (b) $(x - 4)^2 + (y - 3)^2 + (z + 2)^2 = 55/3$

7 (b) $a\mathbf{i} + a\mathbf{j} + (2a - 1)\mathbf{k}$

8 $(\mathbf{i} + \mathbf{j} + \mathbf{k}) \cdot \mathbf{r} = 4$; $\frac{4}{3}(\mathbf{i} + \mathbf{j} + \mathbf{k})$; $1/\sqrt{6}$

9 (a) $x - 5y + 3z = -1, 9$
(i) $3\sqrt{(3/29)}$ (ii) $\frac{1}{2}\sqrt{6}$

10 $-2\mathbf{i} - \mathbf{j} + 3\mathbf{k}, \mathbf{i} - 3\mathbf{j} + 2\mathbf{k}; 7\mathbf{i} + 7\mathbf{j} + 7\mathbf{k}$; $(7\sqrt{3})/2; 2/\sqrt{3}; 7/3$

11 (a) $-4/9$ (b) $\frac{1}{2}\sqrt{65}$ (c) $(6/\sqrt{65}, 2/\sqrt{65}, -5/\sqrt{65})$ (d) $1/6$

12 $\mathbf{a} - \mathbf{b} + \mathbf{c}$ (a) $(1, 2, 6)$
(b) $6x - 3y + 2z = 12$ (c) 7 (d) $37/3$

13 $7(\mathbf{i} - \mathbf{j} - \mathbf{k})$; (a) $1/\sqrt{46}$
(b) $\frac{7}{2}\sqrt{129}$ (c) $49/2$ (d) $90°$

14 (a) $x + 2y + 5z = 30$ (b) 450
(c) $(1, 2, 5)$ (d) $45\sqrt{30}$

15 $(\mathbf{i} + \mathbf{j} + \mathbf{k}) \cdot \mathbf{r} = 1$

16 $6x + 3y + 2z = 6, x/6 = y/3 = z/2$; $(36/49, 18/49, 12/49), 6/7; 1; 7/2$

17 (a) $\mathbf{r} = \mathbf{a} + \lambda(\mathbf{b} - \mathbf{a}) + \mu(\mathbf{c} - \mathbf{a})$
or $(\mathbf{r} - \mathbf{a}) \cdot [(\mathbf{b} - \mathbf{a}) \times (\mathbf{c} - \mathbf{a})] = 0$
(b) $\dfrac{|(\mathbf{d} - \mathbf{a}) \cdot (\mathbf{b} \times \mathbf{c} + \mathbf{c} \times \mathbf{a} + \mathbf{a} \times \mathbf{b})|}{|\mathbf{b} \times \mathbf{c} + \mathbf{c} \times \mathbf{a} + \mathbf{a} \times \mathbf{b}|}$
(c) $\mathbf{r} = \mathbf{a} + \lambda(\mathbf{b} - \mathbf{a}) + \mu(\mathbf{d} - \mathbf{c})$
or $(\mathbf{r} - \mathbf{a}) \cdot [(\mathbf{b} - \mathbf{a}) \times (\mathbf{d} - \mathbf{c})] = 0$
(d) $\dfrac{|(\mathbf{d} - \mathbf{a}) \cdot [(\mathbf{b} - \mathbf{a}) \times (\mathbf{d} - \mathbf{c})]|}{|(\mathbf{b} - \mathbf{a}) \times (\mathbf{d} - \mathbf{c})|}$

18 $\frac{1}{6}\mathbf{a} \cdot (\mathbf{b} \times \mathbf{c})$

19 (c) $\mathbf{b}_1 \times \mathbf{b}_2 = \pm 4\mathbf{a}$

20 $\frac{1}{4}\sqrt{885}, 20/3; 120(\mathbf{i} - \mathbf{j} + \mathbf{k}), 18\mathbf{b} + 3\mathbf{c}$
$(40/7)(\mathbf{i} - \mathbf{j} + \mathbf{k})$;
$\mathbf{r} \cdot (47\mathbf{i} + 18\mathbf{j} + 34\mathbf{k}) = 360$

21 $\mathbf{a} = (4\mathbf{i} + 4\mathbf{j} + 7\mathbf{k})/9$
$\mathbf{b} = (8\mathbf{i} - \mathbf{j} - 4\mathbf{k})/9$

22 (a) O, A, B, C coplanar; O, C, P collinear
(b) (i) $\mathbf{r} = \mathbf{c} + t(\mathbf{a} \times \mathbf{b})$,
$\dfrac{x - c_1}{a_2 b_3 - a_3 b_2} = \dfrac{y - c_2}{a_3 b_1 - a_1 b_3}$
$= \dfrac{z - c_3}{a_1 b_2 - a_2 b_1}$;
(ii) $(\mathbf{b} - \mathbf{a}) \cdot (\mathbf{r} - \mathbf{c}) = 0$,
$(b_1 - a_1)x + (b_2 - a_2)y + (b_3 - a_3)z = (b_1 - a_1)c_1 + (b_2 - a_2)c_2 + (b_3 - a_3)c_3$

Index

ratio test 199
rectangular hyperbola 343
reductio ad absurdum 18
reduction formulae 240
relationship analysis 6
repeated roots 273
residue classes 398
rhombus 401
ring 412
root mean square value 257
roots of equations 272
rotation of axes 375
row transformation 454

scalar triple product 508
second-order convergence 333
second-order differential equation
    121, 323
sequences
  convergent 187, 330
  divergent 187
  monotonic 187
series
  convergent 192
  divergent 192
$\sinh x$ 76
$\sinh^{-1} x$ 85
solution of equations 15
spanning 426
square roots 25
step-by-step method 311
step length 310
straight lines 38, 162

subgroup 403
subspace 427
sufficient condition 5
symmetry group 401

tangent 344, 356, 367
  at origin 384
$\tanh x$ 80
$\tanh^{-1} x$ 87
Taylor series 215
Taylor series method 319
tetrahedron 509
theorems of Pappus 261
transformation of z-plane
  $w = az + b$ 48
  $w = 1/z$ 51
  $w = (az + b)/(cz + d)$ 57
translation 48
transpose 447
truth set 2
truth table 1
truth value 1

unit matrix 451
unit vector 497

vector
  differential equation 150
  product 496
  space 425
  triple product 511

Wallis's formulae 245